Also from Wiley...

EMAG SOLUTIONS POWERED BY JUSTASK!

A website with answers!

EMAG Solutions invites you to be a part of the solution as it walks you step-by-step through a total of 120 end-of-chapter problems and 15 interactive applications from the text. This powerful online problem-solving tool takes select end-of-chapter problems and provides you with more than just the answers. EMAG Solutions is your passport to greater understanding of key concepts. Save time in your studies while building valuable analysis skills.

Wherever you see ▶ $\underset{\text{SOLUTIONS}}{\text{EMAG}}$ in front of an end-of-chapter problem, you know that the EMAG Solutions will provide you with your choice of:

- ▶ Step-by-Step, detailed solutions and answers

- ▶ Detailed hints if you choose to solve the problem yourself

- ▶ Convenient pop up windows that highlight relevant concept, background theory methods, and laws that should be applied when solving each problem

- ▶ Solution guidelines that illustrate the steps you need to take in order to solve the problem, while allowing you to solve the problem on your own

- ▶ A complete glossary of key terms and definitions.

Use the registration code on the opposite page to find out more about EMAG Solutions

www.wiley.com/college/wentworth

Electromagnetics like you have never seen it!

Fundamentals of
Electromagnetics with
Engineering Applications

Fundamentals of Electromagnetics with Engineering Applications

STUART M. WENTWORTH

Auburn University

WILEY
JOHN WILEY & SONS, INC.

EXECUTIVE EDITOR   Bill Zobrist
ASSISTANT EDITOR   Catherine Mergen
SENIOR PRODUCTION EDITOR   Valerie A. Vargas
MARKET DEVELOPMENT MANAGER   Jennifer Powers
MARKETING MANAGER   Julie Lindstrom
NEW MEDIA EDITOR   Tom Kulsea
PRODUCTION SERVICES   ARGOSY PUBLISHING
COVER DESIGNER   Madelyn Lesure
COVER IMAGE   Photo Disc/Getty Image

This book was set in 10/12 Times Roman by Argosy Publishing and printed and bound by R.R. Donnelley-C. The cover was printed by Lehigh Press.

This book is printed on acid free paper.

Library of Congress Cataloging in Publication Data:
Wentworth, Stuart, M.
Fundamentals of electromagnetics with engineering applications / Stuart M. Wentworth
 1st ed
   p. cm
   ISBN 0-471-26355-9
Electric engineering. 2. Electromagnetism. I. Title

TK146.W435 2005

621.3—dc22

                                                                2004049216

ISBN 0-471-26355-9
ISBN 0-471-66132-5 (WIE)

Printed in the United States of America

10 9 8 7 6 5 4 3 2 1

To my wife, Julie, and our son, Austin

Preface

Why is it important for an electrical engineer to understand electromagnetics?

Beyond the fact that electrical circuits can only be understood superficially without an understanding of electromagnetics, consider that as microelectronic circuits continue to get smaller and faster, simple circuit theory breaks down. Only by application of electromagnetic principles can microelectronic circuits be understood and designed. As a second example, consider that future power needs may be partially met by beamed solar power. Orbiting solar panels would capture electromagnetic radiation from the sun and beam the power to receiving antennas on the ground. Electromagnetic theory will be applied to design these systems. Finally, consider the explosive growth of wireless communications. The circuits, antennas, and signal transmission all depend on electromagnetic principles. The anticipated continued growth in wireless technology increases the demand for electrical engineers with a solid electromagnetics background.

This text can be used in a one- or two-semester electromagnetics sequence for electrical engineering students at the junior and senior level. The students are assumed to have completed the freshman and sophomore physics and calculus courses and, therefore, are expected to be competent at integration and differentiation and at least have some familiarity with vectors.

The text begins with an introductory chapter describing the role that electromagnetics has in various aspects of a wireless communications system. After this motivational beginning, the text features a classical organization, easing the student into electromagnetics and vector algebra, beginning with electrostatic fields. Vector concepts such as dot product and gradient are introduced where needed. It is hoped that students will achieve some level of comfort working with vectors and different coordinate systems before progressing to the slightly more complicated topic of magnetostatic fields, featuring cross product and curl operations. Then, variation with time is introduced with dynamic fields, culminating in Maxwell's equations. This is followed by coverage of plane-wave propagation.

The concepts covered in the first half of the text are all fundamental to electromagnetics and form the foundation for understanding and appreciating the applied topics of the second half. The applied topics are as follows:

Transmission lines—Introducing such concepts as impedance matching and signal reflection, the subject of transmission lines is applicable to high-frequency circuits such as the connecting wires used in high-speed microprocessors.

Waveguide—Although rectangular waveguide is itself an interesting and useful topic, in this text it also acts as a stepping stone to the increasingly important topic of optical fibers.

Antennas—This is certainly the most visible manifestation of electromagnetics. An understanding of antennas is critical for the understanding of wireless communications.

Electromagnetic interference—The impact of noise on performance of an electrical system grows more important as the circuits get smaller and faster. This is especially true for digital circuits.

Microwave engineering—Here we describe many of the circuit components used in wireless communication transceivers (e.g., power dividers, filters, and amplifiers) that transmit and receive high-frequency signals.

Features of Wentworth, Fundamentals of Electromagnetics with Engineering Applications, 1e

Who will benefit from using this text?

This text is designed for use in a one or two-semester electromagnetics sequence for electrical engineering students at the junior and senior level.

Unique Approach

This text features practical applications for Wireless Systems, Transmission Lines, Waveguide, Antennas, Electromagnetic Interference, Microwave Engineering.

Pedagogical Features

- **Worked-out example problems**—numerous worked-out example problems give students hands-on experience in how to solve electromagnetic problems.

- **Drill problems**—many relatively simple drill problems are included for reinforcement of the course material.

- **End-of-chapter problems**—plentiful end-of-chapter problems including problems on MATLAB are arranged by chapter section, and many of the odd-numbered problems have answers provided in the appendix.

- **MATLAB**—there are many detailed MATLAB examples that provide for deeper illumination of the subject matter.

- **Practical Applications**—a number of practical applications are provided that show how electromagnetic theory is put into practice.

- **End-of-chapter summaries**—a concise summary at the end of each chapter captures the key points.

Logo used to indicate MAT-LAB Problem

Media Features

EMAG SOLUTIONS

Logo used to indicate EMAG Solution Problem

EMAG Solutions—students may also wish to enhance their understanding by using EMAG Solutions, an innovative student-oriented website that provides the problem-solving approach and step-by-step solution to marked problems from the text. Included in the website are interactive visualizations of electromagnetic theory that is impossible to easily convey on the printed page. Users rave as to the simplicity and usefulness of the website. Access to the site is free with purchase of the textbook. Students who do not have the textbook can purchase access to this site on www.wiley.com/college/wentworth.

Supplements

- **Solutions Manual**—a Solutions Manual for all problems will be available for instructors on the Instuctor Companion site at: www.wiley.com/college/wentworth.
- **Lecture Slides and Illustrations** of all figures and tables are available on the Instructor Companion Site at: www.wiley.com/college/wentworth.

Motivation for the Text

One area emphasized by Auburn University's College of Engineering is wireless communications technology. This emphasis provided part of the motivation for me to compose a new text on electromagnetics. Although many of the topics in this text are covered elsewhere, no other text has the particular collection of topics that we deem necessary as adequate background for the study of wireless communications.

A second motivation for this text was my department's decision to emphasize MATLAB[1] in our undergraduate curriculum. There are very good reasons for doing so. MATLAB is *the* programming language for most electrical engineering applications. It has supplanted the C programming language for many of the problem-solving tasks our engineers encounter in the field. Numerous industry representatives have strongly encouraged that our students be well acquainted with MATLAB.

No presently available electromagnetics text that I am aware of spends significant time with MATLAB. This text does so. It has numerous examples, many of which help illustrate some of the fine points in electromagnetics. There are also a large number of MATLAB-specific end-of-chapter problems to solve.

A third motivation for this text, shared I am sure by anyone who undertakes the attempt, is to provide a clearer explanation of what electromagnetics is all about. Professors of electrical engineering who do not teach electromagnetics often quip "Why does it take so long for you guys to cover electromagnetics? After all, there are only four equations!" The answer is that these four equations (Maxwell's equations) are elegant and subtle and take no little effort to fully appreciate.

Suggested Coverage

As stated previously, the text is designed for a two-semester sequence. Table P.1 shows the topic coverage for a conventional approach, the one I use in my classroom, along with an approach that starts with transmission lines. A number of instructors feel that this alternate approach is a better way to begin the study of electromagnetics. It is certainly possible to start with transmission lines with this text, as long as the students understand that the mysterious field components they are exposed to in the transmission lines chapter will be explained later when the fundamentals section of the text is covered. The one caveat is that the MATLAB

[1]MATLAB is a registered trademark of The MathWorks, Inc. For MATLAB product information, please contact The MathWorks, Inc., 3 Apple Hill Drive, Natick, MA 01760-2098, Tel: (508) 647-7000, Fax: (508) 647-7101.

TABLE P1   Suggested Coverage for Two 44 Contact Hour Courses

Conventional Approach			Transmission Lines First Approach		
Topic	Sections	Hrs	Topic	Sections	Hrs
Semester 1			**Semester 1**		
Introduction	1.1–1.3	1	Introduction	1.1–1.3	1
Electrostatics	2.1–2.14	12	Wave Fundamentals & Phasors	4.2, 4.8	1
Magnetostatics	3.1–3.9	10	Transmission Lines	6.1–6.5	7
Dynamic Fields	4.1–4.9	7	Electrostatics	2.1–2.14 (skim 2.11,2.12)	10
Plane Waves	5.1–5.9	9	Magnetostatics	3.1–3.9 (skim 3.7)	9
Scheduled exams / review		5	Dynamic Fields	4.1, 4.3–4.7, 4.9	6
			Plane Waves	5.1–5.5	5
			Scheduled exams / review		5
Total		44	Total		44
Semester 2			**Semester 2**		
Introduction / Review		1	Introduction / Review		1
Transmission Lines	6.1–6.7, 6.9	10	More Transmission Lines	6.6–6.9	4
Waveguide	7.1–7.7 (skim 7.2)	6	Waveguide & Optics	7.1–7.2, 5.6–5.8, 7.3–7.7	10
Antennas	8.1–8.4, 8.6–8.9	9	Antennas	8.1–8.9	10
Electromagnetic Interference	9.1–9.7	4	Electromagnetic Interference	9.1–9.7	4
Microwave Engineering	10.1–10.3, 10.5–10.7	9	Microwave Engineering	10.1–10.7	11
Scheduled exams / review		5	Scheduled exams / review		5
Total		44	Total		44

examples and assignments in the transmission lines chapter assume the student has climbed the learning curve with the simpler examples from the fundamentals chapters. If students already have the programming expertise, this does not pose a problem; however, if they aren't yet comfortable with MATLAB, the instructor must take this into account.

If a particular curriculum only requires one electromagnetics course, then the suggested coverage for the first semester of either presented approach in the table is appropriate. The remaining subject matter could be used in an advanced course or technical elective in electromagnetics.

Success in Learning Electromagnetics (a Note to the Students)

There are some aspects of electromagnetics that many students find daunting. The use of vectors and coordinate systems other than Cartesian, along with frequent use of derivatives and integrals, are likely to frighten a number of students. There is really no need for fear, only a need for resolve! The only way to learn electromagnetics is to study the material and to apply it to as many problems as possible. Students should certainly work through all of the example and drill problems. Successful students will also read (and reread) the text, will work more end-of-the-chapter problems than required of the homework assignments, and will rework the best problems.

Supplements

EMAG Solutions—students may also wish to enhance their understanding by using EMAG Solutions, an innovative student-oriented website that provides the problem-solving approach and step-by-step solution to marked problems from the text. Included in the website are interactive visualizations of electromagnetic theory that is impossible to easily convey on the printed page. Users rave as to the simplicity and usefulness of the website. Access to the site is free with purchase of the textbook. Students who do not have the textbook can purchase access to this site on the Book Companion Site at: www.wiley.com/college/wentworth.

Solutions Manual—a Solutions Manual for all problems will be available for instructors on the Instuctor Companion site at: www.wiley.com/college/wentworth.

Lecture Slides and Illustrations of all figures and tables are available on the Instructor Companion Site at: www.wiley.com/college/wentworth.

Both students and instructors using this text have access to all of the **MATLAB example program files** on the Student Companion Site at www.wiley.com/college/wentworth.

Acknowledgments

My sincere appreciation goes to Auburn University for granting me a semester of Professional Improvement Leave to work on this book. My colleagues Mike Baginski, Hulya Kirkici, Sadasiva Rao, Lloyd Riggs, and Thomas Shumpert have provided much useful discussion as well as pointers on how to teach electromagnetics. Professor Martial Honnell has in particular given me a number of good suggestions for improving the text. Thanks also to Dr. John Henderson of Harris Corporation for sharing his electromagnetics philosophy with me, and for the use of several photographs. I'd also like to thank the electrical and computer engineering students at Auburn University for having the patience to work with a brand new text, errors and all. Their feedback has been most valuable.

My gratitude also goes to the reviewers of this book for their many useful comments and suggestions. They are

Mani Mina, Iowa State University,

Richard Selfridge, Brigham Young University,

Rajendra Arora, Florida State University,

Lloyd Riggs, Auburn University,

Svetla Jivkova, Penn State University,

Masoud Mostafavi, San Jose State University, and

P. Robert Kotiuga, Boston University.

I am extremely grateful for the guidance provided by Bill Zobrist, my Executive Editor at John Wiley & Sons. In particular, I'd like to thank Assistant Editor Katie Mergen, Designer Madeline Lesure, Senior Production Editor Valerie Vargas, and Sally Boylan of Argosy. The support from Bill and his team has been invaluable in preparing this text.

I also appreciate the moral support provided by friends that I've joined "by the Banks of the Withywindle" for many a discussion.

Finally, I could never have done this without the support and patience of my loving wife, Julie.

Stuart M. Wentworth

About the Author

Stuart M. Wentworth was born and grew up in Pensacola, Florida, USA. In 1982 he received his B.S degree in Chemical Engineering from Auburn University, Alabama, and his M.S. and Ph.D. degrees in Electrical Engineering from the University of Texas at Austin were obtained in 1987 and 1990, respectively. He has been a member of the Electrical & Computer Engineering faculty at Auburn University since 1990. Dr. Wentworth's research has focused on the high-frequency characterization of materials used for electronics packaging. Dr. Wentworth has received numerous teaching awards at Auburn University, including the Birdsong Merit Teaching Award in 1999. He is a Senior Member of the Institute of Electrical and Electronics Engineers (IEEE).

Brief Contents

Contents

CHAPTER 1

Introduction

Learning Objectives

▶ Introduce the electromagnetic spectrum

▶ Explain how electromagnetics is fundamental to wireless communications

▶ Provide guidelines for numeric precision and the handling of dimensions

We are immersed in electromagnetic fields. They are everywhere, being generated naturally (e.g., solar radiation and lightning) and by us (e.g., radio stations, cell phones, and power lines). The modern office, kitchen, and automobile are all stuffed full of devices that rely on electricity, and magnetic fields are in action anywhere an electric motor is running. The wireless communications revolution has electromagnetics at its very core: Voice and data information is transmitted and received via antennas and high-frequency electronics, components requiring knowledge of electromagnetics to design and understand. The study of electromagnetics is necessary for understanding even simple electronic components such as resistors, capacitors, and inductors.

Human beings have been aware of magnetic materials for as long as there has been recorded history, and the Greek Thales of Miletus recorded evidence of both static electricity and magnetic attraction around 600 B.C. But it wasn't until the latter half of the 18th century, and the 19th century in particular, that progress was made in recognizing and understanding electromagnetic phenomena. The timeline in Figure 1.1 shows some of the key advances.[1.1] The real era of understanding began after Alessandro Volta invented the voltaic cell, allowing research to be conducted with controlled currents. From there, Oersted's discovery that electric currents create magnetic fields, and Faraday's discovery that magnetic fields changing with time create electric fields, culminated in James Clerk Maxwell's unification of electricity and magnetism in the concise four equations known as *Maxwell's equations*. Development and understanding of these four equations is the task of the next four chapters of this text.

Before getting started on the first fundamental chapter of the text on electrostatics, we address several topics. First, we present a brief overview of the electromagnetic spectrum, showing the basic relationships between frequency and wavelength and identifying specific frequency bands of interest in wireless communications. Then, we'll describe how

---

[1.1]A torrent of engineering applications followed the development of Maxwell's equations, led by the prolific engineers Thomas Alva Edison and Nikola Tesla.

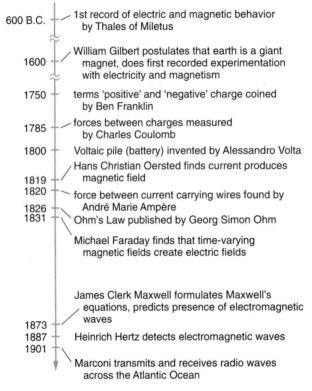

Figure 1.1   The key historical events in electromagnetics.

electromagnetics is pivotal to the application of wireless communications. Finally, we'll discuss some odds and ends on dealing with units and unit conversions.

▶ 1.1   THE ELECTROMAGNETIC SPECTRUM

Maxwell predicted that light consists of electric and magnetic fields oscillating in tandem. Such an *electromagnetic wave* can propagate in a vacuum with velocity $c = 2.998 \times 10^8$ m/s. Over a very broad range, a continuous[1.2] spectrum of electromagnetic radiation is possible. The spectrum shown in Figure 1.2 ranges from 0.1 Hz up to 10^{23} Hz, where a hertz[1.3] (Hz) is equal to one cycle per second. In vacuum, the frequency f and wavelength λ are related by the speed of light,

$$c = \lambda f \tag{1.1}$$

so the spectrum can also be indicated in terms of wavelength.

Below 300 GHz, it is customary to refer to the electromagnetic waves in terms of frequencies. For instance, the microwave bands employed in radar and communications applications are in terms of frequency, as shown in the figure. Above 300 GHz, or at wavelengths

[1.2]The spectrum is continuous, that is, down to the level of a discrete quantum.

[1.3]The hertz is named in honor of German physicist Heinrich R. Hertz (1857–1874).

Figure 1.2   The electromagnetic spectrum.

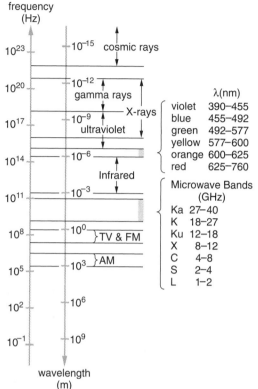

below 1 mm, waves are more likely to be referred to in terms of wavelength. Thus, the spectrum of visible light is listed by wavelength in the figure.

For wireless communications, a high frequency is desirable since the amount of information communicated directly scales with frequency. However, it may be recalled from physics that the energy U of a photon is proportional to frequency as

$$U = hf \qquad (1.2)$$

where h is Planck's constant ($h = 6.63 \times 10^{-34}$ J-s). At very high frequencies (i.e., X rays), the energy of the radiation can cause damage to materials (and people). At somewhat lower frequencies, for instance ultraviolet and visible light, the signal is severely attenuated[1.4] by material media and clouds. Fiber optic and line-of-site communication schemes are utilized at such frequencies. Wireless communication is undertaken at certain microwave frequencies, between 1 and 100 GHz, where there are windows of relatively low signal attenuation in the atmosphere. Some of the important low-attenuation windows are < 18 GHz, 26–40 GHz, and 94 GHz.

Also, efficient transmission of signals requires an antenna of dimension roughly on the order of the signal wavelength. Some AM radio stations require antennas as long as 100 m, and wireless transmission at lower frequencies rapidly becomes impractical.

[1.4]If the amplitude of a wave decreases in the direction of propagation, we say it *attenuates*.

▶ **1.2 WIRELESS COMMUNICATIONS**

Cellular telephones are truly sophisticated feats of engineering. In addition to basic phone service, these gadgets also allow their users Internet and e-mail access. They may also have global positioning satellite (GPS) capability and personal digital assistant (PDA) capability; some even come with games.

There is a limited number of frequency channels available to handle cellular communications—far fewer than would seem necessary to handle the millions of transactions that take place daily. The way a cellular system is able to handle all of these transactions is to break up a town or city into multiple small sections, or *cells*, each serviced by its own cell tower. These cells are typically arranged in a hexagonal grid, as shown in Figure 1.3. Each cell, because it has six neighbors, can use one-seventh of the available frequency channels. Since the transmit and receive power is not strong enough to communicate with a tower two cells away, nonadjacent cells can use the same frequency channels.

Communication between the cell phone and tower is represented in Figure 1.4. The towers are tied into the conventional phone grid. The signal being transmitted by the cell phone is at a different frequency than the received signal frequency. This allows simultaneous transmission and reception, unlike, for instance, a pair of walkie-talkies that use a single frequency. In the cell phone, the analog voice signal is converted to a digital signal via the analog-to-digital (A/D) converter. Digital signals can be compressed and broadcast in a variety of communication schemes that enable many users to utilize the same system at once. The digital signal processing (DSP) block handles very high speed signal calculations, typically as many as 40 million instructions per second. The microprocessor handles other operations, including the user interface (liquid crystal display for output and key pad for input) and accessing memory (perhaps using a "flash card" containing phone numbers or other information). The radio frequency (RF) front end amplifies the weak received RF signal and down-converts it to the frequency required by the rest of the electronics in the cell phone. The front end must also up-convert an output signal to the desired RF transmission frequency. Finally, the front end must separate the functions of reception and transmission, which use the same antenna. The cell phone antenna must be small and unobtrusive. At the tower, each vertical bar is typically an array of patch antennas.

Understanding the signal processing and the information handling required for a cellular system is a worthy objective; it is expected that these topics are covered in some of the students' other courses. But where does electromagnetics come into play in a cellular system?

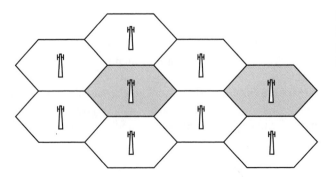

Figure 1.3 Hexagonal cell tower grid. The two shaded grids can use the same frequencies.

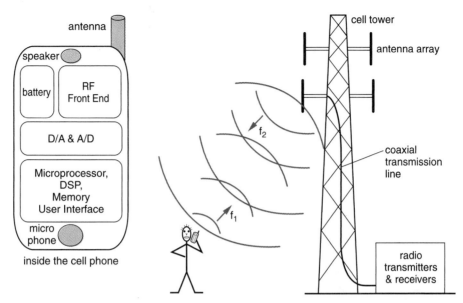

Figure 1.4 Typical cellular telephone system.

Certainly the physical operation of the microelectronic devices is governed by the laws of electromagnetics. Here are some other ways in which electromagnetics plays a role:

- Waves propagate in space and through material media (Chapters 4 & 5).
- Waves are radiated and received by antennas (Chapter 8).
- Waves propagate in transmission lines such as coaxial cables (Chapter 6).
- Efficient signal handling requires impedance matching of transmission lines (Chapter 6 and 10).
- RF components, such as those in the RF front end and in the tower's box, are typically designed and understood via electromagnetics (Chapter 10).
- Communications between towers may employ fiber optics and optical components (Chapter 7).
- Noise and interference between electronic components impact system performance (Chapter 9).

Besides cellular phone systems, other wireless communications systems include direct broadcast satellite (DBS) systems for television, GPS service for navigation, and radio frequency identification (RFID) tags for inventory control and item tracking. Understanding any of these wireless systems requires a firm foundation in electromagnetics.

▷ 1.3 DEALING WITH UNITS

In many of the problems to be solved in the upcoming chapters, students will be required to calculate a numerical solution. Rarely is this number, by itself, correct unless the appropriate units are applied. For instance, suppose 12 V is dropped across a resistor when 0.2 A passes through it. The resistance is not 60; rather, it is 60 Ω.

When reporting large or small numbers, scientific notation gives way to engineering notation, whereby appropriate prefixes are used for multiples of 10^3 (or submultiples of 10^{-3}).[1.5] Common prefix multipliers are listed in Table 1.1. Why use engineering notation instead of scientific notation? Consider an 18-GHz frequency. I can say "eighteen times ten to the nine hertz" or I can say "eighteen gigahertz." An efficient engineer prefers brevity, and students must get in the habit of speaking in engineering language before assuming their careers.

Students should also report numbers to the appropriate precision. Too often, the answer for a calculation like the one we did to find resistance is reported as something like "60.127 Ω." This has too many significant digits. The number of significant digits in a reported answer represents the precision with which the number is known. Significant digits for a number with no decimal point are counted from the first nonzero number on the left side to the last nonzero number on the right. For instance, the number "4030" has three significant digits, whereas the number "4000" has only one. If there is a decimal point, you count to the last digit reported. So "4000." has four significant digits, and "2.1×10^6" has two. It is further assumed that the last number reported can be off by as much as a half. So if I say the resistance is "60.000 Ω," I am reporting a number to five significant digits, and what I mean is that the resistance must lie between 59.9995 Ω and 60.0005 Ω. Likewise, a "60-Ω" resistor has only one significant digit, and it can have resistance between 55 and 65 Ω. A "60.-Ω" resistor has two significant digits with a value between 59.5 and 60.5 Ω.

Some rules concerning the use of significant digits in calculations are as follows:

- When multiplying two numbers, the answer can have no more significant digits than the lowest number of significant digits of either multiplicand. This goes for division as well.

- When adding (or subtracting) numbers, precision of the answer is determined by whichever number's last significant digit is further left. For example,

$$60 + 0.001 = 60$$

Table 1.1   Engineering Notation Prefixes

Prefix	Pronunciation	Multiplier
T	tera	10^{12}
G	giga	10^9
M	mega	10^6
k	kilo	10^3
m	milli	10^{-3}
μ	micro	10^{-6}
n	nano	10^{-9}
p	pico	10^{-12}
f	femto	10^{-15}

[1.5]A notable exception is the ubiquitous use of the centimeter, or a hundredth of a meter.

where the first number has its last significant digit well to the left of the second number. As a second example, consider

$$60.0000 + 0.001 = 60.001$$

where now the second number's last significant digit is the leftmost and determines the outcome.

- For lengthy calculations, it is a good idea to retain more significant digits than are needed until the calculation is complete, at which time the final answer should be properly reported.
- Note that pure integers or counted quantities are known to infinite precision. For instance, if you count 5 cows, you can report 5.000000000 etc. cows!

As a final topic in this section, there will be many calculations with multiple quantities and a variety of units involved. Students must use care to make sure their answers, including the units, follow from the calculation. Conversion factors (also called "unity ratios") are used to convert a quantity in terms of one unit into the terms of another. We can use a horizontal line to separate numerator and denominator quantities, and vertical lines separate quantities to be multiplied together. This *dimensional equation* approach is a convenient way to include the conversion factors and prevents common mistakes like dividing by a conversion factor when it should be multiplied or improperly accounting for number prefixes.

▷ **EXAMPLE 1.1**

Suppose we want to find the energy associated with a photon at 100 GHz. We use the equation $U = hf$ from before and have

$$U = \frac{6.63 \times 10^{-34} \text{ J s}}{} \left| \frac{100 \text{ GHz}}{} \right| \frac{10^9 \text{ Hz}}{\text{GHz}} \left| \frac{1}{\text{s} \cdot \text{Hz}} \right| = 6.63 \times 10^{-23} \text{ J}$$

After the 100 GHz, the next two items are unity ratios used to convert the units. You can ensure the proper units in the final answer by canceling like units in the numerator and denominator.

Drill 1.1 How much voltage is dropped across a 1.1-kΩ resistor when 10.6 mA is passed through it? (*Answer:* 12 V)

Drill 1.2 A voltage of 1.08 V is measured across a resistor that has 7.43 μA passing through it. How much power is dissipated in the resistor? (*Answer:* 8.02 μW)

Drill 1.3 What frequency ranges are associated with (a) orange light and (b) blue light? (*Answer:* (a) 500–480 THz, (b) 659–609 THz)

PART I

Fundamental
Electromagnetics

CHAPTER 2

Electrostatics

Learning Objectives

▷ Introduce Cartesian, cylindrical, and spherical coordinate systems

▷ Discuss vectors, vector addition, dot product, and divergence

▷ Describe electric field intensity and electric flux density

▷ Define and utilize Coulomb's Law

▷ Determine the electric field resulting from various charge distributions

▷ Use Gauss's Law to find the electric field for symmetrical charge distributions

▷ Describe electric potential and its relation to electric field intensity

▷ Present Ohm's Law and explain current in conductors

▷ Describe the features of dielectric materials

▷ Compare the electric fields across material boundaries

▷ Define capacitance and calculate it for various geometries

It is convenient to begin the study of electromagnetics by looking at electrostatic fields. Such fields are easier to visualize than static magnetic fields, and they are certainly easier to understand than time-varying electromagnetic fields. Electrostatics provides the simplest platform for the introduction of coordinate systems, vectors, and vector algebra. In addition, there are numerous practical applications of electrostatics. It is fundamental to the operation of copying machines and ink-jet printers, is a basic prerequisite for understanding lightning in the global electric circuit, and controls the operation of many electronic devices, including field effect devices, CCD imaging cameras, and liquid crystal displays. Electrostatics also has application in pollution-control filters and industrial electrostatic separation.

Following a discussion of vectors in Cartesian coordinates, the study of electrostatics begins with Coulomb's law and electric field intensity. The field intensity resulting from points of charge provides the appropriate setting to introduce the spherical coordinate system. Then, electric field intensity is determined for various distributions of charge. For lines of charge, the cylindrical coordinate system is introduced. The concepts of flux and electric flux density are discussed, leading to Gauss's law and the concept of divergence. Then, the electric potential is described as it relates to charges and electric fields, followed by a discussion of materials issues (i.e., conductors and dielectrics). In the last part of the chapter we describe the relation between the fields across the boundary separating two materials, leading to a description of capacitance.

▶ 2.1 VECTORS IN THE CARTESIAN COORDINATE SYSTEM

To this point, students have mostly dealt with *scalar* quantities: those quantities represented by a single number such as length or time. A *scalar field* represents the mapping of a scalar quantity as it varies with position. For instance, a topographical map is a scalar field of altitude. In electromagnetics, we deal with *vector* quantities, which have both a magnitude and a direction. For instance, at a point in space relative to earth, there is a gravity vector that decreases in magnitude as you venture further out into space, but it always points toward the center of the earth. A *vector field* can represent a mapping of these vectors in space.

Determining the precise location of a point in three-dimensional space can be accomplished using the Cartesian coordinate system, named in honor of the mathematician René Descartes. A Cartesian coordinate system as shown in Figure 2.1 is a *right-handed* system of orthogonal coordinate axes *x, y,* and *z.* The term "orthogonal" means that each axis is mutually perpendicular to the other two. Right-handed refers to the sequence *x, y, z.* These coordinates are right-handed because if the fingers of the right hand point in the *x* direction and are curled toward the *y* direction, the thumb will point in the *z* direction. Another way to visualize a right-handed system is as a right-handed screw. If a screw pointing in the +*z* direction is turned clockwise (moving from the positive *x*-axis to the positive *y*-axis), it moves in the positive *z* direction.

Although the Cartesian coordinate system is the most common and easiest to visualize, many situations call for other coordinate systems that can provide easier problem solving or can enhance understanding. The cylindrical coordinate system and the spherical coordinate system, along with the Cartesian system, round out the three most prevalent systems. Cylindrical and spherical coordinates will be introduced in later sections of this chapter.

The (*x, y, z*) coordinates define a point in the Cartesian coordinate system. For instance, Figure 2.2 shows the points P(0, 1, 0), Q(0, 4, 0), and R(3, 4, 0). A vector drawn from the

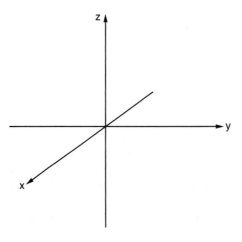

Figure 2.1 The Cartesian coordinate system.

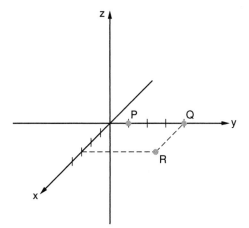

Figure 2.2   Cartesian coordinate points P(0, 1, 0), Q(0, 4, 0), and R(3, 4, 0).

origin O(0, 0, 0) to point P in Figure 2.3 has a magnitude of 1 and is pointing in the positive *y* direction. Since it has a magnitude of 1, this vector is known as a *unit vector* and may be written

$$\mathbf{P} = 1\mathbf{a}_y \tag{2.1}$$

It is customary to indicate vector quantities in bold-faced type. However, when handwriting vectors (for instance, on one of your many homework problems) it is easier to indicate a vector by drawing a small arrow over the letter. A bold lowercase **a** is typically used to represent unit vectors, with the direction indicated by a subscript.[2.1] Handwritten unit vectors are commonly written as a lowercase "a" with a hat on top. Using this approach, (2.1) would be rewritten as

$$\vec{P} = 1\hat{a}_y$$

A vector may be multiplied by a scalar that will change its magnitude but not its direction. Thus, a vector **Q** drawn from the origin to Q in Figure 2.2 is equivalent to placing four **a**$_y$ unit vectors end to end, or

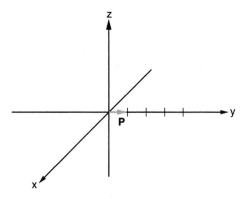

Figure 2.3   A unit vector $\mathbf{P} = \mathbf{a}_y$.

[2.1] Some texts use **x, y, z** to represent unit vectors in the Cartesian coordinate system. Many physicists like to use **i, j,** and **k.**

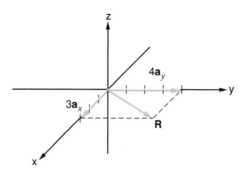

Figure 2.4   A vector from the origin to point R(3, 4, 0) separated into its components along each axis.

$$\mathbf{Q} = 4\mathbf{P} = 4\mathbf{a}_y$$

Likewise, a vector from the origin to point R(3, 4, 0) can be separated into its components along each axis, as shown in Figure 2.4, and written

$$\mathbf{R} = 3\mathbf{a}_x + 4\mathbf{a}_y$$

This vector can also be expressed as

$$\mathbf{R} = R\mathbf{a}_R \tag{2.2}$$

where R is the magnitude of $\mathbf{R}$ determined by the Pythagorean theorem, or

$$R = |\mathbf{R}| = \sqrt{3^2 + 4^2} = 5$$

and the unit vector in the direction of $\mathbf{R}$, $\mathbf{a}_R$, is then

$$\mathbf{a}_R = \frac{\mathbf{R}}{R} = \frac{3}{5}\mathbf{a}_x + \frac{4}{5}\mathbf{a}_y$$

In each of these cases, the vectors are drawn from the origin to a particular point and are termed *position vectors* since they relate to the location of the terminal point. But a general vector can go from any point to another. To find the vector $\mathbf{A}_{mn}$ from point $M(x_m, y_m, z_m)$ to point $N(x_n, y_n, z_n)$, the starting point is subtracted from the ending point for each component, or

$$\mathbf{A}_{mn} = (x_n - x_m)\mathbf{a}_x + (y_n - y_m)\mathbf{a}_y + (z_n - z_m)\mathbf{a}_z \tag{2.3}$$

For instance, a vector from point R to point P of Figure 2.2 gives

$$\mathbf{R}_{RP} = -3\mathbf{a}_x - 3\mathbf{a}_y$$

In general, a vector in the Cartesian coordinate system is written

$$\mathbf{A} = A_x\mathbf{a}_x + A_y\mathbf{a}_y + A_z\mathbf{a}_z \tag{2.4}$$

The magnitude of this vector is found by extending the Pythagorean theorem to three dimensions:

$$A = |\mathbf{A}| = \sqrt{A_x^2 + A_y^2 + A_z^2} \tag{2.5}$$

Vectors may be added or subtracted from each other by adding or subtracting the component values. For instance, if

$$\mathbf{B} = B_x\mathbf{a}_x + B_y\mathbf{a}_y + B_z\mathbf{a}_z$$

then

$$\mathbf{A} + \mathbf{B} = (A_x + B_x)\mathbf{a}_x + (A_y + B_y)\mathbf{a}_y + (A_z + B_z)\mathbf{a}_z \qquad (2.6)$$

Graphically, for vectors **A** and **B** shown in Figure 2.5a, **A** + **B** can be visualized by placing the tail of **B** at the head of **A** (Figure 2.5b) or by starting each vector at a common point and completing the parallelogram (Figure 2.5c). Likewise,

$$\mathbf{A} - \mathbf{B} = (A_x - B_x)\mathbf{a}_x + (A_y - B_y)\mathbf{a}_y + (A_z - B_z)\mathbf{a}_z \qquad (2.7)$$

which is represented by Figure 2.5d.

▷ **EXAMPLE 2.1**

Given the points P(0.0, –4.0, 0.0), Q(0.0, 0.0, 5.0), R(1.0, 0.0, 0.0), and S(0.0, 5.0, 0.0):

    a. Find and sketch the vector **A** from P to Q.

    b. Find and sketch the vector **B** from R to S.

    c. Find the direction of **A** + **B**.

    We begin by using (2.3) to find **A**:

$$\mathbf{A} = (0 - 0)\mathbf{a}_x + (0 - -4)\mathbf{a}_y + (5 - 0)\mathbf{a}_z = 4.0\mathbf{a}_y + 5.0\mathbf{a}_z \quad (a)$$

Likewise,

$$\mathbf{B} = -1.0\mathbf{a}_x + 5.0\mathbf{a}_y \quad (b)$$

To find the direction of **A** + **B**, we first need **A** + **B** from (2.6):

$$\mathbf{A} + \mathbf{B} = (0 - 1)ax + (4 + 5)ay + (5 + 0)az = -1ax + 9ay + 5az$$

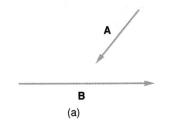

(a)

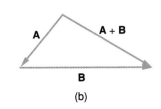

(b)

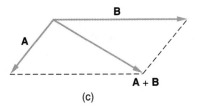

(c)

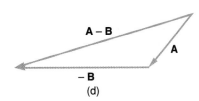

(d)

Figure 2.5 A pair of vectors **A** and **B** shown in (a) are added by the head-to-tail method (b) and by completing the trapezoid (c). In (d), the vector **B** is subtracted from **A**.

The direction of **A** + **B** is the unit vector

$$\mathbf{a} = \frac{\mathbf{A}+\mathbf{B}}{|\mathbf{A}+\mathbf{B}|}$$

where $|A+B| = \sqrt{1^2 + 9^2 + 5^2} = 10.34$, and therefore

$$\mathbf{a} = -0.097\mathbf{a}_x + 0.87\mathbf{a}_y + 0.48\mathbf{a}_z \quad (c)$$

Drill 2.1 Given two points M(−2.0, 3.0, 0.0) and N(3.0, 0.0, 4.0): (a) Find the vector $\mathbf{A}_{MN}$ from point M to N. (b) Determine the magnitude of the vector $\mathbf{A}_{MN}$. (c) Find the unit vector direction of $\mathbf{A}_{MN}$. (*Answer*: (a) $\mathbf{A}_{MN} = 5.0\mathbf{a}_x - 3.0\mathbf{a}_y + 4.0\mathbf{a}_z$, (b) $\mathbf{A}_{MN} = 7.1$, (c) $\mathbf{a}_{MN} = 0.71\mathbf{a}_x - 0.42\mathbf{a}_y + 0.57\mathbf{a}_z$)

Drill 2.2 Given the three points A(2, 0, 0), B(0, 4, 0), and C(0, 0, 3), find (a) the vector $\mathbf{A}_{AB}$, (b) the vector $\mathbf{A}_{AC}$, (c) the sum of the two vectors, and (d) the vector difference $\mathbf{A}_{AB} - \mathbf{A}_{AC}$. (*Answer*: (a) $\mathbf{A}_{AB} = -2\mathbf{a}_x + 4\mathbf{a}_y$, (b) $\mathbf{A}_{AC} = -2\mathbf{a}_x + 3\mathbf{a}_z$, (c) $\mathbf{A}_{AB} + \mathbf{A}_{AC} = -4\mathbf{a}_x + 4\mathbf{a}_y + 3\mathbf{a}_z$, (d) $\mathbf{A}_{AB} - \mathbf{A}_{AC} = 4\mathbf{a}_y - 3\mathbf{a}_z$.)

▶ **MATLAB 2.1**

Vectors are expressed in MATLAB using brackets. For instance, the vector $\mathbf{A}_{MN} = 5\mathbf{a}_x - 3\mathbf{a}_y + 4\mathbf{a}_z$ in Drill 2.1 is expressed as[2.2]

A= [5 − 3 4]

with a space between each number.

To find the magnitude of **A**, we must take each of its components (for instance, $A(1)$ = 5), square it, add it to the other squared components, and finally take the square root.

» magA=sqrt(A(1)^2+A(2)^2+A(3)^2)

magA =
7.071

Finally, the unit vector is simply

» unitvectorA=A/magA

unitvectorA =
 0.7071 −0.4243 0.5657

If we think we will frequently need to calculate a vector magnitude, we can create a MATLAB *function*, similar to MATLAB's built-in functions like sqrt() and sin().

[2.2]We'll use MATLAB's arial font for command-line window text and `courier new` font for the M-file editor text.

In the M-file editor, enter

```
function y=magvector(R)
% Calculates the magnitude of a Cartesian vector R
y=sqrt(R(1)^2+R(2)^2+R(3)^2);
```

and save this as "magvector.m."

It is good practice to add commentary immediately below the function statement. Such comments are used in the "help" command. For instance, on the command line,

```
» help magvector
Calculates the magnitude of a cartesian vector R.
```

Likewise, a function can be created to calculate the unit vector. In the M-file editor, enter

```
function y=unitvector(R)
% Calculates the unit vector of a Cartesian vector R
y=R/magvector(R);
```

and save this as "unitvector.m."

Now, on the command line enter

```
» magvector(A)

ans =
7.071

» unitvector(A)

ans =
 0.7071 −0.4243 0.5657
```

Now suppose we add a second vector $\mathbf{B} = -1\mathbf{a}_x - 2\mathbf{a}_y + 3\mathbf{a}_z$. In MATLAB, we make the following entries in the command-line window:

```
» A=[5 −3 4]; B=[−1 −2 3]
» A+B

ans =
4 −5 7
```

We would write this as $\mathbf{A} + \mathbf{B} = 4\mathbf{a}_x - 5\mathbf{a}_y + 7\mathbf{a}_z$. Note that we didn't really need to add the vector $\mathbf{A}$ in MATLAB again, but if $\mathbf{A}$ didn't exist already we could write it on the same line with $\mathbf{B}$.

For a user-defined function to be usable, it must exist in your MATLAB work file. Some functions, like "magvector," will be used in the MATLAB routines of later chapters.

For many problems to come it will be necessary to perform line, surface, and volume integrations. These integrations require differential elements. A differential box in the Cartesian coordinate system is shown in Figure 2.6 along with an expanded inset showing the sides dx, dy, and dz. A differential volume is the product of these three differential lengths, or

$$dv = dx\, dy\, dz \tag{2.8}$$

Differential surfaces are the product of the appropriate differential lengths. For instance, the differential surface closest to the reader in Figure 2.6 is $dS = dy\, dz$. In many instances, a

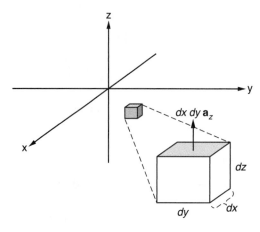

Figure 2.6 A differential volume in Cartesian coordinates shows an expanded inset with sides dx, dy, and dz. The differential surface vector $d\mathbf{S} = dxdy\mathbf{a}_z$ is shaded.

direction is desired for the differential surface. Since there are an infinite number of vector directions in the plane of the surface, a surface vector is only unique if it is normal to the surface. But, you say, how can the vector be unique if it can point away from the surface on either side? The convention is to choose the vector direction that points away from the volume element. So, for the shaded area inset in Figure 2.6, the differential surface vector would be

$$d\mathbf{S} = dx\, dy\, \mathbf{a}_z \tag{2.9}$$

In some problems, a surface will be given without a volume, and the choice of the surface vector direction will then be based on other considerations.

▶ 2.2 COULOMB'S LAW

The Greek philosopher Thales noted that amber, after being rubbed on silk, could pick up pieces of lint or straw. This was one of the earliest documented accounts of static electricity,[2.3] where a positive charge induced on the amber would attract the negative charge of the lint. In the late 18th century, Colonel Charles Augustus Coulomb of the French Army Engineers invented a sensitive torsion balance that he used to experimentally determine the force exerted on one charge by another. He found that the force is proportional to the product of the two charges, inversely proportional to the square of the distance between the charges, and acts in a line containing the two charges. Furthermore, the force is repulsive if both charges are of like sign (i.e., both positive or both negative[2.4]) and attractive if the charges are of different sign. Experimentally, the proportionality constant was found to be $1/(4\pi\varepsilon_o)$, where ε_o (pronounced "epsilon-naught") is the *free space permittivity* in units of farads (F) per meter with a value given by

$$\boxed{\varepsilon_o = 8.854 \times 10^{-12}\, \frac{F}{m} \approx \frac{10^{-9}}{36\pi}\, \frac{F}{m}} \tag{2.10}$$

[2.3]"Electricity" is the Greek word for amber, a rubbery tree secretion.

[2.4]Ben Franklin was the first to use the terms positive and negative to denote the two kinds of charge.

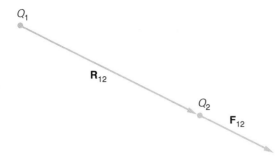

Figure 2.7 Coulomb's law example showing the vector force $\mathbf{F}_{12}$ acting on Q_2 from Q_1.

Consider Figure 2.7, where Q_1 and Q_2 have charge quantities given in coulombs (C). A distance vector $\mathbf{R}_{12} = R_{12}\,\mathbf{a}_{12}$ of magnitude R_{12}(m) and direction $\mathbf{a}_{12}$ can be drawn between the charges. Charge Q_1 exerts a vector force $\mathbf{F}_{12}$, in newtons (N), on charge Q_2 that is given by Coulomb's law:[2.5]

$$\boxed{\mathbf{F}_{12} = \frac{Q_1 Q_2}{4\pi\varepsilon_o R_{12}^2}\,\mathbf{a}_{12}} \tag{2.11}$$

▷ **EXAMPLE 2.2**

Suppose we have a 10. nC charge[2.6] Q_1 located at (0.0, 0.0, 4.0 m) and a 2.0 nC charge Q_2 located at (0.0, 4.0 m, 0.0) as shown in Figure 2.8. We wish to find the force acting on Q_2 from Q_1. For short, we say "Find the force exerted by $Q_1(0, 0, 4\ \text{m}) = 10$ nC on $Q_2(0, 4\ \text{m}, 0) = 2$ nC."

To employ Coulomb's law (2.11), we must first find vector $\mathbf{R}_{12}$, which is

$$\mathbf{R}_{12} = 4\mathbf{a}_y - 4\mathbf{a}_z$$

or

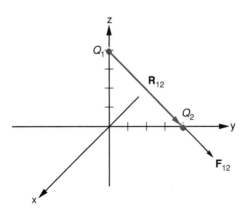

Figure 2.8 Coulomb's law example where $Q_1(0, 0, 4\ \text{m}) = 10$ nC and $Q_2(0, 4\ \text{m}, 0) = 2$ nC.

[2.5] The units for (2.11) can be verified using the following conversions: farad = coulomb/volt (F = C/V), coulomb = joule/volt (C = J/V), and joule = newton-meter (J = N-m).

[2.6] A nanocoulomb, or nC, is 10^{-9} C. Other commonly used numerical prefixes are listed in Table 1.1.

$$R_{12}\mathbf{a}_{12} = 4\sqrt{2}\left(\frac{1}{\sqrt{2}}\mathbf{a}_y - \frac{1}{\sqrt{2}}\mathbf{a}_z\right)$$

Then

$$\mathbf{F}_{12} = \frac{\left(10\times10^{-9}\text{C}\right)\left(2\times10^{-9}\text{C}\right)\left(\frac{1}{\sqrt{2}}\mathbf{a}_y - \frac{1}{\sqrt{2}}\mathbf{a}_z\right)}{4\pi\left(\frac{10^{-9}}{36\pi}\frac{\text{F}}{\text{m}}\right)\left(4\sqrt{2}\text{m}\right)^2}\times\left(\frac{\text{F-V}}{\text{C}}\right)\left(\frac{\text{J}}{\text{C-V}}\right)\left(\frac{\text{N-m}}{\text{J}}\right)$$

$$= 4.0\left(\mathbf{a}_y - \mathbf{a}_z\right)\text{nN}$$

Suppose there are two charges fixed at different points, each exerting force on a third charge. By the *principle of superposition* the total force on the third charge can be calculated by vector addition of the forces from the other two charges.

▶ **EXAMPLE 2.3**

Figure 2.9 shows a third charge, Q_3 (0.0, 0.0, –4.0 m) = 10. nC, added to the previous example. We want to find the total force exerted on charge Q_2 from the charges Q_1 and Q_3.

Observe that, in this case, addition of the two force vectors results in a canceling of the $\mathbf{a}_z$ component and a doubling of the $\mathbf{a}_y$ component to give $\mathbf{F}_{tot} = 8.0\ \mathbf{a}_y$ nN.

Drill 2.3 Find the force exerted by $Q_1(0.00, 0.00, 0.00) = 100.$ nC on $Q_2(4.00$ m, 3.00 m, 0.00) = 3.00 nC. (*Answer:* $\mathbf{F}_{12} = 86.4\mathbf{a}_x + 64.8\mathbf{a}_y$ nN)

Electric Field Intensity

Now suppose we fix a charge Q_1 at, say, the origin. We know that a second charge Q_2 will have force acting on it from Q_1 that can be calculated by Coulomb's law. In fact, we could calculate the force vector that would act on some test charge Q_2 at every point in space and generate a *field* of such predicted force values.

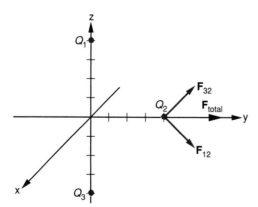

Figure 2.9 Vector addition is used to combine the coulombic forces.

It becomes convenient to define the *electric field intensity* $\mathbf{E}_1$ (force per unit charge) as

$$\boxed{\mathbf{E}_1 = \frac{\mathbf{F}_{12}}{Q_2}}$$

(2.12)

This field from charge Q_1 fixed at the origin results from the force vector $\mathbf{F}_{12}$ for any arbitrarily chosen value of Q_2.[2.7] The electric field concept is useful because the fields can be predicted for any number of charges, and the total field at a particular point in space can be found by adding the fields from each charge.

Coulomb's law can be rewritten as

$$\boxed{\mathbf{E} = \frac{Q}{4\pi\varepsilon_o R^2}\, \mathbf{a}_R}$$

(2.13)

to find the electric field intensity in volts per meter[2.8] at any point in space resulting from a fixed charge Q.

▷ **EXAMPLE 2.4**

Let's find the electric field intensity for the Coulomb's law examples 2.2 and 2.3 by dividing the force vector by the 2-nC "test" charge.

For the first case the field becomes

$$\mathbf{E}_{12} = \frac{\mathbf{F}_{12}}{Q_2} = \frac{4.0\left(\mathbf{a}_y - \mathbf{a}_z\right)\mathrm{nN}}{2.0\ \mathrm{nC}}\ \frac{\mathrm{J}}{\mathrm{N\text{-}m}}\ \frac{\mathrm{C-V}}{\mathrm{J}} = 2.0\left(\mathbf{a}_y - \mathbf{a}_z\right)\frac{\mathrm{V}}{\mathrm{m}}$$

For the second case we have

$$\mathbf{E}_{\mathrm{tot}} = \frac{\mathbf{F}_{\mathrm{tot}}}{Q_2} = \frac{8.0\mathbf{a}_y\,\mathrm{nN}}{2.0\ \mathrm{nC}} = 4.0\mathbf{a}_y\ \frac{\mathrm{V}}{\mathrm{m}}$$

Drill 2.4   For the charges in Drill 2.3, find the electric field intensity at point 2 from the charge at point 1. (*Answer*: $\mathbf{E}_1 = 28.8\mathbf{a}_x + 21.6\mathbf{a}_y$ V/m)

Field Lines

The behavior of the fields can be visualized by using *field lines*.[2.9] These are lines that follow the direction of field vectors at convenient points in space, as illustrated in Figures 2.10a & 2.10b. In Figure 2.10a, the field vectors are found within a regular grid in

---

[2.7]As a fine point in this discussion, when defining the fields the charge Q_2 is considered to be small enough that it doesn't significantly alter the field in which it is placed. Ideally, it will be infinitesimal. A single electron (with charge $q = -1.6 \times 10^{-19}$ C) makes an excellent test charge.

[2.8]The V/m units for $\mathbf{E}$ can be found by reducing the units on the right side of (2.13) to C/(F-m) and then by applying the conversion F = C/V.

[2.9]Field lines are also called *lines of force* since they represent the direction and magnitude of the force component on a test charge placed at a particular point. Other terms used are *streamlines* and *flux lines*.

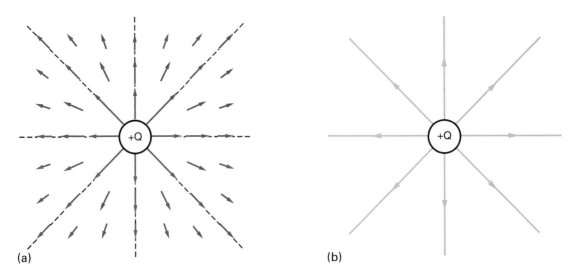

(a)   (b)

Figure 2.10 (a) Field vectors plotted within a regular grid in two-dimensional space surrounding a point charge. Some of these field vectors can easily be joined by field lines, as shown in (b), that emanate from the positive charge.

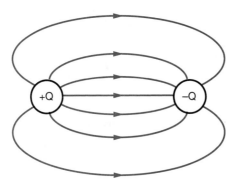

Figure 2.11 Field lines for a pair of opposite charges.

two-dimensional space surrounding a point charge. Some of these field vectors can easily be joined by field lines, as shown in Figure 2.10b, that radiate away from the positive charge. Field strength at a point is related to how close together the field lines are near that point. The field lines provide a convenient visual aid to understand what the fields are doing. For example, Figure 2.11 shows the field lines plotted for a pair of opposite charges.

► 2.3 THE SPHERICAL COORDINATE SYSTEM

Consider a point charge located at the origin. The electric field intensity resulting from this charge can be written for any point in the Cartesian coordinate system as

$$\mathbf{E} = \frac{Q}{4\pi\varepsilon_o} \frac{\left(x\mathbf{a}_x + y\mathbf{a}_y + z\mathbf{a}_z\right)}{\left[x^2 + y^2 + z^2\right]^{3/2}} \tag{2.14}$$

Although not readily apparent from (2.14), inspection shows that the fields are everywhere directed radially away from the point charge and that the magnitude of the field depends only on this radial distance. A serious challenge arises when we use (2.14) to find the total field resulting from a distribution of point charges. A multivariable integration would be required.

A spherical coordinate system offers relief. As a point in the Cartesian system is represented by the orthogonal points (x, y, z), a point in the spherical system is represented by the orthogonal points (r, θ, ϕ), as shown in Figure 2.12. Here, r is the radial distance, or *range*, from the origin to the point. The angle θ (theta) is measured from the positive z-axis and is sometimes referred to as the *colatitude*. This angle can have values ranging from 0 to π radians (0° to 180°). The angle ϕ (phi) is the angle as you move around the z-axis starting from the positive x-axis and is often referred to as the *azimuthal* angle. It can have values ranging from 0 to 2π radians (0° to 360°). Spherical coordinates find extensive use in radar and antenna applications.

Conversion from the point $P(x, y, z)$ to $P(r, \theta, \phi)$ begins with determination of r using the Pythagorean theorem in three dimensions:

$$r = \sqrt{x^2 + y^2 + z^2} \tag{2.15}$$

The angles θ and ϕ are found from trigonometry to be

$$\theta = \cos^{-1}\left(\frac{z}{r}\right) \tag{2.16}$$

and

$$\phi = \tan^{-1}\left(\frac{y}{x}\right) \tag{2.17}$$

To convert from spherical to Cartesian coordinates it is easy to see that

$$z = r \cos \theta \tag{2.18}$$

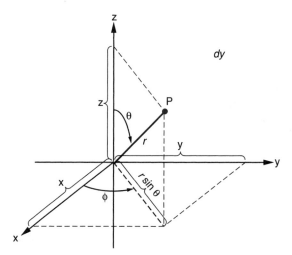

Figure 2.12 The spherical coordinate system is represented by the orthogonal points (r, θ, ϕ).

To find x and y, a line from the origin to P is projected onto the x–y plane. This projected line, of length $r \sin \theta$, is used with trigonometry to give

$$x = r \sin \theta \cos \phi$$
$$y = r \sin \theta \sin \phi$$

(2.19)

A vector in the spherical coordinate system can be written

$$\boxed{\mathbf{A} = A_r \mathbf{a}_r + A_\theta \mathbf{a}_\theta + A_\phi \mathbf{a}_\phi}$$

(2.20)

Converting a Cartesian vector to a spherical vector (and vice versa) requires that the location of the vector in space is known. Some formulas for handling the transformation are given in Appendix B, but a simple example is illustrated in Figure 2.13. Suppose a vector located at (0, 5, 0) is given by $2\mathbf{a}_x + 3\mathbf{a}_y$ in the Cartesian coordinate system. At that location, the $3\mathbf{a}_y$ converts to $3\mathbf{a}_r$, and $2\mathbf{a}_x$ converts to $-2\mathbf{a}_\phi$. So the vector becomes $3\mathbf{a}_r - 2\mathbf{a}_\phi$. Note that moving the Cartesian vector to a different point would result in a different spherical vector. For instance, if the vector were at (4, 0, 0), then it would convert to $2\mathbf{a}_r + 3\mathbf{a}_\phi$. This example is relatively simple because we are restricted to the x–y plane ($\theta = \pi/2$).

Returning to the expression for a field from a point charge located at the origin, we can write (2.14) much more elegantly in terms of spherical coordinates as

$$\boxed{E = \frac{Q}{4\pi\varepsilon_o r^2} a_r}$$

(2.21)

Figure 2.14 shows a differential element in the spherical coordinate system. If the element is considered to be extremely small (which it should be as it is differential) then it can be treated as a box where the sides are the differential lengths indicated in the figure inset. These lengths are found in Appendix B by trigonometric considerations. The volume of the differential element is

$$dv = r^2 \sin \theta \, dr \, d\theta \, d\phi$$

(2.22)

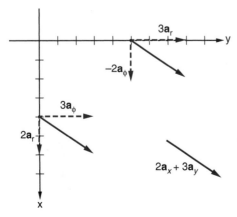

Figure 2.13   An illustration showing the conversion between Cartesian and spherical vectors. $2\mathbf{a}_x + 3\mathbf{a}_y$ at (0, 5, 0) converts to $3\mathbf{a}_r - 2\mathbf{a}_\phi$, and the same vector located at (4, 0, 0) converts to $2\mathbf{a}_r + 3\mathbf{a}_\phi$.

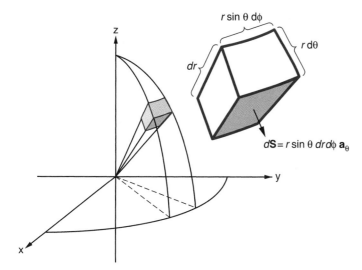

Figure 2.14 A differential element in the spherical coordinate system. One of the element's six surfaces is shaded with the differential surface vector indicated.

There are also six differential surfaces. The one that is shaded in the inset is the product of sides dr and $r \sin \theta \, d\phi$. Since by convention the direction of a surface vector is away from the volume, this differential surface vector is

$$dS = r \sin \theta \, dr \, d\phi \, \mathbf{a}_\theta \qquad (2.23)$$

The other differential surfaces and their directions can be determined in a like manner.

▶ **EXAMPLE 2.5**

As a simple example of the use of dv and dS, consider a solid sphere of radius R. We can determine the total volume by

$$\int dv = \iiint r^2 \sin \theta \, dr \, d\theta \, d\phi = \int_0^R r^2 \, dr \int_0^\pi \sin \theta \, d\theta \int_0^{2\pi} d\phi = \frac{4}{3} \pi R^3$$

The total surface area can be found by

$$\int dS = \iint r^2 \sin \theta \, d\theta \, d\phi = R^2 \int_0^\pi \sin \theta \, d\theta \int_0^{2\pi} d\phi = 4\pi R^2$$

Drill 2.5 Convert the Cartesian coordinate point P(3.00, 5.00, 9.00) to its equivalent point in spherical coordinates. (*Answer*: P(10.7, 32.9°, 59.0°))

Drill 2.6 Convert the spherical coordinate point M(5, 60°, 135°) to its equivalent point in Cartesian coordinates. (*Answer*: M(−3.06, 3.06, 2.5))

Drill 2.7 Determine the volume bounded by $1.00 \le r \le 4.00$ m, $45.0° \le \theta \le 135.°$. (*Answer*: $v = 187$ m^3)

▶ 2.4 LINE CHARGES AND THE CYLINDRICAL COORDINATE SYSTEM

We have seen that the spherical coordinate system is very useful for problems having spherical symmetry. The *cylindrical* coordinate system[2.10] is good for solving problems that have cylindrical symmetry, such as those involving long wires and transmission lines.

The cylindrical coordinate system uses the orthogonal components ρ, ϕ, and z, as illustrated in Figure 2.15, where point P (ρ, ϕ, z) is shown. Here, ρ ("rho") is the radial distance from the z-axis to the point P. The angle ϕ goes from the +x-axis to a projection of the radial line onto the x–y plane. This azimuthal angle is the same as the angle ϕ for the spherical coordinate system. Moreover, z is the same as in the Cartesian system.

In converting P(x, y, z) to its equivalent point in the cylindrical coordinate system, P(ρ, ϕ, z), it is helpful to project the radial line onto the x–y plane. Then, the radial distance ρ is easily seen to be

$$\rho = \sqrt{x^2 + y^2} \tag{2.24}$$

The angle ϕ makes from the +x-axis to the projected radial line is related to x and y by

$$\phi = \tan^{-1}\left(\frac{y}{x}\right) \tag{2.25}$$

Finally,

$$z = z \tag{2.26}$$

As an example, Figure 2.16 shows the point P(3, 4, 5). Using the conversions of (2.24)–(2.26), which are readily verified by inspection of the figure, the point in cylindrical coordinates is P(5, 53°, 5).

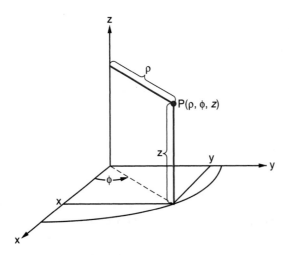

Figure 2.15 Cylindrical coordinate system.

[2.10]This system is also called *circular cylindrical* to distinguish it from rarely used types of cylindrical coordinate systems such as elliptical, hyperbolic, and parabolic.

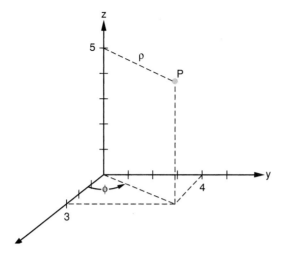

Figure 2.16 Shows conversion of the point P(3,4,5) in Cartesian coordinates to its equivalent point in cylindrical coordinates.

The formulas to convert from cylindrical to Cartesian coordinates are easily derived by inspection of Figure 2.15 to be

$$x = \rho \cos \phi$$
$$y = \rho \sin \phi \qquad\qquad (2.27)$$
$$z = z$$

A vector in cylindrical coordinates is written as

$$\boxed{\mathbf{A} = A_\rho \mathbf{a}_\rho + A_\phi \mathbf{a}_\phi + A_z \mathbf{a}_z} \qquad\qquad (2.28)$$

Just as for spherical coordinates, a vector in cylindrical coordinates is a function of its location. Suppose you have a unit vector $\mathbf{a}_y$ in the Cartesian system. This would convert to a unit vector $\mathbf{a}_\phi$ if the point is located on the +x-axis, and to $\mathbf{a}_\rho$ if the point is located on the +y-axis. Formulas for converting from one system to another are provided in Appendix B.

A differential element in the cylindrical coordinate system is shown in Figure 2.17. The differential volume is given by

$$dv = \rho \, d\rho \, d\phi \, dz \qquad\qquad (2.29)$$

and one of the differential surface vectors is indicated in the figure inset as

$$d\mathbf{S} = \rho \, d\rho \, d\phi \, \mathbf{a}_z$$

▷ **EXAMPLE 2.6**

Consider a volume bounded by radius ρ from 3.00 to 4.00 cm, height from 0.00 to 6.00 cm, and angle from 90.0° to 135.0°. To determine the volume of this component, the integral is

$$V = \iiint \rho \, d\rho \, d\phi \, dz = \int_3^4 \rho \, d\rho \int_{\pi/2}^{3\pi/4} d\phi \int_0^6 dz = 16.5 \text{ cm}^3$$

Note that we had to convert the limits on ϕ from degrees to radians (i.e., 180° = π radians).

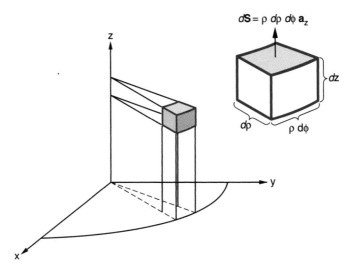

Figure 2.17 A differential element in cylindrical coordinates.

Drill 2.8 Convert the Cartesian coordinate point P(3.0, 5.0, 9.0) to its equivalent point in cylindrical coordinates. (*Answer*: P(5.8, 59°, 9))

Drill 2.9 Convert the cylindrical coordinate point M(5, 180°, 2) to its equivalent point in Cartesian coordinates. (*Answer*: M(−5, 0, 2))

▶ **MATLAB 2.2**

Although MATLAB does have some limited functions to handle cylindrical and spherical coordinates, it is instructive to come up with our own program (a MATLAB *script file*) to perform a conversion, for instance, from cylindrical to Cartesian coordinates.

In the M-Editor window, the following program (saved as ML0202) is entered:

```
%   M-File: ML0202
%
%   This program converts a cylindrical coordinate point
%   to a Cartesian point.
%
%   Wentworth, 7/5/02
%
%   Variables:
%   r           radial distance from z-axis
%   thetad      angle with x-axis in degrees
%   z           z location
%   theta       angle in radians
%   x,y,z       the cartesian coordinates
%   cart        the cartesian coord. point

clc             %clears the command window
clear           %clears variables
```

```
%Prompt for input values
r=input('enter the value of r: ');
thetad=input('enter the value of theta (in degrees): ');
z=input('enter the value of z: ');

%Perform conversion
theta=thetad*pi/180; %convert deg to rad
x=r*cos(theta);
y=r*sin(theta);
z=z;
cart=[x,y,z]
```

Now, we'll run this program to solve Drill 2.9. In the command-line window, we type

ML0202

In this window we get several prompts to answer:

```
enter the value of r: 5
enter the value of theta: 180
enter the value of z: 2


cart =
 -5.0000  0.0000  2.0000
```

Formal commands in MATLAB to do coordinate system conversions are cart2pol, cart2sph, pol2cart, and sph2cart. The reader should use the "help" feature to determine how to use these commands. For instance,

```
» help cart2pol
```

CART2POL Transform Cartesian to polar coordinates.
   [TH,R] = CART2POL(X,Y) transforms corresponding elements of data stored in Cartesian coordinates X,Y to polar coordinates (angle TH and radius R). The arrays X and Y must be the same size (or either can be scalar). TH is returned in radians.

   [TH,R,Z] = CART2POL(X,Y,Z) transforms corresponding elements of data stored in Cartesian coordinates X,Y,Z to cylindrical coordinates (angle TH, radius R, and height Z). The arrays X,Y, and Z must be the same size (or any of them can be scalar). TH is returned in radians.

   See also CART2SPH, SPH2CART, POL2CART.

Notice that the order of the angles may be different from what we've been using.

Infinite Length Line of Charge

Now we wish to derive the electric field at any point in space resulting from an infinite length line of charge placed conveniently along the z-axis. Many students complain at this point about the unreality of an infinite length line. However, this can actually be a good approximation for many practical applications. For instance, a test charge placed a couple of centimeters from an elevated transmission line will see what appears to be an infinite length line.

   To create this line of charge, we will place an amount of charge, in coulombs, evenly along every meter of the z-axis. The *linear charge density* is the coulombs of charge per meter length and is given by ρ_L (C/m). Using ρ with an appropriate subscript is a standard way to denote charge density and is not to be confused with the ρ used for radial distance in the cylindrical coordinate system. L, S, and V are the subscripts used for line (ρ_L), surface (ρ_S), and volume (ρ_V) charge densities, respectively.

Consider Figure 2.18a showing the line charge along the z-axis and choose an arbitrary point P(ρ, ϕ, z) where we want to find the electric field intensity. If we fix ρ and ϕ, we notice that the line looks exactly the same to us no matter where we are in z because the line is infinitely long. That is, if we go up 20 m, we expect the field acting on us to be exactly the same as it was 20 m lower. Likewise, if we only fix ρ, and move around to some new angle ϕ, the line continues to appear exactly the same to us. From this we can argue that, whatever the electric field intensity, it will only vary with the radial distance from the line. Mathematically we can say

$$\mathbf{E}(\rho) = E_\rho\mathbf{a}_\rho + E_\phi\mathbf{a}_\phi + E_z\mathbf{a}_z \qquad (2.30)$$

where the (ρ) indicates that **E** is only a function of ρ. Since only the ρ coordinate affects our answer, we are free to reorient our problem and test a point on the more convenient ρ–z axes shown in Figure 2.18b. Now let's simplify (2.30). Selecting an arbitrary segment of charge dQ at a distance z below the radial axis, we note that it will give us E_ρ and E_z components of the field, but not E_ϕ In fact, we can find no segment of charge dQ anywhere on the z-axis that will give us E_ϕ. So (2.30) becomes

$$\mathbf{E}(\rho) = E_\rho\mathbf{a}_\rho + E_z\mathbf{a}_z \qquad (2.31)$$

Now consider a dQ segment a distance z above the radial axis. The additional field components are given in Figure 2.18c. Notice that the E_z components cancel each other, and the E_ρ components add. In fact, for any arbitrary charge segment on this infinite length line of

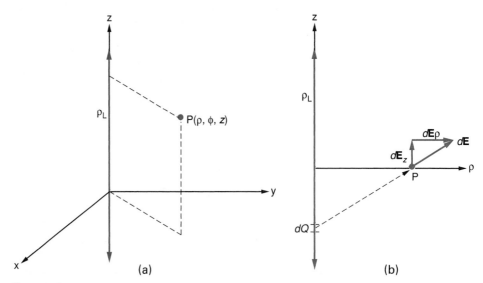

Figure 2.18 (a) An infinite length line of charge, ρ_L, is placed along the z-axis. We wish to find **E** at an arbitrary point P(ρ, ϕ, z). (b) Redrawn problem in the ρ–z plane with field components for a charge element dQ as indicated.

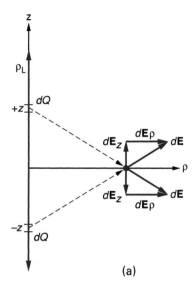

(a)

Figure 2.18 (c) Adding the field components for a second charge element dQ above the radial axis.

charge, you can always find another segment that will cancel out the E_z component. We say that the E_z component cancels *by symmetry*. So now our equation becomes

$$\mathbf{E}(\rho) = E_\rho \mathbf{a}_\rho \tag{2.32}$$

Next, we recall the expression for the electric field intensity for a point charge is

$$\mathbf{E} = \frac{Q}{4\pi\varepsilon_o R^2}\mathbf{a}_R$$

If we consider a collection of such charges, the total field will be a summation of the vector field for each charge. For a continuous charge distribution, the summation becomes an integral written as

$$\boxed{E = \int \frac{dQ}{4\pi\varepsilon_o R^2}\mathbf{a}_R} \tag{2.33}$$

The common procedure for solving charge distribution problems of this sort is to start with (2.33), make the proper substitutions for each component, and finally perform the integration. Figure 2.19 guides us in making substitutions for each component of (2.33). The differential charge is $dQ = \rho_L\, dz$. The vector drawn from the source to the test point is

$$\mathbf{R} = R\mathbf{a}_R = \rho\mathbf{a}_\rho - z\mathbf{a}_z$$

This has a magnitude $R = \sqrt{\rho^2 + z^2}$ and unit vector $\mathbf{a}_R = (\rho\mathbf{a}_\rho - z\mathbf{a}_z)/R$. Placing all this into (2.33) yields

$$E = \int \frac{\rho_L dz}{4\pi\varepsilon_o} \frac{\rho\mathbf{a}_\rho - z\mathbf{a}_z}{\left(\rho^2 + z^2\right)^{3/2}} \tag{2.34}$$

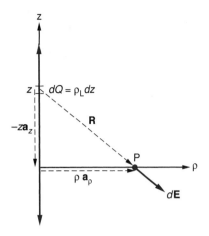

Figure 2.19 Sketch to show where all the components come from to set up the integral.

Now, because we've already established that no $\mathbf{a}_z$ component will be present in the final answer, we can simplify (2.34) to

$$E = \frac{\rho_L \rho \mathbf{a}_\rho}{4\pi\varepsilon_o} \int_{-\infty}^{+\infty} \frac{dz}{\left(\rho^2 + z^2\right)^{3/2}}$$

(2.35)

where the constants have been pulled outside of the integral, and the limits of integration over z go from $-\infty$ to $+\infty$. Solving the integral[2.11] gives the electric field intensity at any point ρ away from an infinite length line of charge as

$$\boxed{E = \frac{\rho_L}{2\pi\varepsilon_o \rho} \mathbf{a}_\rho}$$

(2.36)

Note that (2.34) and (2.35) can be solved for finite lengths of line by appropriately changing the limits on the integral. We must use (2.34) rather than (2.35) unless we are sure that symmetry will cause the $\mathbf{a}_z$ component to cancel.

► **EXAMPLE 2.7**

Suppose an infinite length line of charge $\rho_{L1} = 4.00$ nC/m exists at $x = 2.00$ m, $z = 4.00$ m. We want to find the electric field intensity at the origin.

Since this is an infinite length line of charge, we want to figure out the various components in (2.36). A sketch of the problem is a good starting point. In Figure 2.20a the line charge is sketched in three dimensions. The problem is easiest to solve with the cross section of Figure 2.20b. Here the vector $\rho \mathbf{a}_\rho$ is clearly

$$\rho \mathbf{a}_\rho = -2\mathbf{a}_x - 4\mathbf{a}_z$$

from which we see that $\rho = \sqrt{20}$, and the unit vector is

[2.11]The solution of this and other common integrals encountered in electromagnetics can be found in Appendix D.

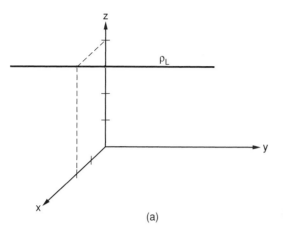

(a)

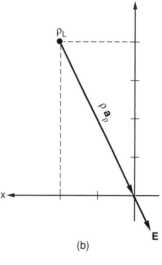

(b)

Figure 2.20 (a) An infinite length line charge is located at $x = 2$, $z = 4$. (b) Cross section at $y = 0$.

$$\mathbf{a}_\rho = -\frac{2}{\sqrt{20}}\mathbf{a}_x - \frac{4}{\sqrt{20}}\mathbf{a}_z$$

Inserting this into (2.36), we obtain

$$\mathbf{E} = \frac{4\times10^{-9}\,\text{C/m}}{2\pi}\frac{36\pi}{10^{-9}}\frac{\text{m}}{\text{F}}\frac{1}{\sqrt{20}\text{m}}\frac{\left(-2\mathbf{a}_x - 4\mathbf{a}_z\right)}{\sqrt{20}}\frac{\text{F-V}}{\text{C}} = -7.2\mathbf{a}_x - 14.4\mathbf{a}_z\,\frac{\text{V}}{\text{m}}$$

Drill 2.10 Suppose to Example 2.7 we add a second infinite length line of charge ρ_{L2} = 8.00 nC/m at $x = -2.00$ m, $y = 3.00$ m. Determine the total electric field intensity at the origin resulting from the presence of both lines of charge. (*Answer:* $\mathbf{E} = 15.0\mathbf{a}_x - 33.2\mathbf{a}_y - 14.4\mathbf{a}_z$ V/m)

Drill 2.11 A segment of line charge $\rho_L = 10.$ nC/m exists on the y-axis from $y = -3.0$ m to $y = +3.0$ m. Determine $\mathbf{E}$ at the point $(3.0, 0.0, 0.0)$m. (*Answer:* $\mathbf{E} = 42\mathbf{a}_x$ V/m)

Ring of Charge

As another illustration of the procedure for using Coulomb's law to find $\mathbf{E}$, consider the ring of charge, of charge density ρ_L, centered at the origin in the x–y plane. The task is to find $\mathbf{E}$ at a point on the z-axis, at $(0, 0, h)$. The problem is illustrated in Figure 2.21a. We work this problem in cylindrical coordinates, and our first task is to determine which components of $\mathbf{E}$ are present. Grabbing a dL section of charge, as shown in Figure 2.21b, we notice by inspection that it delivers $d\mathbf{E}_\rho$ and $d\mathbf{E}_z$ contributions to the field, but not $d\mathbf{E}_\phi$. We also see that a dL section of charge grabbed on the opposite side of the ring will also deliver $d\mathbf{E}_\rho$ and

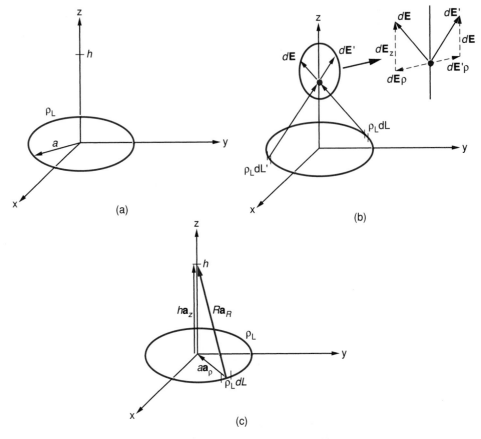

Figure 2.21 (a) **E** is sought at a point h on the z-axis from a ring of charge centered on the x–y plane. (b) $d\mathbf{E}$ components from a pair of charge segments dL on opposite sides of the ring. (c) Terms for setting up the integral.

$d\mathbf{E}_z$, but the $d\mathbf{E}_\rho$ component will be in the opposite direction as the first one and will there- fore cancel by symmetry. So we see that only $\mathbf{E}_z$ will be present in the solution.

As before, we need to make the proper substitutions into Eq. (2.33),

$$\mathbf{E} = \int \frac{dQ}{4\pi\varepsilon_o R^2}\,\mathbf{a}_R$$

Figure 2.21c guides us in the proper determination of each component for (2.33). The dif- ferential charge is $dQ = \rho_L\, a\, d\phi$. The vector drawn from the source to the test point is

$$\mathbf{R} = R\mathbf{a}_R = -a\mathbf{a}_\rho + h\mathbf{a}_z$$

The integral then becomes

$$\mathbf{E} = \int \frac{\rho_L\, a\, d\phi}{4\pi\varepsilon_o \left(a^2 + h^2\right)^{3/2}}\left(-a\mathbf{a}_\rho + h\mathbf{a}_z\right)$$

Rearranging the integral and considering that the $\mathbf{a}_\rho$ components cancel by symmetry, we have

$$E = \frac{\rho_L a h \mathbf{a}_z}{4\pi\varepsilon_o \left(a^2 + h^2\right)^{3/2}} \int\limits_0^{2\pi} d\phi$$

This is easily solved, resulting in

$$E = \frac{\rho_L a h \mathbf{a}_z}{2\varepsilon_o \left(a^2 + h^2\right)^{3/2}}$$

The problem would be considerably more difficult if we were asked to find **E** somewhere off of the z-axis. In such a case, a numerical integration in the Cartesian coordinate system would be the best approach.

2.5 SURFACE AND VOLUME CHARGE

The electric charge on a conductor will tend to distribute itself on the surface. Such a thin layer of charge is considered a continuous distribution of surface charge[2.12] and is given by the surface charge density ρ_s (C/m²).

To begin, we'll determine the field at a height h above a charge sheet of infinite area, for instance occupying the x–y plane. Note that since the sheet is infinite in both the x and y directions, the field at a point P is independent of the particular x and y location. So, we can conveniently locate the point P at a height h on the +z-axis as shown in Figure 2.22. If we consider a differential charge $dQ = \rho_s\, dx\, dy$, we see that the differential field vector $d\mathbf{E}$ at P will have E_x, E_y, and E_z field components. But we can always choose other differential charges to cancel the E_x and E_y components, leaving us with only E_z. Drawing the vector from the given source element to the test point, we manipulate

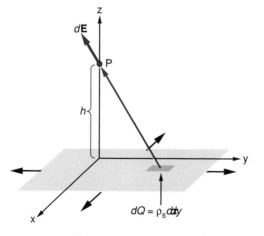

Figure 2.22 The point P is shown h above a current sheet of infinite extent occupying the x–y plane.

[2.12]At the atomic level, the charge distribution appears grainy. However, most practical problems occur at the macroscopic level where the charge distribution does indeed appear to be continuous.

$$\mathbf{E} = \int \frac{dQ}{4\pi\varepsilon_o R^2}\mathbf{a}_R$$

to get

$$\mathbf{E} = \iint \frac{\rho_s dx\, dy\, h\mathbf{a}_z}{4\pi\varepsilon_o\left(x^2 + y^2 + h^2\right)^{3/2}} \tag{2.37}$$

where the limits on the x and y integrals both go from $-\infty$ to $+\infty$. This is a very difficult integral to solve. We could certainly perform a numerical analysis (i.e., use a computer algorithm) to estimate $\mathbf{E}$ for a very large sheet. Fortunately, there are other, easier ways to solve this problem. Let's consider that the infinite extent sheet of charge is broken up into a continuous series of infinite length line charges, such that a single line of charge in the x-direction would have a value $\rho_L = \rho_s\, dy$ as shown in Figure 2.23. We previously found that, for an infinite length line of charge,

$$\mathbf{E} = \frac{\rho_L \mathbf{a}_\rho}{2\pi\varepsilon_o\rho}$$

In our infinite-extent sheet problem, each line would contribute $d\mathbf{E}$ and we'll have to integrate over all y to get the contributions from all the lines. Again referring to Figure 2.23, we see that the vector from source to test point is $\rho\mathbf{a}_\rho = -y\mathbf{a}_y + h\mathbf{a}_z$. Realizing that the E_y component is zero by the problem's symmetry, we have

$$\mathbf{E} = \int \frac{\rho_s dy\, h\mathbf{a}_z}{2\pi\varepsilon_o\left(y^2 + h^2\right)} = \frac{\rho_s h\mathbf{a}_z}{2\pi\varepsilon_o}\int_{-\infty}^{+\infty} \frac{dy}{\left(y^2 + h^2\right)} \tag{2.38}$$

From the table of integrals in Appendix D, we find

$$\int_{-\infty}^{+\infty} \frac{dy}{\left(y^2 + h^2\right)} = \frac{1}{h}\tan^{-1}\left(\frac{y}{h}\right)\Bigg|_{-\infty}^{+\infty} = \frac{\pi}{h}$$

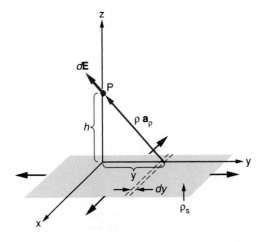

Figure 2.23 Infinite sheet of charge broken up into a continuous series of lines of charge.

Therefore,

$$E = \frac{\rho_s \mathbf{a}_z}{2\varepsilon_o} \tag{2.39}$$

A general expression for the field from a sheet charge is

$$\boxed{E = \frac{\rho_s}{2\varepsilon_o} \mathbf{a}_n} \tag{2.40}$$

where $\mathbf{a}_n$ is the unit vector normal from the sheet to the test point. Notice that the field does not depend on the distance from the sheet! As an analogy, consider a uniform, featureless, white sheet of infinite extent. Let it be back-lit and provide the only source of light (so you won't cast a shadow on the sheet). Equation 2.40 says you will not be able to tell whether you are standing a meter away from the sheet or 100 km. The intensity of illumination you will see is the same in either case.

This approach can be extended to find the field at a point directly above the center of a ribbon of charge of infinite length by changing the limits on the integral in (2.38).

Another way to solve the infinite extent sheet of charge problem is by considering the sheet to consist of a continuous series of concentric rings. From the derivation for a ring charge we have the contribution to $\mathbf{E}$ for each ring of

$$d\mathbf{E} = \frac{(\rho_s d\rho)\rho h \mathbf{a}_z}{2\varepsilon_o \left(\rho^2 + h^2\right)^{3/2}}$$

Here, the ring charge density ρ_L has been replaced with $\rho_s d\rho$. To find the total field, we must integrate the radius ρ from 0 to ∞, or

$$\mathbf{E} = \frac{\rho_s h \mathbf{a}_z}{2\varepsilon_o} \int_0^\infty \frac{\rho d\rho}{\left(\rho^2 + h^2\right)^{3/2}}$$

This is easily integrated by substitution to yield (2.39). This approach can also be adapted to the problem where the field is sought at a point above the middle of a circular disk of finite radius by changing the upper limit on the integration.

▷ **EXAMPLE 2.8**

An infinite extent sheet of charge with $\rho_s = 10.0$ nC/m² exists at the plane $y = -2.00$ m. We want to find the electric field intensity at the point P(0.00, 2.00 m, 1.00 m).

We can use Figure 2.24 to determine the components to use in (2.40). The unit vector directed away from the sheet and toward the point P is $\mathbf{a}_y$. Then we have

$$\mathbf{E}_p = \frac{\rho_s}{2\varepsilon_o} \mathbf{a}_n = \frac{10 \times 10^{-9}\,\text{C/m}^2}{2} \frac{36\pi}{10^{-9}} \frac{\text{m}}{\text{F}} \mathbf{a}_y = 565 \mathbf{a}_y \frac{\text{V}}{\text{m}}$$

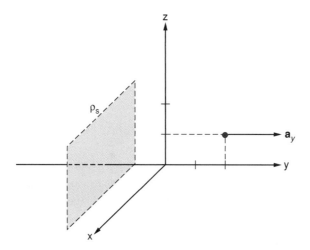

Figure 2.24 Infinite extent sheet of charge at $y = -2$ m for Example 2.8.

Drill 2.12 A charged sheet with $\rho_{s1} = 100.$ nC/m^2 occupies the $z = -3.00$ m plane, and a second charged sheet with $\rho_{s2} = -100.$ nC/m^2 occupies the $z = +3.00$ m plane. Find the electric field intensity at (a) the origin, (b) M(0.00, 0.00, 6.00 m), (c) N(6.00 m, 6.00 m, 6.00 m), and (d) P(0.00, 0.00, −6.00 m). (*Answer:* (a) $\mathbf{E} = 11.3\ \mathbf{a}_z$ kV/m, (b) $\mathbf{E} = 0$, (c) $\mathbf{E} = 0$, (d) $\mathbf{E} = 0$.)

Drill 2.13 A ribbon of charge in the x–z plane with $\rho_s = 10.0$ nC/m^2 exists in the range -2.00 m $\leq x \leq 2.00$ m. Find the electric field intensity at the point (0.00, 3.00 m, 10.0 m). (*Answer:* $\mathbf{E} = 212\ \mathbf{a}_z$ V/m)

Drill 2.14 A circular disk in the x–y plane with $\rho_s = 10.0$ nC/m^2 exists in the range $0 \leq \rho \leq 5.00$ m. Find the electric field intensity at the point (0.00, 0.00, 5.00 m). (*Answer:* $\mathbf{E} = 166\ \mathbf{a}_z$ V/m)

MATLAB 2.3

The closer you get to a finite radius disk of charge, the more you would expect it to resemble an infinite sheet. In the following example, we consider a charged disk of radius a and we want to find the electric field intensity at a point h above the center of the disk. Considering Drill 2.14, we expect

$$\mathbf{E} = \frac{\rho_s}{2\varepsilon_o}\left[1 - \frac{h}{\sqrt{a^2 + h^2}}\right]\mathbf{a}_z$$

So, the ratio of this E_z to the E_z for an infinite sheet will just be the portion of the equation inside the brackets. If we plot this ratio against a factor $k = a/h$, we can manipulate the ratio to be

$$\frac{(E_z)_{\text{actual}}}{(E_z)_{\text{ideal}}} = \left[1 - \frac{1}{\sqrt{1+k^2}}\right]$$

and plotting the ratio of the fields versus k will give us some idea of when the sheet starts to look infinite.

In the M-Editor window, the following program (saved as ML0203) is entered:

```
%  M-File: ML0203
%
%  This program compares the E-field from a finite
%  radius disk of charge to the E-field from an
%  infinite sheet of charge. The ratio (E from disk to
%  E from sheet) is plotted versus the ratio k=a/h,
%  where a is the disk radius and h is the height from
%  the disk center.
%
%  Wentworth, 7/6/02
%
%  Variables:
%  k         the ratio a/h
%  Eratio    ratio of E from disk to E from sheet

clc           %clears the command window
clear         %clears variables

%  Initialize k array and calculate Eratio
k=0.1:0.1:100;
Eratio=1-(1./(sqrt(1+(k.^2))));

%  Plot Eratio versus k
plot(k,Eratio)
grid on
xlabel('k=a/h')
ylabel('E ratio: finite to infinite disk')
```

Notice the "./" and ".^" operations in this program. The dot is needed so that the operation will be an array operation (operating on each individual element) instead of a linear algebra–based matrix operation.

We run this program by typing ML0203 in the command-line window, or we can select "Run" under the Debug menu of the editor window. When the plot comes up, in the figure window under tools, we'll select "edit plot" and then "edit axis properties." Setting the x-axis to a log scale (rather than linear), we get the plot shown in Figure 2.25. Note that, instead of using the edit feature, we could also have replaced the "plot" command with a "semilogx" command in the program.

Notice that when the radius is 10 times the height, the field is 90% of what we would get for an infinite sheet.

Volume Charge

A volume charge is distributed over a volume and is characterized by its volume charge density ρ_v in C/m^3. Plasma, a charged gaslike form of matter, can be contained in a magnetic field and is an example of a volume charge. Very high temperature plasmas have been

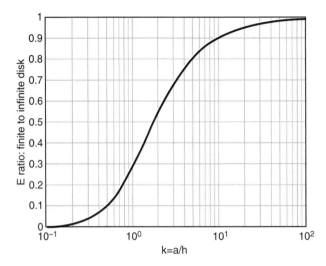

Figure 2.25 The electric field intensity at a height h above a circular disk of radius a is divided by the electric field intensity for a charged sheet of infinite extent. This ratio is plotted versus a/h using MATLAB.

under study for years in the quest to make nuclear fusion a feasible energy source. As another example of volume charge, consider a *doped* semiconductor (i.e., a semiconductor to which specific impurities are added). When a semiconductor such as silicon is doped with atoms containing one extra electron than silicon (arsenic, for instance), this extra electron is very loosely held to the donor atom and is free to move around. These mobile electrons leave behind a distribution of positive charged ions fixed in the crystal lattice. Formulas governing semiconductor operation are derived based on electrostatics with consideration of the volume charge distribution for ions fixed in the lattice.

The total charge in a volume containing a charge distribution ρ_v is found by integrating over the volume: $Q = \int \rho_v dv$.

▶ **EXAMPLE 2.9**

Consider a spherical collection of charge with density ρ_v. The total charge for this sphere of radius a would be simply

$$Q = \iiint \rho_v r^2 \sin\theta\, dr\, d\theta\, d\phi = \rho_v \int_0^a r^2 dr \int_0^\pi \sin\theta\, d\theta \int_0^{2\pi} d\phi = \frac{4}{3}\pi a^3 \rho_v$$

More often, the charge density will vary with position. Suppose the charge density for the sphere is a function of the squared radius, or $\rho_v = kr^2$, where k is a constant. Then the total charge would be

$$Q = \int_0^a kr^4 dr \int_0^\pi \sin\theta\, d\theta \int_0^{2\pi} d\phi = \frac{4}{5}\pi ka^5$$

To find the electric field intensity resulting from a volume charge, we must solve the integral

$$\mathbf{E} = \int \frac{\rho_v dv\, \mathbf{a}_R}{4\pi\varepsilon_o R^2}$$

Since the vector **R** from source to test point will vary over the volume, and since in general ρ_v will also vary over the volume, this triple integral can be extremely difficult to solve analytically. It may instead be necessary to employ numerical integration to solve for this field. An example of this approach is given in the MATLAB example that follows.

In problems with sufficient symmetry, it can be much simpler to determine **E** using *Gauss's law*, which is the subject of an upcoming section.

Drill 2.15 In cylindrical coordinates, $\rho_v = 4\sin\phi/\rho$ C/m^3 for the volume $0.0 \le \rho \le 2.0$ m, $0 \le \phi \le \pi$ and -2.0 m $\le z \le 2.0$ m. Find the total charge in this volume. (*Answer:* $Q = 64$ C)

MATLAB 2.4

The following program was constructed to find the field at point (0, 10 m, 0) for a solid spherical charge of density $\rho_v = 4r^2$ nC/m^3 extending out to radius of 4 m. Figure 2.26 illustrates the problem. The program was saved as ML0204.

```
%    M-File: ML0204
%
%    This program finds the field at point P(0,10m,0)
%    from a spherical distribution of charge given by
%    rhov=4*r^2 nC/m^3 from 0 < r < 4m.
%
%    Wentworth, 7/6/02
%
%    Variables:
%    d               y-axis dist. to test point(m)
%    a               sphere radius(m)
%    dV              diff. charge volume where
%                    dV=delta_r*delta_theta*delta_phi
%    eo              free space permittivity (F/m)
%    r,theta,phi
%                    spherical coordinate location of center of a
%                    differential charge element
%    x,y,z           Cartesian coord location of charge element
%    R               vector from charge elem. to P
%    Rmag            magnitude of R
%    aR              unit vector of R
%    dr,dtheta,dphi
%                    differential spherical elements
%    dEi,dEj,dEk     partial field values
%    Etot            total field at P resulting from charge

clc                  %clears the command window
clear                %clears variables

%   Initialize variables
eo=8.854e-12;
d=10;a=4;
delta_r=40;delta_theta=72;delta_phi=144;

%   Perform calculation
for k=(1:delta_phi)
```

```
        for j=(1:delta_theta)
            for i=(1:delta_r)
                r=i*a/delta_r;
                theta=j*pi/delta_theta;
                phi=k*2*pi/delta_phi;
                x=r*sin(theta)*cos(phi);
                y=r*sin(theta)*sin(phi);
                z=r*cos(theta);
                R=[-x,(d-y),-z];
                Rmag=magvector(R);
                aR=R/Rmag;
                dr=a/delta_r;
                dtheta=pi/delta_theta;
                dphi=2*pi/delta_phi;
                dV=r^2*sin(theta)*dr*dtheta*dphi;
                dQ=4e-9*r^2*dV;
                dEi(i)=dQ*aR(2)/(4*pi*eo*Rmag^2);
            end
            dEj(j)=sum(dEi);
        end
        dEk(k)=sum(dEj);
end
Etot=sum(dEk)
```

In the command-line window, enter ML0204:

```
Etot =
 983.9
```

Note that the exact solution for the magnitude of **E** found by applying Gauss's law in Section 2.7 will be 925 V/m, rather than the 984 V/m found here. The solution gets closer and closer to the exact one as the differential volume element is made smaller. Note also that, because the field only depends on radius, the E_r calculated is valid for any 10-m radius, not just on the y-axis.

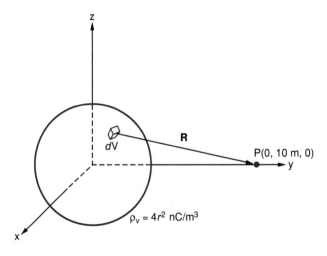

Figure 2.26 Volume charge has a charge density that varies with the radius. We wish to find the electric field intensity at point P in ML0204.

Practical Application: Laser Printer

Laser printers were introduced in 1984 as a method for producing crisp, quality print much sharper than the dot-matrix printer. Featuring up to 1200 dpi (dots per inch) resolution, laser printers cost more on the front end than ink-jet printers, their chief rival. But ink-jet printers require expensive ink cartridges whereas laser printers use much cheaper toner powder. Laser printers are therefore the workhouse printers in most office environments.

Laser printer operation is best described with the aid of Figure 2.27. The heart of the printer is the *organic photoconductive cartridge*, or OPC drum. On the surface of this drum is a special coating that will hold an electrostatic charge. The surface is also photoconductive, meaning a spot on the drum hit by light will be discharged. To begin, a portion of the drum passes under a negatively charged wire known as the *charge corona wire*. This wire's large negative charge induces a positive charge on the drum just under the wire. Next, the image to be printed is delivered to this charged region of the drum by a finely crafted laser and spinning mirror combination. Wherever the laser light strikes the OPC drum, the photoconductive material is discharged. The drum then rolls past a toner dispenser. The toner is a fine black powder that is given a positive charge. It is thus drawn to those portions of the drum that have been discharged by the laser. Meanwhile, paper is fed through the printer at the same speed as the drum. It passes over a positively charged wire (the *transfer corona wire*) that gives the paper a strong negative charge just prior to its contact with the drum. The positively charged toner powder on the drum is transferred to the stronger negative charge on the paper. The paper is then passed near a negatively charged wire (the *detac corona wire*) that removes the negative charge from the paper and therefore prevents it from statically clinging to the drum. The paper and loose toner powder are then passed through heated fuser rollers, where the powder melts into the paper fiber. The warm paper then exits the printer. The drum continues rolling, passing through a high-intensity light that discharges all the photoconductors, thereby erasing the image from the drum and making it ready for application of a positive charge again from the charge corona wire.

The technology of laser printers continues to improve and their prices continue to drop. Advanced printers are now available that can print in color by running the page through separate printing operations, each with a different color toner powder.

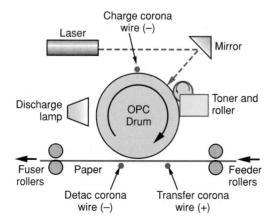

Figure 2.27 General schematic of a laser printer.

▷ 2.6 ELECTRIC FLUX DENSITY

Suppose we apply an amount of charge $+Q$ to a metallic sphere of radius a. As depicted in Figure 2.28a, we enclose this charged sphere using a pair of connecting hemispheres with radius b ($b > a$), being very careful not to ever let any part of the outer sphere come in contact with the inner sphere. Next, as shown in the cross section of Figure 2.28b we briefly ground the outer sphere. We then remove the ground connection and find in the Figure 2.28c cross section that $-Q$ of charge has accumulated on the outer sphere. Somehow, the $+Q$ charge of the inner sphere has induced the $-Q$ charge on the outer sphere.[2.13]

It becomes convenient to define an *electric flux* that extends from the positive charge and casts about for a negative charge (through the brief ground connection), pulling in the negative charge as close as it can to the positive charge. We say that the *electric flux* Ψ (psi), in coulombs, begins at the $+Q$ charge and terminates at the $-Q$ charge. These lines will be radially directed away from the inner sphere to the outer and will spread themselves out to get maximum separation between the like charges on each sphere. Considering that the flux lines pass through a spherical surface in the region between the spheres, we can define an *electric flux density* $\mathbf{D}$, in C/m², as

$$\mathbf{D} = \frac{\Psi}{4\pi r^2}\, \mathbf{a_\rho} \qquad \text{. (2.41)}$$

Note that this is very similar to the electric field intensity expression for a point charge (Eq. 2.21)

$$\mathbf{E} = \frac{Q}{4\pi\varepsilon_o r^2}\, \mathbf{a_\rho}$$

In fact, this expression also holds for the region between the spheres. Since the amount of flux Ψ emanating from the inner sphere is equal to the charge Q on the sphere, we can conclude from (2.41) and (2.21) that

$$\boxed{\mathbf{D} = \varepsilon_o \mathbf{E}} \qquad (2.42)$$

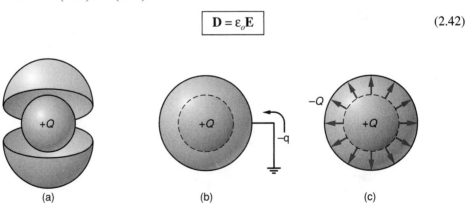

(a)   (b)   (c)

Figure 2.28   (a) Metallic sphere of radius a with charge $+Q$ is enclosed by a pair of hemispherical shells of radius b. (b) The outer shell is briefly grounded, (c) allowing a charge $-Q$ to accumulate on the outer shell.

[2.13]This experiment was carried out by Michael Faraday to determine the effect of different insulative materials on the electric field.

This is the relation between **D** and **E** in free space (or, for most practical purposes, air). In Section 2.11 we will see that, for general media, $\mathbf{D} = \varepsilon\mathbf{E}$, where ε is the material's permittivity. The advantage of using electric flux density rather than electric field intensity is that the former relates to the number of flux lines emanating from one set of charge and terminating on the other, independent of the media.

The amount of flux passing through a surface is given by the product of **D** and the amount of surface normal to **D**. Inspecting Figure 2.29, we can see that the flux is given by

$$\psi = |\mathbf{D}|\,|\mathbf{S}|\cos\theta$$

where θ is the angle between the electric flux density vector and the surface vector. In vector algebra, this relation is known as the *dot product*.[2.14] For a general pair of vectors **A** and **B**,

$$\boxed{\mathbf{A}\cdot\mathbf{B} = |\mathbf{A}|\,|\mathbf{B}|\cos\theta_{AB}} \tag{2.43}$$

In general it is difficult to find the angle θ_{AB} to evaluate the dot product. Fortunately, there is an easier way.

Consider that the dot product obeys both the commutative law ($\mathbf{A}\cdot\mathbf{B} = \mathbf{B}\cdot\mathbf{A}$) and the distributive law ($\mathbf{A}\cdot(\mathbf{B}+\mathbf{C}) = \mathbf{A}\cdot\mathbf{B} + \mathbf{A}\cdot\mathbf{C}$). Because of the distributive law, the dot product of a pair of vectors **A** and **B** can be expanded as

$$\begin{aligned}
\mathbf{A}\cdot\mathbf{B} &= (A_x\mathbf{a}_x + A_y\mathbf{a}_y + A_z\mathbf{a}_z)\cdot(B_x\mathbf{a}_x + B_y\mathbf{a}_y + B_z\mathbf{a}_z) \\
&= A_x\mathbf{a}_x\cdot B_x\mathbf{a}_x + A_x\mathbf{a}_x\cdot B_y\mathbf{a}_y + A_x\mathbf{a}_x\cdot B_z\mathbf{a}_z \\
&\quad + A_y\mathbf{a}_y\cdot B_x\mathbf{a}_x + A_y\mathbf{a}_y\cdot B_y\mathbf{a}_y + A_y\mathbf{a}_y\cdot B_z\mathbf{a}_z \\
&\quad + A_z\mathbf{a}_z\cdot B_x\mathbf{a}_x + A_z\mathbf{a}_z\cdot B_y\mathbf{a}_y + A_z\mathbf{a}_z\cdot B_z\mathbf{a}_z
\end{aligned}$$

Note that $\mathbf{a}_x\cdot\mathbf{a}_x = 1$ by (2.43) since the angle between parallel directions is $0°$ ($\cos(0°) = 1$). But $\mathbf{a}_x\cdot\mathbf{a}_y = 0$ since these two directions are orthogonal ($\cos(90°) = 0$). The result is that the dot product for this pair of vectors is found by adding the products of the like components. For Cartesian, spherical, and cylindrical coordinates, respectively,

$$\boxed{\begin{aligned}
\mathbf{A}\cdot\mathbf{B} &= A_x B_x + A_y B_y + A_z B_z \\
\mathbf{A}\cdot\mathbf{B} &= A_r B_r + A_\theta B_\theta + A_\phi B_\phi \\
\mathbf{A}\cdot\mathbf{B} &= A_\rho B_\rho + A_\phi B_\phi + A_z B_z
\end{aligned}} \tag{2.44}$$

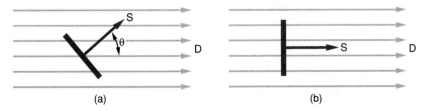

(a) (b)

Figure 2.29   The flux through a surface that is at an angle to the direction of flux (a) is less than the flux through an equivalent surface normal to the direction of flux (b).

[2.14]Also, but not as often, it is called a *scalar product*.

▶ **EXAMPLE 2.10**

Suppose we have two vectors $\mathbf{A} = 5\mathbf{a}_x + 3\mathbf{a}_z$ and $\mathbf{B} = 2\mathbf{a}_y - 4\mathbf{a}_z$. The dot product is

$$\mathbf{A} \cdot \mathbf{B} = (5)(0) + (0)(2) + (3)(-4) = -12$$

The dot product $\mathbf{A} \cdot \mathbf{B}$ represents the amount of $\mathbf{A}$ that is in the direction of $\mathbf{B}$, multiplied by the magnitude of $\mathbf{B}$. It is sometimes referred to as the projection of $\mathbf{A}$ onto the direction of $\mathbf{B}$. By the commutative law, it is also the projection of $\mathbf{B}$ onto the direction of $\mathbf{A}$.

We can thus find the flux through a surface by

$$\psi = |\mathbf{D}|\,|\mathbf{S}|\cos\theta = \mathbf{D} \cdot \mathbf{S} \tag{2.45}$$

However, what if $\mathbf{D}$ varies over the surface for which Ψ is desired? Finding the total flux is then a matter of integrating over the surface:

$$\boxed{\psi = \int \mathbf{D} \cdot d\mathbf{S}} \tag{2.46}$$

▶ **EXAMPLE 2.11**

Suppose $\mathbf{D} = 3xy\,\mathbf{a}_x + 4x\,\mathbf{a}_z$ C/m² and we wish to find the amount of electric flux through the surface at $z = 0$ with $0 \le x \le 5$ m, $0 \le y \le 3$ m. The differential surface vector is $d\mathbf{S} = dx\,dy\,\mathbf{a}_z$. Note that we could have chosen $d\mathbf{S} = dx\,dy\,(-\mathbf{a}_z)$, but our first choice has the differential surface vector pointing in the same direction as the flux, which will conveniently give us a positive answer. Equation 2.46 becomes

$$\psi = \int (3xy\mathbf{a}_x + 4x\mathbf{a}_z) \cdot dx\,dy\mathbf{a}_z = \int_0^5 4x\,dx \int_0^3 dy = 150 \text{ C}$$

In solving this integral, the $3xy\mathbf{a}_x$ component didn't factor into the calculation for Ψ since its dot product with $\mathbf{a}_z$ is zero.

▶ **EXAMPLE 2.12**

Suppose we have an electric flux density given as $\mathbf{D} = 3r\mathbf{a}_r - 9r\mathbf{a}_\theta + 6\mathbf{a}_\phi$ C/m² and we wish to find the electric flux passing through a spherical surface at $r = 2$ m. The differential surface vector in this case is $d\mathbf{S} = r^2\sin\theta\,d\theta\,d\phi\,\mathbf{a}_r$, and the dot product equation becomes

$$\psi = \int (3r\mathbf{a}_r - 9r\mathbf{a}_\theta + 6\mathbf{a}_\phi) \cdot r^2 \sin\theta\,d\theta\,d\phi\,\mathbf{a}_r = 3r^3 \int_0^\pi \sin\theta\,d\theta \int_0^{2\pi} d\phi = 96\pi \text{ C}$$

Drill 2.16 Suppose $\mathbf{A} = 6\mathbf{a}_x - 4\mathbf{a}_y + 2\mathbf{a}_z$ and $\mathbf{B} = -3\mathbf{a}_x - 24\mathbf{a}_y + 6\mathbf{a}_z$. Use (2.43) and (2.44) to find the angle between the two vectors. (*Answer:* 61°)

Drill 2.17 Given $\mathbf{D} = 10\mathbf{a}_\rho + 5\mathbf{a}_\phi$ C/m², find the electric flux passing through the surface defined by $\rho = 6$ m, $0 \le \phi \le 90°$, and $-2 \le z \le +2$ m. (*Answer:* 120π C)

> **MATLAB 2.5**

Given a pair of vectors **A** and **B**, the dot product in MATLAB is dot(A,B):

```
» A=[1 2 3];B=[0 3 1];
» dot(A,B)

ans =
   9
```

A short routine may also be written to calculate the interior angle between a pair of vectors using the dot product. It assumes you have saved the function "magvector" from MATLAB 2.1 in your work file. The program is saved as "interiorangle":

```
function y=interiorangle(A,B)
%   Calculates the interior angle between a pair of
%   vectors A and B.
y=(180/pi)*acos(dot(A,B)/(magvector(A)*magvector(B)));
% 180/pi converts radians to degrees

» A=[1 2 3];B=[0 3 1];
» interiorangle(A,B)

ans =
   40.4795
```

▶ 2.7 GAUSS'S LAW AND APPLICATIONS

If we completely enclose a charge, then the net flux passing through the enclosing surface must be equal to the charge enclosed, Q_{enc}. A formal statement of Gauss's law is as follows:

> The net electric flux through any closed surface is equal to the total charge enclosed by that surface.

In mathematical form, this is written as

$$\boxed{\oint \mathbf{D} \cdot d\mathbf{S} = Q_{enc}} \tag{2.47}$$

where the circle on the integral indicates the integration is performed over a closed surface. Equation 2.47 is called the *integral form of Gauss's law* and is one of the four Maxwell's equations.

Gauss's law, although very simple, is quite useful in finding the fields for problems that have a high degree of symmetry. From the symmetry of the problem, we determine what variables influence **D** and what components of **D** are present. Then we select an enclosing surface, called a *Gaussian surface*, whose differential surface vector is directed outward from the enclosed volume and is everywhere either tangential to **D** (in which case $\mathbf{D} \cdot d\mathbf{S} = 0$) or normal to **D** (in which case $\mathbf{D} \cdot d\mathbf{S} = DdS$). Over the portion of the surface that is normal to **D**, we also require that the magnitude of **D** be constant so that it can be pulled out of the integral:

$$\int \mathbf{D} \cdot d\mathbf{S} = D \int dS$$

It is then a simple matter to calculate

$$D = Q / \int dS$$

▶ **EXAMPLE 2.13**

As the first of several examples, let's use Gauss's law to determine the electric field intensity resulting from a point charge at the origin. This will give us a chance to try out our technique in a case
where we already know the answer.

The problem has spherical coordinate symmetry, and we see that the field is everywhere directed
radially away from the origin and is not a function of θ or ϕ (i.e., if we are a fixed radius away from
the point charge, it looks exactly the same to us no matter what our position in θ or ϕ). From this
inspection of symmetry we conclude that

$$\mathbf{D} = D_r(r)\mathbf{a}_r$$

where $D_r(r)$ means the D_r component is only a function of r. For a Gaussian surface, we choose a
sphere centered at the origin as seen in Figure 2.30. The differential surface vector is

$$dS = r^2 \sin\theta \, d\theta \, d\phi \, \mathbf{a}_r$$

so

$$\oint \mathbf{D} \cdot dS = \oint D_r \mathbf{a}_r \cdot r^2 \sin\theta \, d\theta \, d\phi \, \mathbf{a}_r = \oint D_r \, r^2 \sin\theta \, d\theta \, d\phi$$

Since our Gaussian surface has a fixed radius from the origin, D_r will be constant over the surface, and
so it can be taken outside the integral to yield

$$\oint \mathbf{D} \cdot dS = D_r r^2 \oint \sin\theta \, d\theta \, d\phi = D_r r^2 \int_0^\pi \sin\theta \, d\theta \int_0^{2\pi} d\phi = 4\pi r^2 D_r$$

Now, since the total net charge enclosed by the Gaussian surface is Q, we have

$$4\pi r^2 D_r = Q$$

and

$$D_r = \frac{Q}{4\pi r^2}$$

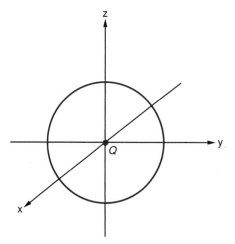

Figure 2.30 A spherical Gaussian surface surrounding a point charge at the origin.

If we consider the point charge to exist in free space, this leads to the expected result

$$\mathbf{E} = \frac{Q}{4\pi\varepsilon_o r^2}\,\mathbf{a}_r$$

▷ **EXAMPLE 2.14**

Now let us find the electric field intensity resulting from an infinite length line of charge density ρ_L on the z-axis. We wish to find $\mathbf{D}$ at any point P(ρ, ϕ, z).

An element of charge dQ along the line will give D_ρ and D_z components at point P, but not D_ϕ. Also, proper selection of a second dQ element will result in cancellation of D_z. So $\mathbf{D} = D_\rho\,\mathbf{a}_\rho$, and from the problem's symmetry it is clear that D_ρ is only a function of ρ, or $\mathbf{D} = D_\rho(\rho\,)\mathbf{a}_\rho$.

We choose a cylindrical Gaussian surface centered along the z-axis with radius ρ and length h as shown in Figure 2.31 and containing the point P somewhere on its side. The flux through the closed surface can be written

$$\oint \mathbf{D}\cdot d\mathbf{S} = \int \mathbf{D}\cdot d\mathbf{S}_{\text{top}} + \int \mathbf{D}\cdot d\mathbf{S}_{\text{bottom}} + \int \mathbf{D}\cdot d\mathbf{S}_{\text{side}}$$

where

$$d\mathbf{S}_{\text{top}} = \rho\,d\rho\,d\phi\,\mathbf{a}_z$$

$$d\mathbf{S}_{\text{bottom}} = \rho\,d\rho\,d\phi\,(-\mathbf{a}_z)$$

and

$$d\mathbf{S}_{\text{side}} = \rho\,d\phi\,dz\,\mathbf{a}_\rho$$

The top and bottom integrals are zero since $\mathbf{a}_\rho \cdot \mathbf{a}_z = 0$. Then, because we know that D_ρ is constant on the side of the Gaussian surface,

$$\int \mathbf{D}\cdot d\mathbf{S}_{\text{side}} = \int D_\rho\mathbf{a}_\rho \cdot \rho\,d\phi\,dz\,\mathbf{a}_\rho = D_\rho\rho\int_0^{2\pi}\!\!d\phi\int_0^{h}\!\!dz = 2\pi h\rho D_\rho$$

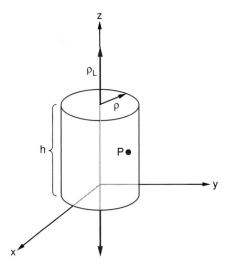

Figure 2.31 A Gaussian surface containing the point P is placed around a section of an infinite length line of charge density ρ_L occupying the z-axis.

The charge enclosed by the Gaussian surface is

$$Q_{enc} = \int_0^h \rho_L dz = \rho_L h$$

Equating the flux to the enclosed charge, we solve for D_ρ and find

$$D_\rho = \frac{\rho_L}{2\pi\rho}$$

▷ **EXAMPLE 2.15**

For a third Gauss's law example, let's determine the field everywhere resulting from an infinite extent sheet of charge density ρ_s placed on the x–y plane at $z = 0$. From examination of the problem's symmetry, we see that only a D_z component will be present, and it will clearly not be a function of either x or y. We did find earlier that it is also not a function of z, but we'll pretend we don't know this yet. Now we'll situate the point at which we want to find the field along the z-axis at a height h. Our Gaussian surface must contain this point and also surround some portion of the charged sheet. We may use a rectangular box of sides $2x$, $2y$, and $2h$ in the x, y, and z directions, respectively, as shown in Figure 2.32.[2.15] Here, the charge enclosed is simply

$$Q = \int \rho_s dS = \rho_s \int_{-x}^{x} dx \int_{-y}^{y} dy = 4\rho_s xy$$

Because there is no net flux through the sides of the box, we must find the flux through the top and bottom surfaces:

$$\oint \mathbf{D} \cdot d\mathbf{S} = \int_{top} \mathbf{D} \cdot d\mathbf{S} + \int_{bottom} \mathbf{D} \cdot d\mathbf{S}$$

$$= \int_{top} D_z \mathbf{a}_z \cdot dxdy\mathbf{a}_z + \int_{bottom} D_z(-\mathbf{a}_z) \cdot dxdy(-\mathbf{a}_z) = 2(4xy)D_z$$

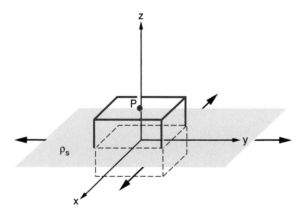

Figure 2.32 A rectangular box is employed as the Gaussian surface surrounding a section of sheet charge to determine the field resulting from an infinite extent sheet of charge density ρ_s.

[2.15]It would be no more challenging to use a cylinder of some arbitrary radius centered along the z-axis with length 2h. However, as we have just used a cylindrical Gaussian surface for the infinite length line of charge, let's try something new.

Notice that this answer is independent of the height of the box. Equating the net flux to the charge enclosed, we see that the $4xy$ portion cancels and we have

$$D_z = \frac{\rho_s}{2}$$

at our point on the positive z-axis.

The general expression for the electric flux density resulting from the infinite extent sheet of charge is

$$\mathbf{D} = \frac{\rho_s}{z}\, \mathbf{a_n}$$

where $\mathbf{a}_n$ is a unit vector normal directed away from the charged sheet.

Coaxial Cable

We now turn to a more complicated example. Figure 2.33a shows a section of an infinitely long coaxial cable with an inner conductor of radius a surrounded by a thin conductive shell at radius b. Actual coaxial cable generally has a solid inner conductor, but in operation the charge concentrates at the conductor surface (see Section 2.10). For practice in using Gauss's law, however, we will assume the charge on the inner conductor has a uniformly distributed charge density ρ_v. The outer conductor is grounded. Our task is to find the fields everywhere.

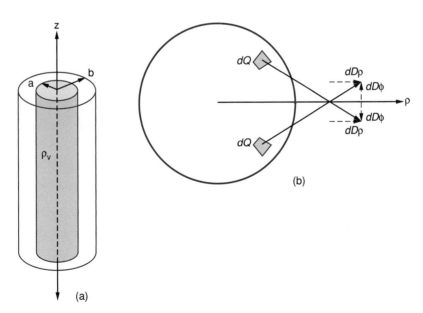

Figure 2.33 (a) Coaxial cable with a solid inner conductor of radius a and charge density ρ_v surrounded by a grounded thin conductive shell at radius b. (b) Cross section of the inner conductor with field components for a pair of differential charge elements.

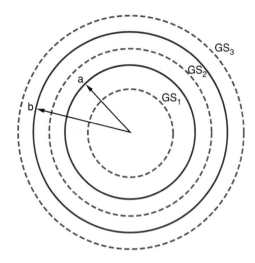

Figure 2.33 (continued) (c) Cross section of the coaxial cable showing the three Gaussian surfaces.

To begin, we notice from the symmetry of the problem that $\mathbf{D}$ only appears to be a function of ρ. To find the components of $\mathbf{D}$ it is helpful to look at the cross section of the inner conductor given in Figure 2.33b. From this, it is apparent the D_ϕ components of the field will cancel by symmetry. The D_z components will cancel from a similar argument, leaving only $\mathbf{D} = D_\rho(\rho)$.

There are three basic regions where we must calculate the field: $\rho \le a$, $a \le \rho \le b$, and $\rho > b$. Three Gaussian surfaces are therefore required. We choose cylinders centered along the z-axis with a height h and radii as indicated in the cross section of Figure 2.33c. There is no flux through the top and bottom surfaces of the cylinder, so the flux through a Gaussian surface is

$$\oint \mathbf{D} \cdot d\mathbf{S} = \int D_\rho \mathbf{a}_\rho \cdot \rho \, d\phi \, dz \, \mathbf{a}_\rho = D_\rho \rho \int_0^{2\pi} d\phi \int_0^h dz = 2\pi \rho h D_\rho$$

This determination of the flux will be valid for all three Gaussian surfaces.

The charge enclosed by the first Gaussian surface is found by integration:

$$Q_{\text{enc}} = \int \rho_v \, dV = \rho_v \int_0^\rho \rho \, d\rho \int_0^{2\pi} d\phi \int_0^h dz = \pi \rho_v h \rho^2$$

So now the field in the first region ($\rho \le a$) is found as

$$D_\rho = \frac{\pi \rho_v h \rho^2}{2\pi \rho h} = \frac{\rho_v}{2} \rho$$

The charge enclosed by the second Gaussian surface is found by taking the previous Q_{enc} integration from 0 to a on ρ, resulting in $Q_{\text{enc}} = \pi \rho_v h a^2$. This yields

$$D_\rho = \frac{\pi \rho_v h a^2}{2\pi \rho h} = \frac{\rho_v}{2} \frac{a^2}{\rho}$$

Finally, since the outer conductor is tied to ground, we know from Faraday's experiment that the charge on this conductor is $-Q$. The total net charge enclosed by this third Gaussian surface is therefore zero, and $D_\rho = 0$ for $\rho > b$.

MATLAB 2.6

Consider the coaxial example to have an inner radius of 3 cm and an outer radius of 6 cm. The inner conductor has a charge density 8 nC/cm^3. You are to use MATLAB to generate a plot of D_ρ versus ρ out to 12 cm.

```
%   M-File: ML0206
%
%   This program plots D vs radius of a coaxial cable.
%
%   Wentworth, 7/7/02
%
%   Variables:
%   a           radius (cm) of solid inner conductor
%   b           radius (cm) of outer conductive shell
%   N           number of points to plot per cm
%   Qdens       charge density (nC/cm^3)
%   rho(n)      radial point
%   Drho(n)     flux density at rho(n)

clc             %clears the command window
clear           %clears variables
%Initialize variables
a=3;b=6;
Qdens=8;
N=20;

%   Perform calculation
for n=1:a*N
    rho(n)=n/N;
    Drho(n)=Qdens*rho(n)/2;
end

for n=1+a*N:b*N;
    rho(n)=n/N;
    Drho(n)=Qdens*a^2/(2*rho(n));
end

for n=1+b*N:12*N;
    rho(n)=n/N;
    Drho(n)=0;
end

%   Plot Drho vs rho
plot(rho,Drho)
grid on
xlabel('rho(cm)')
ylabel('Magnitude of Elect Flux Density (nC/cm^2)')
Title('Coaxial Cable Example')
```

The result is plotted in Figure 2.34. It is worth noting that the step in D_ρ at $\rho = b$ is equivalent to the sheet charge density on the outer conductor.

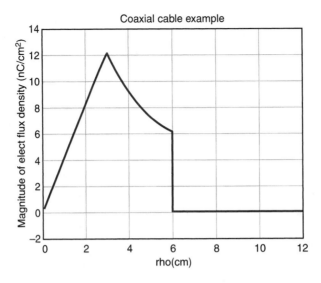

Figure 2.34 D_ρ versus radius for the coaxial cable in MATLAB 2.6.

Drill 2.18 Derive the formulas to characterize the electric flux density everywhere resulting from a sphere of charge, of radius a, containing a uniform charge density ρ_v (C/m³). (*Answer:* $\mathbf{D} = (\rho_v/3)r\,\mathbf{a}_r$ for $0 \leq r \leq a$, $\mathbf{D} = (\rho_v/r)(a^3/r^2)\,\mathbf{a}_r$ for $r \geq a$)

▶ 2.8 DIVERGENCE AND THE POINT FORM OF GAUSS'S LAW

Related to Gauss's law, where net flux is evaluated exiting a closed surface, is the concept of *divergence*. The divergence of a vector field at a particular point in space is a spatial derivative of the field indicating to what degree the field emanates (or *diverges*) from the point. Its value, a scalar quantity, says whether the point contains a source or a sink of field. The concept of divergence will also lead to the very useful differential form of Gauss's law.

We can derive an expression for divergence by applying Gauss's law to the small cubic element shown in Figure 2.35, which has at its center point $P(x_0, y_0, z_0)$ a known flux density vector

$$D_P = D_{x_0}\mathbf{a}_x + D_{y_0}\mathbf{a}_y + D_{z_0}\mathbf{a}_z$$

The total flux through the surface of the cube is equal to the net charge enclosed, by Gauss's law, and is the sum of the flux through each face:

$$\oint \mathbf{D}\cdot d\mathbf{S} = Q_{\text{enc}} = \left(\underset{\text{front}}{\int} + \underset{\text{back}}{\int} + \underset{\text{left}}{\int} + \underset{\text{right}}{\int} + \underset{\text{top}}{\int} + \underset{\text{bottom}}{\int}\right)\mathbf{D}\cdot d\mathbf{S} \qquad (2.48)$$

Looking at just the front face, we have

$$\underset{\text{front}}{\int} \mathbf{D}\cdot d\mathbf{S} = D_{x_0+\Delta x/2}\mathbf{a}_x \cdot \Delta y \Delta z\mathbf{a}_x \qquad (2.49)$$

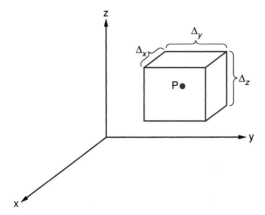

Figure 2.35 A differential volume element used to derive divergence.

The flux density at the front face $D_{x_0 + \Delta x/2}$ is related to the flux at point P by the approximation

$$D_{x_0 + \Delta x/2} \cong D_{x_0} + \frac{\Delta x}{2} \frac{\partial D_x}{\partial x} \qquad (2.50)$$

where the change in D_x from D_{x_0} is the product of the spatial rate of change $\partial D_x/\partial x$ and the distance from P to the front face, $\Delta x/2$. This approximation becomes exact as Δx is shrunk to zero. So we have

$$\int_{\text{front}} \mathbf{D} \cdot d\mathbf{S} \cong D_{x_0} \Delta y \Delta z + \frac{\Delta x \Delta y \Delta z}{2} \frac{\partial D_x}{\partial x} \qquad (2.51)$$

We can find the flux through the back face in a similar manner. The differential surface vector directed away from the volume is

$$d\mathbf{S} = \Delta y \Delta z (-\mathbf{a}_x)$$

and the flux density at the back face $D_{x_0 - \Delta x/2}$ is

$$D_{x_0 - \Delta x/2} \cong D_{x_0} - \frac{\Delta x}{2} \frac{\partial D_x}{\partial x}$$

resulting in

$$\int_{\text{back}} \mathbf{D} \cdot d\mathbf{S} \cong -D_{x_0} \Delta y \Delta z + \frac{\Delta x \Delta y \Delta z}{2} \frac{\partial D_x}{\partial x} \qquad (2.52)$$

Adding the flux from the front and back faces, we see that the D_{x_0} terms cancel and we're left with

$$\int_{\text{front+back}} \mathbf{D} \cdot d\mathbf{S} \cong \frac{\partial D_x}{\partial x} \Delta x \Delta y \Delta z = \frac{\partial D_x}{\partial x} \Delta v \qquad (2.53)$$

where a differential volume Δv replaces $\Delta x \Delta y \Delta z$.

In a like manner, the sum of the flux through the left and right faces is

$$\int_{\text{left+right}} \mathbf{D} \cdot d\mathbf{S} \cong \frac{\partial D_y}{\partial y} \Delta v \tag{2.54}$$

and that through the top and bottom faces is

$$\int_{\text{top+bottom}} \mathbf{D} \cdot d\mathbf{S} \cong \frac{\partial D_z}{\partial z} \Delta v \tag{2.55}$$

Summing all these flux terms and recalling Gauss's law gives

$$\oint \mathbf{D} \cdot d\mathbf{S} = Q_{\text{enc}} \cong \left(\frac{\partial D_x}{\partial x} + \frac{\partial D_y}{\partial y} + \frac{\partial D_z}{\partial z} \right) \Delta v \tag{2.56}$$

Because our approximations for the flux density at each face are most accurate when the differential elements are made very small, we can obtain an exact solution by dividing each side of (2.56) by Δv and taking the limit as Δv goes to zero:

$$\lim_{\Delta v \to 0} \frac{\oint \mathbf{D} \cdot d\mathbf{S}}{\Delta v} = \lim_{\Delta v \to 0} \frac{Q_{\text{enc}}}{\Delta v} = \frac{\partial D_x}{\partial x} + \frac{\partial D_y}{\partial y} + \frac{\partial D_z}{\partial z} \tag{2.57}$$

The first term in (2.57) is known as the *divergence of* $\mathbf{D}$, written

$$\text{div } \mathbf{D} = \lim_{\Delta v \to 0} \frac{\oint \mathbf{D} \cdot d\mathbf{S}}{\Delta v} \tag{2.58}$$

This equation holds for any vector field and states that the divergence of a vector field is equal to the net flow out of a closed surface per unit volume enclosed by the surface as the volume shrinks to zero.

From inspection of the second term in (2.57) it is apparent that

$$\lim_{\Delta v \to 0} \frac{Q_{\text{enc}}}{\Delta v} = \rho_v \tag{2.59}$$

The third term in (2.57) is related to the flux density vector by the *del operator*,[2.16] which for Cartesian coordinates is written

$$\nabla = \frac{\partial}{\partial x} \mathbf{a}_x + \frac{\partial}{\partial y} \mathbf{a}_y + \frac{\partial}{\partial z} \mathbf{a}_z \tag{2.60}$$

and

$$\boxed{\nabla \cdot \mathbf{D} = \frac{\partial D_x}{\partial x} + \frac{\partial D_y}{\partial y} + \frac{\partial D_z}{\partial z}} \tag{2.61}$$

[2.16]In addition to divergence of a vector field $\mathbf{A}$ ($\nabla \cdot \mathbf{A}$), the del operator will also be used for the *curl* of $\mathbf{A}$ ($\nabla \times \mathbf{A}$), for the *gradient* of a scalar field A (∇A), and for the *Laplacian* of a scalar field A($\nabla^2 A$).

is pronounced "del dot D" or more commonly "the divergence of D." Notice that $\nabla \cdot D$ returns a scalar quantity. It indicates how much flux is leaving the small closed surface without imparting any information about the flux direction.

From Eqs. 2.57, 2.59, and 2.61, the differential form of Gauss's law is

$$\boxed{\nabla \cdot D = \rho_v} \tag{2.62}$$

This is also called the *point form* of Gauss's law since it occurs at some particular point in space.

A common example used to give a physical picture of divergence is to consider the expansion of a gas as pressure is lowered. Consider the motion of air molecules in a small, fixed volume within the apparatus shown in Figure 2.36. If the plunger is stationary, there is no net movement of molecules[2.17] through the closed surface. Now, as the plunger moves up, lowering the pressure in the chamber, there is a net movement of molecules out of the closed surface. At any point in the chamber, the air molecules are diverging; that is, the air is expanding. If the plunger is pushed in, the net flux out of the closed surface would be negative, as would be the divergence, which would indicate the air is compressing.

In terms of electric flux density, a positive divergence at a point indicates the presence of a source of flux at that point (i.e., a positive charge). A negative divergence indicates the presence of a flux sink (i.e., a negative charge).

It should be noted that in our derivation of (2.61) we assumed a differential volume element from the Cartesian coordinate system. For cylindrical and spherical coordinate systems, the volume elements are different and our derivation would lead to the following equations:

$$\boxed{\begin{aligned} \nabla \cdot D_{\text{cyl}} &= \frac{1}{\rho} \frac{\partial}{\partial \rho}\left(\rho D_\rho\right) + \frac{1}{\rho} \frac{\partial D_\phi}{\partial \phi} + \frac{\partial D_z}{\partial z} \\ \nabla \cdot D_{\text{spher}} &= \frac{1}{r^2} \frac{\partial}{\partial r}\left(r^2 D_r\right) + \frac{1}{r\sin\theta} \frac{\partial}{\partial \theta}\left(D_\theta \sin\theta\right) + \frac{1}{r\sin\theta} \frac{\partial D_\phi}{\partial \phi} \end{aligned}} \tag{2.63}$$

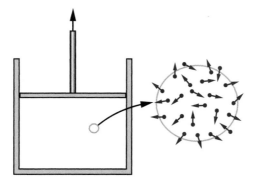

Figure 2.36 Expansion of air in a sealed chamber with increasing volume is demonstrated by the net flow of air molecules out of a small volume element.

[2.17]Except at absolute zero temperature, gas molecules are always skittering about, bumping into each other. For a small closed surface, the statistical average of molecules leaving the surface equals the average going back in. Thus, there is no *net* movement through the closed surface.

Recalling once again the integral form of Gauss's law, we can write

$$\oint \mathbf{D} \cdot d\mathbf{S} = Q_{enc} = \int \rho_v dv$$

Using the point form of Gauss's law, ρ_v can be replaced by $\nabla \cdot \mathbf{D}$, leading to the *divergence theorem*:

$$\boxed{\oint \mathbf{D} \cdot d\mathbf{S} = \int \nabla \cdot \mathbf{D} \, dv} \qquad (2.64)$$

This very handy relation says that integrating the normal component of a vector field over a closed surface is equivalent to integrating the divergence of the vector field at every point in the volume enclosed by that surface.

▶ **EXAMPLE 2.16**

Suppose $\mathbf{D} = \rho^2 \, \mathbf{a}_\rho$. Find the flux through the surface of a cylinder with $0 \leq z \leq h$ and $\rho = a$ by evaluating

    a. the left side of the divergence theorem and

    b. the right side of the divergence theorem.

A sketch of this cylinder is shown in Figure 2.37 along with differential surface vectors for each of the cylinder's three surfaces. We can first evaluate the left side of the divergence theorem by considering

$$\psi = \oint \mathbf{D} \cdot d\mathbf{S} = \underbrace{\int \mathbf{D} \cdot d\mathbf{S}}_{top} + \underbrace{\int \mathbf{D} \cdot d\mathbf{S}}_{side} + \underbrace{\int \mathbf{D} \cdot d\mathbf{S}}_{bottom}$$

The integrals over the top and bottom surfaces are each zero, since $\mathbf{a}_\rho \cdot \mathbf{a}_z = 0$. So we have

$$\psi = \int \mathbf{D} \cdot d\mathbf{S} = \int_{z=0}^{h} \int_{\phi=0}^{2\pi} \rho^2 \mathbf{a}_\rho \cdot \rho \, d\phi \, dz \, \mathbf{a}_\rho = 2\pi h \rho^3 \Big|_{\rho=a} = 2\pi h a^3$$

For evaluation of the right side of the divergence equation, we can first find the divergence in cylindrical coordinates (from Eq. 2.63 or from Appendix A). We have

$$\nabla \cdot \mathbf{D}_{cyl} = \frac{1}{\rho} \frac{\partial}{\partial \rho} (\rho D_\rho) = \frac{1}{\rho} \frac{\partial}{\partial \rho} (\rho^3) = 3\rho$$

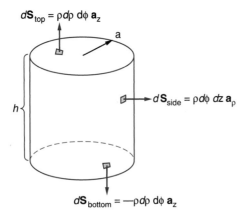

$d\mathbf{S}_{top} = \rho \, d\rho \, d\phi \, \mathbf{a}_z$

a

h

$d\mathbf{S}_{side} = \rho \, d\phi \, dz \, \mathbf{a}_\rho$

$d\mathbf{S}_{bottom} = -\rho \, d\rho \, d\phi \, \mathbf{a}_z$

Figure 2.37 Cylindrical volume of Example 2.16 with differential surface vectors shown.

Performing a volume integration on this divergence, we have

$$\psi = \int_{vol} \nabla \cdot \mathbf{D}\,dv = \int (3\rho)\rho\,d\rho\,d\phi\,dz = 3\int_0^a \rho^2 d\rho \int_0^{2\pi} d\phi \int_0^h dz = 2\pi h a^3$$

This is the same as that calculated for the other side of the divergence equation.

It is worth commenting on the units in this problem. We are given that

$$\mathbf{D}\left(\frac{C}{m^2}\right) = \rho(m)^2\,\mathbf{a}_\rho$$

Clearly these units do not match up. We must assume a unity conversion in the equation, or

$$\mathbf{D}\left(\frac{C}{m^2}\right) = \left(1\frac{C}{m^4}\right)\rho(m)^2\,\mathbf{a}_\rho$$

This can be carried through to the final answer:

$$\psi(C) = \left(1\frac{C}{m^4}\right)2\pi h(m)a(m)^3$$

Drill 2.19 If the electric flux density is given by $\mathbf{D} = x\,\mathbf{a}_x + y^2\,\mathbf{a}_y$ C/m², determine the volume charge density at the point P(2, 3, 4). (*Answer:* 7 C/m³)

2.9 ELECTRIC POTENTIAL

Students of electrical engineering are very familiar with the concept of electric potential (*voltage* in circuit analysis classes). In this section we develop the concept of electric potential and show its relationship to electric field intensity.

When force is applied to move an object, work is the product of the force and the distance the object travels in the direction of the force. Mathematically, in moving the object from point a to point b, the work can be expressed as

$$W = \int_a^b \mathbf{\Phi} \cdot d\mathbf{L} \tag{2.65}$$

where $d\mathbf{L}$ is a differential length vector along some portion of the path between a and b. We know from Coulomb's law that the force exerted on a charge Q by an electric field $\mathbf{E}$ is $F = QE$. The work done *by the field* in moving the charge from point a to b is then

$$W_{E-field} = Q\int_a^b \mathbf{E} \cdot d\mathbf{L} \tag{2.66}$$

If an external force moves the charge against the field, the work done is the negative of $W_{E-field}$, or

$$W = -Q\int_a^b \mathbf{E} \cdot d\mathbf{L} \tag{2.67}$$

▶ **EXAMPLE 2.17**

Let us calculate the work required to move a 10–nC charge from the origin to point P(1, 1, 0) against the static field $\mathbf{E} = 5\mathbf{a}_x$ V/m.

We notice in Figure 2.38 that several paths are indicated; all will require the same amount of work in moving the charge from a to b.[2.18] We choose path i and divide the problem into a pair of integrals

$$W = -(10 \text{ nC}) \int_0^{1\text{ m}} 5\mathbf{a}_x (\text{V/m}) \cdot dy\,\mathbf{a}_y - (10 \text{ nC}) \int_0^{1\text{ m}} 5\mathbf{a}_x (\text{V/m}) \cdot dx\,\mathbf{a}_x$$

The first integral takes us from the origin to the point (0, 1, 0). The amount of work for this portion of the path is zero since $\mathbf{a}_x \cdot \mathbf{a}_y = 0$. The second integral takes us from (0, 1, 0) to (1, 1, 0) and results in

$$W = -(10 \text{ nC}) (5 \text{ V/m}) (1 \text{ m}) = -50 \text{ nJ}$$

where the conversion $J = C \cdot V$ has been employed. Since the work expended to move the charge is negative, the field is doing 50 nJ of work.

Now we can define the *electric potential difference V_{ba}* as the work done by an external source to move a charge from point a to point b in an electric field divided by the amount of charge moved:

$$V_{ba} = \frac{W}{Q} = -\int_a^b \mathbf{E} \cdot d\mathbf{L} \tag{2.68}$$

This potential difference can also be related to absolute potentials, or *electrostatic potentials*,[2.19] at points a and b:

$$V_{ba} = V_b - V_a \tag{2.69}$$

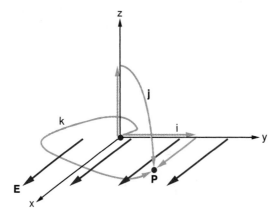

Figure 2.38 Three different paths to calculate work moving from the origin to a point P against an electric field.

[2.18]Although the student may quip that it would take much more *mathematical* work to use path k in Figure 2.38, the actual work done will be the same as the other, simpler paths.

[2.19]When we do work to move a pair of like point charges toward each other (moving one against the field of the other), the energy is stored as *potential* energy that can be recovered if we allow the charges to return to their starting positions.

Finding the absolute potential at some point requires that we have a reference potential. Often a ground plane, or plate, is chosen as the zero potential reference. In the case of coaxial cable, the zero potential reference is chosen to be the grounded outer conductor. For a collection of point charges near the origin, the zero potential reference is often selected at infinite radius. The reference can also be a known or assumed potential at some point.

▷ **EXAMPLE 2.18**

Let's find the potential difference V_{PO} between the origin and point P in Example 2.17.
The potential difference V_{PO} is calculated as

$$V_{PO} = \frac{-50\,\text{nJ}}{10\,\text{nC}} = -5\,\text{V}$$

If we know that the absolute potential at the origin is, for instance, 8 V, or $V_O = 8$ V, then $V_P = V_{PO} + V_O$, or $V_P = 3$ V.

It is interesting to see that, if a closed path is chosen, the integral will return zero potential difference:

$$\boxed{\oint \mathbf{E} \cdot d\mathbf{L} = 0} \tag{2.70}$$

This result[2.20] is the very familiar *Kirchhoff's voltage law*.

Let's calculate the potential difference between two points in space resulting from the field of a point charge located at the origin. Since the electric field intensity is radially directed, only movement in the radial direction will influence the potential. If we move from radius a to radius b, we have

$$V_{ba} = -\int_a^b \mathbf{E} \cdot d\mathbf{L} = -\int_a^b \frac{Q}{4\pi\varepsilon_o r^2} \mathbf{a}_r \cdot dr\,\mathbf{a}_r$$

which upon evaluating the integral yields

$$V_{ba} = \left. \frac{Q}{4\pi\varepsilon_o r} \right|_{r=a}^{r=b} = \frac{Q}{4\pi\varepsilon_o}\left(\frac{1}{b} - \frac{1}{a}\right) = V_b - V_a$$

Now, if we set a reference voltage of zero at an infinite radius, then the absolute potential at some finite radius from a point charge fixed at the origin is

$$V = \frac{Q}{4\pi\varepsilon_o r} \tag{2.71}$$

We can define this as the work per coulomb required to pull a test charge from infinity to the radius r. If we have a collection of N charges, the total potential can be found by adding the potential for each charge, or

---

[2.20]Equation (2.70) is valid for *static fields*, but not for time-varying fields. One of the key discoveries in electromagnetics was that time-varying magnetic fields produce electric fields (and that time-varying electric fields produce magnetic fields). This will be discussed further in Chapter 4.

$$V = \sum_{i=1}^{N} \frac{Q_i}{4\pi\varepsilon_o r_i} \tag{2.72}$$

If the collection of charges becomes a continuous distribution, we can find the total potential by integrating:

$$\boxed{V = \int \frac{dQ}{4\pi\varepsilon_o r}} \tag{2.73}$$

Drill 2.20 The field for an infinite length line of charge on the *z*-axis is

$$\mathbf{E} = \frac{\rho_L}{2\pi\varepsilon_o \rho}\mathbf{a}_\rho$$

Suppose the charge density ρ_L is 100 nC/m. (a) Find the work done moving a 10-nC charge from $\rho = 3$ m (point a) to $\rho = 1$ m (point b). (b) Find the potential difference V_{ba}. (*Answer: W = 20 μJ, V_{ba} = 2 kV*)

Drill 2.21 Three 1-nC charges exist at points (1 m, 0, 0), (–1 m, 0, 0), and (0, 0, 1 m), respectively. Determine the absolute electrostatic potential at the point (0, 1 m, 0) assuming a zero-potential reference at infinite distance from the origin. (*Answer: 19.1 V*)

MATLAB 2.7

There is a point charge $Q = 10$ nC at the origin, and the potential difference V_{ba} is to be calculated going from a point A(3 m, 4 m, 0) to a point B(3 m, 0, 0). Now, the solution is not hard to derive and gives you an answer of 12.00 V. However, you want to see how to get the same solution using MATLAB so that you'll have some confidence in solving more difficult problems where the answer is not so easy to derive.

Here's one approach:

```
%    M-File: ML0207
%
%    This program calculates potential difference going
%    from point A(3,4,0) to point B(3,0,0), given a point
%    charge Q=10nC at the origin. The approach will be to
%    break up the distance from A to B into k sections.
%    The field E will be found at the center of each
%    section (located at point P) and then dot(Ep,dLv)
%    will give the potential drop across the kth section.
%    Total potential is found by summing the potential
%    drops.
%
%    Wentworth, 7/7/02
%
%    Variables:
%    Q            point charge, in nC
```

```
%   k           number of num. integration steps
%   dL          magnitude of one step
%   dLv         vector for a step
%   y(n)        y location at center of section at P
%   R           vector from Q to P
%   E           electric field at P
%   V(n)        portion of dot(E,dL) at P

clc             %clears the command window
clear           %clears variables

%   Initialize variables
k=32;
dL=4/k;
dLv=dL*[0 -1 0];

%   Perform calculation
for n=1:k
    y(n)=(n-1)*dL+dL/2;
    R=[3 y(n) 0];
    Rmag=magvector(R);
    E=90*R/Rmag^3;
    V(n)=dot(E,dLv);
end
Vtot=sum(-V)
```

Sufficient steps must be chosen to converge on the correct solution. Here are the potentials calculated for various numbers of steps:

$$K = 2 \qquad V = 12.76$$
$$K = 4 \qquad V = 12.17$$
$$K = 8 \qquad V = 12.04$$
$$K = 16 \qquad V = 12.01$$
$$K = 32 \qquad V = 12.003$$

The program therefore requires less than 32 integration steps to achieve convergence.

Gradient

Figure 2.39 is a plot of the electrostatic potential superimposed over the field lines for a point charge. The electrostatic potential contours form *equipotential surfaces* surrounding the point charge. All points on such a surface have the same potential. It is evident that these surfaces are always orthogonal to the field lines. In fact, if the behavior of the potential is known, the electric field can be determined by finding the maximum rate and direction of spatial change of the potential field. We can use the del operator again, this time in the *gradient equation*

$$\boxed{\mathbf{E} = -\nabla V}$$

(2.74)

where the negative sign indicates that the field is pointing in the direction of decreasing potential. In Cartesian, cylindrical, and spherical coordinates, respectively, the gradient equations are

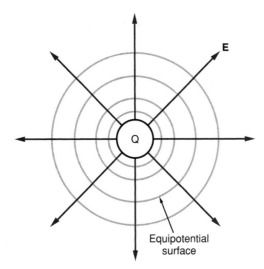

Equipotential
surface

Figure 2.39 Equipotential lines are shown orthogonal to field lines for a point charge.

$$
\nabla V_{\text{Cart}} = \frac{\partial V}{\partial x}\mathbf{a}_x + \frac{\partial V}{\partial y}\mathbf{a}_y + \frac{\partial V}{\partial z}\mathbf{a}_z
$$

$$
\nabla V_{\text{cyl}} = \frac{\partial V}{\partial \rho}\mathbf{a}_\rho + \frac{1}{\rho}\frac{\partial V}{\partial \phi}\mathbf{a}_\phi + \frac{\partial V}{\partial z}\mathbf{a}_z \qquad (2.75)
$$

$$
\nabla V_{\text{spher}} = \frac{\partial V}{\partial r}\mathbf{a}_r + \frac{1}{r}\frac{\partial V}{\partial \theta}\mathbf{a}_\theta + \frac{1}{r\sin\theta}\frac{\partial V}{\partial \phi}\mathbf{a}_\phi
$$

We can demonstrate the gradient equation by applying it to the potential field of (2.71). We have

$$
\mathbf{E} = -\nabla V = -\frac{\partial}{\partial r}\frac{Q}{4\pi\varepsilon_o r}\mathbf{a}_r = \frac{Q}{4\pi\varepsilon_o r^2}\mathbf{a}_r
$$

as expected for a point charge at the origin.

So now we have three ways to calculate **E**. First, if there is sufficient symmetry, we can employ Gauss's law. Second, we can use the Coulomb's law approach. But this approach can be very difficult to implement for some charge distributions. The gradient equation provides us with a powerful, third technique for finding electric field intensity. Here, we can do an integration (without worrying about vectors) followed by a relatively simple differentiation. The electrostatic potential then acts as a stepping stone, or intermediate step, that replaces a single difficult mathematical operation with two simpler ones.

▶ **EXAMPLE 2.19**

Consider a disk of charge density ρ_s as shown in Figure 2.40 extending to a radius a. We want to first find the potential at point h on the z-axis and then find **E** at that point. We start with (2.73) and find that $dQ = \rho_s \, \rho \, d\rho \, d\phi$ and $r = \sqrt{h^2 + \rho^2}$, giving us

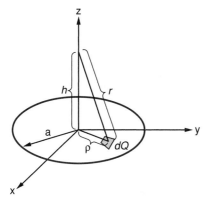

Figure 2.40 Disk of charge to find V and $\mathbf{E}$ on the z-axis.

$$V = \frac{\rho_s}{4\pi\varepsilon_o} \int_0^a \frac{\rho\,d\rho}{\sqrt{h^2+\rho^2}} \int_0^{2\pi} d\phi$$

The integral over $d\phi$ is simple and the other integral is accomplished by substitution (i.e., letting $u = h^2+\rho^2$ and $du = 2\rho\,d\rho$ leads to the integral $\int u^{-1/2}du$). The result is

$$V = \frac{\rho_s}{2\varepsilon_o}\sqrt{h^2+\rho^2}\,\bigg|_{\rho=0}^{\rho=a} = \frac{\rho_s}{2\varepsilon_o}\left(\sqrt{h^2+a^2} - h\right)$$

Now, to find $\mathbf{E}$ we have to know how V is changing with position. In this case, where we want to know how $\mathbf{E}$ varies along the z-axis, we can simply replace h with z in the answer for V, and proceed with the gradient expression:

$$\mathbf{E} = -\nabla V = -\frac{\partial V}{\partial z}\mathbf{a}_z$$

$$= -\frac{\rho_s}{2\varepsilon_o}\left[\frac{1}{2}\frac{2z}{\sqrt{z^2+a^2}} - 1\right]\mathbf{a}_z = \frac{\rho_s}{2\varepsilon_o}\left[1 - \frac{z}{\sqrt{z^2+a^2}}\right]\mathbf{a}_z$$

In many practical problems, the charge distribution is unknown but we are provided the potentials. For example, we might be given the potentials on each plate of a capacitor. We can use the gradient equation, along with Laplace's or Poisson's equation, to find the field. This will be explained further in Section 2.12.

Drill 2.22 Given the field $V = x^2yz$ (V), find $\mathbf{E}$ at (2 m, 3 m, 0). (*Answer:* $\mathbf{E} = -12\,\mathbf{a}_z$ V/m)

MATLAB 2.8

In this example we compare the electrostatic potential to the electric field intensity as a function of radial distance from a 1-nC point charge.

The SUBPLOT(m,n,p) command breaks the figure window into an *m*-by-*n* matrix of small panes and selects the *p*th pane for the current plot. The SEMILOGX command plots a linear *y* against a logarithmic *x*-axis. Running this program returns the plots in Figure 2.41.

```
%   M-File: ML0208
%
%   This program compares V and Er versus radial
%   distance from a point charge.
%
%   Wentworth, 7/7/02
%
%   Variables:
%   Q    point charge (nC)
%   eo   free space permittivity (F/m)
%   r    radial distance from point charge (m)
%   V    electrostatic potential (V)
%   Er   radial electric field (V/m)

clc %clears the command window
clear    %clears variables

%   Initialize variables
Q=1e-9;
eo=8.854e-12;
r=0.1:0.01:10;

%   Perform calculation
V=Q./(4*pi*eo*r);
Er=V./r;

Subplot(2,1,1)
semilogx(r,V,'-r')
ylabel('potential (V)')
subplot(2,1,2)
semilogx(r,Er,'k')
ylabel('electric field intensity (V/m)')
xlabel('radial distance (m)')
```

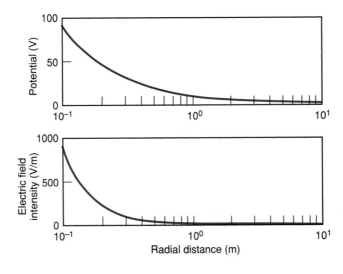

Figure 2.41 *E* and *V* for a point charge in MATLAB 2.8.

▷ **EXAMPLE 2.20**

Consider a pair of point charges of equal magnitude and opposite sign, $+Q$ and $-Q$, in close proximity as depicted by Figure 2.42. Such a pair, termed an *electric dipole*, is useful for describing the behavior of dielectric materials, which will be encountered in Section 2.11. We wish to find an expression for the potential field of such a dipole at a distance r that is large compared to the charge separation d and then apply the gradient equation to find the resulting electric field.

From (2.72), the total potential at point P is

$$V_P = \frac{+Q}{4\pi\varepsilon_o R^+} + \frac{-Q}{4\pi\varepsilon_o R^-} = \frac{+Q}{4\pi\varepsilon_o}\left(\frac{R^- - R^+}{R^+ R^-}\right)$$

We are interested in the potential at a distant point such that $r \gg d$. In this case, the two lines R^+ and R^- are approximately the same length, and the product in the denominator can be estimated as $R^+ R^- = r^2$. In the numerator, the difference $R^- - R^+$ is required. From the lower part of the figure we see that the two lines are approximately parallel, leading to $R^- - R^+ = d\cos\theta$. So our electric potential becomes

$$V_p = \frac{Qd\cos\theta}{4\pi\varepsilon_o r^2}$$

The electric field for $r \gg d$ can be found using the gradient operation on V_p. We have

$$\mathbf{E} = -\nabla V_p = \frac{-\partial V_p}{\partial r}\mathbf{a}_r + \frac{-1}{r}\frac{\partial V_p}{\partial\theta}\mathbf{a}_\theta + \frac{-1}{r\sin\theta}\frac{\partial V_p}{\partial\phi}\mathbf{a}_\phi$$

Since V_p only varies with r and θ, the E_ϕ term is zero. The E_r term is

$$-\frac{\partial V_p}{\partial r}\mathbf{a}_r = \frac{Qd\cos\theta}{4\pi\varepsilon_o}\frac{\partial}{\partial r}\left(r^{-2}\right)\mathbf{a}_r = \frac{2Qd\cos\theta}{4\pi\varepsilon_o r^3}\mathbf{a}_r$$

and the E_θ term is

$$-\frac{1}{r}\frac{\partial V_p}{\partial\theta}\mathbf{a}_\theta = -\frac{1}{r}\frac{Qd}{4\pi\varepsilon_o r^2}\frac{\partial}{\partial\theta}(\cos\theta)\mathbf{a}_\theta = \frac{Qd\sin\theta}{4\pi\varepsilon_o r^3}\mathbf{a}_\theta$$

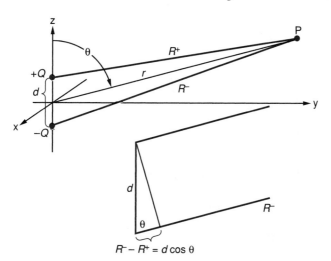

Figure 2.42 An electric dipole.

Combining these terms, we have

$$\mathbf{E} = \frac{Qd}{4\pi\varepsilon_o r^3}\left(2\cos\theta\,\mathbf{a}_r + \sin\theta\,\mathbf{a}_\theta\right)$$

▷ 2.10 CONDUCTORS AND OHM'S LAW

In the next few sections we turn to how fields behave in material space. In electromagnetics, materials are defined by their *constitutive parameters*: permeability μ ("mu"), permittivity ε ("epsilon"), and conductivity σ ("sigma"). Permeability, associated with magnetic fields and inductance, will be described in Chapter 3. Permittivity, discussed in the next section, describes dielectric materials used, for instance, in capacitors. Conductivity relates to a material's ability to conduct electricity and is the subject of this section. In our treatment of the constitutive parameters, we will generally assume that the materials are *homogeneous*,[2.21] meaning the properties are the same at every point in the material, and that the materials are *isotropic*,[2.22] meaning the properties are independent of direction.

The units for conductivity are most often expressed in terms of siemens per meter (S/m), or by the equivalent terms $1/\Omega$-m and, less commonly, $(\mho/\text{m})$ (pronounced "*mhos per meter*"). Figure 2.43 shows the range of conductivity exhibited by good conductors ($\sigma > 10^4$ S/m), good insulators ($\sigma < 10^{-4}$ S/m), and semiconductors.

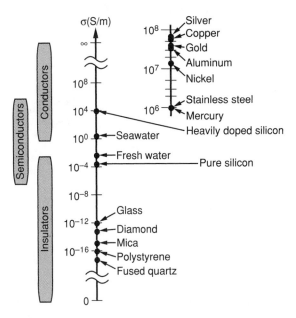

Figure 2.43 Conductivity chart (at room temperature).

[2.21]In some problems, parameters will be given as a function of position, clearly indicating an *inhomogeneous* material.

[2.22]Constitutive properties for semiconductor crystals are anisotropic, meaning they depend on direction in the crystal.

Most conductors are metals with an abundance of electrons available for conduction. The conductivity in metals depends on the charge density and on the scattering of electrons by their interactions with the crystal lattice. Conductivity decreases with increasing temperature for metals since more lattice vibration and hence more scattering occurs at higher temperatures. Most metals commonly used for electrical wiring have conductivities between 10^7 and 10^8 S/m (see Figure 2.43 and Appendix E). Note that an electrical conductor does not have to be solid. Mercury, a liquid at room temperature and pressure, conducts well enough to be used as an electrical probe for delicate surfaces. Also, seawater is somewhat conductive.

A perfect conductor has infinite conductivity and is termed a *superconductor*. Until the 1980s, only a limited number of metals existed that would superconduct at very low temperatures (in the vicinity of 10 K or so). Then, a startling discovery was made that certain ceramic-layered structures, beginning with yttrium-barium-copper-oxide, would superconduct at significantly higher temperatures. The most advanced superconductors today are limited to modest field intensity levels and still are only operable at well below room temperatures.

In a good insulator, or dielectric, the electrons are tightly bound by their parent atoms and so are not available to conduct. In the presence of a very strong field (or large potential across the insulator), the electrons may be stripped from their orbits and conduction will ensue, sometimes with unfortunate results. The insulator's breakdown voltage is therefore a critical parameter when large fields are involved.

In semiconductors, electrons are loosely bound to their parent atoms, and with the addition of thermal energy they are made available for conduction. When the electron is pulled from its parent atom it leaves behind a vacancy, or *hole*, which for all practical purposes behaves as a mobile positive charge. For pure silicon at room temperature, the number of mobile charge carriers (electrons and holes) is modest and conductivity is low, as indicated in Figure 2.43. With increasing temperature, however, the number of charge carriers increases rapidly and the conductivity increases in spite of the increased lattice vibrations. Another way to increase a semiconductor's conductivity is to intentionally add, or *dope*, the semiconductor with impurities that can easily donate a charge carrier. As seen in Figure 2.43, heavily doped silicon can achieve a conductivity as high as 10^4 S/m.

Current and Current Density

An ampere (A) or amp of *current* is defined as the amount of charge (in coulombs) that passes through a reference plane in a given amount of time (in seconds). Thus, 1 A = 1 C/s. The *current density* is the current divided by the area through which the current passes. The current density is expressed as a vector quantity $\mathbf{J}$ (A/m^2) and is related to current I by

$$\boxed{I = \int \mathbf{J} \cdot d\mathbf{S}} \tag{2.76}$$

There are three types of current density: convection, conduction, and displacement. Displacement current density is a time-varying field phenomenon that allows current to flow between the plates of a capacitor and will be discussed in Chapter 4.

Convection current density involves the movement of charged particles through vacuum, air, or other nonconductive media. An example is the beam of electrons in a cathode ray tube (i.e., a conventional television picture tube). If a charge density ρ_v is moving at velocity $\mathbf{u}$, the convection current density is

$$\mathbf{J} = \rho_v \mathbf{u} \tag{2.77}$$

Drill 2.23   A 4.00-mm-diameter columnar beam of electrons with charge density -0.200 nC/m^3 moves with velocity $\mathbf{u} = 6.00 \times 10^6$ m/s. Determine the current. (*Answer:* 15.1 nA)

Conduction current density involves the movement of electrons through conductive media in response to an applied electric field. It is given by the *point form of Ohm's law*,

$$\boxed{\mathbf{J} = \sigma\mathbf{E}} \tag{2.78}$$

Suppose charge is introduced into a good conductor. The charges repel each other and quickly accumulate very near or on the conductor surface.[2.23] The result is that within a good conductor the charge density is zero, and by Gauss's law considerations the electric field intensity inside the conductor is zero. With $\mathbf{E} = 0$, the potential difference

$$V_{ab} = -\int_b^a \mathbf{E}\cdot d\mathbf{L} \tag{2.79}$$

is zero between any two points in the conductor. The conductor is considered to be an *equipotential* medium. Another way to see that $\mathbf{E} = 0$ inside a good conductor is to consider the field within a perfect conductor where $\sigma = \infty$. The only way $\mathbf{J}$ can be a finite quantity is if $\mathbf{E}$ approaches zero.

Although perfect conductors are equipotential volumes, real conductors and resistive materials will encounter a potential difference in the direction of current. The relationship between the potential difference and the current is the resistance R, given by the version of Ohm's law we've seen before in circuit theory, $R = V/I$.

As an example, and to show the derivation of the circuit-theory form of Ohm's law from the point form, let us consider a cylinder of material with conductivity σ, length L, and cross-sectional area S as indicated in Figure 2.44. A potential difference V_{ab} between the two ends of the cylinder will establish an electric field intensity $\mathbf{E} = (V_{ab}/L)\,\mathbf{a}_z$. The current density will be related to the field by the point form of Ohm's law, $\mathbf{J} = \sigma\mathbf{E}$. Because the current is related to the current density by $\mathbf{J} = (I/S)\,\mathbf{a}_z$, the point form of Ohm's law becomes

$$\frac{I}{S} = \sigma\frac{V_{ab}}{L}$$

Manipulating this equation to arrive at an expression for resistance, we have

$$\boxed{R = \frac{V_{ab}}{I} = \frac{1}{\sigma}\frac{L}{S}} \tag{2.80}$$

[2.23]We assume, since this chapter concerns static electric fields, that the charges have already reached the surface.

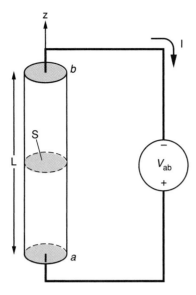

Figure 2.44 Bar with conductivity σ for deriving the circuit-theory form of Ohm's law.

Drill 2.24 Determine the resistance of a 1.00-m-long, 1.00-mm-diameter nichrome wire. (*Answer:* 1.27 Ω)

A more general expression for resistance is

$$R = \frac{V_{ab}}{I} = \frac{-\int_b^a \mathbf{E} \cdot d\mathbf{L}}{\int \sigma \mathbf{E} \cdot d\mathbf{S}} \tag{2.81}$$

▶ **EXAMPLE 2.21**

As an example of how we can apply (2.81), let us find the resistance between the inner conductive shell (radius a) and outer conductive shell (radius b) for a length L of coaxial cable filled with material of conductivity σ. Assuming a charge Q on the inner shell, we have by Gauss's law a field for $a \le \rho \le b$ of

$$\mathbf{E} = \frac{Q}{2\pi\varepsilon_o \rho L} \mathbf{a}_\rho$$

Then, using this field we can determine potential difference between the inner and outer conductors as

$$V_{ab} = -\int_b^a \frac{Q}{2\pi\varepsilon_o \rho L} \mathbf{a}_\rho \cdot d\rho \, \mathbf{a}_\rho = \frac{Q}{2\pi\varepsilon_o L} \ln\left(\frac{b}{a}\right) \tag{2.82}$$

The current is found from (2.76) and (2.78) to be

$$I = \int \mathbf{J} \cdot d\mathbf{S} = \int \sigma \mathbf{E} \cdot d\mathbf{S}$$
$$= \int \sigma \frac{Q}{2\pi\varepsilon_o \rho L} \mathbf{a}_\rho \cdot \rho \, d\phi \, dz = \frac{\sigma Q}{2\pi\varepsilon_o L} \int_0^{2\pi} d\phi \int_0^L dz = \frac{\sigma Q}{\varepsilon_o} \tag{2.83}$$

Dividing V_{ab} by I to find resistance, we have

$$R = \frac{1}{2\pi\sigma L} \ln\left(\frac{b}{a}\right) \tag{2.84}$$

The resistance between the conductors of coaxial cable given by (2.84) can also be expressed in terms of a *conductance G*, where $G = 1/R$. So, the conductance for the coaxial cable is

$$G = \frac{2\pi\sigma L}{\ln(b/a)} \tag{2.85}$$

and the conductance per unit length, G', is simply

$$G' = \frac{G}{L} = \frac{2\pi\sigma}{\ln(b/a)} \tag{2.86}$$

This value is one of the distributed parameters to be employed when studying transmission lines in Chapter 6.

Joule's Law

The electric field does work in moving charges through a material. Some of the energy of the moving charges is given up in collisions with atoms of the material. The amount of energy given up per unit time is called the *dissipated power*, denoted P.

The differential force exerted by the electric field to move a differential charge $dQ = \rho_v \, dv$ is

$$d\mathbf{F} = dQ\,\mathbf{E} = \rho_v dv \mathbf{E} \tag{2.87}$$

The incremental work done is simply

$$dW = d\mathbf{F} \cdot d\mathbf{L} = \rho_v dv\, \mathbf{E} \cdot d\mathbf{L} \tag{2.88}$$

The increment of power dissipated is this work divided by the increment of time, or

$$dP = \frac{dW}{dt} = \rho_v dv\, \mathbf{E} \cdot \frac{d\mathbf{L}}{dt} = \rho_v dv\, \mathbf{E} \cdot \mathbf{u} \tag{2.89}$$

where $\mathbf{u}$ is the velocity vector $d\mathbf{L}/dt$. Now, this can be rearranged to give

$$dP = \mathbf{E} \cdot \rho_v \mathbf{u}\, dv \tag{2.90}$$

which, since $\mathbf{J} = \rho_v\, \mathbf{u}$, is equal to

$$dP = \mathbf{E} \cdot \mathbf{J}\, dv \tag{2.91}$$

Finally, we can integrate over the volume to find the total dissipated power,

$$\boxed{P = \int \mathbf{E} \cdot \mathbf{J}\, dv = \int \sigma E^2 dv} \tag{2.92}$$

Equation 2.92 is known as *Joule's law*.

Considering again the cylinder of conductive material in Figure 2.44, we can write (2.92) as

$$P = \int \mathbf{E} \cdot \mathbf{J}\, dv = \int \frac{V_{ab}}{L}\, d\mathbf{L} \int \frac{I}{S}\, d\mathbf{S} \tag{2.93}$$

which is easily evaluated to get the more familiar electric-circuit-theory form of Joule's law,

$$P = VI \tag{2.94}$$

▷ **EXAMPLE 2.22**

The electric field intensity in a 10-cm-diameter cylinder of material with conductivity $\sigma = 10^{-3}$ S/m is measured as $\mathbf{E} = 12\,\rho\,\mathbf{a}_z$ V/cm, where ρ is in centimeters. What is the dissipated power in a 1-m length of this cylinder?

To maintain the correct units, we'll convert σ to 10^{-5} S/cm. Then using (2.92) we have

$$P = \sigma \int_0^{5\ \text{cm}} (12\rho)^2\, \rho\, d\rho \int_0^{2\pi} d\phi \int_0^{100\ \text{cm}} dz = \left(10^{-5}\right)(144)\frac{\left(5^4\right)}{4}(2\pi)(100) = 140\ \text{W}$$

Drill 2.25   How much power is dissipated if 1 V is placed across the wire of Drill 2.24? (*Answer:* 0.79 W)

▷ 2.11   DIELECTRICS

In the previous section we saw that insulators (dielectrics) are different from conductors in that they have few, if any, free charges available for conduction. Such materials do have fixed, or *bound*, charges that influence the field within the material. At the atomic level, where an electron cloud surrounds a positively charged nucleus, an external applied field will cause some shifting of the electron cloud such that *electric dipoles* are formed and aligned, as indicated in Figure 2.45. Each dipole consists of a positive charge at the center of the nucleus separated by a small distance from a negative charge at the center of the electron cloud. We say that the material is *polarized* by the electric field, meaning the dipoles are aligned, and the degree and direction of alignment are given by

$$\mathbf{P} = \chi_e \varepsilon_o \mathbf{E} \tag{2.95}$$

where the polarization vector $\mathbf{P}$ aligns with the electric field and χ_e ("chi") is the material's *electric susceptibility*.

Molecules like the fictitious ones depicted in Figure 2.45 are *nonpolar*, meaning they have no dipole until influenced by an electric field. However, some molecules are *polar*, meaning they have a built-in electric dipole. A water molecule is an example of a polar molecule. Because the hydrogen atoms are not placed on opposite sides of the oxygen,[2.24] and the shared electrons spend a bit more time in the vicinity of the oxygen nucleus than with

---

[2.24]Think of the silhouette of a very famous cartoon mouse; the round ears are the hydrogen atoms and the head is the oxygen atom.

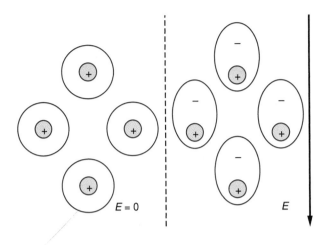

Figure 2.45 A collection of atoms before and after application of an electric field shows the formation of aligned dipoles.

the hydrogen nuclei, the water molecule is an electric dipole. These polar molecules are randomly arranged, but under the influence of an electric field they are polarized, just like the ones in Figure 2.45. The degree to which a material is susceptible to polarization is termed the electric susceptibility χ_e.

The polarization creates an opposing electric field within the material, where the net effect is a decrease of the field. The lines of flux, however, remain the same, and we can relate the flux density to polarization by the equation

$$\mathbf{D} = \varepsilon_o \mathbf{E} + \mathbf{P} \tag{2.96}$$

The susceptibility and polarization terms are useful for giving us some insight into what is going on in the material to influence the fields, but in practice we combine (2.95) and (2.96) to give a more compact relation between $\mathbf{D}$ and $\mathbf{E}$:

$$\boxed{\mathbf{D} = \varepsilon_o(1 + \chi_e)\mathbf{E} = \varepsilon_r\varepsilon_o\mathbf{E} = \varepsilon\mathbf{E}} \tag{2.97}$$

where ε is the material's permittivity, related to free-space permittivity by the factor ε_r, called the *relative permittivity* or *dielectric constant*. [2.25] The relative permittivity is clearly related to the electric susceptibility by

$$\varepsilon_r = 1 + \chi_e \tag{2.98}$$

Appendix E lists a number of dielectrics with their relative permittivities and breakdown voltages.

Drill 2.26 The relative permittivity for polystytrene is 2.6. Determine the electric susceptibility and the permittivity. (*Answer:* 1.6, 23 pF/m)

[2.25]The term dielectric constant, though widely used, is a bit deceptive in that it is not really constant. It can change with temperature and frequency.

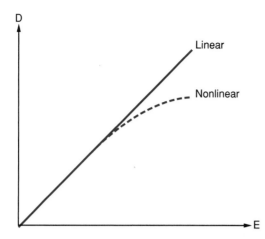

Figure 2.46 The *D* versus *E* relation is compared for a linear and a nonlinear material.

The relative permittivity may itself be a function of electric field intensity. In such a case, the material is considered *nonlinear*, meaning that a plot of D versus E would not yield a straight line, as evidenced in Figure 2.46. Whereas (2.96) holds for all materials, (2.97) is true only for linear, isotropic, homogeneous materials. We will restrict our treatment to linear materials.

The many equations developed thus far for fields in free space are simply modified by replacing ε_o with ε.

Drill 2.27 A 10-nC point charge exists at the origin in free space. (a) Find **D** and **E** at a point 1 m away from the origin. (b) Repeat the problem with the point charge embedded in a large volume of distilled water ($\varepsilon_r = 81$). (*Answer:* (a) **D** = 0.796 $\mathbf{a}_r$ nC/m², **E** = 90 $\mathbf{a}_r$ V/m; (b) **D** = 0.796 $\mathbf{a}_r$ nC/m², **E** = 1.11 $\mathbf{a}_r$ V/m)

Application of a sufficiently strong field can strip the electrons from the parent atoms and allow conduction in the dielectric. This can cause a runaway effect, whereby the collision of a stripped electron with another atom can lead to further generation of charges and breakdown of the dielectric. The *dielectric strength* is the maximum electric field a dielectric can handle before breakdown. The spark between the charged poles in a Jacob's ladder is an example in which the breakdown voltage in air has been exceeded. A more common example of dielectric breakdown is the lightning bolt, where sufficient charge has accumulated to overcome the dielectric strength in air (about 3 MV/m!).

In high-power or high-voltage applications involving dielectrics, high-voltage cables for instance, the dielectric strength is an important design criterion. In the design of capacitors, as discussed in Section 2.13, care must be taken not to exceed the dielectric's breakdown voltage.

▷ **EXAMPLE 2.23**

Suppose a pair of capacitive plates is to support a 6-kV potential difference. We want to bring the plates as close together as we can with a mica separation layer without the dielectric breaking down.

Mica has $\varepsilon_r = 6$ and a dielectric breakdown value of 200 MV/m. The field in the dielectric is $E = 6$ kV/d, where d is the separation distance, and this value of E must be kept below 200 MV/m. Dividing 6 kV by 200 MV/m, we see that d must be at least 30-µm thick.

Drill 2.28 Suppose a 30-µm-thick layer of polystyrene is substituted for mica in Example 2.23. What is the maximum potential difference that can be supported across the plates? (*Answer:* 600 V)

Practical Application: Electret Microphone

An electret is a dielectric material specially treated to sustain an electric field. This is analogous to a magnet that sustains a magnetic field. Electrets find use primarily in electroacoustic transducers, such as microphones. They are also employed to produce miniature electric motors and a variety of sensors.

The first electrets were formed by aligning dipoles within a dielectric. The dielectric must contain dipolar molecules, and at sufficient heat where the dipoles are able to twist and turn, a strong electric field is externally applied, thus aligning the dipoles. The heat is removed while maintaining the electric field, and the dipoles are frozen into place. The first such *thermal electret* was made in 1919 using a combination of carnauba wax and beeswax.

A second, much more widely used approach is to inject charge into a good dielectric using irradiation. When the charges are placed within the dielectric, they are locked into place and maintain a very long lasting electric field. These electrets are routinely made of polypropylene, Mylar, or Teflon. The irradiation may be accomplished in a number of ways, including direct injection of an electron beam. A general approach is to use a *corona discharge*, the dielectric breakdown of air in a strong electric field. Rolls of polymer film electrets are heavily irradiated by passing the film through a corona discharge.

A very common type of microphone uses a variable capacitor, where one of the electrodes is a thin, taut membrane (or diaphragm) that is deflected by sound waves. Such a microphone is known as a *condensor microphone*. In an electret microphone, shown in Figure 2.47, the membrane is made of a metallized electret film used as one side of the variable capacitor. As sound waves impinge on the membrane, the distance between the plates is varied. An output signal is produced that is proportional to the diaphragm deflection. The

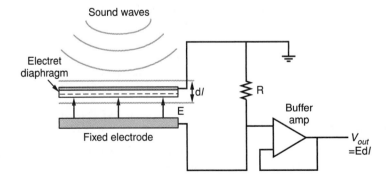

Figure 2.47 Simplified schematic of an electret microphone circuit.

big advantage of the electret microphone is that it doesn't require an external power source. Some power is, however, required for the support electronics.

Thin electret films can also be used in microelectromechanical systems (MEMS). Not only can MEMS-based electret microphones be very small with high sensitivity, but they can also be mass-produced and integrated with electronics. Their construction can be much simpler than that of a conventional condensor-based microphone. One application is to build microphonic arrays, capable of directional reception of sound (identical to beam steering in antenna arrays).

▶ 2.12 BOUNDARY CONDITIONS

We have looked at fields in conductors and in dielectrics. Now we want to see how the fields behave at the boundary between a pair of dielectrics or between a dielectric and a conductor. Beginning with the boundary between two dielectrics, we perform a line integral of **E** around a closed rectangular path, as indicated in Figure 2.48. In the figure, fields are shown in each medium along with their normal and tangential components. For static fields we have Kirchhoff's voltage law

$$\oint \mathbf{E} \cdot d\mathbf{L} = 0 \tag{2.99}$$

We integrate in the loop clockwise starting at a and have

$$\int_a^b \mathbf{E} \cdot d\mathbf{L} + \int_b^c \mathbf{E} \cdot d\mathbf{L} + \int_c^d \mathbf{E} \cdot d\mathbf{L} + \int_d^a \mathbf{E} \cdot d\mathbf{L} = 0 \tag{2.100}$$

Evaluating each segment of this integral we have

$$\int_a^b \mathbf{E} \cdot d\mathbf{L} = \int_0^{\Delta w} E_{T1}\mathbf{a_T} \cdot dL\,\mathbf{a_T} = E_{T1}\Delta w$$

$$\int_b^c \mathbf{E} \cdot d\mathbf{L} = \int_{\Delta h/2}^0 E_{N1}\mathbf{a_N} \cdot dL\,\mathbf{a_N} + \int_0^{-\Delta h/2} E_{N2}\mathbf{a_N} \cdot dL\,\mathbf{a_N} = -(E_{N1} + E_{N2})\frac{\Delta h}{2}$$

$$\int_c^d \mathbf{E} \cdot d\mathbf{L} = \int_{\Delta w}^0 E_{T2}\mathbf{a_T} \cdot dL\,\mathbf{a_T} = -E_{T2}\Delta w \tag{2.101}$$

$$\int_d^a \mathbf{E} \cdot d\mathbf{L} = \int_{-\Delta h/2}^0 E_{N2}\mathbf{a_N} \cdot dL\,\mathbf{a_N} + \int_0^{\Delta h/2} E_{N1}\mathbf{a_N} \cdot dL\,\mathbf{a_N} = (E_{N1} + E_{N2})\frac{\Delta h}{2}$$

Now summing the results for each segment, we see that the normal components cancel[2.26] and we have $E_{T1} = E_{T2}$, or

$$\boxed{\mathbf{E}_{T1} = \mathbf{E}_{T2}} \tag{2.102}$$

[2.26]But, you say, what if the normal components vary over the distance Δw? Then the normal components wouldn't cancel, right? We can avoid this problem two ways. Let's make sure that Δw is small so that there isn't much change, but let's also shrink Δh to close to zero. This ensures that the normal components cancel.

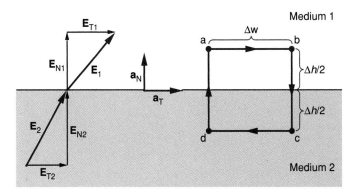

Figure 2.48 Boundary condition for a line integral.

So we see that the tangential component of the electric field intensity must be continuous across the boundary.

We can get a second boundary condition by applying Gauss's law,

$$\oint \mathbf{D} \cdot d\mathbf{S} = Q_{\text{enc}}$$

over a very small pillbox-shaped Gaussian surface[2.27] enclosing a portion of the surface, as shown in Figure 2.49. The left side of Gauss's law becomes

$$\oint \mathbf{D} \cdot d\mathbf{S} = \underbrace{\int \mathbf{D} \cdot d\mathbf{S}}_{\text{Top}} + \underbrace{\int \mathbf{D} \cdot d\mathbf{S}}_{\text{Bottom}} + \underbrace{\int \mathbf{D} \cdot d\mathbf{S}}_{\text{Side}} \qquad (2.103)$$

If we make the pillbox short enough, negligible flux passes through the side, leaving us with just the flux through the top and bottom, or

$$\underbrace{\int \mathbf{D} \cdot d\mathbf{S}}_{\text{Top}} = \int D_{N1}\mathbf{a_N} \cdot d S \mathbf{a_N} = D_{N1}\Delta S$$

$$\underbrace{\int \mathbf{D} \cdot d\mathbf{S}}_{\text{Bottom}} = \int D_{N2}\mathbf{a_N} \cdot d S(-\mathbf{a_N}) = -D_{N2}\Delta S \qquad (2.104)$$

which sums to

$$(D_{N1} - D_{N2})\Delta S = Q_{\text{enc}} \qquad (2.105)$$

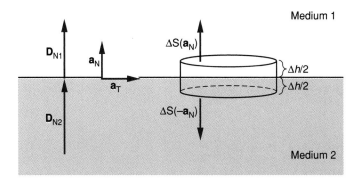

Figure 2.49 Boundary condition for Gauss's law.

[2.27]*Pillbox* is the historical term used since it is hollow and has the correct shape. As pillboxes are rather out of fashion, it would be easier to think of a Gaussian surface with the shape of a penny.

The right side of Gauss's law, the charge enclosed, is

$$Q_{enc} = \int \rho_s dS = \rho_s \Delta S \tag{2.106}$$

Equations (2.105) and (2.106) lead to the second boundary condition

$$D_{N1} - D_{N2} = \rho_s \tag{2.107}$$

The sign of (2.107) is a consequence of the normal direction being chosen to go from medium 2 to medium 1. Had we chosen the normal direction from medium 1 to medium 2, we would have gotten $D_{N2} - D_{N1} = \rho_s$. To always get the sign correct we can use a general expression for this boundary condition:

$$\boxed{\mathbf{a}_{21} \cdot (\mathbf{D}_1 - \mathbf{D}_2) = \rho_s} \tag{2.108}$$

where $\mathbf{a}_{21}$ is the unit vector normal from medium 2 to medium 1, and the dot product ensures that we consider only the normal components of $\mathbf{D}_1$ and $\mathbf{D}_2$.

We see that, if there is no surface charge, (2.108) says the normal component of the electric flux density must be continuous across the surface. However, the presence of a surface charge indicates an abrupt change in D_N at the boundary.

If the boundary between two dielectrics is not normal to the field, we can use (2.102) and (2.108) to relate the field components on each side and determine how much the field bends from one medium to the other.

▷ **EXAMPLE 2.24**

Consider that the field $\mathbf{E}_1$ is known for one of a pair of dielectrics as shown in Figure 2.50 and we wish to find the field $\mathbf{E}_2$ in the other dielectric and also the angles that the fields in each dielectric make with a normal to the surface.

We follow a bookkeeping approach as indicated in the figure. By inspection, in step 1 we see that the component of $\mathbf{E}_1$ normal to the boundary is just $5\mathbf{a}_z$. Mathematically, we can find $\mathbf{E}_{N1}$ by finding out how much of $\mathbf{E}_1$ is in the normal direction ($\mathbf{a}_N = \mathbf{a}_z$), and then multiplying this by $\mathbf{a}_N$, or $\mathbf{E}_{N1} = (\mathbf{E}_1 \cdot \mathbf{a}_z)\mathbf{a}_z$. The tangential portion of $\mathbf{E}_1$ is simply $\mathbf{E}_1 - \mathbf{E}_{N1}$ (step 2). In step 3, we get the angle $\mathbf{E}_1$ makes with a normal by employing routine trigonometry. Then, by (2.102) the tangential component in the second dielectric is known (step 4). Using the permittivity information, we are able to determine the normal component of $\mathbf{D}_1$ (step 5), and by (2.108) this is equivalent to $\mathbf{D}_{N2}$ since $\rho_s = 0$ (step 6). The known permittivity in medium 2 allows calculation of $\mathbf{E}_{N2}$ (step 7), which is used with $\mathbf{E}_{T2}$ to find the angle to the normal (step 8) and the total field $\mathbf{E}_2$ (step 9).

Using our boundary conditions for a pair of dielectrics, we can also find the boundary conditions between a dielectric and a good conductor. Realizing that in a good conductor $\mathbf{E} = 0$, we see that the first boundary condition becomes

$$\boxed{E_T = 0} \tag{2.109}$$

There is no tangential electric field intensity at the boundary of a good conductor. For the second boundary condition, the electric flux density is also zero inside the conductor, so

$$\boxed{D_N = \rho_s} \tag{2.110}$$

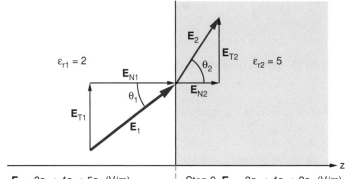

$\mathbf{E}_1 = 3\mathbf{a}_x + 4\mathbf{a}_y + 5\mathbf{a}_z$ (V/m)

Step 1. $\mathbf{E}_{N1} = 5\mathbf{a}_z$

Step 2. $\mathbf{E}_{T1} = 3\mathbf{a}_x + 4\mathbf{a}_y$

Step 3. $\theta_1 = \tan^{-1}\left(\dfrac{|\mathbf{E}_{T1}|}{|\mathbf{E}_{N1}|}\right) = 45°$

Step 5. $\mathbf{D}_{N1} = \varepsilon_{r1}\varepsilon_o\mathbf{E}_{N1} = 10\varepsilon_o\mathbf{a}_z$

Step 9. $\mathbf{E}_2 = 3\mathbf{a}_x + 4\mathbf{a}_y + 2\mathbf{a}_z$ (V/m)

Step 7. $\mathbf{E}_{N2} = \mathbf{D}_{N2}/\varepsilon_{r1}\varepsilon_o = 2\mathbf{a}_z$

Step 4. $\mathbf{E}_{T2} = \mathbf{E}_{T1} = 3\mathbf{a}_x + 4\mathbf{a}_y$

Step 8. $\theta_2 = \tan^{-1}\left(\dfrac{|\mathbf{E}_{T2}|}{|\mathbf{E}_{N2}|}\right) = 68.2°$

Step 6. $\mathbf{D}_{N2} = \mathbf{D}_{N1} = 10\varepsilon_o\mathbf{a}_z$

Figure 2.50 Procedure for evaluating the fields on both sides of a boundary separating a pair of dielectrics.

Drill 2.29   For $\rho \le 2$ m, $\varepsilon_{r1} = 2$ and $\mathbf{E}_1 = 3\mathbf{a}_\rho + 6\mathbf{a}_\phi + 9\mathbf{a}_z$ V/m. For $\rho > 2$ m, $\varepsilon_{r2} = 3$. Determine $\mathbf{E}_2$. (Answer: $\mathbf{E}_2 = 2\mathbf{a}_\rho + 6\mathbf{a}_\phi + 9\mathbf{a}_z$ V/m)

Drill 2.30   A surface charge $\rho_s = 3$ nC/m^2 exists at the plane $y = 0$. For $0 \le y \le 3$ m, $\varepsilon_{r1} = 9$; for $y > 3$ m, $\varepsilon_{r2} = 12$. Determine $\mathbf{E}_1$ and $\mathbf{E}_2$. (*Answer:* $\mathbf{E}_1 = 37.7\ \mathbf{a}_y$ V/m, $\mathbf{E}_2 = 28.3\ \mathbf{a}_y$ V/m)

Boundary Value Problems

The boundary conditions are very useful for finding the electric fields when some of the field quantities are known. In many instances, however, only the potentials are known along with, perhaps, some information about charge distribution. Commonly we will know the potential difference across a pair of conductors of some geometrical configuration separated by a known dielectric and will want to determine the potential everywhere along with the electric field. We can employ Poisson's and Laplace's equations to help us find the potential function when conditions at the boundaries are specified.

We have, from the divergence expression,

$$\nabla \cdot \mathbf{D} = \rho_v \qquad (2.111)$$

which we can rewrite by considering $\mathbf{D} = \varepsilon\mathbf{E}$ and dividing both sides by ε (since we are assuming it is not a variable with position) so that

$$\nabla \cdot \mathbf{E} = \frac{\rho_v}{\varepsilon} \qquad (2.112)$$

We also know that the electric field is a function of the potential field by the gradient expression

$$\mathbf{E} = -\nabla V \tag{2.113}$$

which upon insertion into (2.112) gives

$$\boxed{\nabla \cdot \nabla V = \nabla^2 V = -\frac{\rho_v}{\varepsilon}} \tag{2.114}$$

The divergence of a gradient is expressed by the Laplacian operator (∇^2), and (2.114) is known as *Poisson's equation*. For a charge-free medium in which $\rho_v = 0$, the equation becomes *Laplace's equation*,

$$\boxed{\nabla^2 V = 0} \tag{2.115}$$

The Laplacian operator is easily expanded in Cartesian coordinates to give

$$\boxed{\nabla^2 V_{\text{Cart}} = \frac{\partial^2 V}{\partial x^2} + \frac{\partial^2 V}{\partial y^2} + \frac{\partial^2 V}{\partial z^2}} \tag{2.116}$$

The expansion in cylindrical and spherical coordinate systems, although not so obvious are

$$\boxed{\nabla^2 V_{\text{cyl}} = \frac{1}{\rho}\frac{\partial}{\partial \rho}\left(\rho \frac{\partial V}{\partial \rho}\right) + \frac{1}{\rho^2}\left(\frac{\partial^2 V}{\partial \phi^2}\right) + \frac{\partial^2 V}{\partial z^2}} \tag{2.117}$$

and

$$\boxed{\nabla^2 V_{\text{spher}} = \frac{1}{r^2}\frac{\partial}{\partial r}\left(r^2 \frac{\partial V}{\partial r}\right) + \frac{1}{r^2 \sin\theta}\frac{\partial}{\partial \theta}\left(\sin\theta \frac{\partial V}{\partial \theta}\right) + \frac{1}{r^2 \sin^2\theta}\frac{\partial^2 V}{\partial \phi^2}} \tag{2.118}$$

Solution of Poisson's and Laplace's equations to determine the potential field requires that we know the potential on the boundaries (i.e., the *boundary conditions*). A useful application is finding the resistance for a material of nonuniform cross section. We assume a conductive surface at each end of the material and then employ Laplace's equation to arrive at an expression for the potential, which is solved based on the boundary conditions (i.e., the potentials at each end). The electric field is determined using the gradient equation, which allows calculation of the current. The resistance is then the ratio of the potential difference to the current, which only depends on the geometry and the conductivity. In this section we restrict our discussion to simple, symmetric geometries where application of the equations is straightforward. The fields for more complicated geometries can be solved via Poisson's and Laplace's equations by employing numerical analysis techniques.

▷ **EXAMPLE 2.25**

Let's derive the resistance over a length d of a block of material that has conductivity σ and cross-sectional area S as shown in Figure 2.51.

We'll put a potential difference across the block by placing V_d at $z = d$ and by grounding the block at $z = 0$. Equation (2.115) reduces to

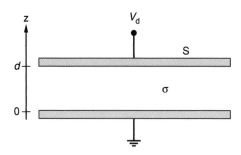

Figure 2.51 Cross section of a block of material for which resistance is to be calculated.

$$\frac{\partial^2 V(z)}{\partial z^2} = 0$$

where the z in parentheses for $V(z)$ indicates that the potential is only a function of z. Integrating once, we have

$$\frac{\partial V(z)}{\partial z} = A$$

where A is a constant of integration. A second integration gives

$$V(z) = Az + B$$

where B is a second constant of integration. These constants are determined by employing the boundary conditions, namely,

$$V(d) = V_d$$

and

$$V(0) = 0$$

Applying $V(0) = 0$ returns a value of 0 for B, and applying $V(d) = V_d$ returns a value of V_d/d for A. So now we have

$$V(z) = \frac{V_d}{d} z$$

The electric field intensity can now be found by the gradient equation,

$$\mathbf{E} = -\nabla V = -\frac{\partial}{\partial z}\left(\frac{V_d}{d} z\right)\mathbf{a}_z = -\frac{V_d}{d}\mathbf{a}_z$$

The current, found by integrating over a cross-sectional surface, is then

$$I = \int \sigma \mathbf{E} \cdot d\mathbf{S} = -\sigma \frac{V_d}{d}\mathbf{a}_z \cdot S(-\mathbf{a}_z) = \frac{\sigma S}{d} V_d$$

So we see that the resistance is the expected result

$$R = \frac{V_d}{I} = \frac{V_d}{\dfrac{\sigma S}{d} V_d} = \frac{1}{\sigma}\frac{d}{S}$$

▶ **EXAMPLE 2.26**

Let's determine the electric potential in the dielectric region between a pair of concentric spheres that have a potential difference V_{ab} as indicated in Figure 2.52.

In this problem we'll also assume a charge distribution $\rho_v = \rho_o/r$ C/m^3 and employ Poisson's equation to find the fields. From the problem's symmetry we know the potential is only a function of r, so (2.114) with (2.118) reduces to

$$\frac{1}{r^2}\frac{\partial}{\partial r}\left(r^2\frac{\partial V(r)}{\partial r}\right) = -\frac{\rho_o}{r\varepsilon_r\varepsilon_o}$$

Multiplying both sides by r^2 and integrating we obtain

$$r^2\frac{\partial V(r)}{\partial r} = -\frac{\rho_o r^2}{2\varepsilon_r\varepsilon_o} + A$$

where A is a constant of integration. Dividing both sides by r^2 and integrating again gives

$$V(r) = -\frac{\rho_o r}{2\varepsilon_r\varepsilon_o} - \frac{A}{r} + B$$

where B is a second constant of integration. To solve for the constants, we'll employ boundary conditions. Let's assume that the potential difference V_{ab} consists of a voltage V_a on the inner conductor and we'll ground the outer conductor. The conditions result in

$$B = \frac{\rho_o(a+b)}{2\varepsilon_r\varepsilon_o} + \frac{V_a a}{a-b}$$

and

$$A = \frac{V_a ab}{a-b} + \frac{\rho_o ab}{2\varepsilon_r\varepsilon_o}$$

Hence the potential between the spheres is given by

$$V(r) = \frac{\rho_o}{2\varepsilon_r\varepsilon_o}\left(\frac{a}{r} - r\right) + \frac{V_a a}{r}$$

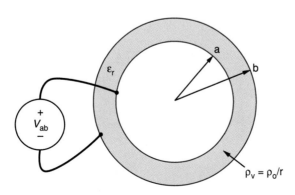

Figure 2.52 Concentric conductive spheres with a charge distribution between them.

Drill 2.31 Find an expression for the electric field intensity for Example 2.26.

$$\left(Answer:\ \mathbf{E} = \left[\left(\frac{\rho_o}{2\varepsilon_r\varepsilon_o} + \frac{V_a}{b-a} \right) ab \ln r - \frac{\rho_o}{2\varepsilon_r\varepsilon_o} \right] \mathbf{a}_r \right)$$

Drill 2.32 Use Laplace's equation to find V and $\mathbf{E}$ as a function of ρ within the dielectric of a coaxial cable of inner conductor radius a and outer conductor radius b.

$$\left(Answer:\ V(\rho) = V_{ab} \frac{\ln\rho - \ln b}{\ln a - \ln b}, \quad \mathbf{E} = \frac{V_{ab}}{\rho \ln(b/a)} \mathbf{a}_\rho \right)$$

▷ **2.13 CAPACITANCE**

Suppose a potential difference is applied across a pair of conductors separated by a dielectric, as shown in Figure 2.53. In the configuration shown, positive charge ($+Q$) will accumulate on the bottom surface of the top plate and an equal amount of negative charge ($-Q$) will accumulate on the top surface of the bottom plate. The amount of charge that accumulates as a function of potential difference is called the *capacitance*. The formula for capacitance C is

$$\boxed{C = \frac{Q}{V}}$$
(2.119)

where Q is the charge on the positive plate and V is the potential difference between the top and bottom plates.[2.28] The capacitance unit is the farad (F), defined as a coulomb per volt.

A device used to store charge, and hence electrical energy, is known as a capacitor.[2.29] Capacitors are frequently employed in electrical circuits for DC blocking, AC bypassing, filtering, tuning, and noise suppression. Typically their values are less than a microfarad (μF, or 10^{-6} F), and picofarad (pF, or 10^{-12} F) values are not uncommon, while some state-of-the-art supercapacitors have values in the multifarad range.

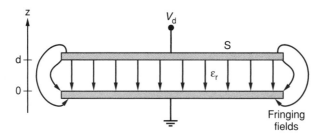

Fringing fields

Figure 2.53 Cross section of a parallel-plate capacitor.

[2.28]You could also say capacitance is the magnitude of the charge on one of the plates divided by the magnitude of the potential difference.

[2.29]The capacitance is a capacitor's capacity for storing charge.

The value of capacitance depends on the conductor–dielectric configuration and the dielectric's permittivity. It does not depend on Q or V, as the ratio of Q and V will be constant.[2.30] There are two basic methods for determining capacitance, which we will call the Q-method and the V-method:

Q-Method

- Assume a charge $+Q$ on plate a and a charge $-Q$ on plate b.
- Solve for **E** using the appropriate method (Coulomb's law, Gauss's law, or boundary conditions).
- Solve for the potential difference V_{ab} between the plates.
- $C = Q/V_{ab}$ (the assumed Q will divide out).

V-Method

- Assume V_{ab} between the plates.
- Find **E** (and then **D**) using Laplace's equation.
- Find ρ_s (and then Q) at each plate using conductor–dielectric boundary conditions ($D_N = \rho_s$).
- $C = Q/V_{ab}$ (the assumed V_{ab} will divide out).

▷ **EXAMPLE 2.27**

Let's use the Q-method to find the capacitance for the parallel-plate capacitor shown in Figure 2.53.

We start by placing charge $+Q$ on the inner surface of the top plate and $-Q$ on the upper surface of the bottom plate. The charge density $\rho_s = Q/S$, and we can relate this to the electric flux density by using the conductor–dielectric boundary conditions to obtain $\mathbf{D} = Q/S\,(-\mathbf{a}_z)$. Now, the electric field intensity is found by dividing **D** by the permittivity, or $\mathbf{E} = -Q\,\mathbf{a}_z/(\varepsilon_r\varepsilon_o S)$. The potential difference across the plates is

$$V_{ab} = -\int_b^a \mathbf{E}\cdot d\mathbf{L} = -\int_0^d \frac{-Q}{\varepsilon_r\varepsilon_o S}\mathbf{a}_z\cdot dz\,\mathbf{a}_z = \frac{Qd}{\varepsilon_r\varepsilon_o S}$$

Finally, we divide Q by V_{ab} to get C:

$$C = \frac{Q}{V_{ab}} = \frac{Q}{\left(\dfrac{Qd}{\varepsilon_r\varepsilon_o S}\right)} = \frac{\varepsilon_r\varepsilon_o S}{d}$$

In the derivation of the parallel-plate capacitance we neglected the *fringing fields*, which are shown in Figure 2.53 at the very edge of the capacitor. These fringing field lines increase the capacitance. But if the surface S is large compared to the separation distance d, it is common practice to ignore the fringing fields.

[2.30]This conclusion can only be made with the important assumption that the permittivity is independent of field strength.

It is interesting to compare the parallel-plate capacitance with the resistance calculated between the plates in the previous section. We see that

$$RC = \frac{\varepsilon}{\sigma}$$

(2.120)

This is actually a general relationship that is quite useful since if the capacitance of a structure is known, it can be used to find the resistance.

▶ **EXAMPLE 2.28**

Let's use the V-method to determine the capacitance for a length L of coaxial line of inner conductor radius a and outer radius b, filled with dielectric of permittivity $\varepsilon_r\varepsilon_o$ as shown in Figure 2.54.

V_{ab} is applied across the dielectric, and we want to employ Laplace's equation to find the potential field everywhere in the dielectric. Here, we'll make the simplifying assumptions that fringing fields may be neglected and that the field is only a function of ρ. Then Laplace's equation reduces to

$$\frac{\partial}{\partial \rho}\left(\rho\frac{\partial V}{\partial \rho}\right) = 0$$

Integrating twice we obtain

$$V(\rho) = A\ln \rho + B$$

where A and B are constants of integration determined by applying the boundary conditions. We can let $V(b) = 0$ and $V(a) = V_{ab}$, leading to

$$A = \frac{-V_{ab}}{\ln(b/a)}, \quad B = A\ln b$$

and

$$V(\rho) = \frac{-V_{ab}\ln(\rho b)}{\ln(b/a)}$$

Next, we find the electric field intensity by applying the gradient:

$$E = -\nabla V = -\frac{\partial V(\rho)}{\partial \rho}a_\rho$$

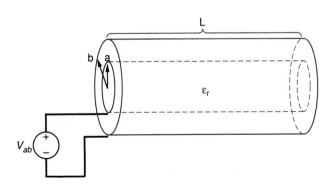

Figure 2.54   Coaxial capacitor.

which yields

$$\mathbf{E} = \frac{V_{ab}}{\rho \ln(b/a)} \mathbf{a}_\rho$$

The electric flux density is then

$$\mathbf{D} = \frac{\varepsilon_r \varepsilon_o V_{ab}}{\rho \ln(b/a)} \mathbf{a}_\rho$$

which we can use to find the surface charge density at the conductive plates. At the inner conductor, the flux is directed outward, indicating a positive surface charge density

$$\rho_s = \frac{\varepsilon_r \varepsilon_o V_{ab}}{a \ln(b/a)}$$

We can find Q on the inner conductor by multiplying ρ_s by the surface area S, which gives

$$Q = \rho_s S = \left(\frac{\varepsilon_r \varepsilon_o V_{ab}}{a \ln(b/a)}\right)(2\pi a L) = \frac{2\pi L \varepsilon_r \varepsilon_o V_{ab}}{\ln(b/a)}$$

Now the capacitance is found by dividing Q by V_{ab}, leaving us with

$$C = \frac{Q}{V_{ab}} = \frac{2\pi L \varepsilon_r \varepsilon_o}{\ln(b/a)}$$

Drill 2.33 Use the Q-method to find the capacitance for the coaxial capacitor.

Drill 2.34 Use the V-method to find the capacitance for the parallel-plate capacitor.

Electrostatic Potential Energy

Work is required to assemble a collection of like charges, and if held in place this collection of charges constitutes potential energy. By considering how much work it would take to assemble a collection of charges, we can arrive[2.31] at the relationship

$$\boxed{W_E = \frac{1}{2}\int \mathbf{D} \cdot \mathbf{E}\, dv = \frac{1}{2}\int \varepsilon_r \varepsilon_o E^2\, dv} \qquad (2.121)$$

where W_E is the electrostatic potential energy.

For a parallel-plate capacitor where fringing fields are neglected, the field is constant over the Sd volume and (2.121) becomes

$$W_E = \frac{1}{2}\varepsilon_r \varepsilon_o E^2 Sd \qquad (2.122)$$

[2.31]A different sort of proof for this relation will be seen in Chapter 5 in the discussion of the Poynting theorem.

By using $E = V/d$ and $C = \varepsilon_r \varepsilon_o S/d$, (2.122) becomes

$$\boxed{W_E = \frac{1}{2} CV^2}$$

(2.123)

Although this expression is derived for a parallel-plate capacitor, it holds for any capacitor configuration.

▷ **EXAMPLE 2.29**

A parallel-plate capacitor has a plate area of 4 m² and a separation distance of 0.01 m and is filled with a dielectric of $\varepsilon_r = 10$ and $\sigma = 10^{-8}$ S/m. We apply 12 V to the top plate and ground the bottom plate. We want to find the electrostatic potential energy stored in this capacitor as well as the amount of power dissipated.

We can first calculate the capacitance:

$$C = \frac{\varepsilon_r \varepsilon_o S}{d} = \frac{(10)(8.854 \times 10^{-12})(4)}{(0.01)} = 35 \text{ nF}$$

Then, the electrostatic potential energy from (2.123) is

$$W_E = \frac{1}{2} CV^2 = \frac{1}{2}(35 \times 10^{-9})(12) = 210 \text{ nJ}$$

Finally, with 12 V applied across 0.01 m, we know $E = 1.2$ kV, and we can apply (2.92) to get

$$P = \sigma E^2 Sd = (10^{-8})(1200)^2(4)(.01) = 0.58 \text{ mW}$$

Drill 2.35 Suppose a coaxial capacitor has inner and outer radii of 2 and 4 cm, respectively, a dielectric with $\varepsilon_r = 4$, and a 50-cm length. (a) Calculate the capacitance. (b) Calculate the electrostatic potential energy stored in this capacitor if a 9-V potential is applied across the conductors. (*Answer:* (a) 160 pF, (b) 6.5 nJ)

Practical Application: Electrolytic Capacitors

For circuit applications that require high values of capacitance, electrolytic capacitors are often used. The very high capacitance available for electrolytic capacitors (up to 220 mF) results from a very large electrode surface area combined with a very thin insulative dielectric. This is represented by Figure 2.55. One of the electrodes for an electrolytic capacitor is a porous metallic slug formed by compressing and baking a metal powder. The metal used to form the slug is generally either aluminum or tantalum. Although aluminum is far cheaper, tantalum electrolytic capacitors perform better. The nooks and crannies of the porous slug can deliver surface area on the order of a square meter for a cubic centimeter of slug. The other electrode is a conductive electrolyte separated from the slug by a very thin insulative layer. This layer can be formed either by oxidation or by anodization. The quality and thickness of this insulative layer determines the capacitor's maximum voltage rating.

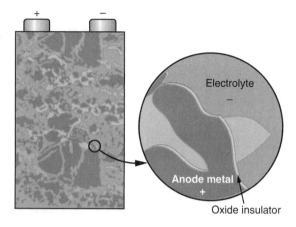

Figure 2.55 Electrolytic capacitor.

If the insulative layer is formed by oxidation, the baking step is performed in air, leading to an oxidized surface. The oxide quality is not very good, however, and the capacitor will not be able to sustain very large voltage drops. In layer formation by anodization, an electrochemical reaction results in a high-quality insulative oxide of thickness between 0.01 and 0.1 μm.

The electrolytes should be chemically and thermally stable with appropriate conductivity. One type of electrolyte is manganese dioxide. After forming the insulative layer, the porous slug is heated while immersed in manganese nitrate solution. Semiconducting manganese dioxide is formed in the porous region.

With an anodized insulative layer, it is important that the anode metal always be biased positive with respect to the electrolyte. Otherwise, performance is that of a rectifier. With a reverse-bias voltage of more than a volt or two, large currents can result that damage the capacitor. Operation is therefore limited to DC applications. If AC signals are present, the total instantaneous voltage should never be negative.

▷ SUMMARY

- Vectors in Cartesian, cylindrical, and spherical coordinate systems, respectively, are

$$\mathbf{A}_{\text{Cart}} = A_x \mathbf{a}_x + A_y \mathbf{a}_y + A_z \mathbf{a}_z$$

$$\mathbf{A}_{\text{cyl}} = A_\rho \mathbf{a}_\rho + A_\phi \mathbf{a}_\phi + A_z \mathbf{a}_z$$

and

$$\mathbf{A}_{\text{spher}} = A_r \mathbf{a}_r + A_\theta \mathbf{a}_\theta + A_\phi \mathbf{a}_\phi$$

- The force exerted by charge Q_1 on charge Q_2 in a medium of permittivity ε is given by Coulomb's law,

$$\mathbf{F}_{12} = \frac{Q_1 Q_2}{4\pi\varepsilon R_{12}^2} \mathbf{a}_{12}$$

where $\mathbf{R}_{12} = R_{12} \mathbf{a}_{12}$ is a vector from charge Q_1 to Q_2.

- Electric field intensity $\mathbf{E}_1$ is related to force $\mathbf{F}_{12}$ by

$$\mathbf{E}_1 = \frac{\mathbf{F}_{12}}{Q_2}$$

and the Coulomb's law expression can be written

$$\mathbf{E} = \frac{Q}{4\pi\varepsilon R^2} \mathbf{a}_R$$

For a continuous charge distribution, $\mathbf{E}$ is found by integrating:

$$\mathbf{E} = \int \frac{dQ}{4\pi\varepsilon R^2} \mathbf{a}_R$$

- For a point charge Q at the origin,

$$\mathbf{E} = \frac{Q}{4\pi\varepsilon r^2} \mathbf{a}_r$$

For an infinite length line charge ρ_L on the z-axis,

$$E = \frac{\rho_L}{2\pi\varepsilon\rho}\mathbf{a}_\rho$$

For an infinite extent sheet of charge ρ_s,

$$E = \frac{\rho_s}{2\varepsilon}\mathbf{a}_N$$

- For a pair of vectors **A** and **B**, the dot product $\mathbf{A}\cdot\mathbf{B}$ is a measure of how much of **A** is in the direction of **B**, multiplied by the magnitude of **B**. In Cartesian coordinates,

$$\mathbf{A}\cdot\mathbf{B} = |\mathbf{A}|\,|\mathbf{B}|\cos\theta_{AB} = A_xB_x + A_yB_y + A_zB_z$$

- Electric flux density is related to field intensity by

$$\mathbf{D} = \varepsilon_r\varepsilon_o\mathbf{E}$$

where ε_r is the relative permittivity in a linear, isotropic, homogeneous material. Electric flux passing through a surface is given by

$$\psi = \int \mathbf{D}\cdot d\mathbf{S}$$

- The divergence theorem relates a surface integral to a volume integral and is given by

$$\oint \mathbf{D}\cdot d\mathbf{S} = \int \nabla\cdot\mathbf{D}\,dv$$

where $\nabla\cdot\mathbf{D}$ is the divergence of **D**, given in Cartesian coordinates by

$$\nabla\cdot\mathbf{D} = \frac{\partial D_x}{\partial x} + \frac{\partial D_y}{\partial y} + \frac{\partial D_z}{\partial z}$$

- Gauss's law states that the net electric flux through any closed surface is equal to the total charge enclosed by that surface:

$$\oint \mathbf{D}\cdot d\mathbf{S} = Q_{enc}$$

The point form of Gauss's law is

$$\nabla\cdot\mathbf{D} = \rho_v$$

- The electric potential difference V_{ba} between a pair of points a and b in an electric field is given by

$$V_{ba} = -\int_a^b \mathbf{E}\cdot d\mathbf{L} = V_b - V_a$$

where V_b and V_a are the electrostatic potentials at b and a, respectively. For a distribution of charge in the vicinity of the origin, where a zero reference voltage is taken at infinite radius,

$$V = \int \frac{dQ}{4\pi\varepsilon r}$$

- **E** is related to V by the gradient equation,

$$\mathbf{E} = -\nabla V$$

which for Cartesian coordinates is

$$\nabla V = \frac{\partial V}{\partial x}\mathbf{a}_x + \frac{\partial V}{\partial y}\mathbf{a}_y + \frac{\partial V}{\partial z}\mathbf{a}_z$$

- The point form of Ohm's law relates the current density **J** to the electric field intensity **E** by the material conductivity σ as

$$\mathbf{J} = \sigma\mathbf{E}$$

The current through an area is

$$I = \int \mathbf{J}\cdot d\mathbf{S}$$

and the resistance is

$$R = \frac{-\int \mathbf{E}\cdot d\mathbf{L}}{\int \sigma\mathbf{E}\cdot d\mathbf{S}}$$

- Joule's law indicates amount of dissipated power P by

$$P = \int \mathbf{E}\cdot\mathbf{J}\,dv$$

- The conditions for fields at the boundary between a pair of dielectrics is given by

$$\mathbf{E}_{T1} = \mathbf{E}_{T2}$$

and

$$\mathbf{a}_{21}\cdot(\mathbf{D}_1 - \mathbf{D}_2) = \rho_s$$

where $\mathbf{E}_{T1}$ and $\mathbf{E}_{T2}$ are the electric field components tangential to the boundary, $\mathbf{a}_{21}$ is a unit vector from medium 2 to medium 1, and ρ_s is the surface charge at the boundary. If no surface charge is present, the components of **D** normal to the boundary are equal:

$$\mathbf{D}_{N1} = \mathbf{D}_{N2}$$

At the boundary between a conductor and a dielectric, the conditions are

$$\mathbf{E}_T = 0$$

and

$$\mathbf{D}_N = \rho_s$$

- Poisson's equation is

$$\nabla^2 V = -\frac{\rho_v}{\varepsilon}$$

where the Laplacian of V in Cartesian coordinates is given by

$$\nabla^2 V = \frac{\partial^2 V}{\partial x^2} + \frac{\partial^2 V}{\partial y^2} + \frac{\partial^2 V}{\partial z^2}$$

In a charge-free medium, Poisson's equation reduces to Laplace's equation

$$\nabla^2 V = 0$$

These equations are used in conjunction with values at the boundaries to solve for V in a region.

- Capacitance is a measure of charge storage capability and is given by

$$C(\text{F}) = \frac{Q(\text{C})}{V(\text{V})}$$

where Q is the magnitude of charge on one of a pair of conductive plates and V is the magnitude of the potential difference across the plates. For a parallel-plate capacitor of plate area S and separation distance d, filled with a medium of permittivity ε,

$$C = \frac{\varepsilon S}{d}$$

where fields fringing at the edges are ignored.

- Electrostatic potential energy is a measure of the energy stored in the electric field and is given by

$$W_E = \frac{1}{2} \int \varepsilon E^2 \, dv = \frac{1}{2} C V^2$$

- The use of MATLAB to perform vector calculations was demonstrated and several examples demonstrating basic programming and plotting techniques were presented.

▶ PROBLEMS

2.1 Vectors in the Cartesian Coordinate System

2.1 Given P(4, 2, 1) and $\mathbf{A}_{PQ} = 2\mathbf{a}_x + 4\mathbf{a}_y + 6\mathbf{a}_z$, find the point Q.

2.2 Given the points P(4 m, 1 m, 0) and Q(1 m, 3 m, 0), construct a table showing the vector, vector magnitude, and unit vector for the following: (a) the vector **A** from the origin to P, (b) the vector **B** from the origin to Q, (c) the vector **C** from P to Q, (d) the vectors **A** + **B**, **C** − **A**, and **B** − **A**. (e) Make a sketch of the vectors found in (a) through (d).

2.3 Write a program that will find the vector between a pair of arbitrary points in the Cartesian coordinate system.

2.2 Coulomb's Law, Electric Field Intensity, and Field Lines

2.4 Suppose $Q_1(0.0, -3.0 \text{ m}, 0.0) = 4.0$ nC, $Q_2(0.0, 3.0 \text{ m}, 0.0) = 4.0$ nC, and $Q_3(4.0 \text{ m}, 0.0, 0.0) = 1.0$ nC. (a) Find the total force acting on the charge Q_3. (b) Repeat the problem after changing the charge of Q_2 to −4.0 nC. (c) Find the electric field intensity for parts (a) and (b).

2.5 Find the force exerted by $Q_1(3.0 \text{ m}, 3.0 \text{ m}, 3.0 \text{ m}) = 1.0 \ \mu\text{C}$ on $Q_2(6.0 \text{ m}, 9.0 \text{ m}, 3.0 \text{ m}) = 10$. nC.

2.6 Suppose 10.0-nC point charges are located on the corners of a square of side 10.0 cm. Locating the square in the x–y plane (at $z = 0.00$) with one corner at the origin and one corner at P(10.0 cm, 10.0 cm, 0.00), find the total force acting at point P.

2.7 Four 1.00-nC point charges are located at (0.00, −2.00 m, 0.00), (0.00, 2.00 m, 0.00), (0.00, 0.00, −2.00 m), and (0.00, 0.00, +2.00 m), respectively. Find the total force acting on a 1.00-nC charge located at (2.00 m, 0.00, 0.00).

2.8 A 20.0–nC point charge exists at P(0.00, 0.00, −3.00 m). Where must a 10.0-nC charge be located to make the total field zero at the origin?

2.3 The Spherical Coordinate System

2.9 Convert the following points from Cartesian to spherical coordinates:

(a) P(6.0, 2.0, 6.0)

(b) Q(0.0, −4.0, 3.0)

(c) R(−5.0, −1.0, −4.0)

2.10 Convert the following points from spherical to Cartesian coordinates:

(a) P(3.0, 30.°, 45.°)

(b) Q(5.0, π/4, 3π/2)

(c) R(10., 135°, 180°)

2.11 Given a volume defined by $1.0 \text{ m} \leq r \leq 3.0 \text{ m}$, $0° \leq \theta \leq 90°$, and $0° \leq \phi \leq 90°$, (a) sketch the volume, (b) perform the integration to find the volume, and (c) perform the necessary integrations to find the total surface area.

2.4 Line Charges and the Cylindrical Coordinate System

2.12 Convert the following points from Cartesian to cylindrical coordinates:

(a) P(0.0, 4.0, 3.0)

(b) Q(−2.0, 3.0, 2.0)

(c) R(4.0, −3.0, −4.0)

2.13 Convert the following points from cylindrical to Cartesian coordinates:

(a) P(2.83, 45.0°, 2.00)

(b) P(6.00, 120.°, −3.00)

(c) P(10.0, −90.0°, 6.00)

2.14 A 20.0-cm-long section of copper pipe has a 1.00-cm-thick wall and outer diameter of 6.00 cm.

(a) Sketch the pipe, conveniently overlaying the cylindrical coordinate system and lining up the length direction with the z-axis.

(b) Determine the total surface area. (This could actually be useful if, say, you needed to do an electroplating step on this piece of pipe.)

(c) Determine the weight of the pipe given the density of copper of 8.96 g/cm^3.

2.15 A line charge with charge density 2.00 nC/m exists at $y = -2.00 \text{ m}$, $x = 0.00$. (a) A charge $Q = 8.00 \text{ nC}$ exists somewhere along the y-axis. Where must you locate Q so that the total electric field is zero at the origin? (b) Suppose instead of the 8.00-nC charge of part (a) that you locate a charge Q at (0.00, 6.00 m, 0.00). What value of Q will result in a total electric field intensity of zero at the origin?

2.16 You are given two z-directed line charges of charge density +1.0 nC/m at $x = 0$, $y = -1.0 \text{ m}$ and of charge density −1.0 nC/m at $x = 0$, $y = 1.0 \text{ m}$. Find **E** at P(1, 0, 0).

2.17 Suppose you have a segment of line charge of length $2L$ centered on the z-axis and having a charge distribution ρ_L. Compare the electric field intensity at a point on the y-axis a distance d from the origin with the electric field at that point assuming the line charge is of infinite length. Plot the ratio of E for the segment to E for the infinite line versus the ratio L/d using MATLAB.

2.18 A segment of line charge $\rho_L = 10.$ nC/m exists on

the y-axis from the origin to $y = +3.0 \text{ m}$. Determine **E** at the point (3.0 m, 0, 0).

2.5 Surface and Volume Charge

2.19 In free space, there is a point charge $Q = 8.0 \text{ nC}$ at (−2.0 m, 0, 0), a line charge $\rho_L = 10.$ nC/m at $y = -9.0 \text{ m}$, $x = 0 \text{ m}$, and a sheet charge $\rho_s = 12.$ nC/m² at $z = -2.0 \text{ m}$. Determine **E** at the origin.

2.20 An infinitely long line charge ($\rho_L = 21\pi$ nC/m) lies along the z-axis. An infinite area sheet charge ($\rho_s = 3$ nC/m²) lies in the x–z plane at $y = 10 \text{ m}$. Find a point on the y-axis where the electric field intensity is zero.

2.21 Sketch the following surfaces and find the total charge on each surface given a surface charge density of $\rho_s = 1\text{nC/m}^2$. Units (other than degrees) are meters.

(a) $-3 \leq x \leq 3$, $0 \leq y \leq 4$, $z = 0$

(b) $1 \leq r \leq 4$, $180° \leq \phi \leq 360°$, $\theta = \pi/2$

(c) $1 \leq \rho \leq 4$, $180° \leq \phi \leq 360°$, $z = 0$

2.22 Consider a circular disk in the x–y plane of radius 5.0 cm. Suppose the charge density is a function of radius such that $\rho_s = 12\rho$ nC/cm² (when ρ is in centimeters). Find the electric field intensity a point 20.0 cm above the origin on the z-axis.

2.23 Suppose a ribbon of charge with density ρ_s exists in the y–z plane of infinite length in the z direction and extending from $-a$ to $+a$ in the y direction. Find a general expression for the electric field intensity at a point d along the x-axis.

2.24 Sketch the following volumes and find the total charge for each given a volume charge density of $\rho_v = 1$ nC/m³. Units (other than degrees) are meters.

(a) $0 \leq x \leq 4$, $0 \leq y \leq 5$, $0 \leq z \leq 6$

(b) $1 \leq r \leq 5$, $0 \leq \theta \leq 60°$

(c) $1 \leq \rho \leq 5$, $0° \leq \phi \leq 90°$, $0 \leq z \leq 5$

2.25 You have a cylinder of 4.00-in diameter and 5.00-in length (imagine a can of tomatoes) that has a charge distribution that varies with radius as $\rho_v = (6\,\rho, \text{ nC/in}^3)$ where ρ is in inches. [It may help you with the units to think of this as $\rho_v \text{ (nC/in}^3) = 6 \text{ (nC/in}^4) \rho\text{(in)}$.] Find the total charge contained in this cylinder.

2.26 Consider a rectangular volume with $0.00 \leq x \leq 4.00$ m, $0.00 \leq y \leq 5.00$ m, and $-6.00 \text{ m} \leq z \leq 0.00$ with charge density $\rho_v = 40.0$ nC/m³. Find the electric field intensity at the point P(0.00, 0.00, 20.0 m).

2.27 Consider a sphere with charge density $\rho_v = 120$ nC/m^3 centered at the origin with a radius of 2.00 m. Now, remove the top half of the sphere, leaving a hemisphere below the *x–y* plane. Find the electric field intensity at the point P(8.00 m, 0.00, 0.00). (*Hint: See MATLAB 2.4, and consider that your answer will now have two field components.*)

2.6 Electric Flux Density

2.28 Use the definition of dot product to find the three interior angles for the triangle bounded by the points P(–3.00, –4.00, 5.00), Q(2.00, 0.00, –4.00), and R(5.00, –1.00, 0.00).

2.29 Given $\mathbf{D} = 2\rho\mathbf{a}_\rho + \sin\phi\ \mathbf{a}_z$ C/m^2, find the electric flux passing through the surface defined by $2.0 \le \rho \le 4.0$ m, $90.°$ $\le \phi \le 180°$, and $z = 4.0$ m.

2.30 Suppose the electric flux density is given by $\mathbf{D} = 3r$ $\mathbf{a}_r - \cos\phi\ \mathbf{a}_\theta + \sin^2\theta\ \mathbf{a}_\phi$ C/m^2. Find the electric flux through both surfaces of a hemisphere of radius 2.00 m and $0.00° \le \theta \le 90.0°$.

2.7 Gauss's Law and Applications[2.32]

2.31 Given a 3.00-mm-radius solid wire centered on the *z*-axis with an evenly distributed 2.00 C of charge per meter length of wire, plot the electric flux density D_ρ versus radial distance from the *z*-axis over the range $0 \le \rho \le 9$ mm.

2.32 Given a 2.00-cm-radius solid wire centered on the *z*-axis with a charge density $\rho_v = 6\rho$ C/cm^3 (when ρ is in centimeters), plot the electric flux density D_ρ versus radial distance from the *z*-axis over the range $0 \le \rho \le 8$ cm.

2.33 A cylindrical pipe with a 1.00-cm wall thickness and an inner radius of 4.00 cm is centered on the *z*-axis and has an evenly distributed 3.00 C of charge per meter length of pipe. Plot D_ρ as a function of radial distance from the *z*-axis over the range $0 \le \rho \le 10$ cm.

2.34 An infinitesimally thin metallic cylindrical shell of radius 4.00 cm is centered on the *z*-axis and has an evenly distributed charge of 100. nC per meter length of shell. (a) Determine the value of the surface charge density on the conductive shell and (b) plot D_ρ as a function of radial distance from the *z*-axis over the range $0 \le \rho \le 12$ cm.

2.35 A spherical charge density is given by $\rho_v = \rho_o\ r/a$ for $0 \le r \le a$ and $\rho_v = 0$ for $r > a$. Derive equations for the electric flux density for all *r*.

2.36 A thick-walled spherical shell, with inner radius

2.00 cm and outer radius 4.00 cm, has an evenly distributed 12.0 nC charge. Plot D_r as a function of radial distance from the origin over the range $0 \le r \le 10$ cm.

2.37 Given a coaxial cable with solid inner conductor of radius *a*, an outer conductor that goes from radius *b* to *c* (so $c > b > a$), a charge $+Q$ that is assumed evenly distributed throughout a meter length of the inner conductor, and a charge $-Q$ that is assumed evenly distributed throughout a meter length of the outer conductor, derive equations for the electric flux density for all ρ. You may orient the cable in any way you wish.

2.8. Divergence and the Point Form of Gauss's Law

2.38 Determine the charge density at the point P(3.0 m, 4.0 m, 0.0) if the electric flux density is given as $\mathbf{D} = xyz\ \mathbf{a}_z$ C/m^2.

2.39 Given $\mathbf{D} = 3\mathbf{a}_x + 2xy\mathbf{a}_y + 8x^2y^3\mathbf{a}_z$ C/m^2, (a) determine the charge density at the point P(1, 1, 1). Find the total flux through the surface of a cube with $0.0 \le x \le 2.0$ m, $0.0 \le y \le 2.0$ m, and $0.0 \le z \le 2.0$ m by evaluating (b) the left side of the divergence theorem and (c) the right side of the divergence theorem.

2.40 Suppose $\mathbf{D} = 6\rho\cos\phi\ \mathbf{a}_\phi$ C/m^2. (a) Determine the charge density at the point (3 m, 90°, –2 m). Find the total flux through the surface of a quartered cylinder defined by $0 \le \rho \le 4$ m, $0 \le \phi \le 90°$, and -4 m $\le z \le 0$ by evaluating (b) the left side of the divergence theorem and (c) the right side of the divergence theorem.

2.41 Suppose $\mathbf{D} = r^2\sin\theta\ \mathbf{a}_r + \sin\theta\cos\phi\ \mathbf{a}_\phi$ C/m^2. (a) Determine the charge density at the point (1.0 m, 45°, 90°). Find the total flux through the surface of a volume defined by $0.0 \le r \le 2.0$ m, $0.0° \le \theta \le 90.°$, and $0.0 \le \phi \le 180°$ by evaluating (b) the left side of the divergence theorem and (c) the right side of the divergence theorem.

2.9 Electric Potential

2.42 A sheet of charge density $\rho_s = 100$ nC/m^2 occupies the *x–z* plane at $y = 0$. (a) Find the work required to move a 2.0-nC charge from P(–5.0 m, 10. m, 2.0 m) to M(2.0 m, 3.0 m, 0.0). (b)Find V_{MP}.

2.43 A surface is defined by the function $2x + 4y^2 - \ln z = 12$. Use the gradient equation to find a unit vector normal to the plane at the point (3.00 m, 2.00 m, 1.00 m).

2.44 For the following potential distributions, use the gradient equation to find **E**.

[2.32]It would be very good practice to use MATLAB for generating the plots required in this section.

(a) $V = x + y^2 z$ (V)

(b) $V = \rho^2 \sin\phi$ (V)

(c) $V = r \sin\theta\cos\phi$ (V)

2.45 A 100-nC point charge is located at the origin. (a) Determine the potential difference V_{BA} between the point A(0.0, 0.0, –6.0 m) and point B(0.0, 2.0 m, 0.0). (b) How much work would be done to move a 1.0-nC charge from point A to point B against the electric field generated by the 100-nC point charge?

2.46 Suppose you have a pair of charges Q_1(0.0, –5.0 m, 0.0) = 1.0 nC and Q_2(0.0, 5.0 m, 0.0) = 2.0 nC. Write a MATLAB routine to calculate the potential V_{RO} moving from the origin to the point R(5.0 m, 0.0, 0.0). Your numerical integration will involve choosing a step size ΔL and finding the field at the center of the step. You should try several different step sizes to see how much this affects the solution.

2.47 For an infinite length line of charge density $\rho_L = 20$ nC/m on the z-axis, find the potential difference V_{BA} between point B(0, 2 m, 0) and point A(0, 1 m, 0).

2.48 Find the electric field at point P(0.0, 0.0, 8.0 m) resulting from a surface charge density $\rho_s = 5.0$ nC/m² existing on the $z = 0$ plane from $\rho = 2.0$ m to $\rho = 6.0$ m. Assume $V = 0$ at a point an infinite distance from the origin.

2.49 Suppose a 6.0-m-diameter ring with charge density 5.0 nC/m lies in the x–y plane with the origin at its center. Determine the potential difference V_{HO} between the point H(0.0, 0.0, 4.0 m) and the origin. *(Hint: First find an expression for E on the z-axis as a general function of z.)*

2.10 Conductors and Ohm's Law

2.50 A columnar beam of electrons from $0 \le \rho \le 1$ mm has a charge density $\rho_v = -0.1 \cos(\pi\rho/2)$ nC/mm³ (where ρ is in millimeters) and a velocity of 6×10^6 m/s in the $+\mathbf{a}_z$ direction. Find the current.

2.51 Two spherical conductive shells of radii a and b ($b > a$) are separated by a material with conductivity σ. Find an expression for the resistance between the two spheres.

2.52 The typical length of each piece of jumper wire on a student's protoboard is 5.0 cm. Assuming AWG-20 (with a wire diameter of 0.812 mm) copper wire, (a) determine the resistance for this length of wire. (b) Determine the power dissipated in the wire for 10. mA of current.

2.53 A 150-m length of AWG-22 (0.644 mm diameter) copper magnet wire with a very thin insulative sheath is used to make a tightly wrapped coil. Determine the resistance for this length of wire.

2.54 Determine an expression for the power dissipated per unit length in coaxial cable of inner radius a, outer radius b, and conductivity between the conductors σ if a potential difference V_{ab} is applied.

2.55 Find the resistance per unit length of a stainless steel pipe of inner radius 2.5 cm and outer radius 3.0 cm.

2.56 A nickel wire of diameter 5.0 mm is surrounded by a 0.50-mm-thick layer of silver. What is the resistance per unit length for this wire? Assuming 1.0 m of this wire carries 1.0 A of current, determine the power dissipated in the nickel portion and in the silver portion of the wire.

2.11 Dielectrics

2.57 A material has 12.0 V/m $\mathbf{a}_x$ field intensity with permittivity 194.5 pF/m. Determine the electric flux density.

2.58 A 20-nC point charge at the origin is embedded in Teflon ($\varepsilon_r = 2.1$). Find and plot the magnitudes of the polarization vector, the electric field intensity, and the electric flux density at a radial distance from 0.1 cm out to 10 cm.

2.59 Suppose the force is very carefully measured between a pair of point charges separated by a dielectric material and is found to be 20 nN. The dielectric material is removed without changing the position of the point charges, and the force has increased to 100 nN. What is the relative permittivity of the dielectric?

2.60 The potential field in a material with $\varepsilon_r = 10.2$ is $V = 12 xy^2$ (V). Find **E**, **P**, and **D**.

P2.61 In a mineral oil dielectric, with breakdown voltage of 15 MV/m, the potential function is $V = x^3 - 6x^2 - 3.1x$ (MV). Is the dielectric likely to break down, and if so, where?

2.12 Boundary Conditions

2.62 For $y < 0$, $\varepsilon_{r1} = 4.0$ and $\mathbf{E}_1 = 3\mathbf{a}_x + 6\pi\mathbf{a}_y + 4\mathbf{a}_z$ V/m. At $y = 0$, $\rho_s = 0.25$ nC/m². If $\varepsilon_{r2} = 5.0$ for $y > 0$, find $\mathbf{E}_2$.

2.63 For $z \le 0$, $\varepsilon_{r1} = 9.0$, and for $z > 0$, $\varepsilon_{r2} = 4.0$. If $\mathbf{E}_1$ makes a 30° angle with a normal to the surface, what angle does $\mathbf{E}_2$ make with a normal to the surface?

2.64 A plane defined by $3x + 2y + z = 6$ separates two dielectrics. The first dielectric, on the side of the plane containing the origin, has $\varepsilon_{r1} = 3.0$ and $\mathbf{E}_1 = 4.0\mathbf{a}_z$ V/m. The other dielectric has $\varepsilon_{r2} = 6.0$. Find $\mathbf{E}_2$.

2.65 Consider a dielectric–dielectric charge-free boundary at the plane $z = 0$. Construct a program that will allow the user to enter ε_{r1} (for z < 0), ε_{r2}, and $\mathbf{E}_1$ and will then calculate $\mathbf{E}_2$. (Just for fun, you may want to have the program calculate the angles that $\mathbf{E}_1$ and $\mathbf{E}_2$ make with a normal to the surface).

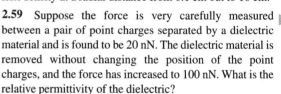

2.66 A 1.0-cm-diameter conductor is sheathed with a 0.50-cm thickness of Teflon and then a 2.0-cm (inner) diameter outer conductor. (a) Use Laplace's equation to find an expression for the potential as a function of ρ in the dielectric. (b) Find **E** as a function of ρ. (c) What is the maximum potential difference that can be applied across this coaxial cable without breaking down the dielectric?

2.67 A 1.0-m-long carbon pipe of inner diameter 3.0 cm and outer diameter 5.0 cm is cut in half lengthwise. Determine the resistance between the inner surface and the outer surface of one of the half sections of pipe.

2.68 For a coaxial cable of inner conductor radius a and outer conductor radius b and a dielectric ε_r in between, assume a charge density $\rho_v = \rho_o/\rho$ is added in the dielectric region. Use Poisson's equation to derive an expression for V and **E**. Calculate ρ_s on each plate.

2.69 For the parallel-plate capacitor given in Figure 2.51, suppose a charge density

$$\rho_v = \rho_o \sin\left(\frac{\pi z}{2d}\right)$$

is added between the plates. Use Poisson's equation to derive a new expression for V and **E**. Calculate ρ_s on each plate.

2.13 Capacitance

2.70 A parallel-plate capacitor is constructed such that the dielectric can be easily removed. With the dielectric in place, the capacitance is 48 nF. With the dielectric removed, the capacitance drops to 12 nF. Determine the relative permittivity of the dielectric.

2.71 A parallel-plate capacitor with a 1.0 m^2 surface area for each plate, a 2.0-mm plate separation, and a dielectric with relative permittivity of 1200 has a 12.-V potential difference across the plates. (a) What is the minimum allowed dielectric strength for this capacitor? Calculate (b) the capacitance and (c) the magnitude of the charge density on one of the plates.

2.72 A conical section of material extends over the range 2.0 cm $\le r \le$ 9.0 cm for $0 \le \theta \le 30°$ with $\varepsilon_r = 9.0$ and $\sigma = 0.020$ S/m. Conductive plates are placed at each radial end of the section. Determine the resistance and capacitance of the section. ►EMAG SOLUTIONS

2.73 An inhomogeneous dielectric fills a parallel-plate capacitor of surface area 50. cm^2 and thickness 1.0 cm. You are given $\varepsilon_r = 3(1 + z)$, where z is measured from the bottom plate in centimeters. Determine the capacitance.

2.74 Given **E** = $5xy\mathbf{a}_x + 3z\mathbf{a}_z$ V/m, find the electrostatic potential energy stored in a volume defined by $0 \le x \le 2$ m, $0 \le y \le 1$ m, and $0 \le z \le 1$ m. Assume $\varepsilon = \varepsilon_o$.

2.75 Suppose a coaxial capacitor with inner radius 1.0 cm, outer radius 2.0 cm, and length 1.0 m is constructed with two different dielectrics. When oriented along the z-axis, ε_r for $0° \le \phi \le 180°$ is 9.0, and ε_r for $180° \le \phi \le 360°$ is 4.0. (a) Calculate the capacitance. (b) If 9.0 V is applied across the conductors, determine the electrostatic potential energy stored in each dielectric for this capacitor.

CHAPTER 3

Magnetostatics

Learning Objectives

▷ Describe magnetic field intensity and magnetic flux density

▷ Define Biot–Savart's law and use it to determine the magnetic field resulting from various current distributions

▷ Use Ampère's circuital law to find magnetic field intensity for symmetrical current distributions

▷ Introduce magnetic forces, torque, and moment

▷ Describe the features of magnetic materials

▷ Compare the magnetic field across material boundaries

▷ Define inductance and calculate it for various geometries

▷ Describe magnetic circuits and electromagnets

It is thought that 4500 years ago the Chinese discovered that certain types of iron ore could attract each other and certain metals. Carefully suspended slivers of this metal were found to always point in the same direction, and as such they could be used as compasses for navigation. The first compass is thought to have been used by the Chinese around 376 B.C. Greeks found this iron ore near Magnesia, in what is present-day Turkey. It contained magnetite (Fe_3O_4) and came to be known as *magnetic lodestone*.

In 1600, William Gilbert of England postulated that magnetic lodestones, or compasses, work because the earth is one big magnet. The magnetic field is generated by the spin of the molten inner core. The north end of a compass needle points to the geographic North Pole, which corresponds to the earth's south magnetic pole.

Magnetism and electricity were considered distinct phenomena until 1820 when Hans Christian Oersted conducted an experiment that showed a compass needle deflecting when in proximity to a current-carrying wire. The principle of magnetism is now used in a host of applications, including magnetic memory, motors and generators, microphones and speakers, and magnetically levitated high-speed vehicles.[3.1] So the study of magnetostatics is

---

[3.1]Some people claim that magnets can help ease pain if placed at strategic locations about the body. A similar claim was made about electric fields early in the 20th century. Neither claim appears to have any scientific validity.

important in its own right, but it is also a prerequisite for the understanding of dynamic electromagnetic fields.

The action at a distance exhibited by magnets suggests the presence of a *magnetic field* analogous to electric fields. It has just been mentioned that magnetic fields are produced by permanent magnets and by steady electric currents. They are also produced from time varying electric fields, a key aspect of Maxwell's equations, which will be discussed in Chapter 4. The field from a permanent magnet is very difficult to understand; it is thought to arise from quantum mechanical electron "spin," which can be considered charge in motion (i.e., current). It is much more straightforward to base our magnetostatics theory on steady electric currents.[3.2]

So in this chapter, magnetic fields will be introduced in a manner paralleling our treatment of electric fields. The key difference between the two types of fields is that action is at right angles to magnetic fields. Handling this in terms of vectors requires the use of a vector *cross product*, to be described in Section 3.1. Also, we will see that, unlike electric flux, magnetic flux has no starting or ending point, instead existing as continuous loops.

▶ 3.1 MAGNETIC FIELDS AND THE CROSS PRODUCT

Magnetic fields are easily visualized by sprinkling iron filings on a piece of paper suspended over a magnet, for instance over the bar magnet in Figure 3.1a. The iron filings align themselves with the direction of the field lines. A schematic of the bar magnet is shown in Figure 3.1b, where the field lines are in terms of the *magnetic field intensity* **H**, in units of amps per meter. This is analogous to the volts per meter units for electric field intensity **E**. The field lines are directed from the magnetic north end to the magnetic south end of the bar magnet.

A loosely held conductor with current I is shown in Figure 3.2a. The wire is deflected in the presence of a magnetic field as shown in Figure 3.2b. The force of deflection depends on the velocity of the moving charge in the conductor (i.e., the current) and on the strength of the field.[3.3] Note that the force on the wire is in the $\mathbf{a}_y$ direction, normal to both the

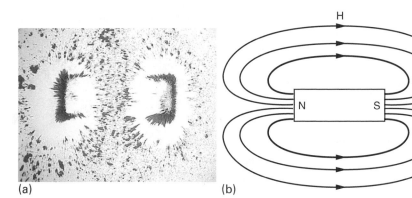

(a)　　　　　　　(b)

Figure 3.1 (a) Iron filings "map" of a bar magnet's field. (b) Schematic view of a bar magnet showing the magnetic field.

[3.2]Although the charges are in motion for DC current, the magnetic field established by the moving charges is static.

[3.3]We will return to this concept in Section 3.6.

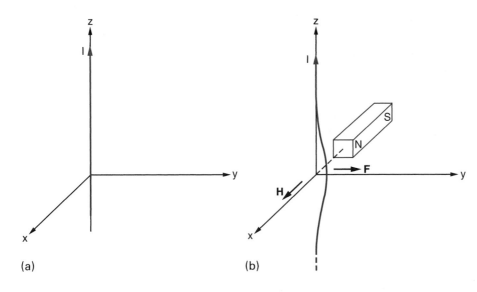

Figure 3.2 (a) A loosely held conductor carrying current. (b) Upon application of a magnetic field, the wire is deflected in a direction normal to both the field and the direction of current.

magnetic field and the direction of the current. If the current is reversed, the deflecting force would be in the $-\mathbf{a}_y$ direction.

This action at right angles is a key difference between magnetic and electric fields, and we need some mathematical way to handle this. When we dealt with vectors in Chapter 2, we found the scalar dot product of a pair of vectors. Now, we wish to multiply a pair of vectors and return a vector product. The *cross product* of a pair of vectors $\mathbf{A}$ and $\mathbf{B}$ is

$$\mathbf{A} \times \mathbf{B} = |\mathbf{A}|\,|\mathbf{B}|\,\sin\theta_{AB}\mathbf{a}_N \qquad (3.1)$$

which is the product of the magnitude of the vectors multiplied by the sine of the angle between the vectors, and $\mathbf{a}_N$ is a unit vector in the normal direction of $\mathbf{A} \times \mathbf{B}$ taken by the *right-hand rule*. Here, the fingers of the right hand point in the direction of the first vector, $\mathbf{A}$, and as they curl toward the second vector, $\mathbf{B}$, the thumb points in the direction of the cross product. Notice that, although the magnitude of $\mathbf{A} \times \mathbf{B}$ is the same as that for $\mathbf{B} \times \mathbf{A}$, the unit vector normal will be in the opposite direction.

Figure 3.3 shows the unit vector directions for the Cartesian coordinate system. For a right-handed system, the coordinates of a point or the coordinates of a vector are listed in a "right-handed" sequence, (x, y, z). If we take the cross product of the first two components in order, we get the third:

$$\mathbf{a}_x \times \mathbf{a}_y = (1)(1)\sin 90°\mathbf{a}_z = \mathbf{a}_z$$

This result is easily confirmed using the right-hand rule with Figure 3.3. If we extend the sequence (x, y, z, x, y), and take the cross product of the second and third components (moving in the *right* direction of the sequence), we get the fourth component:

$$\mathbf{a}_y \times \mathbf{a}_z = \mathbf{a}_x$$

If we reverse the order (move in the left direction of the sequence) we get negative results:

$$\mathbf{a}_z \times \mathbf{a}_y = -\mathbf{a}_x$$

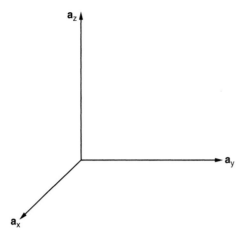

Figure 3.3 Cartesian coordinate unit vectors.

A physical picture of the cross product is given by considering the torque exhibited at the pivot point of a lever arm that has some force applied to it as shown in Figure 3.4. The torque vector $\boldsymbol{\tau}$ (tau), in newton-meters, is given by

$$\boldsymbol{\tau} = \mathbf{r} \times \mathbf{F} = |\mathbf{r}| \, |\mathbf{F}| \sin\theta \, \mathbf{a}_N \tag{3.2}$$

where $\mathbf{F}$ is a force vector in newtons applied at the end of the lever-arm vector $\mathbf{r}$ with length in meters. The most torque is generated when $\mathbf{F}$ is normal to $\mathbf{r}$.

▷ **EXAMPLE 3.1**

Let us consider the cross product $\mathbf{A} \times \mathbf{B}$, where $\mathbf{A} = 3\mathbf{a}_x + 4\mathbf{a}_y$ and $\mathbf{B} = 3\mathbf{a}_y$.
   We can easily calculate the magnitude of these vectors, and in Figure 3.5 it is apparent that $\sin\theta$ is equal to 3/5. The right-hand rule indicates a unit vector normal of $\mathbf{a}_z$, so

$$\mathbf{A} \times \mathbf{B} = (5)(3)\left(\tfrac{3}{5}\right)\mathbf{a}_z = 9\mathbf{a}_z$$

This example is simple because the sine of the angle between the vectors is easy to calculate. However, this is rarely the case in most problems. Fortunately, we can also perform a vector cross product using a determinant,

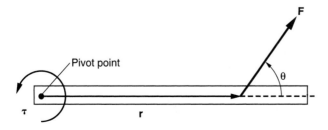

Figure 3.4 Illustration of torque on a lever arm.

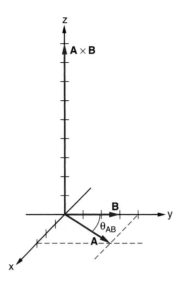

Figure 3.5 Cross product example.

$$\mathbf{A} \times \mathbf{B} = \begin{vmatrix} \mathbf{a}_x & \mathbf{a}_y & \mathbf{a}_z \\ A_x & A_y & A_z \\ B_x & B_y & B_z \end{vmatrix} = \left(A_y B_z - A_z B_y\right)\mathbf{a}_x - \left(A_x B_z - A_z B_x\right)\mathbf{a}_y + \left(A_x B_y - A_y B_x\right)\mathbf{a}_z \qquad (3.3)$$

In our example, we have

$$\mathbf{A} \times \mathbf{B} = \begin{vmatrix} \mathbf{a}_x & \mathbf{a}_y & \mathbf{a}_z \\ 3 & 4 & 0 \\ 0 & 3 & 0 \end{vmatrix} = 9\mathbf{a}_z$$

The cross products for cylindrical and spherical coordinate systems are identical in form. For each case, we can look at the unit vectors at a convenient point away from the origin.[3.4] As with the Cartesian system, the coordinates for cylindrical and spherical points are listed in right-handed order. That is, for cylindrical coordinates we have (ρ, ϕ, z) and we see from Figure 3.6a that

$$\mathbf{a}_\rho \times \mathbf{a}_\phi = \mathbf{a}_z$$

For spherical coordinates (Figure 3.6b) we have (r, θ, ϕ), and inspecting Figure 3.6b we have

$$\mathbf{a}_r \times \mathbf{a}_\theta = \mathbf{a}_\phi$$

[3.4]At the origin, we have no $\mathbf{a}_\phi$ component in cylindrical coordinates and no $\mathbf{a}_\theta$ nor $\mathbf{a}_\phi$ component in spherical coordinates.

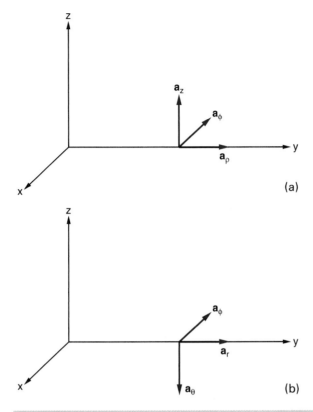

Figure 3.6 Demonstrating the cross product relation for unit vectors in (a) cylindrical coordinates and (b) spherical coordinates.

▷ **MATLAB 3.1**

The cross product $\mathbf{A} \times \mathbf{B}$ is accomplished in MATLAB using cross(A,B). For our example with $\mathbf{A} = 3\mathbf{a}_x + 4\mathbf{a}_y$ and $\mathbf{B} = 3\mathbf{a}_y$, we have

```
» A=[3 4 0]; B=[0 3 0];
» cross(A,B)

ans =

   0   0   9
```

Drill 3.1 The points O(0, 0, 0), P(3, 0, 3), and Q(0, 4, 2) are the three vertices of a triangle. (a) Find the interior angles for the triangle and (b) find a unit vector normal to the plane containing the triangle. (*Answer:* (a) $\theta_O = 72°$, $\theta_P = 56°$, $\theta_Q = 52°$; (b) $\mathbf{a}_n = -0.67\mathbf{a}_x - 0.33\mathbf{a}_y + 0.67\mathbf{a}_z$)

Oersted's Experiment

A compass is simply a weak magnet that aligns itself to the local magnetic field. In the presence of a bar magnet, an obvious source of magnetic field, a compass will align itself with the fields. A meter or so away from the typical bar magnet, the earth's magnetic field dominates.

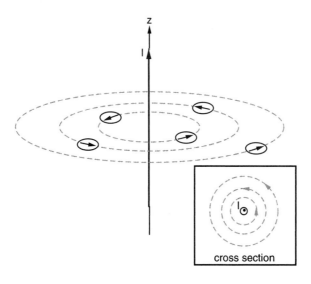

Figure 3.7 Oersted's experiment with a compass placed in several positions in close proximity to a current-carrying wire. The inset shows ⊙ used to represent the cross section for current coming out of the paper; this represents the head of an arrow. A ⊗ symbol would represent the feathered end of an arrow and would correspond to a current heading into the paper.

In 1820, Hans Christian Oersted (1777–1851) used a compass to show that current produces magnetic fields that loop around the conductor, as indicated in Figure 3.7. As one moves away from the source of current, the field grows weaker. Prior to this tremendous discovery by Oersted, electricity and magnetism were thought to be two separate entities. Oersted's discovery released a flood of study that culminated in Maxwell's equations in 1865.

▶3.2 BIOT–SAVART'S LAW

Shortly following Oersted's discovery that currents produce magnetic fields, Jean Baptiste Biot (1774–1862) and Felix Savart (1791–1841) arrived at a mathematical relation between the field and current. The *law of Biot–Savart* is

$$d\mathbf{H}_2 = \frac{I_1 d\mathbf{L}_1 \times \mathbf{a}_{12}}{4\pi R_{12}^2}$$

(3.4)

where Figure 3.8 identifies each term in the equation. Subscripts are included in this introduction to the Biot–Savart law to clarify the location of each element. However, in general the subscripts are left out for simplicity. Equation 3.4 is analogous to the Coulomb's law equation for the electric field resulting from a differential charge,

$$d\mathbf{E}_2 = \frac{dQ_1 \mathbf{a}_{12}}{4\pi \varepsilon R_{12}^2}$$

(3.5)

Of course, you never really have an isolated differential segment of current. To get the total field resulting from a current, you can sum the contributions from each segment by integrating,

$$\mathbf{H} = \int \frac{I d\mathbf{L} \times \mathbf{a}_R}{4\pi R^2}$$

(3.6)

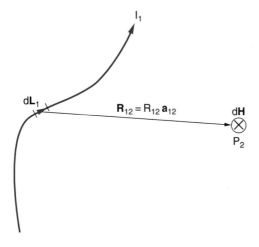

Figure 3.8 Illustration of the law of Biot–Savart showing magnetic field arising from a differential segment of current.

Progression from (3.4) to (3.6) is possible because, just like electric fields, magnetic fields can be added by superposition.

▷ **EXAMPLE 3.2**

Consider an infinite length line along the z-axis conducting current I in the $+\mathbf{a}_z$ direction as displayed in Figure 3.9. We want to find the magnetic field everywhere.

We first inspect the symmetry and see that the field will be independent of z and ϕ and only dependent on ρ. So we consider a point a distance ρ from the line along the ρ-axis. Now we must determine each component in (3.6) before integrating. The term $Id\mathbf{L}$ is simply $Idz\mathbf{a}_z$, and the vector from the source to the test point is

$$R\mathbf{a}_R = -z\mathbf{a}_z + \rho\mathbf{a}_\rho$$

Combining these terms, we have

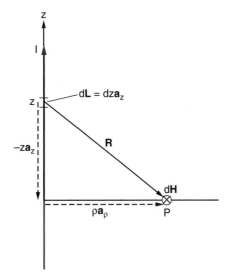

Figure 3.9 Component values for the equation to find the magnetic field intensity resulting from an infinite length line of current on the z-axis.

$$\mathbf{H} = \int_{-\infty}^{\infty} \frac{Idz\mathbf{a}_z \times \left(-z\mathbf{a}_z + \rho\mathbf{a}_\rho\right)}{4\pi\left(z^2 + \rho^2\right)^{\frac{3}{2}}}$$

Pulling the constants outside of the integral and realizing that $\mathbf{a}_z \times \mathbf{a}_z = 0$ and $\mathbf{a}_z \times \mathbf{a}_\rho = \mathbf{a}_\phi$, we have

$$\mathbf{H} = \frac{I\rho\mathbf{a}_\phi}{4\pi} \int_{-\infty}^{\infty} \frac{dz}{\left(z^2 + \rho^2\right)^{\frac{3}{2}}}$$

An interesting aspect of this problem is that no matter what differential current element we choose, there will only be an $\mathbf{a}_\phi$ element of field at the test point. Finding the solution to this integral in Appendix D, we have

$$\mathbf{H} = \frac{I\rho\mathbf{a}_\phi}{4\pi} \left[\frac{z}{\rho^2 \sqrt{z^2 + \rho^2}} \right]_{-\infty}^{+\infty} \tag{3.7}$$

Solving, we find the magnetic field intensity resulting from an infinite length line of current:

$$\boxed{\mathbf{H} = \frac{I\mathbf{a}_\phi}{2\pi\rho}} \tag{3.8}$$

The solution for a segment would also be straightforward, just requiring a change to the limits on the integral. Equation 3.8 suggests another version of the right-hand rule. If you grip the conducting wire with your right hand,[3.5] with your thumb in the direction of current, then the fingers will curl around the wire in the direction of the magnetic field.

▶ **EXAMPLE 3.3**

Let us now consider a ring of current with radius a lying in the x–y plane with a current I in the $+\mathbf{a}_\phi$ direction as shown in Figure 3.10a. The objective is to find an expression for the field at an arbitrary point a height h on the z-axis. Venturing away from the z-axis makes the problem significantly harder, but it is still solvable as we will see in MATLAB 3.2.

Our first task is to solve for each term in the Biot–Savart equation. Figure 3.10b shows what these component values are, leading to

$$\mathbf{H} = \int_{\phi=0}^{2\pi} \frac{Iad\phi\mathbf{a}_\phi \times \left(h\mathbf{a}_z - a\mathbf{a}_\rho\right)}{4\pi\left(h^2 + a^2\right)^{\frac{3}{2}}}$$

We can further simplify this expression by considering the symmetry of the problem in Figure 3.10c. A particular differential current element will give a field with an $\mathbf{a}_\rho$ component (from $\mathbf{a}_\phi \times \mathbf{a}_z$) and an $\mathbf{a}_z$ component (from $\mathbf{a}_\phi \times -\mathbf{a}_\rho$). By taking the field from a differential current element on the opposite side of the ring, it is apparent that the radial components cancel whereas the $\mathbf{a}_z$ components add. Taking this into consideration, and pulling all the constants outside of the integral, we have

$$\mathbf{H} = \frac{Ia^2\mathbf{a}_z}{4\pi\left(h^2 + a^2\right)^{\frac{3}{2}}} \int_0^{2\pi} d\phi$$

[3.5]If you actually perform such an experiment, you may wish to wear insulative gloves.

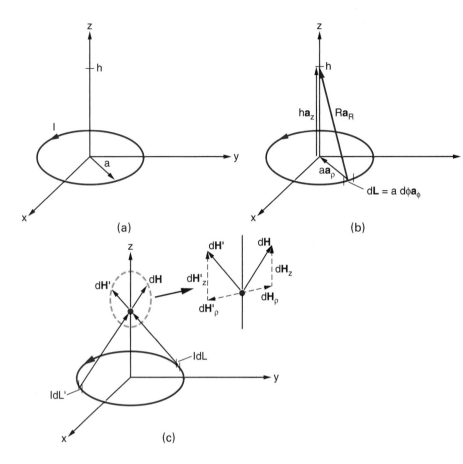

Figure 3.10 (a) We want to find **H** a height h above a ring of current centered in the x–y plane. (b) The component values are shown for use in the Biot–Savart equation. (c) The radial components of **H** cancel by symmetry.

which is easily solved to get

$$\mathbf{H} = \frac{Ia^2}{2\left(h^2 + a^2\right)^{3/2}} \mathbf{a}_z \qquad (3.9)$$

At $h = 0$, the center of the loop, this equation reduces to

$$\mathbf{H} = \frac{I}{2a} \mathbf{a}_z \qquad (3.10)$$

Drill 3.2 A segment of conductor is $2h$ long, centered on the z-axis. If this segment conducts current I in the $+\mathbf{a}_z$ direction, find $\mathbf{H}(\rho, \phi, 0)$.

$$\left(\textit{Answer: } \mathbf{H} = \frac{Ih\mathbf{a}_\phi}{2\pi\rho\sqrt{h^2 + \rho^2}} \right)$$

Drill 3.3   An infinite length line with a 4π A current in the $+\mathbf{a}_x$ direction exists along the x-axis. Find $\mathbf{H}(0, 1\text{ m}, 0)$. (*Answer:* 2 A/m $\mathbf{a}_z$)

Drill 3.4   A ring of radius a is centered in the x–y plane at $z = 0$ and has 1 A of current in the $+\mathbf{a}_\phi$ direction. An infinite length line of current exists at $z = 0$, $y = 2a$. Determine the magnitude and direction of current needed in this infinite length line to make the magnetic field at the origin equal to 0. (*Answer:* 2π A in the $+\mathbf{a}_x$ direction)

> **MATLAB 3.2**

Let us now find the field inside a circular loop of current-carrying wire. We can find the field at any point by summing the $d\mathbf{H}$ contributions (using (3.4)) from each differential current element in the ring.

Figure 3.11 shows the key parameters used in the MATLAB routine. One of the parameters we need is the vector direction of each $d\mathbf{L}$ element. We get this by finding a unit vector from the origin to the location of the $d\mathbf{L}$ element, and crossing this unit vector with $-\mathbf{a}_z$.

Note that this method can be modified to find $\mathbf{H}$ at any point, not just in the x–y plane (see Problem 3.10).

```
%   M-File: ML0302
%%  Magnetic Field Inside a Ring of Current
%
%   This program determines and plots the magnetic field
%   intensity at a location on the x-axis between the
%   center and the periphery of a ring of current.
%
%   Wentworth, 7/15/02
%
%   Variables:
%   I           current(A) in +phi direction on ring
%   a           ring radius (m)
%   Ndeg        number of increments for phi
%   f           angle of phi in radians
%   df          differential change in phi
%   dL          differential length vect. on the ring
%   dLmag       magnitude of dL
%   dLuv        unit vector in direction of dL
%   [xL,yL,0]   location of source point
%   Ntest       number of test points
%   Rsuv        unit vector from origin to source point
%   R           vector from source to test point
%   Ruv         unit vector for R
%   Rmag        magnitude of R
%   dH          differential portion of H
%   dHmag       magnitude of dH
%   radius      radial distance from origin
%   Hz          total mag. field at test point
```

```
clc           %clears the command window
clear         %clears variables

%   Initialize Variables
a=1;
I=1;
Ndeg=90;
Ntest=40;
df=360/Ndeg;
dLmag=(df*pi/180)*a;

%   Perform Calculation
for j=1:Ntest
    x=(j-1)*a/Ntest;
    for i=1:df:360
        f=i*pi/180;
        xL=a*cos(f);
        yL=a*sin(f);
        Rsuv=[xL yL 0]/a;
        dLuv=cross([0 0 1],Rsuv);
        dL=dLmag*dLuv;
        R=[x-xL -yL 0];
        Rmag=magvector(R);
        Ruv=R/Rmag;
        dH=I*cross(dL,Ruv)/(4*pi*Rmag^2);
        dHmag(i)=magvector(dH);
    end
    radius(j)=x;
    Hz(j)=sum(dHmag);
end
```

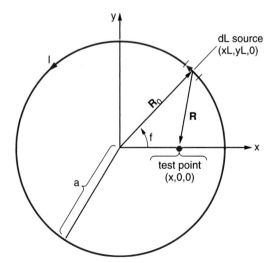

Figure 3.11   The ring with key parameters used in MATLAB 3.2.

```
%   Generate Plot
plot(radius,Hz)
grid on
xlabel('radius(m)')
ylabel('Hz(A/m)')
```

The plot is given in Figure 3.12. The magnetic field is seen to be fairly constant near the center of the loop.

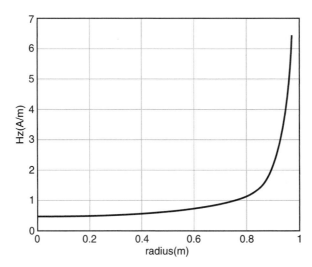

Figure 3.12 The field inside a ring of current.

Solenoid

Figure 3.13a shows many turns of insulated wire coiled in the shape of a cylinder. Such constructs are known as *solenoids* and are commonly used as relays and switches. We can use our solution for the field at the center of a single loop of current to find the field at any point along the center axis of a solenoid.

Suppose the solenoid has a length h and a radius a and is made up of N turns of current-carrying wire. For tight wrapping, we can consider the solenoid to be made up of N loops of current. To find the magnetic field intensity from a single loop at a point P along the axis of the solenoid, from (3.9) we have

$$\mathbf{H_P} = \frac{Ia^2}{2\left(z'^2 + a^2\right)^{3/2}} \mathbf{a}_z$$

We could also consider that this is a differential amount of field resulting from a differential amount of current, or

$$d\mathbf{H_P} = \frac{dIa^2}{2\left(z'^2 + a^2\right)^{3/2}} \mathbf{a}_z$$

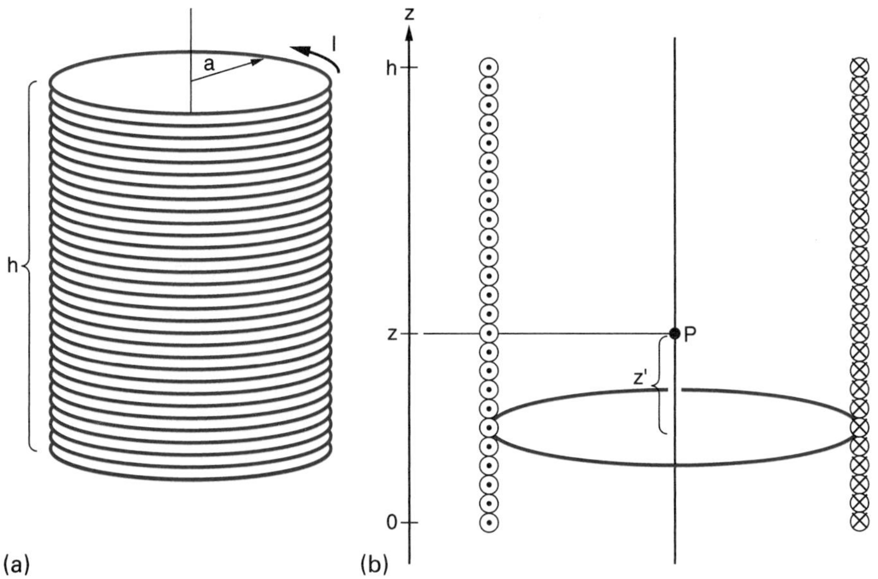

Figure 3.13 (a) A solenoid. (b) Schematic with parameters for solving the magnetic field intensity at P.

where the differential amount of current can be considered a function of the number of loops and the length of the solenoid as

$$dI = \frac{N}{h} I dz'$$

Fixing the point P where the field is desired (see Figure 3.13b), we have that z' will range from $-z$ to $h - z$, or

$$\mathbf{H} = \int_{-z}^{h-z} \frac{NIa^2 dz'}{2h(z'^2 + a^2)^{3/2}} \mathbf{a}_z = \frac{NIa^2}{2h} \int_{-z}^{h-z} \frac{dz'}{(z'^2 + a^2)^{3/2}} \mathbf{a}_z$$

This integral is found from Appendix D, leading to the solution

$$\mathbf{H} = \frac{NI}{2h} \left[\frac{h-z}{\sqrt{(h-z)^2 + a^2}} + \frac{z}{\sqrt{z^2 + a^2}} \right] \mathbf{a}_z \qquad (3.11)$$

At the very center of the solenoid ($z = h/2$), with the assumption that the length is considerably bigger than the loop radius ($h \gg a$), the equation reduces to

$$\mathbf{H} = \frac{NI}{h} \mathbf{a}_z \qquad (3.12)$$

Surface and Volume Current Densities

In addition to linear current I (A) and volume current density $\mathbf{J}$ (A/m²), we can also consider a *surface current density* $\mathbf{K}$ (A/m). This vector, also called *sheet current*, is considered to flow in an infinitesimally thin layer. Surface current density can be a good approximation for the current in a thin metallic conductive layer for a circuit board, for instance. In the solenoid example, the coils were considered so tightly wound that the current could be assumed as an evenly distributed sheet current. We could write

$$\mathbf{K} = \frac{NI}{h}\mathbf{a}_\phi$$

for the solenoid.

The Biot–Savart law can also be written in terms of surface and volume current densities by replacing $Id\mathbf{L}$ with $\mathbf{K}dS$ and $\mathbf{J}dv$:

$$\mathbf{H} = \int \frac{\mathbf{K}dS \times \mathbf{a}_R}{4\pi R^2} \quad \text{and} \quad \mathbf{H} = \int \frac{\mathbf{J}dv \times \mathbf{a}_R}{4\pi R^2} \tag{3.13}$$

Notice that the sheet current's direction is given by the vector quantity $\mathbf{K}$ rather than by a vector direction for dS, since dS would be normal to the direction of current.[3.6]

▶ EXAMPLE 3.4

We wish to find $\mathbf{H}$ at a point centered adjacent to an infinite length ribbon of sheet current as shown in Figure 3.14a.

Using (3.13) with $\mathbf{K}dS = K_z dxdz\mathbf{a}_z$, we have a double integral that is not conveniently solved. Fortunately, we can treat the ribbon as a collection of infinite length lines of current $K_z dx$. Each line of current will contribute $d\mathbf{H}$ of field from (3.8),

$$d\mathbf{H} = \frac{I}{2\pi\rho}\mathbf{a}_\phi$$

where we can use Figure 3.14b to see what each of the components in the equation will be. We know

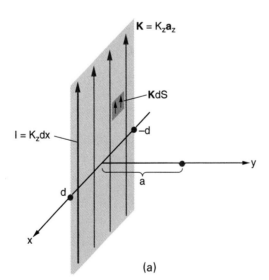

Figure 3.14   (a) A 2d-wide ribbon with current density $\mathbf{K}$.

(a)

[3.6]It might also make sense to treat line current as a vector quantity and have $Id\mathbf{L}$ rather than $Id\mathbf{L}$. But the convention is to treat line current as a scalar quantity, so we will do so.

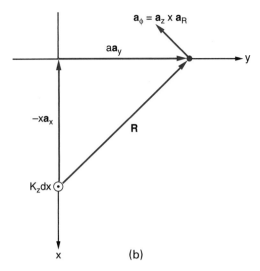

Figure 3.14 (b) Cross section showing parameter values.

$$\mathbf{R} = -x\mathbf{a}_x + a\mathbf{a}_y$$

and $\rho = R$. Also, by the law of Biot–Savart, the $\mathbf{a}_\phi$ direction is given by the cross product of the current element direction and $\mathbf{a}_R$. To find the total field, we integrate from $x = -d$ to $x = +d$:

$$\mathbf{H} = \int_{-d}^{d} \frac{K_z dx \left(\mathbf{a}_z \times \left(-x\mathbf{a}_x + a\mathbf{a}_y\right)\right)}{2\pi\left(x^2 + a^2\right)}$$

Expanding the integral gives us

$$\mathbf{H} = \frac{K_z}{2\pi}\left[\int_{-d}^{d} \frac{-xdx\mathbf{a}_y}{x^2 + a^2} - \int_{-d}^{d} \frac{adx\mathbf{a}_x}{x^2 + a^2}\right]$$

We notice by symmetry arguments that the first term inside the brackets, the $\mathbf{a}_y$ component, is zero. This is confirmed by solving this portion of the integral. If we were to look for the field at a point other than adjacent to the middle of the ribbon, the $\mathbf{a}_y$ component would not cancel. The second bracketed integral is solved using the integral solutions of Appendix D, leading to

$$\mathbf{H} = \frac{-K_z}{\pi}\tan^{-1}\left(\frac{d}{a}\right)\mathbf{a}_x \tag{3.14}$$

Finally, we can find the field resulting from an infinite extent sheet of current by letting d be infinite. In such a case we get

$$\mathbf{H} = -\frac{K_z}{2}\mathbf{a}_x \tag{3.15}$$

For many problems involving surface current densities, and indeed for most problems involving volume current densities, solving for the magnetic field intensity using the law of Biot–Savart can be quite cumbersome and require numerical integration. For some problems that we will encounter with volume charge densities, there will be sufficient symmetry to be able to solve for the fields using *Ampère's circuital law*, the topic of the next section.

▶ ## 3.3 AMPÈRE'S CIRCUITAL LAW

In electrostatics problems that featured considerable symmetry we were able to apply Gauss's law to solve for the electric field intensity much more easily than Coulomb's law. Likewise, in magnetostatic problems with sufficient symmetry we can employ *Ampère's circuital law* more easily than the law of Biot–Savart.

Ampère's circuital law says that the integration of **H** around any closed path is equal to the net current enclosed by that path. This is stated in equation form as

$$\oint \mathbf{H} \cdot d\mathbf{L} = I_{\text{enc}} \qquad\qquad (3.16)$$

This equation can be derived from the law of Biot–Savart, but that requires the use of *Stoke's theorem* and *curl*, concepts to be covered later in this chapter. However, it is verified by experiment, and most students prefer to take the equation on faith rather than slog their way through the derivation.

The line integral of **H** around a closed path is termed the *circulation* of **H**. The path of the circulation doesn't matter in solving for the current enclosed, but in practical application, a symmetrical current distribution is given and you want to solve for **H**, so it is important to make a careful selection of an *Amperian path* (analogous to a Gaussian surface) that is everywhere either tangential or normal to **H**, and over which H is constant.

The direction of the circulation is chosen such that the right-hand rule is satisfied. That is, with the thumb in the direction of the current, the fingers will curl in the direction of the circulation. We'll show how to use Ampère's circuital law to find the well-known field resulting from an infinite length line of current on the z-axis. This will be followed by increasingly complex examples to further illustrate the procedure.

▶ **EXAMPLE 3.5**

Here we want to find the magnetic field intensity everywhere resulting from an infinite length line of current situated on the z-axis as shown in Figure 3.15.

The figure also shows a pair of Amperian paths, a and b. Performing the circulation of **H** about either path will result in the same current I. But we choose path b, which has a constant value of H_ϕ around the circle specified by the radius ρ. In Ampère's circuital law equation, we substitute $H_\phi \mathbf{a}_\phi$ for **H** and $\rho d\phi\, \mathbf{a}_\phi$ for d**L**, giving

$$\oint \mathbf{H} \cdot d\mathbf{L} = I_{\text{enc}} = \int_0^{2\pi} H_\phi \mathbf{a}_\phi \cdot \rho d\phi \mathbf{a}_\phi = 2\pi\rho H_\phi = I$$

Here we have chosen to perform the circulation in the $+\mathbf{a}_\phi$ direction in accordance with the right-hand rule. Solving for H_ϕ, we find that the field resulting from an infinite length line of current is the expected result (3.8)

$$\mathbf{H} = \frac{I}{2\pi\rho}\mathbf{a}_\phi$$

▶ **EXAMPLE 3.6**

Let us now use Ampère's circuital law to find the magnetic field intensity resulting from an infinite extent sheet of current.

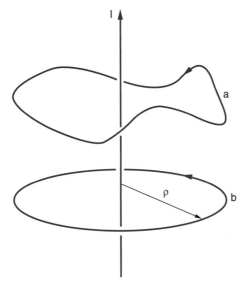

Figure 3.15 Two possible Amperian paths around an infinite length line of current.

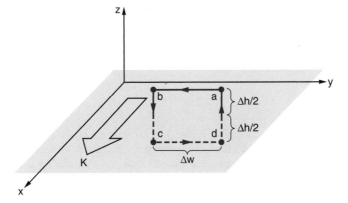

Figure 3.16 Calculating **H** resulting from a current sheet $\mathbf{K} = K_x \mathbf{a}_x$ in the x–y plane.

Figure 3.16 shows a current sheet with uniform current density $\mathbf{K} = K_x \mathbf{a}_x$ in the $z = 0$ plane along with a rectangular Amperian path of height Δh and width Δw. In accordance with the right-hand rule, where the thumb of the right hand points in the direction of the current and the fingers curl in the direction of the field, we'll perform the circulation in the order $a \rightarrow b \rightarrow c \rightarrow d \rightarrow a$. We have

$$\oint \mathbf{H} \cdot d\mathbf{L} = I_{\text{enc}} = \int_a^b \mathbf{H} \cdot d\mathbf{L} + \int_b^c \mathbf{H} \cdot d\mathbf{L} + \int_c^d \mathbf{H} \cdot d\mathbf{L} + \int_d^a \mathbf{H} \cdot d\mathbf{L}$$

From symmetry arguments in Section 3.2 we know that **H** only has an H_y component. Therefore the integrals from $b \rightarrow c$ and from $d \rightarrow a$, with $d\mathbf{L} = dz\, \mathbf{a}_z$, will be zero. Above the sheet $\mathbf{H} = H_y(-\mathbf{a}_y)$ and below the sheet $\mathbf{H} = H_y\mathbf{a}_y$. Also, above the sheet we could write $d\mathbf{L} = dy(-\mathbf{a}_y)$. But instead, we'll adopt the convention of letting $d\mathbf{L} = dy\mathbf{a}_y$ and take care of the sign by integrating from Δw to zero.[3.7] So we have

---

[3.7]Had we used $d\mathbf{L} = -dy\mathbf{a}_y$ and integrated from Δw to 0, we would have accounted for the direction twice and would have had a sign error in our answer.

$$\oint \mathbf{H} \cdot d\mathbf{L} = \int_{\Delta w}^{0} H_y\left(-\mathbf{a}_y\right) \cdot dy\mathbf{a}_y + \int_{0}^{\Delta w} H_y\mathbf{a}_y \cdot dy\mathbf{a}_y = 2H_y\Delta w$$

The current enclosed by the path is just

$$I = \int_{0}^{\Delta w} K_x dy = K_x{}^t$$

and equating the two terms gives

$$H_y = \frac{K_x}{2}$$

The result of Example 3.6 can be extended to give a general equation for the magnetic field intensity resulting from an infinite extent sheet of current,

$$\boxed{\mathbf{H} = \frac{1}{2}\mathbf{K} \times \mathbf{a_N}} \tag{3.17}$$

where $\mathbf{a_N}$ is a normal vector from the sheet current to the test point.

> **Drill 3.5** An infinite extent sheet current $\mathbf{K} = 6\,\mathbf{a}_z$ A/m exists on the x–z plane at $y = 0$. Find $\mathbf{H}(3, 4, 5)$. (*Answer:* $-3\mathbf{a}_x$ A/m)

▶ **EXAMPLE 3.7**

Consider the cylindrical conductor of Figure 3.17 carrying a radially dependent current $\mathbf{J} = J_o\rho\,\mathbf{a}_z$ A/m², where J_o is a constant with units of amperes per cubic meter. We wish to find $\mathbf{H}$ everywhere.

What components of $\mathbf{H}$ will be present in this example? Consider the distributed current to be made up of a bundle of line currents. Grabbing a line current and finding the field at some point P, as shown in Figure 3.18a, we find that the field has both $\mathbf{a}_\rho$ and $\mathbf{a}_\phi$ components. Grabbing a particular second line current in Figure 3.18b, we see that the $\mathbf{a}_\rho$ components cancel and the $\mathbf{a}_\phi$ components add. From this we say that from symmetry arguments $\mathbf{H}$ only has an H_ϕ component.

To calculate $\mathbf{H}$ everywhere, two Amperian paths are required, as shown in Figure 3.17b. Amperian path #1 is for $\rho \leq a$, and path #2 is for $\rho > a$. Evaluating the left side of Ampère's circuital law we have

$$\oint \mathbf{H} \cdot d\mathbf{L} = \int_{0}^{2\pi} H_\phi\mathbf{a}_\phi \cdot \rho d\phi\mathbf{a}_\phi = 2\pi\rho H_\phi$$

This is true for both Amperian paths.

The current enclosed by the first path is found by integrating

$$I = \int \mathbf{J} \cdot d\mathbf{S} = \int J_o\rho\mathbf{a}_z \cdot \rho d\rho d\phi\mathbf{a}_z = J_o\int_{0}^{\rho}\rho^2 d\rho \int_{0}^{2\pi} d\phi = \frac{2\pi J_o\rho^3}{3}$$

Equating both sides of Ampère's circuital law and solving for H_ϕ we have

$$H_\phi = \frac{J_o\rho^2}{3}$$

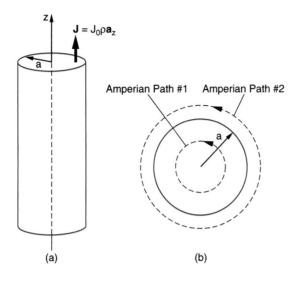

(a)   (b)

Figure 3.17 (a) Infinite length cylindrical conductor with radially dependent current density. (b) Cross section showing location of Amperian paths.

or

$$\mathbf{H} = \frac{J_o}{3}\rho^2 \mathbf{a}_\phi, \text{ for } \rho \leq a$$

For the second Amperian path, the current enclosed is

$$I = \frac{2\pi J_o a^3}{3}$$

and solving Ampère's circuital law we have

$$\mathbf{H} = \frac{J_o}{3}\frac{a^3}{\rho}\mathbf{a}_\phi, \text{ for } \rho > a$$

We note that, for $\rho > a$, the magnetic field intensity falls off as $1/\rho$ just like the field for the line of current.

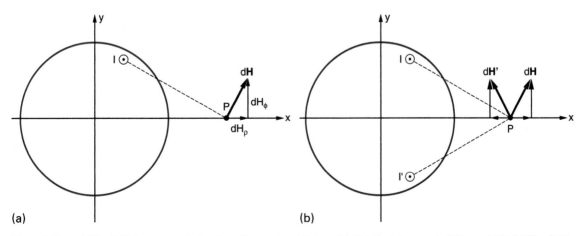

(a)   (b)

Figure 3.18 (a) The field from a particular line of current making up the distributed current of Figure 3.17. (b) The field from a second line of current results in a cancellation of the $\mathbf{a}_\rho$ component.

> **MATLAB 3.3**

In Example 3.7, let's suppose that $J_o = 6.0$ A/m and $a = 2.0$ cm. Let's plot H_ϕ versus ρ out to 6 cm.

```
%  M-File: ML0303
%
%  This program plots the magnetic field intensity
%  versus radial distance from a cylindrical conductor
%  that has a radius-dependent current density.
%
%  Wentworth, 7/15/02
%
%  Variables:
%  Jo      current density constant
%  rho     radial distance (m)
%  Hf      mag. field intensity in phi direction (A/m)
%  N       number of points per cm to plot
%  a       radius of conductor (cm)
%  b       radial plot limit (cm)

clc           %clears the command window
clear         %clears variables

%  Initialize Variables
N=20;
Jo=6;
a=2;
b=6;

%  Perform Calculation
i=1:b*N;
rho=i./N;
Hf(1:a*N)=Jo*(100/3)*rho(1:a*N).^2;
Hf(1+a*N:b*N)=Jo*a^3./(0.03*rho(1+a*N:b*N));

%  Generate Plot
plot(rho,Hf)
grid on
xlabel('rho(cm)')
ylabel('Mag Field Intensity (microamps/m)')
title('ACL: MATLAB 3.3')
```

The result is plotted in Figure 3.19.

▷ **EXAMPLE 3.8**

A fourth example details the very important case of the coaxial cable. A number of configurations are possible, but we'll address the case shown in Figure 3.20a, where even current distributions are assumed in the inner and outer conductors. Other configurations (such as a line of current surrounded by a cylindrical shell of current) will be given as problems at the end of the chapter.

There are four Amperian paths to consider, as shown in the cross section of Figure 3.20b. The left side of Ampère's circuital law will be the same for all four paths. The symmetry of the problem sug-

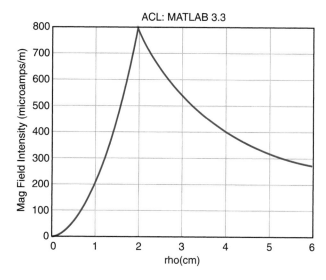

Figure 3.19 Plot of H_ϕ versus ρ from MATLAB 3.3.

gests that only an H_ϕ component is present, and as was seen in the third example of the cylindrical conductor we have

$$\oint \mathbf{H} \cdot d\mathbf{L} = 2\pi\rho H_\phi$$

The current enclosed by the first path is found by integrating

$$I_{enc} = \oint \mathbf{J}_i \cdot d\mathbf{S}$$

but we need to know $\mathbf{J}_i$. Because the problem assumes an even current distribution

$$\mathbf{J}_i = \frac{I}{\pi a^2}\mathbf{a}_z$$

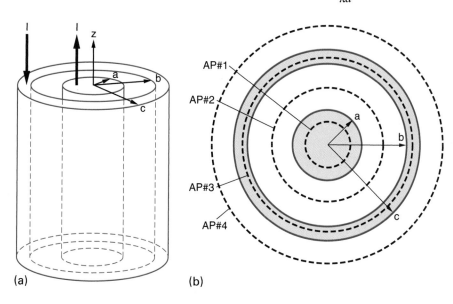

(a)   (b)

Figure 3.20 (a) Coaxial cable. (b) Cross section showing the Amperian paths.

we therefore have

$$I_{enc} = \int \frac{I}{\pi a^2} \mathbf{a}_z \cdot \rho d\rho d\phi \mathbf{a}_z = \frac{I}{\pi a^2} \int_0^\rho \rho d\rho \int_0^{2\pi} d\phi = \frac{I}{a^2} \rho^2$$

Equating both sides of Ampère's circuital law we have

$$H_\phi = \frac{I}{a^2} \frac{\rho^2}{2\pi\rho} = \frac{I\rho}{2\pi a^2} \text{ for } \rho \le a$$

For $a \le \rho \le b$, the current enclosed is just I and we have

$$H_\phi = \frac{I}{2\pi\rho}$$

To find the total current enclosed by Amperian path #3, we need the current density in the outer conductor, given by

$$\mathbf{J}_o = \frac{I(-\mathbf{a}_z)}{\pi(c^2 - b^2)}$$

Now, the total current enclosed by the path is the sum of the inner current and the outer current enclosed by the path, and thus

$$I_{enc} = I + \int \mathbf{J}_o \cdot d\mathbf{S}$$

Evaluating the right side we have

$$\int \mathbf{J}_o \cdot d\mathbf{S} = \int \frac{I(-\mathbf{a}_z)}{\pi(c^2 - b^2)} \cdot \rho d\rho d\phi \mathbf{a}_z = \frac{-I}{\pi(c^2 - b^2)} \int_b^\rho \rho d\rho \int_0^{2\pi} d\phi = -I \frac{\rho^2 - b^2}{c^2 - b^2}$$

Therefore

$$I_{enc} = I\left(1 - \frac{\rho^2 - b^2}{c^2 - b^2}\right) = I \frac{c^2 - \rho^2}{c^2 - b^2}$$

Now we can solve for H_ϕ for Amperian path #3:

$$H_\phi = \frac{I}{2\pi(c^2 - b^2)}\left(\frac{c^2 - \rho^2}{\rho}\right) \text{ for } b \le \rho \le c$$

Finally, at Amperian path #4 the total current enclosed is zero, so $H_\phi = 0$ for $\rho > c$. This further shows the shielding ability exhibited by coaxial cable.

Summarizing the results for the coaxial cable we have the following:

$$\text{for } \rho \le a, \qquad \mathbf{H} = \frac{I\rho}{2\pi a^2} \mathbf{a}_\phi$$

$$\text{for } a < \rho \le b, \qquad \mathbf{H} = \frac{I}{2\pi\rho} \mathbf{a}_\phi$$

$$\text{for } b < \rho \le c, \qquad \mathbf{H} = \frac{I}{2\pi(c^2 - b^2)}\left(\frac{c^2 - \rho^2}{\rho}\right) \mathbf{a}_\phi$$

$$\text{for } c < \rho, \qquad \mathbf{H} = 0$$

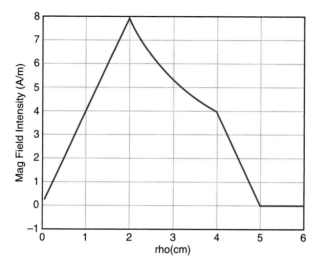

Figure 3.21 Magnetic field intensity plotted against radius for the coaxial cable example of Drill 3.6.

Drill 3.6 Suppose in the coaxial example just examined a = 2.0 cm, b = 4.0 cm, c = 5.0 cm, and I = 1.0 A. Plot H_ϕ versus ρ from 0 to 6 cm using MATLAB. (*Answer:* See Figure 3.21.)

► **EXAMPLE 3.9**

Let us now find the magnetic field intensity within a tightly wrapped solenoid of infinite length.

Cross sections are shown in Figure 3.22. Over a distance h, we have N turns of coil. First, we can make symmetry arguments that the only component of **H** will be H_z.

Now let's look at a cross section as shown in Figure 3.22a. The cross section resembles a pair of infinite extent sheets of current, one directed out of the page and one into the page. The current density for each of these sheets is

$$K = \frac{NI}{h}$$

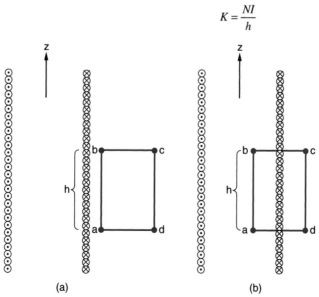

(a)   (b)

Figure 3.22 Cross sections of an infinitely long solenoid.

Using Eq. 3.17, it would be easy to see that the fields from the two sheets cancel each other outside the solenoid and add inside. So from this argument we have inside the solenoid

$$\mathbf{H} = K\mathbf{a}_z = \frac{NI}{h}\mathbf{a}_z$$

Although true, this result isn't entirely satisfactory since we really aren't talking about a pair of infinite extent sheets.

Using an Ampère's circuital law argument in Figure 3.22a, we see that the path $a \rightarrow b \rightarrow c \rightarrow d \rightarrow a$ encloses no current. This can only be true if either $H_z\mathbf{a}_z$ is constant or equal to zero outside the solenoid. If a field exists at all, we would expect it to at least decrease with radial distance away from the solenoid. This suggests that the field outside the solenoid is zero, a result confirmed experimentally.

Now we look at the Amperian path in Figure 3.22b. The only nonzero portion of the circulation integral will be from $a \rightarrow b$, so we have

$$\oint \mathbf{H} \cdot d\mathbf{L} = \int_0^h H_z\mathbf{a}_z \cdot dz\mathbf{a}_z = NI$$

leading to the expression for magnetic field intensity inside a solenoid:

$$\boxed{\mathbf{H} = \frac{NI}{h}\mathbf{a}_z} \tag{3.18}$$

A very interesting aspect of this answer is that we didn't specify a distance from the solenoid wall; thus this is the field *at any point* within the solenoid! This is the same result that we found using the law of Biot–Savart for the very center of a solenoid.

▶3.4 CURL AND THE POINT FORM OF AMPÈRE'S CIRCUITAL LAW

When we studied electrostatic fields, the concept of divergence was employed to find the point form of Gauss's law from the integral form. A nonzero divergence of the electric field indicates the presence of a charge at that point. In this section, *curl* is employed to find the point form of Ampère's circuital law from the integral form. A nonzero curl of the magnetic field will indicate the presence of a current at that point.

To begin, let's apply Ampère's circuital law to a path surrounding a small surface. Dividing both sides by the small surface area, we have the circulation per unit area

$$\frac{\oint \mathbf{H} \cdot d\mathbf{L}}{\Delta S} = \frac{I_{enc}}{\Delta S} \tag{3.19}$$

Taking the limit as ΔS is shrunk to zero, we have

$$\lim_{\Delta S \to 0} \frac{\oint \mathbf{H} \cdot d\mathbf{L}}{\Delta S} = \lim_{\Delta S \to 0} \frac{I_{enc}}{\Delta S} \tag{3.20}$$

Both sides of (3.20) will be maximized if ΔS is chosen normal to the direction of the current, $\mathbf{a}_n$. Multiplying both sides of (3.20) by $\mathbf{a}_n$ we have

$$\lim_{\Delta S \to 0} \frac{\oint \mathbf{H} \cdot d\mathbf{L}}{\Delta S}\mathbf{a}_n = \lim_{\Delta S \to 0} \frac{I_{enc}}{\Delta S}\mathbf{a}_n \tag{3.21}$$

The right side of (3.21) is the current density **J**, and the left side is *the maximum circulation of* **H** *per unit area as the area shrinks to zero*, called *the curl of* **H** for short. So we now have the point form of Ampère's circuital law:

$$\text{curl } \mathbf{H} = \mathbf{J} \tag{3.22}$$

The curl describes the rotation or "vorticity" of a field about a particular point. It gives a measure of the degree to which a field curls around a particular point. It is a position derivative of **H** that returns a vector quantity, and it can be shown that

$$\text{curl } \mathbf{H} = \nabla \times \mathbf{H} \tag{3.23}$$

which, for Cartesian coordinates, can be written

$$\nabla \times \mathbf{H} = \begin{vmatrix} \mathbf{a}_x & \mathbf{a}_y & \mathbf{a}_z \\ \frac{\partial}{\partial x} & \frac{\partial}{\partial y} & \frac{\partial}{\partial z} \\ H_x & H_y & H_z \end{vmatrix} \tag{3.24}$$

The most common way, then, to refer to the point form of Ampère's circuital law is

$$\boxed{\nabla \times \mathbf{H} = \mathbf{J}} \tag{3.25}$$

pronounced "the curl of H equals J."

Divergence and curl are both position derivatives of a vector field. With divergence, the field's magnitude is changing in the direction of the field, so a scalar quantity is adequate. But with curl the change occurs in a direction transverse to that of the field, so a vector quantity is needed to give this direction. For example, the field in Figure 3.23a has a positive divergence but zero curl. In Figure 3.23b, the divergence is zero and the curl is a positive value in the $+\mathbf{a}_z$ direction.

We can confirm this last statement by considering that the field in Figure 3.23b is $\mathbf{A} = A_x \mathbf{a}_x$ and that this field only varies in the y direction. So, expanding (3.24) we see that only the $\partial A_x / \partial y$ term is nonzero, and we have

$$\nabla \times \mathbf{A} = \frac{-\partial A_x}{\partial y} \mathbf{a}$$

Now, because A_x decreases as y increases, $\partial A_x / \partial y$ is negative, leading to a positive $\mathbf{a}_z$ direction for the curl.

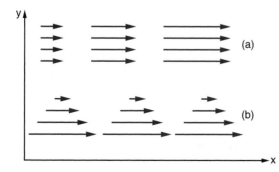

Figure 3.23 Field lines indicating (a) divergence and (b) curl.

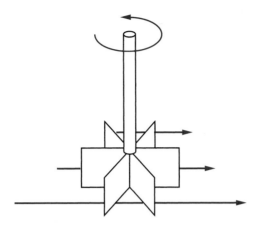

Figure 3.24 A Skilling wheel used to measure curl of the velocity field in flowing water.

A simpler, more intuitive, way to see the direction of the curl in Figure 3.23b is to imagine that the field vectors represent water velocity and to use a *Skilling wheel* (introduced by H. H. Skilling in 1948), as shown in Figure 3.24. This small, imagined paddle wheel can be inserted into the flow and oriented in the direction that provides maximum torque on the axis. Using the right-hand rule with fingers curling in the direction of paddle wheel spin, the thumb will indicate the curl direction. Doing this in Figure 3.24 (for a portion of the field from Figure 3.23b) clearly shows the curl in the $+\mathbf{a}_z$ direction.

As was the case with divergence, the curl in cylindrical and spherical coordinate systems is not as straightforward:

$$\nabla \times \mathbf{H}_{\text{cyl}} =$$

$$\left[\frac{1}{\rho} \frac{\partial H_z}{\partial \phi} - \frac{\partial H_\phi}{\partial z} \right] \mathbf{a}_\rho + \left[\frac{\partial H_\rho}{\partial z} - \frac{\partial H_z}{\partial \rho} \right] \mathbf{a}_\phi + \frac{1}{\rho} \left[\frac{\partial}{\partial \rho} \left(\rho H_\phi \right) - \frac{\partial H_\rho}{\partial \phi} \right] \mathbf{a}_z \tag{3.26}$$

and

$$\nabla \times \mathbf{H}_{\text{spher}} = \frac{1}{r \sin \theta} \left[\frac{\partial}{\partial \theta} \left(H_\phi \sin \theta \right) - \frac{\partial H_\theta}{\partial \phi} \right] \mathbf{a}_r$$

$$+ \frac{1}{r} \left[\frac{1}{\sin \theta} \frac{\partial H_r}{\partial \phi} - \frac{\partial}{\partial r} \left(r H_\phi \right) \right] \mathbf{a}_\theta + \frac{1}{r} \left[\frac{\partial}{\partial r} \left(r H_\theta \right) - \frac{\partial H_r}{\partial \theta} \right] \mathbf{a}_\phi \tag{3.27}$$

Drill 3.7 Find $\mathbf{J}$ at (2 m, 1 m, 3 m) if $\mathbf{H} = 2xy^2 \, \mathbf{a}_z$ A/m. (*Hint: Make sure you perform the curl operation before plugging in the numbers.*) (*Answer:* $8\mathbf{a}_x - 2\mathbf{a}_y$ A/m²)

Drill 3.8 Find $\mathbf{J}$ at (3 m, 90°, 0) if $\mathbf{H} = r^2 \sin\phi \, \mathbf{a}_\theta$ A/m. (*Answer:* $-3\mathbf{a}_r$ A/m²)

Stokes's Theorem

We can rewrite Ampère's circuital law in terms of a current density as

$$\oint \mathbf{H} \cdot d\mathbf{L} = \int \mathbf{J} \cdot d\mathbf{S} \qquad (3.28)$$

and then it is a simple matter to use the point form of Ampère's circuital law to replace $\mathbf{J}$ with $\nabla \times \mathbf{H}$, yielding

$$\oint \mathbf{H} \cdot d\mathbf{L} = \int (\nabla \times \mathbf{H}) \cdot d\mathbf{S} \qquad (3.29)$$

This expression, relating a closed line integral to a surface integral, is known as *Stokes's theorem* (after British mathematician and physicist Sir George Stokes, 1819–1903). Note that the directions $d\mathbf{S}$ and $d\mathbf{L}$ are not independent but are related by the right-hand rule.

That (3.29) is true is apparent by studying Figure 3.25a. The closed line integral is taken around a contour enclosing an area that has been divided into ΔS sections. As ΔS shrinks to zero, the right side of (3.29) is found by summing $\nabla \times \mathbf{H} \cdot \Delta S \mathbf{a}_n$ from each ΔS section. However, recalling that

$$\nabla \times \mathbf{H} = \lim_{\Delta S \to 0} \frac{\oint \mathbf{H} \cdot d\mathbf{L}}{\Delta S} \mathbf{a}_n \qquad (3.30)$$

we see that this is the same as summing the $\oint \mathbf{H} \cdot d\mathbf{L}$ from each section. As shown in the figure, the $\oint \mathbf{H} \cdot d\mathbf{L}$ components at common boundaries will cancel, leaving only those components along the contour (i.e., the left side of (3.29)).

Now suppose we consider that the surface bounded by the contour in Figure 3.25a is actually a rubber sheet. In Figure 3.25b, we can distort the surface while keeping it intact. As long as the surface remains unbroken, Stokes's theorem remains valid.

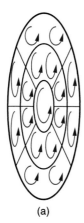

(a)

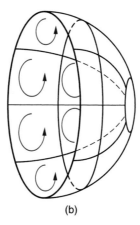

(b)

Figure 3.25 The areas inside the contour divided into ΔS sections.

▶3.5 MAGNETIC FLUX DENSITY

In our dealings with electrostatics we sometimes found it convenient to think in terms of electric flux lines and electric flux density. So too we will find it convenient to work with *magnetic flux density* **B**, which is related to the magnetic field intensity in free space by

$$\mathbf{B} = \mu_o \mathbf{H} \tag{3.31}$$

where μ_o is the *free space permeability*, given in units of *henrys* per meter, or

$$\boxed{\mu_o = 4\pi \times 10^{-7} \text{ H/m}} \tag{3.32}$$

The units of **B** are therefore H-A/m², but it is more instructive to write *webers* per meter squared, or Wb/m², where Wb = (H)(A). Writing the units in terms of webers per square meter is satisfying in that it reminds us of writing electric flux density in terms of coulombs per square meter. We will think of lines of magnetic flux in units of webers. But for brevity, and perhaps to honor a deserving scientist, a *tesla*[3.8] T, equivalent to a Wb/m², is the standard unit adopted by the International System of Units. To further add to your confusion, Wb/m² and T both replace an older unit for magnetic flux density called the *gauss* G, where 10,000 G = 1 T. This unit was used for many years to specify the performance of magnets and is somewhat convenient in that the earth's magnetic flux density is about 0.5 G. Although we will use Wb/m² in the discussions to follow, the student should be conversant with all three terminologies. Table 3.1 lists some typical magnetic flux densities, in terms of both T(Wb/m²) and G.

The amount of magnetic flux ϕ, in webers, from magnetic field passing through a surface is found in a manner analogous to finding electric flux:

$$\phi = \int \mathbf{B} \cdot d\mathbf{S} \tag{3.33}$$

Drill 3.9   (a) Find **B** for an infinite length line of 3.0-A current going in the $+\mathbf{a}_z$ direction along the z-axis in free space, and (b) find the magnetic flux through a surface defined by 1.0 m $\leq \rho <$ 4.0 m, 0 $< z <$ 3.0 m, $\phi = 90°$. (*Answer:* (a) **B** = $(6 \times 10^{-7}/\rho)\mathbf{a}_\phi$ Wb/m², (b) $\phi = 2.5$ μWb)

TABLE 3.1   Approximate Magnetic Flux Densities for Selected Items

Item	G	Wb/m² or T
Human brain	10^{-11}	10^{-15}
Human heart	10^{-4}	10^{-8}
Earth's field	0.5	5×10^{-5}
Refrigerator magnet	100	0.010
Permanent magnet	4000	0.40
Pulsed electromagnet	10^6	100
Neutron star	10^{12}	10^8

[3.8]The unit is named after Nikola Tesla (1856–1943), a Yugoslavian-born American electrical engineer who, among other things, invented the induction motor.

A fundamental feature of magnetic fields that distinguishes them from electric fields is that the field lines form closed loops. Figure 3.26 shows the magnetic field lines in a bar magnet, where the lines are shown extending through the bar forming closed loops. This is quite different from electric field lines, which start on positive charge and terminate on negative charge. You cannot saw the magnet in half to isolate the north and the south poles; as Figure 3.27 shows, if you saw a magnet in half you get two magnets. Put another way, *you cannot isolate a magnetic pole*. From this characteristic of magnetic fields, it is easy to see that the net magnetic flux passing through a Gaussian surface (a closed surface as shown in Figure 3.26) must be zero. What goes into the surface must come back out. Thus we have *Gauss's law for static magnetic fields,*

$$\oint \mathbf{B} \cdot d\mathbf{S} = 0$$

(3.34)

This is also referred to as the law of conservation of magnetic flux.

Applying the divergence theorem to (3.34), we arrive at the point form of Gauss's law for static magnetic fields,

$$\nabla \cdot \mathbf{B} = 0$$

(3.35)

With the addition of Gauss's law for static magnetic fields, we can now present all four of Maxwell's equations for static fields:

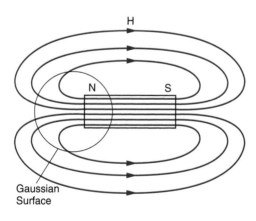

Figure 3.26 Magnetic field lines form closed loops, so the *net flux* through a Gaussian surface is zero.

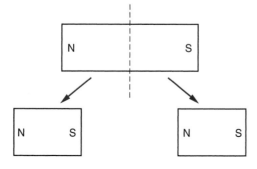

Figure 3.27 Dividing a magnet in two parts results in two magnets. You cannot isolate a magnetic pole.

Integral form	*Differential form*	
$\oint \mathbf{D} \cdot d\mathbf{S} = Q_{enc}$	$\nabla \cdot \mathbf{D} = \rho_v$	
$\oint \mathbf{B} \cdot d\mathbf{S} = 0$	$\nabla \cdot \mathbf{B} = 0$	
$\oint \mathbf{E} \cdot d\mathbf{L} = 0$	$\nabla \times \mathbf{E} = 0$	(3.36)
$\oint \mathbf{H} \cdot d\mathbf{L} = I_{enc}$	$\nabla \times \mathbf{H} = \mathbf{J}$	

The differential, or point, form of Maxwell's equations are easily derived by applying the divergence theorem and Stoke's theorem to the integral form of the equations.

▶ 3.6 MAGNETIC FORCES

We saw in Figure 3.2 that when electric current is passed through a magnetic field, a force is exerted on the wire normal to both the magnetic field and the current direction. This force is actually acting on the individual charges moving in the conductor. The *magnetic force* $\mathbf{F}_m$(N) on a moving charge q(C) is given by

$$\mathbf{F}_m = q\mathbf{u} \times \mathbf{B} \qquad (3.37)$$

where the velocity of the charge is $\mathbf{u}$(m/s) within a field of magnetic flux density $\mathbf{B}$(Wb/m^2). The units are confirmed by using the equivalences Wb = (V)(s) and J = (N)(m) = (C)(V).

By the definition of electric field intensity, the electric force $\mathbf{F}_e$ acting on a charge q within an electric field is given by

$$\mathbf{F}_e = q\mathbf{E} \qquad (3.38)$$

A big difference between the two forces is readily apparent: The electric force acts in the direction of the electric field whereas the magnetic force acts at right angles to the magnetic field. Also, the magnetic force requires that the charged particle be in motion. Finally, recalling the definition of work as

$$W = \int \mathbf{F} \cdot d\mathbf{L}$$

we see that the magnetic force *can do no work*, since $d\mathbf{L}$ is always in a normal direction to the force.

A total force on a charge is given by superposing (3.37) and (3.38), arriving at the *Lorentz*[3.9] *force equation*:

$$\boxed{\mathbf{F} = q\,(\mathbf{E} + \mathbf{u} \times \mathbf{B})} \qquad (3.39)$$

The Lorentz force equation is quite useful in determining the paths charged particles will take as they move through electric and magnetic fields. If we also know the particle mass m, the force is related to acceleration by the equation, fondly remembered from introductory physics,

$$\mathbf{F} = m\mathbf{a} \qquad (3.40)$$

[3.9]The equation is named after Dutch physicist Hendrik Antoon Lorentz (1853–1928).

It should be noted that because the magnetic force acts in a direction normal to the particle velocity, the acceleration is normal to the velocity and the magnitude of the velocity vector is unaffected.

Drill 3.10 At a particular instant in time, in a region of space where $\mathbf{E} = 0$ and $\mathbf{B} = 3\,\mathbf{a}_y$ Wb/m^2, a 2-kg particle of charge 1 C moves with velocity $2\mathbf{a}_x$ m/s. What is the particle's acceleration due to the magnetic field?

$$\left(Answer:\ \mathbf{a} = \frac{q}{m}\mathbf{u} \times \mathbf{B} = 3\mathbf{a}_z\, \frac{m}{s^2};\ \text{to calculate the units:} \right.$$

$$\left. \frac{C}{kg}\frac{m}{s}\frac{Wb}{m^2} \times \left(\frac{kg-m}{N-s^2} \right)\left(\frac{N-m}{J} \right)\left(\frac{J}{C-V} \right)\left(\frac{V-s}{Wb} \right) = \frac{m}{s^2} \right)$$

Force on a Current Element

Consider a line conducting current in the presence of a magnetic field. We wish to find the resulting force on the line. We can look at a small, differential segment dQ of charge moving with velocity $\mathbf{u}$, and we can calculate the differential force on this charge from (3.37) as

$$d\mathbf{F} = dQ\,\mathbf{u} \times \mathbf{B} \tag{3.41}$$

But the velocity can also be written

$$\mathbf{u} = \frac{d\mathbf{L}}{dt} \tag{3.42}$$

and (3.41) can be rearranged as

$$d\mathbf{F} = \frac{dQ}{dt}\,d\mathbf{L} \times \mathbf{B} \tag{3.43}$$

Now, since dQ/dt (in C/s) corresponds to the current I in the line, we have

$$d\mathbf{F} = Id\mathbf{L} \times \mathbf{B} \tag{3.44}$$

Equation 3.44 is often referred to as the *motor equation*.

We can use (3.44) to find the force from a collection of current elements, using the integral

$$\boxed{\mathbf{F}_{12} = \int I_2 d\mathbf{L}_2 \times \mathbf{B}_1} \tag{3.45}$$

where the subscripts are added to indicate that the magnetic force acts from the field $\mathbf{B}_1$ on an element $I_2 d\mathbf{L}_2$.

Let's consider a line of current in the $+\mathbf{a}_z$ direction on the z-axis as shown in Figure 3.28a. For current element $Id\mathbf{L}_a$, we have

$$Id\mathbf{L}_a = Idz_a\mathbf{a}_z$$

We know this element produces magnetic field, but the field cannot exert magnetic force on the element producing it. As an analogy, consider that the electric field of a point charge can exert no electric force on itself.

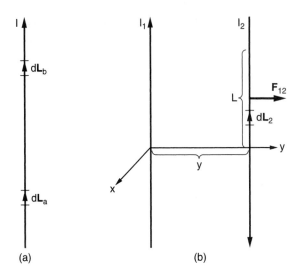

Figure 3.28 (a) Differential current elements on a line. (b) A pair of current-carrying lines will exert magnetic force on each other.

What about the field from a second current element $Id\mathbf{L}_b$ on this line? From Biot–Savart's law, we see that the cross product in this particular case will be zero, since $Id\mathbf{L}$ and $\mathbf{a}_R$ will be in the same direction. So, we can say that a straight line of current exerts no magnetic force on itself.

Now suppose we add a second line of current parallel to the first, as shown in Figure 3.28b. The force $d\mathbf{F}_{12}$ from the magnetic field of line 1 acting on a differential section of line 2 is

$$d\mathbf{F}_{12} = I_2 d\mathbf{L}_2 \times \mathbf{B}_1 \qquad (3.46)$$

The magnetic flux density $\mathbf{B}_1$ for an infinite length line of current is recalled from (3.8) to be

$$\mathbf{B}_1 = \frac{\mu_o I_1}{2\pi\rho} \mathbf{a}_\phi$$

and by inspection of the figure we see that $\rho = y$ and $\mathbf{a}_\phi = -\mathbf{a}_x$. Inserting this into (3.46) and considering that $d\mathbf{L}_2 = dz\mathbf{a}_z$, we have

$$d\mathbf{F}_{12} = I_2 dz\mathbf{a}_z \times \frac{\mu_o I_1}{2\pi y}\left(-\mathbf{a}_x\right) = \frac{\mu_o I_1 I_2}{2\pi y}\, dz\left(-\mathbf{a}_y\right) \qquad (3.47)$$

To find the total force on a length L of line 2 from the field of line 1, we must integrate $d\mathbf{F}_{12}$ from $+L$ to 0. We are integrating in this direction to account for the direction of the current. This gives us

$$\mathbf{F}_{12} = \frac{\mu_o I_1 I_2}{2\pi y}\left(-\mathbf{a}_y\right)\int_L^0 dz = \frac{\mu_o I_1 I_2 L}{2\pi y}\,\mathbf{a}_y \qquad (3.48)$$

which is a repulsive force. Had we instead been seeking $\mathbf{F}_{21}$, the magnetic force acting on line 1 from the field of line 2, we would have found $\mathbf{F}_{21} = -\mathbf{F}_{12}$. So two parallel lines with current in opposite directions experience a force of repulsion. For a pair of parallel lines with current in the same direction, a force of attraction would result.

In the more general case where the two lines are not parallel, or not straight, we could use the law of Biot–Savart to find $\mathbf{B}_1$ and arrive at

$$\mathbf{F}_{12} = \frac{\mu_o}{4\pi} I_2 I_1 \oint\oint \frac{d\mathbf{L}_2 \times (d\mathbf{L}_1 \times \mathbf{a}_{12})}{R_{12}^2} \tag{3.49}$$

This equation is known as *Ampère's law of force* between a pair of current-carrying circuits and is analogous to Coulomb's law of force between a pair of charges. Rather than applying (3.49), it is easier in practice to find the magnetic field $\mathbf{B}_1$ by Biot–Savart's law and then use (3.45) to find $\mathbf{F}_{12}$.

▷ **EXAMPLE 3.10**

Let's find the force on each arm of the square loop of current in Figure 3.29 resulting from the magnetic field created by the infinite length line of current I_1 on the z-axis. Here we will ignore any force that may be acting between current elements in the loop, but in MATLAB 3.4 we will address this sort of problem and you will also consider this force in a homework problem.

As a first step, we find the field $\mathbf{B}_1$ in the y–z plane containing the loop, which we know for this particular infinite length line of current is

$$\mathbf{B}_1 = \frac{\mu_o I_1}{2\pi y}(-\mathbf{a}_x)$$

Next we find the force on each segment of the loop starting with segment a. We have

$$\mathbf{F}_{1a} = \int I_2 d\mathbf{L}_a \times \mathbf{B}_1 = I_2 \int dy \mathbf{a}_y \times \frac{\mu_o I_1}{2\pi y}(-\mathbf{a}_x)$$

$$= \frac{\mu_o I_1 I_2 \mathbf{a}_z}{2\pi} \int_{y_o}^{y_o+w} \frac{dy}{y} = \frac{\mu_o I_1 I_2}{2\pi} \ln\left(\frac{y_o+w}{y}\right)\mathbf{a}_z$$

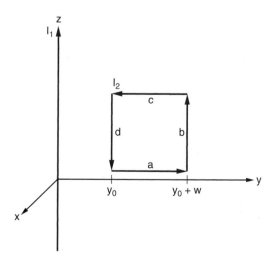

Figure 3.29 Figure used to find the magnetic force exerted by the field of I_1 on each arm of a square loop of current I_2.

For segment b, y is fixed at $y_o + w$ and we have

$$\mathbf{F}_{1b} = I_2 \int_0^w dz \mathbf{a}_z \times \frac{\mu_o I_1}{2\pi(y_o + w)}(-\mathbf{a}_x) = -\frac{\mu_o I_1 I_2 w}{2\pi(y_o + w)}\mathbf{a}_y$$

Likewise, for the other two segments we have

$$\mathbf{F}_{1c} = -\frac{\mu_o I_1 I_2}{2\pi} \ln\left(\frac{y_o + w}{y_o}\right)\mathbf{a}_z$$

and

$$\mathbf{F}_{1d} = \frac{\mu_o I_1 I_2 w}{2\pi y_o}\mathbf{a}_y$$

If we were to sum the fields, we would see that $\mathbf{F}_{1a}$ and $\mathbf{F}_{1c}$ cancel, but the magnitude of force $\mathbf{F}_{1d}$ is bigger than that of $\mathbf{F}_{1b}$ owing to segment d's closer proximity to the infinite length line of current.

Drill 3.11   A pair of parallel infinite length lines each carry current $I = 2$ A in the same direction. Determine the magnitude of the force per unit length between the two lines if their separation distance is (a) 10 cm and (b) 100 cm. Is the force repulsive or attractive? (*Answer*: (a) 8 μN/m, (b) 0.8 μN/m, attractive)

▶ **MATLAB 3.4**

We want to find the magnetic force $d\mathbf{F}$ acting on a differential segment of a current-conducting loop from the field of the rest of the loop as shown in Figure 3.30.

```
%    M-File: ML0304
%
%    This program finds the force on one piece of a loop
%    resulting from the field of the rest of the loop.
%    The loop is broken up into segments subtended by 2
%    degree increments.
%
%    Wentworth, 7/16/02
%
%    Variables:
%    I        loop current (A)
%    a        loop radius (m)
%    F        the angle phi in radians
%    xi,yi    location of ith element
%    Ai       vector from origin to xi,yi
%    ai       unit vector from origin to xi,yi
%    Dli      ith element vector
%    Ri1      vector from ith point to test point
%    ri1      unit vector from ith point to test point
%    DL1      test element vector
%    mu       free space permeability (H/m)
%    az       unit vector in z direction
```

```
clc        %clears the command window
clear      %clears variables

%   Initialize Variables
I=1;
a=1;
mu=pi*4e-7;
az=[0 0 1];
DL1=a*2*(pi/180)*[0 1 0];

%   Perform Calculation
for i=1:179
    F=2*i*pi/180;
    xi=a*cos(F);
    yi=a*sin(F);
    Ai=[xi yi 0];
    ai=unitvector(Ai);
    DLi=(pi*a/90)*cross(az,ai);
    Ri1=[a*(1-cos(F)) -a*sin(F) 0];
    ri1=unitvector(Ri1);
    num=mu*I*cross(DLi,ri1);
    den=4*pi*(magvector(Ri1)^2);
    B=num/den;
    Bx(i)=B(1);
    By(i)=B(2);
    Bz(i)=B(3);
end
Btot=[sum(Bx) sum(By) sum(Bz)];
dF=I*cross(DL1,Btot)
```

Running this program yields

$d\mathbf{F}$= 18.6 nN $\mathbf{a}_x$

In this particular example we see that a force is attempting to expand the loop.

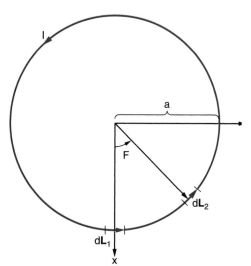

Figure 3.30 In MATLAB 3.4, we want to find the force on $d\mathbf{L}_1$ from the field of the rest of the loop.

Magnetic Torque and Moment

If we place a planar loop of current in a uniform magnetic field, there will be no net magnetic force on the loop. However, there can be a torque, or twisting force, acting on the loop.

We begin by recalling that torque $\boldsymbol{\tau}$ is the cross product of the moment arm $\mathbf{r}$ and a force $\mathbf{F}$:

$$\boldsymbol{\tau} = \mathbf{r} \times \mathbf{F} \tag{3.50}$$

If we have a pair of moment arms and a pair of forces, as shown in Figure 3.31a, the total torque will be a superposition of the individual torques. Here we consider the special case where total net force is zero. Taking the torque about the origin in this figure we have

$$\boldsymbol{\tau} = \mathbf{R}_1 \times \mathbf{F}_1 + \mathbf{R}_2 \times \mathbf{F}_2 = 2\mathbf{a}_y \times F_o \mathbf{a}_z + 4\mathbf{a}_y \times (-F_o \mathbf{a}_z) = -2F_o \mathbf{a}_x \tag{3.51}$$

Now consider taking the torque about the point $(0, 3, 0)$ in Figure 3.31b. Here we see

$$\boldsymbol{\tau} = -1\mathbf{a}_y \times F_o \mathbf{a}_z + 1\mathbf{a}_y \times (-F_o \mathbf{a}_z) = -2F_o \mathbf{a}_x \tag{3.52}$$

Notice that the torque about the center point is the same as the torque about the origin. This is a consequence of the zero net force and means that the torque for this case will be independent of the chosen origin. It will be convenient, then, for us to consider the center point of our conducting loop as the origin, or the axis about which rotation occurs.

Now consider a current-carrying loop in the x–y plane immersed in a uniform magnetic field

$$\mathbf{B} = B_x \mathbf{a}_x + B_y \mathbf{a}_y + B_z \mathbf{a}_z$$

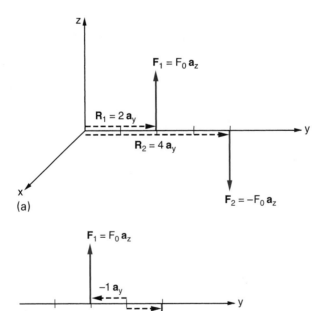

(a)

(b)

Figure 3.31 (a) Finding the total torque about the origin by superposition of a pair of torques. (b) Finding torque at the center point between the forces.

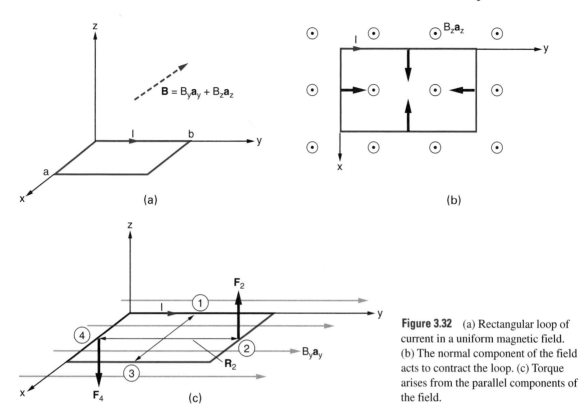

Figure 3.32 (a) Rectangular loop of current in a uniform magnetic field. (b) The normal component of the field acts to contract the loop. (c) Torque arises from the parallel components of the field.

as shown in Figure 3.32a. This magnetic field can be conveniently broken up into a field normal to the plane of the loop, $\mathbf{B}_\perp = B_z\mathbf{a}_z$ (Figure 3.32b), and a field parallel to the plane, $\mathbf{B}_\parallel = B_x\mathbf{a}_x + B_y\mathbf{a}_y$ (Figure 3.32c). For simplicity we'll let $B_x = 0$ for this example. Inspection of Figure 3.32b reveals that the perpendicular magnetic force is acting to contract the loop. If the current had been going in the other direction, the force would be acting to expand the loop. In this case, there is no net force on the loop and no torque. The torque comes from the parallel component of $\mathbf{B}$.

In Figure 3.32c the total torque consists of contributions from each arm,

$$\boldsymbol{\tau} = \mathbf{R}_1 \times \mathbf{F}_1 + \mathbf{R}_2 \times \mathbf{F}_2 + \mathbf{R}_3 \times \mathbf{F}_3 + \mathbf{R}_4 \times \mathbf{F}_4 \qquad (3.53)$$

The force on arm 1 is found by integrating along the length of the arm the differential force $d\mathbf{F}_1$ given by

$$d\mathbf{F}_1 = Idy\mathbf{a}_y \times B_y\mathbf{a}_y \qquad (3.54)$$

which is zero since for this segment the current is in the same direction as the field. The differential force $d\mathbf{F}_3$ will be zero for the same reason, leaving us with

$$\boldsymbol{\tau} = \mathbf{R}_2 \times \mathbf{F}_2 + \mathbf{R}_4 \times \mathbf{F}_4 \qquad (3.55)$$

For arm 2, the differential force is

$$d\mathbf{F}_2 = Idx\mathbf{a}_x \times B_y\mathbf{a}_y = IB_ydx\mathbf{a}_z \qquad (3.56)$$

and the total force $\mathbf{F}_2$ is

$$\mathbf{F}_2 = IB_y \mathbf{a}_z \int_0^a dx = IB_y a a_z \tag{3.57}$$

The moment arm $\mathbf{R}_2$ is $(b/2)\mathbf{a}_y$, resulting in a torque for arm 2

$$\boldsymbol{\tau}_2 = IB_y \frac{ab}{2} \mathbf{a}_x \tag{3.58}$$

Likewise, for arm 4 we have

$$d\mathbf{F}_4 = IB_y dx \mathbf{a}_z \tag{3.59}$$

and

$$\mathbf{F}_4 = IB_y \mathbf{a}_z \int_a^0 dx = -IB_y a a_z \tag{3.60}$$

The moment arm $\mathbf{R}_4$ is $-(b/2)\,\mathbf{a}_y$, so

$$\boldsymbol{\tau}_4 = IB_y \frac{ab}{2} \mathbf{a}_z \tag{3.61}$$

and the total torque on the loop is therefore

$$\boldsymbol{\tau} = IB_y ab \mathbf{a}_x \tag{3.62}$$

It is helpful to define the *magnetic dipole moment* $\mathbf{m}$ of a loop as

$$\boxed{\mathbf{m} = IS\mathbf{a}_N} \tag{3.63}$$

where $\mathbf{m}$ has the dimensions (A-m^2) and $\mathbf{a}_N$ is a unit vector normal to the planar loop of area S. The unique $\mathbf{a}_N$ direction is found via the right-hand rule, where the fingers curl in the direction of the current loop and the thumb points in the direction of $\mathbf{a}_N$. Then, the torque on a magnetic dipole in a field $\mathbf{B}$ is

$$\boxed{\boldsymbol{\tau} = \mathbf{m} \times \mathbf{B}} \tag{3.64}$$

Notice that $\mathbf{B}$ is the general field consisting of both $\mathbf{B}_\perp$ and $\mathbf{B}_\parallel$. However, the $\mathbf{B}_\perp$ portion is in the same direction as $\mathbf{m}$ and will therefore not contribute to the torque. For the problem of Figure 3.32, the surface $S = ab$ and the direction $\mathbf{a}_N = -\mathbf{a}_z$, so $\mathbf{m} = Iab(-\mathbf{a}_z)$, which leads to the same solution as before for $\boldsymbol{\tau}$.

Equations (3.63) and (3.64) will hold for planar loops of any shape, not just rectangular ones. Also, if instead of a single loop there is a winding of N loops of insulated conducting wire then (3.63) is modified as

$$\mathbf{m} = NIS\mathbf{a}_N \tag{3.65}$$

Suppose the loop from Figure 3.32c is free to turn about its axis of rotation. For simplicity, let's ignore the B_z component of the field. In Figure 3.33 we see the loop at three different positions on its axis of rotation. The torque is acting to twist the loop such that the

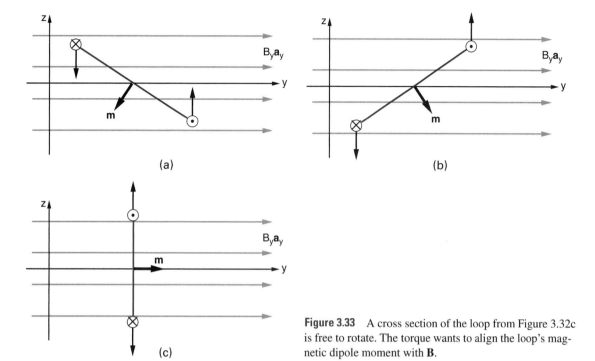

Figure 3.33 A cross section of the loop from Figure 3.32c is free to rotate. The torque wants to align the loop's magnetic dipole moment with **B**.

magnetic dipole moment is in the same direction as the magnetic field. This behavior of magnetic torque is useful for understanding magnetic materials (covered in the next section) as well as simple DC motors.

In a very simple DC motor, the loop (called a rotor) is immersed in the field. The ends of the rotor have conductive brushes that contact a DC current source through a split ring. Current is passed through the loop, resulting in torque. The rotor turns, but just as it reaches the point of zero torque, the brushes pass the split in the ring and the current changes direction. The rotor thus continues to be torqued in the same direction.

> **Drill 3.12** A circular conducting loop of 10.-cm radius lies in the x–y plane and conducts 3.0 A of current in the $\mathbf{a}_\phi$ direction. The loop is immersed in a magnetic field $\mathbf{B} = 3.0\mathbf{a}_x + 4.0\mathbf{a}_z$ Wb/m^2. Determine the loop's (a) magnetic dipole moment and (b) torque. (*Answer:* (a) 0.094 $\mathbf{a}_z$ A-m^2, (b) 0.28 $\mathbf{a}_y$ N-m)

Practical Application: Loudspeakers

A typical loudspeaker is a very simple device consisting of a paper or plastic cone affixed to a voice coil (an electromagnet) suspended in a magnetic field. Alternating current (AC) signals to the voice coil cause it to move back and forth, and the resulting vibration of the cone can reproduce practically any sound.

The cross section of a typical moving-coil speaker is shown in Figure 3.34. The magnetic field comes from a permanent magnet. The front and rear suspensions hold the coil in

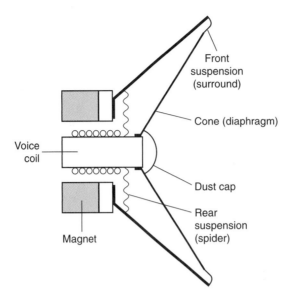

Figure 3.34 Cross section of a moving-coil loudspeaker.

place and allow movement of the cone. The front suspension, often called the *surround*, can be made of rubber. The rear suspension is referred to as the *spider*.

The voice coil is a typical electromagnet consisting of a coil of wire wrapped around an iron cylinder. Current through the coil in one direction will produce a particular polarity in the electromagnet. Since the voice coil is also under the influence of the permanent magnet, the polarity will cause the coil, and the attached cone, to move in a particular direction. That is, the positive pole of the coil is attracted to the negative pole of the permanent magnet and vice versa. When the current is reversed, the coil moves in the opposite direction. An AC signal through the coil therefore produces back-and-forth movement of the coil and diaphragm, thus producing sound waves of the same frequency as the AC signal.

Speaker size determines how well a particular audible frequency can be produced. The diaphragm of a small speaker, called a *tweeter*, can move back and forth (or vibrate) very rapidly and is therefore used to produce high-frequency sound waves. For large speakers, called *woofers*, the diaphragm can move much slower and therefore is used to produce the low-frequency sound waves. In between is the midrange speaker (sometimes called a *squawker*). Typically the speaker system will break up the audio signal into different frequency bands and pass these bands on to the appropriate speaker. The user can choose to amplify a particular band if more bass (lower frequency component) or more treble (higher frequency component) is desired. The speakers are housed in an enclosure specially designed to optimize the sound quality.

Instead of a moving-coil arrangement, a speaker diaphragm can also be vibrated with electrostatic fields. In one design for an electrostatic speaker, a conductive diaphragm is supported between a pair of charged panels, each with opposite polarity. A positive charge on the diaphragm moves it toward the negative charged panel and away from the positive charged panel. A negative charge on the diaphragm moves it in the opposite direction. The electrical signal to the diaphragm is therefore transduced to a sound wave. This approach is most effective for high-frequency sound, where the diaphragm doesn't have to move very far in either direction.

▶3.7 MAGNETIC MATERIALS

We know that current through a coil of wire will produce a magnetic field akin to that of a bar magnet. We also know that we can greatly enhance the field by wrapping the wire around an iron core. The iron is considered a *magnetic material* since it can influence, in this case amplify, the magnetic field.

In Chapter 2 we described the polarization **P** and the electric susceptibility χ_e to arrive at an expression for the ε_r of a dielectric material. In a like manner, for magnetic materials we can consider magnetization **M** and magnetic susceptibility χ_m. The magnetization **M** is the vector sum of all the magnetic dipole moments in a unit volume of the material. It contributes to the total magnetic flux density as

$$\mathbf{B} = \mu_o \mathbf{H} + \mu_o \mathbf{M} \tag{3.66}$$

The magnetization **M** is related to the magnetic field intensity by the material's magnetic susceptibility,

$$\mathbf{M} = \chi_m \mathbf{H} \tag{3.67}$$

The permeability μ and relative permeability μ_r of a material are thus defined in terms of χ_m:

$$\mathbf{B} = \mu_o (1 + \chi_m)\mathbf{H} = \mu \mathbf{H} = \mu_r \mu_o \mathbf{H} \tag{3.68}$$

Rarely do we keep track of **M** and χ_m. Instead we consider the degree to which a material can influence the magnetic field to be given by μ_r. In free space (a vacuum), $\mu_r = 1$ and there is no effect on the field. In most materials, μ_r is either slightly more or slightly less than unity, but in some materials, μ_r can be very large, even as high as 10^6. Table 3.2 lists a variety of materials along with their relative permeabilities.

To understand, at least qualitatively, why a magnetic material behaves as it does requires that we first consider the net magnetic dipole moment for the outermost electrons orbiting their nucleii. Recalling from the previous section that a loop of current will have a

TABLE 3.2   Relative Permeabilities for a Variety of Materials[a]

	Material	μ_r
Diamagnetic	bismuth	0.99983
	gold	0.99986
	silver	0.99998
	copper	0.999991
	water	0.999991
Paramagnetic	air	1.0000004
	aluminum	1.00002
	platinum	1.0003
Ferromagnetic	cobalt	250
(nonlinear)	nickel	600
	iron (99.8% pure)	5000
	iron (99.96% pure)	280,000
	Mo/Ni supermalloy	1,000,000

[a]See also Appendix E.

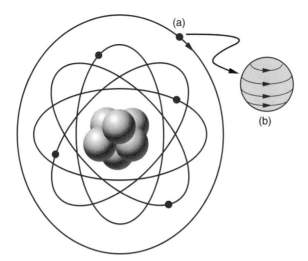

Figure 3.35 The outer electron orbiting a nucleus exhibits (a) an orbital magnetic moment along with (b) a spin magnetic moment.

magnetic moment, we consider the classical orbit of electrons around a nucleus as shown in Figure 3.35. If we assume that all of the inner electron shells are filled, then the net moment of the inner electrons and nucleus tends to be negligible and the majority of the net magnetic dipole moment comes from consideration of the outermost electrons. The outermost electron in Figure 3.35a is a charge in orbital motion that resembles a loop of current, and as such it has an *orbital magnetic dipole moment*, or *orbital moment* for short. The electron itself has a quantum mechanical property known as *spin*, and though not rigorously accurate, it is convenient for us to think of this spin as charge spinning about its own axis as shown in Figure 3.35b. This electron has a *spin magnetic dipole moment*, or *spin moment* for short.

In most materials, the net dipole moment is zero in the absence of a magnetic field. Here, the spin moment cancels the orbital moment. With application of an external magnetic field, the orbital moments are torqued to align with the field whereas the spin moments continue to oppose the orbital moments. This would seem to imply that the net moment remains zero and that μ_r should equal one. But this is not the case.

The externally applied field imparts some energy to the orbiting electron. However, the electron is in a quantized, constant energy state. If energy is added to the electron, it must be decreased some other way to keep the total energy level constant. This reduction in energy is achieved by slowing down the electron. A decrease in the electron's orbital velocity is tantamount to reducing the current in the loop and therefore the orbital moment is decreased. The spin moment is unaffected, however, so the net dipole moment is slightly negative and the material tends to oppose the externally applied field. For such *diamagnetic* materials, μ_r is slightly less than one. Most diamagnetic materials[3.10] have very little influence on the magnetic field. The most diamagnetic material is bismuth, with a $\mu_r = 0.99983$.

In a second class of magnetic materials, known as *paramagnetic*, the orbital and spin moments are not equal and there exists a net magnetic dipole moment. However, the arrangement of the moments is random and so paramagnetic materials exhibit no magnetic

---

[3.10]It should be noted that all materials will exhibit a degree of diamagnetism, but this effect is often overshadowed by other considerations.

behavior in the absence of a magnetic field. But when a magnetic field is applied, the magnetic dipoles experience a torque that tends to align them with the field, and the material can become slightly magnetic. Aluminum is an example of a paramagnetic material, with a relative permeability of 1.000021.

Diamagnetic and paramagnetic materials are two of the three major classes of magnetic material. They can be somewhat understood by considering only the contributions of the orbital and spin moments. The third major class of magnetic material is called *ferromagnetic*. These materials, which are strongly magnetic even in the absence of an externally applied field, are used for permanent magnets. The most common member of this class is iron (whose atomic symbol Fe is from the Latin ferrum). Other common ferromagnetic materials are nickel and cobalt. The mechanism for this third class of magnetic material can be somewhat understood[3.11] by invoking the concepts of *exchange coupling* and the formation of *domains*.

First, consider the energy for a collection of atoms. Assume that each atom has a dipole moment and that they are randomly arranged as in Figure 3.36a. Quantum theory says the system energy will be minimized if the spins on electrons of adjacent atoms are in opposite directions.[3.12] This is indeed the case for most materials that exhibit very little magnetic behavior. However, for ferromagnetic materials, the system energy appears to be *reduced* if adjacent atoms have parallel spin. This is a poorly understood phenomenon, but it is conjectured that a third, "intermediate" electron is involved that tends to have an opposite spin to the electrons on adjacent atoms responsible for the magnetic moment. The mechanism for the alignment of spins on the electrons of adjacent atoms is termed *exchange coupling*.

As more and more of the moments are aligned to reduce the system energy via exchange coupling, as in Figure 3.36b, a stronger and stronger magnetic field is generated by these aligned dipoles. As was the case for electric fields in Section 2.13, magnetic fields store

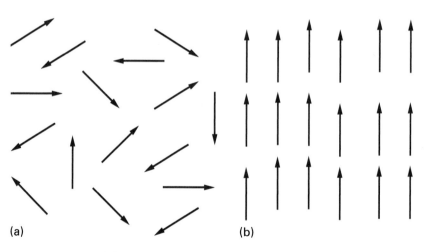

(a)

(b)

Figure 3.36 Randomly oriented dipoles in (a) are all aligned in (b) via exchange coupling.

[3.11]To quote Richard Feynman, "…to the theoretical physicists, ferromagnetism presents a number of very interesting, unsolved, and beautiful challenges. One challenge is to understand why it exists at all" (*The Feynman Lectures on Physics,* Vol. 2, p. 37–13, Addison-Wesley, 1989).

[3.12]As a simple example of a magnetic system seeking a low energy level, consider a collection of 20 or so bar magnets, initially with their north and south poles aligned. Bring them close together and release them and the magnets will repel, flip, and then attract such that the system energy is decreased.

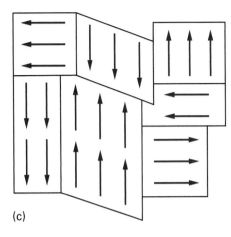

(c)

Figure 3.36 In (c), the dipoles have formed domains (separated by the dashed-line domain walls).

energy. System energy, then, is increased by the increase in the magnetic field if all the moments line up. We can reduce the energy stored in the field if we break up the magnetized regions into *domains* that have random magnetic dipole direction as shown in Figure 3.36c.

Can we continue to break up the sample into smaller and smaller domains to decrease the energy in the magnetic field and thus the system energy? No, because the spins on atoms along the domain walls are in opposite directions, and lower energy is achieved via the exchange coupling phenomenon when the poles are aligned. The formation of domains in conjunction with exchange coupling, as depicted in Figure 3.36c, appears to be a trade-off to minimize energy in the system.

Physically, a magnetized domain contains between 10^{16} and 10^{19} atoms, all with aligned magnetic dipoles. The thin region between a pair of domains is termed a *domain wall* and is on the order of 100 atoms thick. When an external magnetic field is applied, the domains oriented in the direction of the applied field grow larger, with the domain walls moving further into the unaligned domains.[3.13] Sometimes the domain walls get stuck at crystal boundaries or at impurities in the material. Eventually, with more external field applied, the domain wall snaps free, but in doing so there is energy loss to heat from friction.

Because the domain walls have trouble moving past grain boundaries, the relation between *B* and *H* in a ferromagnetic material is nonlinear. This is best seen in a *magnetization curve* as shown in Figure 3.37. We begin at point *a* with a completely demagnetized material. By this we mean that the domains are organized such that there is no net field for the sample. Now we apply a magnetizing field *H* to the ferromagnetic material and measure the resulting flux density *B*. At first, it is relatively easy for the domains in the direction of *H* to grow and *B* proceeds along the initial curve to point *b*. But then it grows harder to align the domains, and finally all the domains are aligned at point *c* and we are at magnetic saturation, denoted by H_s and B_s. Now notice what happens when we reduce the magnetization field *H* to zero, following the path from point *c* to *d*. There remains a *residual flux density* B_r even with no applied *H*. This is the property of ferromagnetic materials employed to make permanent magnets. It actually takes a negative magnetization field to coerce *B* to

---

[3.13]It should be emphasized that although a ferromagnetic material like iron may be composed of many crystals, these crystals are *not* the same as domains.

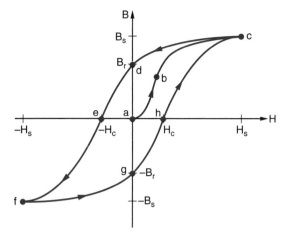

Figure 3.37 Magnetization curve showing the hysteresis loop of a ferromagnetic material.

zero at point *e*. This occurs at H_c, the *coercive field intensity*. If we continue decreasing the magnetization field, eventually we reach magnetic saturation in the other direction (point *f*). We can then start increasing the magnetization field to travel from point *f* to *g*, *h*, and then finally back to point *c*. The complete round trip is termed a *hysteresis*[3.14] *loop*.

Wide hysteresis loops are desirable for permanent magnets, since we want to freeze the domain walls in place over a significant range of external field. For transformers and motors, however, we want the domain walls to move easily so that we can easily change the magnetism. A tall, narrow hysteresis loop is desired in this case. The area enclosed by the hysteresis loop is proportional to the energy dissipated in a round trip around the loop.

If a permanent magnet is subjected to a sufficiently high temperature, known as the *Curie temperature*, the thermal energy will overwhelm the exchange coupling force and the material will lose its magnetization. It will become a paramagnetic material, at least until it cools off again. The Curie temperature for iron is 770°C.

Diamagnetics, paramagnetics, and ferromagnetics account for almost all material types, but there are some exotic materials that should be briefly mentioned.

An *antiferromagnetic* material, for example manganese oxide, has its moments of adjacent atoms locked in opposite directions regardless of the applied field. The coupling of the moments disappears above the material's Curie temperature, at which point the material behaves like a paramagnetic.

A *ferrimagnetic* material, for example iron ferrite, is less magnetic than a ferromagnetic one. Here, the dipole moments of adjacent atoms are aligned opposite, as in an antiferromagnetic material, but the moments are not equal, leaving a small magnetic moment. Some ferrimagnetic materials, known as *ferrites*, have low electrical conductivities and are useful in a variety of high-frequency applications. As with antiferromagnetism, ferrimagnetism disappears above the Curie temperature.

A *superparamagnetic* material comprises simply ferromagnetic particles suspended in a dielectric, as in magnetic audio and video tapes. The particles are far enough apart that their exchange forces do not interact.

[3.14]Hysteresis means "to lag" in Greek.

▶3.8 BOUNDARY CONDITIONS

In Section 2.12 we found how **D** and **E** varied across the boundary separating a pair of dielectric materials. We broke the field into components normal and tangential to the surface, then applied Gauss's law and Kirchhoff's voltage law to derive the boundary conditions. Now we will use a similar approach to find how **B** and **H** vary across a boundary separating a pair of magnetic materials. These *magnetostatic boundary conditions* are derived using Ampère's circuital law

$$\oint \mathbf{H} \cdot d\mathbf{L} = I_{\text{enc}}$$

and Gauss's law for magnetostatic fields

$$\oint \mathbf{B} \cdot d\mathbf{S} = 0$$

Beginning with Ampère's circuital law, we consider a pair of magnetic media separated by a sheet current density **K** as depicted in Figure 3.38. We choose a rectangular Amperian path of width Δw and height Δh, centered at the interface such that half of the path lies in each medium. The path is chosen such that the surface current is normal to the surface enclosed by the path. Then, the current enclosed by the path is

$$I_{\text{enc}} = \int K dW = K\Delta w \tag{3.69}$$

Notice that in this problem we have chosen the sheet current heading into the page ($\otimes$), and we use the right-hand rule to determine the direction of integration around the loop. Pointing the right thumb in the direction of the current, we see that the right-hand fingers curl in a clockwise direction around the closed path. We can break up the circulation of **H** into four integrals:

$$\oint \mathbf{H} \cdot d\mathbf{L} = \int_a^b + \int_b^c + \int_c^d + \int_d^a (\mathbf{H} \cdot d\mathbf{L}) = K\Delta w \tag{3.70}$$

In the first integral, we move from point a to b. We can install our own mini-coordinate system here and say that this corresponds to moving from 0 to Δw. This gives us

$$\int_d^b \mathbf{H} \cdot d\mathbf{L} = \int_0^{\Delta w} H_{T_1} \mathbf{a}_T \cdot dL \mathbf{a}_T = H_{T_1} \Delta w \tag{3.71}$$

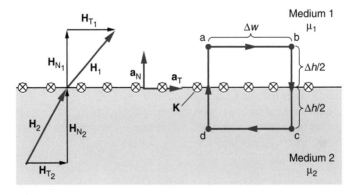

Figure 3.38 Boundary between a pair of magnetic media, and placement of a rectangular path for performing the circulation of **H**.

Moving from b to c involves going from $\Delta h/2$ to 0 in medium 1, and from 0 to $-\Delta h/2$ in medium 2. The second integral is therefore

$$\int_b^c \mathbf{H} \cdot d\mathbf{L} = \int_{\Delta h/2}^0 H_{N_1} \mathbf{a}_N \cdot dL\mathbf{a}_N + \int_0^{-\Delta h/2} H_{N_2} \mathbf{a}_N \cdot dL\mathbf{a}_N = -\left(H_{N_1} + H_{N_2}\right)\frac{\Delta h}{2} \qquad (3.72)$$

Moving from c to d in medium 2 we are going from Δw to 0, so the third integral is

$$\int_c^d \mathbf{H} \cdot d\mathbf{L} = \int_{\Delta w}^0 H_{T_2} \mathbf{a}_T \cdot dL\mathbf{a}_T = -H_{T_2}\Delta w \qquad (3.73)$$

Finally, we complete the path by moving from d to a and the fourth integral is

$$\int_d^a \mathbf{H} \cdot d\mathbf{L} = \int_{-\Delta h/2}^0 H_{N_2} \mathbf{a}_N \cdot dL\mathbf{a}_N + \int_0^{\Delta h/2} H_{N_1} \mathbf{a}_N \cdot dL\mathbf{a}_N = \left(H_{N_1} + H_{N_2}\right)\frac{\Delta h}{2} \qquad (3.74)$$

We see that the second and fourth components cancel. If we are concerned that $\mathbf{H}$ may vary across the small Δw width of the rectangle (i.e., that H_{N_1} at the left side of the box is unequal to H_{N_1} at the right side), then we can simply shrink Δh to zero. Combining our results, we have

$$H_{T_1} - H_{T_2} = K \qquad (3.75)$$

Equation (3.75) requires that the surface of the chosen Amperian path be normal to the surface current, with the direction of circulation about the path determined by the right-hand rule. A more general expression for the first magnetostatic boundary condition can be written as

$$\boxed{\mathbf{a}_{21} \times (\mathbf{H}_1 - \mathbf{H}_2) = \mathbf{K}} \qquad (3.76)$$

where $\mathbf{a}_{21}$ is a unit vector normal going from medium 2 to medium 1.

To find the second boundary condition, we center a Gaussian pillbox across the interface as shown in Figure 3.39. We can shrink Δh such that the flux out of the side of the pillbox is negligible. Then we have

$$\oint \mathbf{B} \cdot d\mathbf{S} = \int B_{N_1} \mathbf{a}_N \cdot dS\mathbf{a}_N + \int B_{N_2} \mathbf{a}_N \cdot dS(-\mathbf{a}_N) = \left(B_{N_1} - B_{N_2}\right)\Delta S = 0 \qquad (3.77)$$

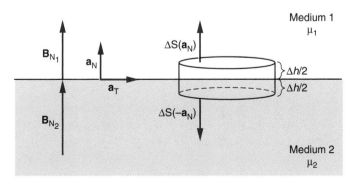

Figure 3.39 Gaussian pillbox placed across the boundary.

TABLE 3.3   Electrostatic and Magnetostatic Boundary Conditions

Electrostatic	Magnetostatic
$\mathbf{E}_{T_1} = \mathbf{E}_{T_2}$	$\mathbf{a}_{21} \times (\mathbf{H}_1 - \mathbf{H}_2) = \mathbf{K}$
$\mathbf{a}_{21} \cdot (\mathbf{D}_1 - \mathbf{D}_2) = \rho_s$	$\mathbf{B}_{N_1} = \mathbf{B}_{N_2}$

Now, since ΔS can be chosen unequal to zero, it follows that

$$\boxed{\mathbf{B}_{N_1} = \mathbf{B}_{N_2}} \tag{3.78}$$

Table 3.3 summarizes the boundary conditions for both electrostatic and magnetostatic fields.

You may recall that, for a conductor–dielectric interface, $E_T = 0$ and $D_N = \rho_s$. A similar situation does not in general exist for magnetostatic fields. The exception is if one of the media is a superconductor. By the *Meissner effect*, the magnetic field rapidly attenuates away from the surface such that $\mathbf{B} = 0$ within a superconductor. If we let medium 2 be the superconductor, the equations at the interface for magnetostatic fields become

$$\mathbf{a}_N \times \mathbf{H}_1 = \mathbf{K} \tag{3.79}$$

and

$$\mathbf{B}_N = 0 \tag{3.80}$$

The second condition is logical since we know magnetic field lines must form closed loops and cannot suddenly terminate, even on a superconductor.

The solutions of many magnetostatic boundary condition problems follow the same kind of bookkeeping approach that we used for electrostatic boundary-condition problems. Example 3.11 and Figure 3.40 detail such a procedure.

▶ **EXAMPLE 3.11**

The magnetic field intensity is given as $\mathbf{H}_1 = 6\mathbf{a}_x + 2\mathbf{a}_y + 3\mathbf{a}_z$ A/m in a medium with $\mu_{r_1} = 6000$ that exists for $z < 0$. We want to find $\mathbf{H}_2$ in a medium with $\mu_{r_2} = 3000$ for $z > 0$.

Following the procedure shown in Figure 3.40, the first step is to break $\mathbf{H}_1$ into its normal component (step 1) and its tangential component (step 2). With no current at the interface, the tangential component is the same on both sides of the boundary (step 3). Next, we find $\mathbf{B}_{N_1}$ by multiplying $\mathbf{H}_{N_1}$ by the permeability in medium 1 (step 4). This normal component $\mathbf{B}$ is the same on both sides of the boundary (step 5), and then we can find $\mathbf{H}_{N_2}$ by dividing $\mathbf{B}_{N_2}$ by the permeability of medium 2 (step 6). The last step is to sum the fields (step 7).

Drill 3.13   A block of iron (99.8% pure) exists for $z < 0$. For $z > 0$, we have air and a magnetic flux density $\mathbf{B}_{air} = 1\mathbf{a}_x + 5\mathbf{a}_y + 12\mathbf{a}_z$ T. Assuming there is no sheet current at the interface, find $\mathbf{B}_{iron}$. (*Answer:* $\mathbf{B}_{iron} = 5000\mathbf{a}_x + 25{,}000\mathbf{a}_y + 12\mathbf{a}_z$ T)

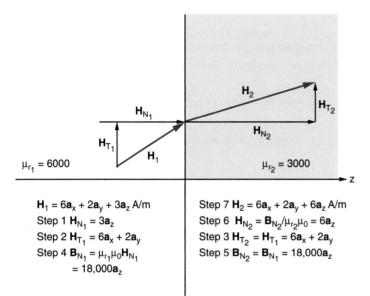

$H_1 = 6a_x + 2a_y + 3a_z$ A/m
Step 1 $H_{N_1} = 3a_z$
Step 2 $H_{T_1} = 6a_x + 2a_y$
Step 4 $B_{N_1} = \mu_{r_1}\mu_0 H_{N_1}$
          $= 18{,}000a_z$

Step 7 $H_2 = 6a_x + 2a_y + 6a_z$ A/m
Step 6 $H_{N_2} = B_{N_2}/\mu_{r_2}\mu_0 = 6a_z$
Step 3 $H_{T_2} = H_{T_1} = 6a_x + 2a_y$
Step 5 $B_{N_2} = B_{N_1} = 18{,}000a_z$

Figure 3.40 Procedure for evaluating the fields on both sides of a boundary separating a pair of magnetic materials.

▶3.9 INDUCTANCE AND MAGNETIC ENERGY

Consider a loop of current I in Figure 3.41a. The flux ϕ_1 that passes through the area S_1 bounded by the loop is

$$\phi_1 = \int_{S_1} \mathbf{B} \cdot d\mathbf{S}$$

Suppose we pass the same current I through two loops, wrapped very close together, as indicated in Figure 3.41b. Each loop generates ϕ_1 of flux, and since they are so closely spaced, the total flux through each loop is $\phi_{tot} = 2\phi_1$. How much flux passes through the total area bounded by the loops, $2S_1$? Because ϕ_{tot} passes through the surface of each loop, the answer is $2\phi_{tot}$, or $4\phi_1$. We say that the two loops of current are *linked* by the total flux ϕ_{tot}.

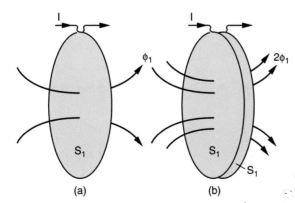

(a)   (b)

Figure 3.41 (a) A single loop of current. (b) Two loops of current.

We define the *flux linkage* λ as the total flux passing through the surface bounded by the contour of the circuit carrying the current. For Figure 3.41a, λ is simply ϕ_1, and for Figure 3.41b, λ is $4\phi_1$. For a tightly wrapped solenoid, the flux linkage is the number of loops multiplied by the total flux linking them. If we have a tightly wrapped solenoid with N turns,

$$\lambda = N\phi_{\text{tot}} = N^2\phi_1 \tag{3.81}$$

where again ϕ_1 is the flux generated by a single loop.

Now we define *inductance L* as the ratio of the flux linkage to the current I generating the flux,

$$\boxed{L = \frac{\lambda}{I} = \frac{N\phi_{\text{tot}}}{I}} \tag{3.82}$$

This has the units of henrys (H), equal to a weber per ampere. Inductors are devices used to store energy in the magnetic field, analogous to the storage of energy in the electric field by capacitors. Inductors most generally consist of loops of wire, often wrapped around a ferrite or ferromagnetic core, and their value of inductance is a function only of the physical configuration of the conductor along with the permeability of the material through which the flux passes.

A procedure for finding the inductance is as follows:

1. Assume a current I in the conductor.
2. Determine **B** using the law of Biot–Savart, or Ampère's circuital law if there is sufficient symmetry.
3. Calculate the total flux ϕ_{tot} linking all the loops.
4. Multiply the total flux by the number of loops to get the flux linkage: $\lambda = N\phi_{\text{tot}}$.
5. Divide λ by I to get the inductance: $L = \lambda/I$. The assumed current will divide out.

▶ **EXAMPLE 3.12**

Let's calculate the inductance for a solenoid with N turns wrapped around a μ_r core as shown in Figure 3.42.

Our first step is to assume a current I going into one end of the conductor. In an earlier section, we found

$$\mathbf{H} = \frac{NI}{h}\mathbf{a}_z$$

inside a solenoid. Technically, near the ends there is a drop off in the field, but here we'll assume that $h \gg a$ so we can neglect any such end effects. Then, within the μ_r core we have

$$\mathbf{B} = \frac{\mu NI}{h}\mathbf{a}_z$$

where $\mu = \mu_r\mu_o$. The cross-sectional area of one loop in the solenoid is πa^2, so the total flux through a loop is

$$\phi_{\text{tot}} = \int \mathbf{B} \cdot d\mathbf{S} = \frac{\mu NI\pi a^2}{h} \tag{3.83}$$

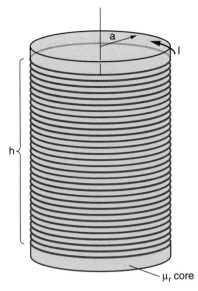

μ$_r$ core   **Figure 3.42**   *N*-turn solenoid.

This flux is linked to the current *N* times, so the flux linkage is

$$\lambda = N\phi_{tot} = \frac{\mu N^2 I \pi a^2}{h} \tag{3.84}$$

Finally, we divide out the current to find the inductance,

$$\boxed{L = \frac{\lambda}{I} = \frac{\mu N^2 \pi a^2}{h}} \tag{3.85}$$

Drill 3.14   Using fine magnet wire (copper wire with a thin insulative enamel coating), you manage to evenly wrap 200 turns along the 10-cm length of a 1-cm-diameter wooden dowel ($\mu \approx \mu_0$). (a) Determine the inductance. (b) Replace the wooden dowel with a 99.8% pure iron core of identical dimensions and recalculate *L*. (*Answer:* (a) 40 μH, (b) 200 mH)

▶ **EXAMPLE 3.13**

Consider a coaxial cable (coax) consisting of a pair of cylindrical metallic shells of inner radius *a* and outer radius *b* as indicated in Figure 3.43. Here we wish to determine the inductance per unit length of the coax.

We can start by assuming a current going in the +**a**$_z$ direction on the inner conductor and returning on the outer conductor. We can easily find the field between the conductors using Ampère's circuital law (see Section 3.3). We can use this field to find the flux through an area of height *h* and width from *a* to *b* as shown in the figure. We have

$$\phi = \int \mathbf{B} \cdot d\mathbf{S} = \iint \left(\frac{\mu I}{2\pi\rho} \mathbf{a}_\phi \right) \cdot d\rho \, dz \, \mathbf{a}_\phi = \frac{\mu I}{2\pi} \int_a^b \frac{d\rho}{\rho} \int_0^h dz = \frac{\mu I h}{2\pi} \ln\left(\frac{b}{a}\right)$$

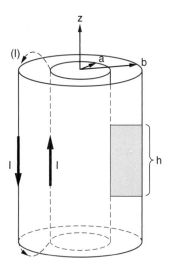

Figure 3.43 Diagram for coaxial cable.

To find the flux linkage, we need to know the number of loops being linked by the flux. Notice in the figure that the inner and outer conductors are shown connected at the ends. These connections are considered to be a long distance from where we are calculating the inductance. It is easy to see that there is only one loop of current. The inductance per unit length is simply

$$\boxed{\frac{L}{h} = \frac{\mu}{2\pi} \ln\left(\frac{b}{a}\right)} \tag{3.86}$$

Mutual Inductance

So far, what we have discussed has been a *self-inductance,*[3.15] where the flux is linked to the circuit containing the current that produced the flux. We could, however, also determine the flux linked to a different circuit than the flux-generating one. In this case we are talking about *mutual inductance*, which is fundamental to the design and operation of transformers.

Consider the pair of coils shown in Figure 3.44. We'll let circuit 1, with N_1 loops, be our driving coil and circuit 2, with N_2 loops, be our receiving coil. When current I_1 is pushed through circuit 1, it produces flux, some of which links the N_2 loops of circuit 2. This flux is common, or *mutual*, to both circuits. We'll call this flux ϕ_{12}, where the subscripts indicate that this is the flux from $\mathbf{B}_1$ of circuit 1 that links circuit 2. We find ϕ_{12} by integrating the dot product of $\mathbf{B}_1$ and the area of a loop in circuit 2:

$$\phi_{12} = \int \mathbf{B}_1 \cdot d\mathbf{S}_2 \tag{3.87}$$

The flux linkage λ_{12} is then the number of times ϕ_{12} links with circuit 2, or

$$\lambda_{12} = N_2 \phi_{12} \tag{3.88}$$

[3.15]When we use the term inductance by itself, we will be talking about self-inductance.

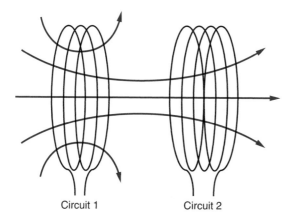

Figure 3.44   Pair of coils used to illustrate mutual inductance.

Circuit 1   Circuit 2

Finally, the *mutual inductance* M_{12} is

$$M_{12} = \frac{\lambda_{12}}{I_1} = \frac{N_2}{I_1} \int \mathbf{B}_1 \cdot d\mathbf{S}_2$$ (3.89)

▶ **EXAMPLE 3.14**

Consider the solenoid from Example 3.12 as having N_1 turns, and add a secondary winding of N_2 turns sharing the same core as shown in Figure 3.45. Our procedure to determine the mutual inductance of this structure is very similar to the approach we took to find self-inductance:

1. Assume a current I_1 in the conductor for the driving circuit.[3.16]
2. Determine the total $\mathbf{B}_1$ from circuit 1.
3. Find the flux ϕ_{12} through one of the loops of circuit 2.
4. Multiply this flux by the number of loops N_2 to get the flux linkage: $\lambda_{12} = N_2\,\phi_{12}$.
5. Divide λ_{12} by I_1 to get the mutual inductance: $M_{12} = \lambda_{12}/I_1$.
6. The assumed driving current will again divide out.

We begin by assuming I_1 in circuit 1 and use this to calculate $\mathbf{B}_1$. We found this for the example of Figure 3.42, and appending circuit 1 subscripts we have

$$\mathbf{B}_1 = \frac{\mu N_1 I_1}{h}\mathbf{a}_z$$

The flux through one of the loops of circuit 2 is

$$\phi_{12} = \frac{\mu N_1 I_1 \pi a^2}{h}$$

This flux links all N_2 loops of circuit 2, so we have

$$\lambda_{12} = N_2\phi_{12} = \frac{\mu N_1 N_2 I_1 \pi a^2}{h}$$

[3.16]Note that we will arrive at the same value for mutual inductance no matter which circuit is the designated driver, but in some problems it will be far easier to find a solution if we make an intelligent selection for the driver.

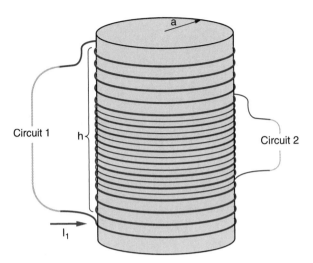

Figure 3.45 Solenoid with a secondary winding.

Dividing out the driving current I_1, we obtain a mutual inductance

$$M_{12} = \frac{\mu N_1 N_2 \pi a^2}{h}$$

Now, since it makes no difference which circuit we designate as the driver for our calculation,

$$M_{12} = M_{21} \tag{3.90}$$

Drill 3.15 For the solenoid described in Drill 3.14, add a secondary winding of 100 turns and calculate the mutual inductance with (a) the wooden dowel and (b) the iron rod. (*Answer:* (a) 20 µH, (b) 99 mH)

Magnetic Energy

In Chapter 3 we found that it takes work to assemble a collection of charges and that the work went into energy stored in the electric field. The field and capacitance quantities were related by the equation

$$W_E = \frac{1}{2}CV^2 = \frac{1}{2}\int \mathbf{D} \cdot \mathbf{E} \, dv$$

Likewise, for inductors it takes work to place a current in the coil and this work is stored as energy in the magnetic field.

From circuit theory, we know that voltage v and current i are related by the inductance as

$$v(t) = L\frac{di(t)}{dt} \tag{3.91}$$

This equation is a consequence of *Faraday's law*, to be discussed in Chapter 4. We also know from circuit theory that the power in an electric component can be found by integrating the product of voltage and current over time,

$$W = \int v(t)i(t)dt \tag{3.92}$$

Rearranging (3.91) as

$$dt = \frac{L}{v(t)} di(t) \tag{3.93}$$

we can rewrite (3.92) as

$$W_M = \int_{i=0}^{I} Li(t)di(t) = \frac{1}{2}LI^2 \tag{3.94}$$

Here we have written W with the subscript M to acknowledge that we're looking for the magnetic energy. This $(1/2)LI^2$ is very similar to the electrostatic energy expression $(1/2)CV^2$, and in fact it can also be shown[3.17] that

$$W_M = \frac{1}{2}\int \mathbf{B} \cdot \mathbf{H}\, dv \tag{3.95}$$

▷ **EXAMPLE 3.15**

Although we will not prove (3.95), we can at least verify the equivalence of (3.94) and (3.95) for the solenoid.

First recall from earlier in this section that the field within the solenoid is

$$\mathbf{B} = \frac{\mu NI}{h}\mathbf{a}_z = B\mathbf{a}_z \tag{3.96}$$

We can solve this for the current,

$$I = \frac{Bh}{\mu N} \tag{3.97}$$

and with our earlier solution for the solenoid inductance

$$L = \frac{\mu N^2 \pi a^2}{h} \tag{3.98}$$

we can rewrite (3.94) as

$$W_M = \frac{1}{2}\left(\frac{\mu N^2 \pi a^2}{h}\right)\left(\frac{Bh}{\mu N}\right)^2 = \frac{1}{2}BH\left(\pi a^2 h\right) \tag{3.99}$$

For (3.95) it is easy to see that

$$W_M = \frac{1}{2}\int \mathbf{B} \cdot \mathbf{H}\, dv = \frac{1}{2}BH\int dv = \frac{1}{2}BH\left(\pi a^2 h\right) \tag{3.100}$$

which is the same result as (3.99).

[3.17]See the discussion of the Poynting theorem in Chapter 5.

Combining (3.94) and (3.95) we have

$$\boxed{W_M = \frac{1}{2}LI^2 = \frac{1}{2}\int \mathbf{B}\cdot\mathbf{H}\,dv} \tag{3.101}$$

This equation gives us a powerful alternative approach to solving for the inductance of a circuit element. If we assume the current in an inductor, and can calculate the field values and integrate over a finite volume to solve for W_M, we can calculate inductance by

$$L = \frac{2W_M}{I^2} \tag{3.102}$$

▷ EXAMPLE 3.16

Let's use the energy approach to find the inductance internal to a length of solid wire with current distributed evenly over the cross section (see Figure 3.46).

From Ampère's circuital law, we can find

$$H_\phi = \frac{I\rho}{2\pi a^2} \tag{3.103}$$

Solving for W_M, we get

$$W_M = \frac{1}{2}\int \mu H_\phi^2\,dv = \frac{1}{2}\frac{\mu I^2}{4\pi^2 a^4}\int_0^a \rho^3 d\rho \int_0^{2\pi} d\phi \int_0^h dz = \frac{\mu I^2 h}{16\pi} \tag{3.104}$$

Inserting this value for W_M into (3.102) we have

$$\frac{L}{h} = \frac{\mu}{8\pi} \tag{3.105}$$

Figure 3.46 A length of solid wire with uniform current I.

The result of Example 3.16 is sometimes referred to as the *internal inductance* per unit length for a wire. This would be extremely difficult to calculate using steps outlined by the flux linkage approach[3.18] and shows the utility of the energy approach. However, consider finding the inductance for a pair of straight parallel wires (see problem 3.55); here, the energy approach would be impractical because of the lack of a finite volume over which to integrate.

Drill 3.16   Use the energy approach to find the inductance per unit length of the coaxial cable of Figure 3.43.

$$\left(Answer:\ \frac{L}{h} = \frac{\mu}{2\pi} \ln\frac{b}{a}\right)$$

Drill 3.17   Consider a coaxial cable with solid inner conductor of radius a and a conductive outer shell at radius b, filled with nonmagnetic material ($\mu_r = 1$). Find the total inductance per unit length.

$$\left(Answer:\ \frac{L}{h} = \frac{\mu_o}{2\pi}\left(\frac{1}{4} + \ln\frac{b}{a}\right)\right)$$

▶ 3.10 MAGNETIC CIRCUITS

Consider the toroid in Figure 3.47a. A ferrite core has been wrapped with N turns of wire, and passing current through this wire results in a flux ϕ through the core. Now consider Figure 3.47b, where instead of being evenly distributed about the toroid, all of the loops are bunched on one side. The same flux is generated in this second case, and in both cases the flux is confined mainly to the ferrite core.

Why does the flux stay in the core? Magnetic field mapping can confirm this behavior, but this topic lies beyond the scope of this chapter. As a qualitative argument, we can consider a magnetic circuit, analogous to an electric circuit, where flux replaces current and permeability replaces conductivity. As current prefers to flow on paths of high conductivity, so the loops of magnetic flux prefer to occupy paths of highest permeability. The flux leakage out of the core is minimal if $\mu_r \gg 1$.

Many magnetic devices such as transformers can be analyzed just like electric circuits. Table 3.4 lists variables used in magnetic circuit analysis that are analogous to those used in electric circuits. Considering the toroid shown in Figure 3.48a, we see that the *magnetomotive force* V_m, analogous to electromotive force V in electric circuits, is the total current enclosed by an Amperian path along the center of the toroid, or

$$\oint \mathbf{H} \cdot d\mathbf{L} = NI = V_m \tag{3.106}$$

The *reluctance* of a magnetic circuit element to hold flux is analogous to resistance for electric circuit elements. In our toroid problem, the reluctance is

$$\mathcal{R} = \frac{\ell}{\mu A} \approx \frac{2\pi\rho_o}{\mu A} \tag{3.107}$$

[3.18]This approach involves a nebulous concept called *differential flux linkage*.

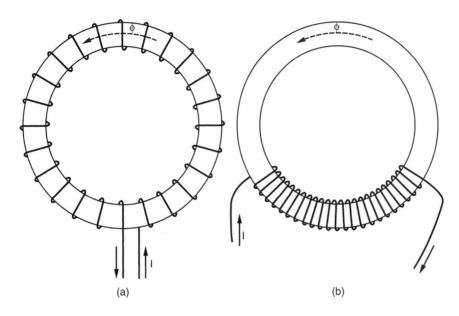

Figure 3.47 For a set N number of loops around a ferrite core, the flux generated is the same even when the loops are bunched together.

(a)   (b)

TABLE 3.4   Analogy between Magnetic and Electric Circuits

Electric circuits		Magnetic circuits	
Electromotive force (volts)	V	V_m	Magnetomotive force (amp-turns)
Current (amps)	I	ϕ	Magnetic flux (webers)
Resistance (ohms)	R	$\mathcal{R}$	Reluctance (amp-turns/webers)
Ohm's law	$V = IR$	$V_m = \phi\mathcal{R}$	Ohm's law for magnetic circuits
Conductivity (siemens/meter)	σ	μ	Permeability (henrys/meter)

where the length of the toroid is estimated using its mean radius ρ_o. The flux for Figure 3.48a can be calculated using the magnetic circuit approach as indicated in Figure 3.48b. By Ohm's law for magnetic circuits we have

$$\phi = \frac{V_m}{\mathcal{R}} \approx \frac{NI\mu A}{2\pi\rho_o} \tag{3.108}$$

As a check on this result, let's calculate the flux by integrating the magnetic flux density over the surface,

$$\phi = \int \mathbf{B} \cdot d\mathbf{S} \tag{3.109}$$

which for our toroidal circuit is simply BA. Since B inside the toroid is given by

$$\mathbf{B} = \frac{\mu NI}{2\pi\rho_o} \tag{3.110}$$

then for this approach we also have

$$\phi = BA = \frac{NI\mu A}{2\pi\rho_o} \tag{3.111}$$

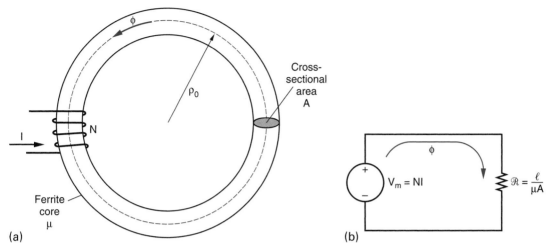

Figure 3.48   A simple toroid wrapped with N turns modeled by a magnetic circuit.

A few caveats need to be mentioned regarding magnetic circuit analysis. First, in electric circuits the conductivity is assumed to be linear with field strength. This is definitely not the case for the analogous quantity permeability in magnetic materials. Using magnetic circuit analysis accompanied by a hysteresis curve can help get around this problem, though it may require an iterative approach for solution. Second, in electric circuits current is restricted to flow in conductors rather than air since the conductivity of air is essentially zero. But the permeability of air is not zero: It is μ_o. So even though ferrite materials can have relative permeabilities in the thousands, some flux will leak from the magnetic circuit and travel in air. In our magnetic circuit calculations, we have made the simplifying assumption of zero flux leakage. And third, because flux can travel in air, we can have air gaps in our ferrite cores that must be treated as magnetic circuit elements.

▷ **EXAMPLE 3.17**

Consider Figure 3.49a, the simple toroidal core from Figure 3.48a, only this time with a small air gap of length ℓ_g. We can analyze this structure by using the magnetic circuit shown in Figure 3.49b.
The reluctance of the core is

$$\mathcal{R}_c \approx \frac{2\pi\rho_o}{\mu A}$$

where we are assuming $\ell_g \ll 2\pi\rho_o$. The reluctance of the short air gap is simply

$$\mathcal{R}_g \approx \frac{\ell_g}{\mu A}$$

where we have assumed the gap is small enough that fringing of the field is negligible.[3.19] The flux looping around the toroid, including the air gap, can be calculated as

[3.19]When inserting the air gap, pretend that there is a "ghost core" with the area A filled with air. This will be the component for which the reluctance of the air gap is calculated.

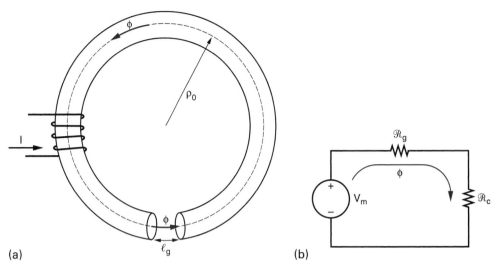

Figure 3.49 Toroid with a small air gap analyzed with a magnetic circuit.

$$\phi = \frac{V_m}{\mathcal{R}_c + \mathcal{R}_g}$$

and then the magnetic flux density within the toroid and air gap can be calculated since

$$B = \frac{\phi}{A}$$

The magnetic field intensity depends on permeability. In the core, it is

$$H_c = \frac{B}{\mu}$$

and in the air gap it is

$$H_g = \frac{B}{\mu_o}$$

The field intensity in the air gap is thus seen to be a factor μ/μ_o greater than the field in the core.

Drill 3.18   In Figure 3.50, half of the 2.0-cm-diameter core consists of magnetic material with $\mu_{r_1} = 3000$, and the other half of material with $\mu_{r_2} = 6000$. The toroid has a mean radius $\rho_o = 50$ cm. For 10.0 A of current driven through 20 loops of wire, find the magnetic field intensity in each material of the toroid. (*Answer: $H_1 = 85$ A/m, $H_2 = 42$ A/m*)

Electromagnets

Consider the electromagnet of Figure 3.51a. We want to determine the magnetic force that is holding an iron bar in place. Our approach will be to displace the bar from the magnet by

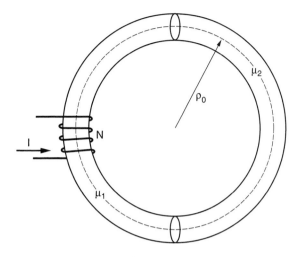

Figure 3.50 Toroid consisting of two different magnetic materials for Drill 3.18.

a differential length $d\ell$ (as shown in Figure 3.51b), and see how the system energy changes. This approach is known as the *principle of virtual work*.

Before we begin, let's recall how energy is related to force. The differential work dW is

$$dW = \mathbf{F} \cdot d\ell \qquad (3.112)$$

or

$$dW = F d\ell \qquad (3.113)$$

if the force and $d\ell$ are in the same direction. The work done by the force in moving the object a distance $d\ell$ is stored as energy. For instance, it takes a force against gravity to raise a bowling ball from the ground to shoulder height. The work done in raising the ball is stored as potential energy.

Now consider that the bar from Figure 3.51a is displaced a distance $d\ell$ from the magnet, as shown in Figure 3.51b. There is magnetic energy in the field between the bar and the magnet, which we know is

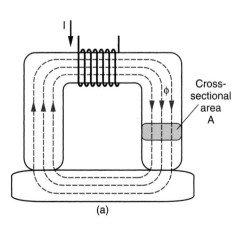

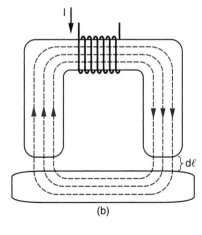

Figure 3.51 (a) An iron bar attached to an electromagnet. (b) The bar displaced by a differential length $d\ell$.

$$W_M = \frac{1}{2}\int \mathbf{B} \cdot \mathbf{H}\, dv = \frac{1}{2}\mu_o \int H^2 dv \qquad (3.114)$$

The differential volume dv is simply $Ad\ell$ for each end of the bar, or $2Ad\ell$ total. Then we have

$$dW_M = 2\left(\frac{1}{2}\mu_o H^2 Ad\ell\right) = \mu_o H^2 Ad\ell \qquad (3.115)$$

This work is also equal to $Fd\ell$, where F is the force of attraction between the magnet and the bar. Equating (3.113) and (3.115) leads to the force

$$F = \mu_o H^2 A \qquad (3.116)$$

▶ **EXAMPLE 3.18**

Let's use the principle of virtual work along with what we know about magnetic circuits to determine how many turns of wire carrying current I for the electromagnet shown in Figure 3.52 are needed to hold up the iron bar of mass m.

We can first calculate the field B by neglecting the presence of the gap. The complete circuit has a reluctance

$$\mathcal{R} = \frac{2(w+h)}{\mu A}$$

The magnetomotive force V_m is simply NI, so the flux in the loop will be

$$\phi = \frac{V_m}{\mathcal{R}} = \frac{NI\mu A}{2(w+h)}$$

The magnetic flux density anywhere in the circuit is then calculated as

$$B = \frac{\phi}{A} = \frac{NI\mu}{2(w+h)}$$

Now we must consider the force of attraction assuming a differential air gap $d\ell$ between the magnet and the bar. The magnetic field H is simply B/μ_o, and so from (3.116) we find a total force of attraction of

$$F = \mu_o H^2 A = \mu_o \left(\frac{NI\mu_r}{2(w+h)}\right)^2 A$$

Now we know that this force must counter the force of gravity pulling on the bar, which is given by $F = mg$. Setting these two force equations equal and solving for N, we find

$$N = \frac{2(w+h)}{I\mu_r}\sqrt{\frac{mg}{\mu_o A}}$$

Drill 3.19 Referring to Figure 3.52, find the number of loops required to lift a 1-kg bar if $h = w = 8$ cm, the cross-sectional area is 1 cm^2, the current is 1 A, and the magnetic material of both the bar and the electromagnet can be assumed to have $\mu_r = 3000$. (*Answer: N = 30 turns*)

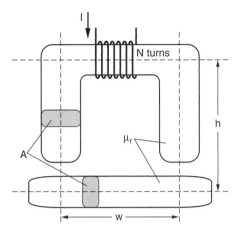

Figure 3.52 Electromagnet supporting a bar of mass m.

Drill 3.20 Suppose $N = 25$ turns in Drill 3.19. Determine the current required to hold up the bar. (*Answer: I* = 1.2 A)

Practical Application: Maglev

Magnetically levitated trains, called Maglevs, are touted as a relatively cheap, fast transportation alternative to conventional rail. They are levitated and propulsed by interaction between electromagnets in the train and current in the guidance-rail cable windings. Both acceleration and braking are handled with electromagnets. There are no moving parts to wear out, and the frictionless support of magnetic levitation offers an extremely smooth, quiet, and fast ride with speeds in excess of 500 km/hr (300 mph). Despite these features, detractors point to the high initial cost of the system and its incompatibility with existing rail.

Research teams in Germany and in Japan have led the development of Maglev trains. China has broken ground on the first Maglev to be put into practical commercial use. Based on the German Transrapid design (Figure 3.53), it is slated for operation in Shanghai in

Figure 3.53 Maglev prototype. Courtesy of Transrapid International.

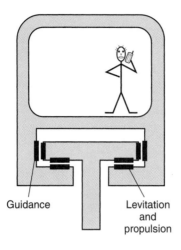

Guidance   Levitation
                and
                propulsion   **Figure 3.54**   The Maglev concept.

2004. Other systems are planned for operation in Pittsburgh, Pennsylvania, in Southern California, and between Baltimore and Washington, D.C.

General operation of a simple Maglev follows the conceptual diagram of Figure 3.54. Interaction between the electromagnets in the train and the current-carrying coils in the guide rail provide levitation. By sending waves along the guide-rail coils, the train magnets are simultaneously pushed and pulled in the direction of travel. The train is guided by magnets on the side of the guide rail. Computer feedback algorithms maintain the separation distance between magnets.

Of special interest are future Maglevs that would utilize superconductors with high transition temperatures. Such superconductors are ceramic composites that can *superconduct* (conduct without resistive losses) above the temperature of liquid nitrogen. Once current gets established in a superconducting loop, it will continue to flow so long as it is kept cold enough. A superconducting Maglev would operate by sending alternating current through superconducting coils on the train. On the guide rails beneath the coils would be a conductive aluminum sheet. As the train moves, the superconducting magnets would induce currents in the conductive sheet. The electromagnetic interaction between these currents and the superconducting electromagnets yields the repulsive force that provides levitation. The only significant power required would be for the cryogenic refrigeration system needed to maintain low temperature.

Even if Maglev is not used for mass transportation, the technology may still find use as an aid in launching rockets. A magnetic *rail-gun* would levitate and accelerate rockets on a sloping track. The rocket could reach speeds of close to 1000 km/hr, before igniting its engines. Rail-gun technology could also be used to launch raw materials from the moon into lunar orbit.

►SUMMARY

- Analysis of magnetic fields is similar in many respects to analysis of electric fields. Analogous values are listed in Table 3.5.

- The *cross product* of a pair of vectors **A** and **B** is

$$\mathbf{A} \times \mathbf{B} = |\mathbf{A}||\mathbf{B}| \sin\theta_{AB}\mathbf{a}_N$$

which is the product of the magnitude of the vectors multiplied by the sine of the angle between the vectors, and $\mathbf{a}_N$ is a unit vector in the normal direction of $\mathbf{A} \times \mathbf{B}$

TABLE 3.5 Analogy between Electric and Magnetic Fields (Static)

Electric fields	Magnetic fields
$\mathbf{E}$(V/m)	$\mathbf{H}$(A/m)
$\mathbf{D}$(C/m^2)	$\mathbf{B}$(Wb/m^2)
ψ(C)	ϕ(Wb)
ε(F/m)	μ(H/m)
$\mathbf{D} = \varepsilon\mathbf{E}$	$\mathbf{B} = \mu\mathbf{H}$
$\nabla\cdot\mathbf{D} = \rho_v$	$\nabla\cdot\mathbf{B} = 0$
$\nabla\times\mathbf{E} = 0$	$\nabla\times\mathbf{H} = \mathbf{J}$
$\psi = \int\mathbf{D}\cdot d\mathbf{S}$	$\phi = \int\mathbf{B}\cdot d\mathbf{S}$
$\mathbf{F}$(N) $= Q\mathbf{E}$	$\mathbf{F}$(N) $= Q\mathbf{u}\times\mathbf{B}$
W_E(J) $= \dfrac{1}{2}\int\mathbf{D}\cdot\mathbf{E}\,dv$	W_M(J) $= \dfrac{1}{2}\int\mathbf{B}\cdot\mathbf{H}\,dv$

taken by the right-hand rule. In Cartesian coordinates, this can also be written

$$\mathbf{A}\times\mathbf{B} = \begin{vmatrix} \mathbf{a}_x & \mathbf{a}_y & \mathbf{a}_z \\ A_x & A_y & A_z \\ B_x & B_y & B_z \end{vmatrix}$$

- For a differential current element $I_1 d\mathbf{L}_1$ at point 1, the magnetic field intensity $\mathbf{H}_2$ at point 2 is given by the law of Biot–Savart,

$$d\mathbf{H}_2 = \frac{I_1 d\mathbf{L}_1 \times \mathbf{a}_{12}}{4\pi R_{12}^2}$$

where $\mathbf{R}_{12} = R_{12}\mathbf{a}_{12}$ is a vector from the source element at point 1 to the location where the field is desired at point 2. By summing the field for all current elements, the total magnetic field intensity can be found as

$$\mathbf{H} = \int\frac{Id\mathbf{L}\times\mathbf{a}_R}{4\pi R^2}$$

- The Biot–Savart law can also be written in terms of surface and volume current densities by replacing $Id\mathbf{L}$ with $\mathbf{K}dS$ and $\mathbf{J}dv$:

$$\mathbf{H} = \int\frac{\mathbf{K}dS\times\mathbf{a}_R}{4\pi R^2}$$

and

$$\mathbf{H} = \int\frac{\mathbf{J}dv\times\mathbf{a}_R}{4\pi R^2}$$

- The magnetic field intensity resulting from an infinite length line of current is

$$\mathbf{H} = \frac{I\mathbf{a}_\phi}{2\pi\rho}$$

From a solenoid with N turns and height h it is

$$\mathbf{H} = \frac{NI}{h}\mathbf{a}_z$$

and from a current sheet of infinite extent it is

$$\mathbf{H} = \frac{1}{2}\mathbf{K}\times\mathbf{a}_N$$

where $\mathbf{a}_N$ is a unit vector normal from the current sheet to the test point.

- An easy way to solve for the magnetic field intensity in problems with sufficient current distribution symmetry is to use Ampère's circuital law, which says that the circulation of $\mathbf{H}$ is equal to the net current enclosed by the circulation path,

$$\oint\mathbf{H}\cdot d\mathbf{L} = I_{\text{enc}}$$

- The point, or differential, form of Ampère's circuital law is

$$\nabla\times\mathbf{H} = \mathbf{J}$$

where the curl operation in Cartesian coordinates is given by

$$\nabla\times\mathbf{H} = \begin{vmatrix} \mathbf{a}_x & \mathbf{a}_y & \mathbf{a}_z \\ \partial/\partial x & \partial/\partial y & \partial/\partial z \\ H_x & H_y & H_z \end{vmatrix}$$

- A closed line integral is related to a surface integral by Stokes's theorem:

$$\oint\mathbf{H}\cdot d\mathbf{L} = \int(\nabla\times\mathbf{H})\cdot d\mathbf{S}$$

- Magnetic flux density $\mathbf{B}$, in Wb/m^2 or T, is related to the magnetic field intensity by

$$\mathbf{B} = \mu\mathbf{H}$$

where the material permeability μ can also be written as

$$\mu = \mu_r\mu_o$$

and the free space permeability is

$$\mu_o = 4\pi\times 10^{-7}\ \text{H/m}$$

- The amount of magnetic flux ϕ, in webers, through a surface is

$$\phi = \int \mathbf{B} \cdot d\mathbf{S}$$

Since magnetic flux forms closed loops, we have Gauss's law for static magnetic fields,

$$\oint \mathbf{B} \cdot d\mathbf{S} = 0$$

- Maxwell's equations for static fields are

integral form	*differential form*
$\oint \mathbf{D} \cdot d\mathbf{S} = Q_{enc}$	$\nabla \cdot \mathbf{D} = \rho_v$
$\oint \mathbf{B} \cdot d\mathbf{S} = 0$	$\nabla \cdot \mathbf{B} = 0$
$\oint \mathbf{E} \cdot d\mathbf{L} = 0$	$\nabla \times \mathbf{E} = 0$
$\oint \mathbf{H} \cdot d\mathbf{L} = I_{enc}$	$\nabla \times \mathbf{H} = \mathbf{J}$

- The total force vector $\mathbf{F}$ acting on a charge q moving through magnetic and electric fields with velocity $\mathbf{u}$ is given by the Lorentz force equation,

$$\mathbf{F} = q(\mathbf{E} + \mathbf{u} \times \mathbf{B})$$

From this equation we can also find that the force $\mathbf{F}_{12}$ from a magnetic field $\mathbf{B}_1$ on a current-carrying line I_2 is

$$\mathbf{F}_{12} = \int I_2 d\mathbf{L}_2 \times \mathbf{B}_1$$

- The magnetic dipole moment $\mathbf{m}$ for N loops of current is

$$\mathbf{m} = NIS\mathbf{a}_N$$

where $\mathbf{a}_N$ is a unit vector normal to the planar loops each of area S with direction found using the right-hand rule. The torque $\boldsymbol{\tau}$ on these loops of current in a magnetic field $\mathbf{B}$ is given by

$$\boldsymbol{\tau} = \mathbf{m} \times \mathbf{B}$$

- The three basic types of magnetic materials are diamagnetic, paramagnetic, and ferromagnetic. Whereas the first two of these have relative permeabilities close to unity, ferromagnetics can have very large, nonlinear permeabilities. Ferromagnetics, like iron, are routinely used as the core material in electromagnets.

- The magnetic fields at the boundary between different materials are given by

$$B_{N_1} = B_{N_2}$$

and

$$\mathbf{a}_{21} \times (\mathbf{H}_1 - \mathbf{H}_2) = \mathbf{K}$$

where $\mathbf{a}_{21}$ is a unit vector normal going from medium 2 to medium 1.

- Inductance L is a measure of an inductor's ability to store magnetic energy and is the ratio of the flux linkage λ to the current I generating the flux. For N loops of current,

$$L = \frac{\lambda}{I} = \frac{N\phi_{tot}}{I}$$

This has the units of henrys (H), equal to a weber per ampere. For N turns or loops of current wrapped around a core of radius a, permeability μ, and height h,

$$L = \frac{\mu N^2 \pi a^2}{h}$$

- For a coaxial cable with inner conductor of radius a separated by an outer conductor of radius b by material with permeability μ, the inductance L per length h is

$$\frac{L}{h} = \frac{\mu}{2\pi} \ln \frac{b}{a}$$

- The mutual inductance M_{12} between a driving coil with current I_1 driven through N_1 loops and a receiving coil with N_2 loops is given by

$$M_{12} = \frac{\lambda_{12}}{I_1} = \frac{N_2}{I_1} \int \mathbf{B}_1 \cdot d\mathbf{S}_2$$

where flux linkage λ_{12} is the number of times the flux generated by the driving coil links with the receiving coil.

- The energy stored in an inductor's magnetic field is related to its inductance and field values by

$$W_M = \frac{1}{2} L I^2 = \frac{1}{2} \int \mathbf{B} \cdot \mathbf{H} \, dv$$

The *energy approach* makes use of this equation to solve for inductance when it is easy to integrate the fields over a volume.

- Transformer and electromagnet design or analysis is often simplified by using a magnetic circuit approach. Here, the magnetomotive force V_m, the reluctance $\mathcal{R}$, and the magnetic flux ϕ are analogous to voltage, resistance, and current, respectively.

▶ PROBLEMS

3.1 Magnetic Fields and Cross Products

3.1 Find $\mathbf{A} \times \mathbf{B}$ for the following:

(a) $\mathbf{A} = 2\mathbf{a}_x - 3\mathbf{a}_y + 4\mathbf{a}_z$, $\mathbf{B} = 5\mathbf{a}_y - 1\mathbf{a}_z$

(b) $\mathbf{A} = \mathbf{a}_\rho + 2\mathbf{a}_\phi + 4\mathbf{a}_z$, $\mathbf{B} = 2\mathbf{a}_\rho + 6\mathbf{a}_z$

(c) $\mathbf{A} = 2\mathbf{a}_r + 5\mathbf{a}_\theta + 1\mathbf{a}_\phi$, $\mathbf{B} = \mathbf{a}_r + 3\mathbf{a}_\phi$

3.2 If a parallelogram has a short side a, a long side b, and an interior angle θ (the smaller of the two interior angles), the area of the parallelogram is given by

$$area = ab \sin \theta$$

Determine how you would use the cross product of a pair of vectors to find the area of a parallelogram defined by the points O(0, 0, 0), P(6, 0, 0), Q(8, 12, 0), and R(2, 12, 0). (Assume dimensions are in meters.)

3.3 Given the vertices of a triangle P(1, 2, 0), Q(2, 5, 0), and R(0, 4, 7), find (a) the interior angles, (b) a unit vector normal to the surface containing the triangle, and (c) the area of the triangle.

3.2 Biot–Savart's Law

3.4 A segment of conductor on the z-axis extends from $z = 0$ to $z = h$. If this segment conducts current I in the $+\mathbf{a}_z$ direction, find $\mathbf{H}(0, y, 0)$. Compare your answer to that of Example 3.2.

3.5 An infinite length line with 2.0-A current in the $+\mathbf{a}_x$ direction exists at $y = -3.0$ m, $z = 4.0$ m. A second infinite length line with 3.0-A current in the $+\mathbf{a}_z$ direction exists at $x = 0$, $y = 3.0$ m. Find $\mathbf{H}(0, 0, 0)$.

3.6 A conductive loop in the shape of an equilateral triangle of side 8.0 cm is centered in the x–y plane. It carries a 20.0-mA current clockwise when viewed from the $+\mathbf{a}_z$ direction. Find $\mathbf{H}(0, 0, 16$ cm$)$.

3.7 A square conductive loop of side 10.0 cm is centered in the x–y plane. It carries a 10.0-mA current clockwise when viewed from the $+\mathbf{a}_z$ direction. Find $\mathbf{H}(0, 0, 10$ cm$)$.

3.8 A conductive loop in the x–y plane is bounded by $\rho = 2.0$ cm, $\rho = 6.0$ cm, $\phi = 0°$, and $\phi = 90°$. A 1.0-A current flows in the loop, going in the $\mathbf{a}_\phi$ direction on the $\rho = 2.0$ cm arm. Determine $\mathbf{H}$ at the origin.

3.9 How close do you have to be to the middle of a finite length of a current-carrying line before it appears infinite in length? Consider $\mathbf{H}_f(0, a, 0)$ to be the field for the finite line of length $2h$ centered on the z-axis and $\mathbf{H}_i(0, a, 0)$ to be the

field for an infinite length line of current on the z-axis. In both cases consider current I in the $+\mathbf{a}_z$ direction. Plot H_f/H_i versus h/a.

3.10 For the ring of current described in MATLAB 3.2, find $\mathbf{H}$ at the following points: (a) (0, 0, 1 m), (b) (0, 2 m, 0), and (c) (1 m, 1 m, 0).

3.11 A solenoid has 200 turns, is 10.0 cm long, and has a radius of 1.0 cm. Assuming 1.0 A of current, determine the magnetic field intensity at the very center of the solenoid. How does this compare with your solution if you make the assumption that 10 cm >> 1 cm?

3.12 For the solenoid of the previous problem, plot the magnitude of the field versus position along the axis of the solenoid. Include the axis 2 cm beyond each end of the solenoid.

3.13 A 4.0-cm-wide ribbon of current is centered about the y-axis on the x–y plane and has a surface current density $\mathbf{K} = 2\pi \, \mathbf{a}_y$ A/cm. Determine the magnetic field intensity at the points (a) P(0, 0, 2 cm) and (b) Q(2 cm, 2 cm, 2 cm).

3.3 Ampère's Circuital Law

3.14 Two infinite extent current sheets exist at $z = -2.0$ m and at $z = +2.0$ m. The top sheet has a uniform current density $\mathbf{K} = 3.0 \, \mathbf{a}_y$ A/m and the bottom one has $\mathbf{K} = -3.0 \, \mathbf{a}_y$ A/m. Find $\mathbf{H}$ at (a) (0, 0, 4 m), (b) (0, 0, 0), and (c) (0, 0, –4 m).

3.15 An infinite extent current sheet with $\mathbf{K} = 6.0 \, \mathbf{a}_y$ A/m exists at $z = 0$. A conductive loop of radius 1.0 m, in the y–z plane centered at $z = 2.0$ m, has zero magnetic field intensity measured at its center. Determine the magnitude of the current in the loop and show its direction with a sketch.

3.16 Given the field $\mathbf{H} = 3y^2 \, \mathbf{a}_x$, find the current passing through a square in the x–y plane that has one corner at the origin and the opposite corner at (2, 2, 0).

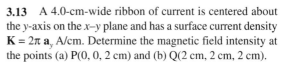

3.17 Given a 3.0-mm-radius solid wire centered on the z-axis with an evenly distributed 2.0 A of current in the $+\mathbf{a}_z$ direction, plot the magnetic field intensity H versus radial distance from the z-axis over the range $0 \le \rho \le 9$ mm.

3.18 Given a 2.0-cm-radius solid wire centered on the z-axis with a current density $\mathbf{J} = 3\rho$ A/cm^2 $\mathbf{a}_z$ (for ρ in centimeters) plot the magnetic field intensity H versus radial distance from the z-axis over the range $0 \le \rho \le 8$ cm.

3.19 An infinitesimally thin metallic cylindrical shell of radius 4.0 cm is centered on the z-axis and carries an evenly distributed current of 10.0 mA in the $+\mathbf{a}_z$ direction. (a) Determine the value of the surface current density on the conductive shell and (b) plot H as a function of radial distance from the z-axis over the range $0 \leq \rho \leq 12$ cm.

▶ EMAG
SOLUTIONS

3.20 A cylindrical pipe with a 1.0-cm wall thickness and an inner radius of 4.0 cm is centered on the z-axis and has an evenly distributed 3.0 A of current in the $+\mathbf{a}_z$ direction. Plot the magnetic field intensity H versus radial distance from the z-axis over the range $0 \leq \rho \leq 10$ cm.

3.21 An infinite length line carries current I in the $+\mathbf{a}_z$ direction on the z-axis, and this is surrounded by an infinite length cylindrical shell (centered about the z-axis) of radius a carrying the return current I in the $-\mathbf{a}_z$ direction as a surface current. Find expressions for the magnetic field intensity everywhere. If the current is 1.0 A and the radius a is 2.0 cm, plot the magnitude of $\mathbf{H}$ versus radial distance from the z-axis from 0.1 to 4 cm.

3.22 Consider a pair of collinear cylindrical shells centered on the z-axis. The inner shell has radius a and carries a sheet current totaling I amps in the $+\mathbf{a}_z$ direction whereas the outer shell of radius b carries the return current I in the $-\mathbf{a}_z$ direction. Find expressions for the magnetic field intensity everywhere. If $a = 2$ cm, $b = 4$ cm and $I = 4$ A, plot the magnitude of $\mathbf{H}$ versus radial distance from the z-axis from 0 to 8 cm.

3.23 Consider the toroid in Figure 3.55 that is tightly wrapped with N turns of conductive wire. For an Amperian path with radius less than a, no current is enclosed and therefore the field is zero. Likewise, for radius greater than c, the net current enclosed is zero and again the field is

zero. Use Ampère's circuital law to find an expression for the magnetic field at radius b, the center of the toroid.

3.4 Curl and the Point Form of Ampere's Circuital Law

3.24 Find $\nabla \times \mathbf{A}$ for the following fields:
(a) $\mathbf{A} = 3xy^2/z\,\mathbf{a}_x$
(b) $\mathbf{A} = \rho\sin^2\phi\,\mathbf{a}_\rho - \rho^2\,z\,\cos\phi\,\mathbf{a}_\phi$
(c) $\mathbf{A} = r^2\sin\theta\,\mathbf{a}_r + r/\cos\phi\,\mathbf{a}_\theta$

3.25 Find $\mathbf{J}$ at (3 m, 60°, 4 m) for $\mathbf{H} = (z/\sin\phi)\,\mathbf{a}_\rho - (\rho^2/\cos\phi)\,\mathbf{a}_z$ A/m.

3.26 Suppose $\mathbf{H} = y^2\mathbf{a}_x + x^2\mathbf{a}_y$ A/m.

▶ EMAG
SOLUTI

(a) Calculate $\oint \mathbf{H} \cdot d\mathbf{L}$ around the path $A \to B \to C \to D \to A$, where A(2 m, 0, 0), B(2 m, 4 m, 0), C(0, 4 m, 0), and D(0, 0, 0).
(b) Divide this $\oint \mathbf{H} \cdot d\mathbf{L}$ by the area S (2 m × 4 m = 8m²).
(c) Evaluate $\nabla \times \mathbf{H}$ at the center point.
(d) Comment on your results for (b) and (c).

3.27 For the coaxial cable in Example 3.8, we found the following:

$$\text{for } \rho \leq a, \qquad \mathbf{H} = \frac{I\rho}{2\pi a^2}\,\mathbf{a}_\phi$$

$$\text{for } a < \rho \leq b, \qquad \mathbf{H} = \frac{I}{2\pi\rho}\,\mathbf{a}_\phi$$

$$\text{for } b < \rho \leq c, \qquad \mathbf{H} = \frac{I}{2\pi(c^2 - b^2)}\left(\frac{c^2 - \rho^2}{\rho}\right)\mathbf{a}_\phi$$

and

$$\text{for } c < \rho, \qquad \mathbf{H} = 0$$

(a) Evaluate the curl in all four regions.
(b) Calculate the current density in the conductive regions by dividing the current by the area. Are these results the same as what you found in (a)?

3.28 Suppose you have the field $\mathbf{H} = r\cos\theta\,\mathbf{a}_\phi$ A/m. Now consider the cone specified by $\theta = \pi/4$, with a height a as shown in Figure 3.56. The circular top of the cone has a radius a.

▶ EMAG
SOLU

(a) Evaluate the right side of Stoke's theorem through the $d\mathbf{S} = dS\mathbf{a}_\theta$ surface.
(b) Evaluate the left side of Stoke's theorem by integrating around the loop.

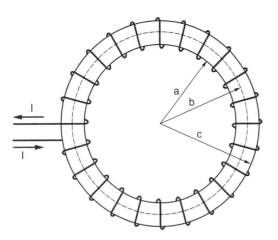

Figure 3.55 Toroid for Problem 3.23.

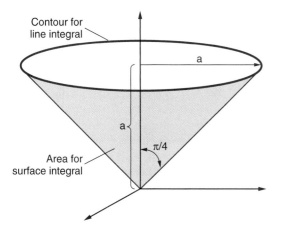

Figure 3.56 Cone for Problem 3.28.

3.5 Magnetic Flux Density

3.29 An infinite length line of 3.0-A current in the $+\mathbf{a}_y$ direction lies on the y-axis. Find the magnetic flux density at P(7.0 m, 0, 0) in (a) teslas, (b) Wb/m², and (c) gauss.

3.30 Suppose an infinite extent sheet of current with $\mathbf{K} = 12\mathbf{a}_x$ A/m lies on the x–y plane at z = 0. Find **B** for any point above the sheet. Find the magnetic flux passing through a 2m² area in the x–z plane for z > 0.

3.31 An infinite length coaxial cable exists along the z-axis, with an inner shell of radius a carrying current I in the $+\mathbf{a}_z$ direction and an outer shell of radius b carrying the return current. Find the magnetic flux passing through an area of length h along the z-axis bounded by radius between a and b.

3.6 Magnetic Forces

3.32 A 1.0-nC charge with velocity 100. m/s in the y direction enters a region where the electric field intensity is 100. V/m $\mathbf{a}_z$ and the magnetic flux density is 5.0 Wb/m² $\mathbf{a}_x$. Determine the force vector acting on the charge.

3.33 A 10.-nC charge with velocity 100. m/s in the z direction enters a region where the electric field intensity is 800. V/m $\mathbf{a}_x$ and the magnetic flux density is 12.0 Wb/m² $\mathbf{a}_y$. Determine the force vector acting on the charge.

3.34 A 10.-nC charged particle has a velocity $\mathbf{v} = 3.0\mathbf{a}_x + 4.0\mathbf{a}_y + 5.0\mathbf{a}_z$ m/s as it enters a magnetic field $\mathbf{B} = 1000.$ T $\mathbf{a}_y$ (recall that T = Wb/m²). Calculate the force vector on the charge.

3.35 What electric field is required so that the velocity of the charged particle in the previous problem remains constant?

3.36 An electron (with rest mass $M_e = 9.11 \times 10^{-31}$ kg and charge $q = -1.6 \times 10^{-19}$ C) has a velocity of 1.0 km/s as it enters a 1.0-nT magnetic field. The field is oriented normal to the velocity of the electron. Determine the magnitude of the acceleration on the electron caused by its encounter with the magnetic field.

3.37 Suppose you have a surface current $\mathbf{K} = 20.$ $\mathbf{a}_x$ A/m along the z = 0 plane. About a meter or so above this plane, a 5.0-nC charged particle is moving along with velocity $\mathbf{v} = -10.\mathbf{a}_x$ m/s. Determine the force vector on this particle.

3.38 A meter or so above the surface current of the previous problem there is an infinite length line conducting 1.0 A of current in the $-\mathbf{a}_x$ direction. Determine the force per unit length acting on this line of current.

3.39 Recall that the gravitational force on a mass m is

$$\mathbf{F} = m\mathbf{g}$$

where, at the earth's surface, $\mathbf{g} = 9.8$ m/s² $(-\mathbf{a}_z)$. A line of 2.0-A current with 100. g mass per meter length is horizontal with the earth's surface and is directed from west to east. What magnitude and direction of uniform magnetic flux density would be required to levitate this line?

3.40 Suppose you have a pair of parallel lines each with a mass per unit length of 0.10 kg/m. One line sits on the ground and conducts 200. A in the $+\mathbf{a}_x$ direction, and the other one, 1.0 cm above the first, has sufficient current to levitate. Determine the current and its direction for line 2.

3.41 In Figure 3.57, a 2.0-A line of current is shown on the z-axis with the current in the $+\mathbf{a}_z$ direction. A current loop exists on the x–y plane (z = 0) that has four wires (labeled 1 through 4) and carries 1.0 mA as shown. Find the force on each arm and the total force acting on the loop from the field of the 2.0-A line.

3.42 Modify MATLAB 3.4 to find the differential force acting from each individual differential segment on the loop. Plot this force against the phi location of the segment.

3.43 Consider a circular conducting loop of radius 4.0 cm in the y–z plane centered at (0, 6 cm, 0). The loop conducts a 1.0-mA current clockwise as viewed from the +x-axis. An infinite length line on the z-axis conducts a 10.-A current in the $+\mathbf{a}_z$ direction. Find the net force on the loop.

3.44 A square loop of 1.0-A current of side 4.0 cm is centered on the x–y plane. Assume 1 mm diameter wire, and

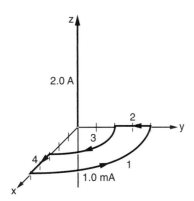

Figure 3.57 A current loop in the x–y plane under the influence of the field from the current on the z-axis used in Problem 3.41. Dimensions are in meters.

estimate the force vector on one arm resulting from the field of the other three arms.

3.45 A current sheet $\mathbf{K} = 100\mathbf{a}_x$ A/m exists at $z = 2.0$ cm. A 2.0-cm-diameter loop centered in the x–y plane at $z = 0$ conducts a 1.0-mA current in the $+\mathbf{a}_\phi$ direction. Find the torque on this loop.

3.46 Ten turns of insulated wire in a 4.0-cm-diameter coil are centered in the x–y plane. Each strand of the coil conducts 2.0 A of current in the $\mathbf{a}_\phi$ direction. (a) What is the magnetic dipole moment of this coil? Now suppose this coil is in a uniform magnetic field $\mathbf{B} = 6.0\mathbf{a}_x + 3.0\mathbf{a}_y + 6.0\mathbf{a}_z$ Wb/m²; (b) what is the torque on the coil?

3.47 A square conducting loop of side 2.0 cm is free to rotate about one side that is fixed on the z-axis. There is 1.0-A current in the loop, flowing in the $-\mathbf{a}_z$ direction on the fixed side. A uniform magnetic flux density exists such that, when the loop is positioned at $\phi = 90°$, no torque acts on the loop, and when the loop is positioned at $\phi = 180°$, a maximum torque of 8.0 $\mathbf{a}_z$ μN-m occurs. Determine the magnetic flux density.

3.7 Magnetic Materials

3.48 A solid nickel wire of diameter 2.0 mm evenly conducts 1.0 A of current. Determine the magnitude of the magnetic flux density $\mathbf{B}$ as a function of radial distance from the center of the wire. Plot to a radius of 2 mm.

3.8 Boundary Conditions

3.49 A planar interface separates two magnetic media. The magnetic field in medium 1 (with μ_{r_1}) makes an angle

α_1 with a normal to the interface. (a) Find an equation for α_2, the angle the field in medium 2 (with μ_{r_2}) makes with a normal to the interface, in terms of α_1 and the relative permeabilities in the two media. (b) Suppose medium 1 is nickel and medium 2 is air and that the magnetic field in the nickel makes an 80° angle with a normal to the surface; find α_2.

3.50 Suppose the $z = 0$ plane separates two magnetic media and that no surface current exists at the interface. Construct a program that prompts the user for μ_{r_1} (for $z < 0$), μ_{r_2} (for $z > 0$), and one of the fields, either $\mathbf{H}_1$ or $\mathbf{H}_2$. The program is to calculate the unknown $\mathbf{H}$. Verify the program using Example 3.11.

3.51 The plane $y = 0$ separates two magnetic media. Medium 1 ($y < 0$) has $\mu_{r_1} = 3.0$ and medium 2 ($y > 0$) has $\mu_{r_2} = 9.0$. A sheet current $\mathbf{K} = (1/\mu_o)\mathbf{a}_x$ A/m exists at the interface, and $\mathbf{B}_1 = 4.0\mathbf{a}_y + 6.0\mathbf{a}_z$ Wb/m².

(a) Find $\mathbf{B}_2$.

(b) What angles do $\mathbf{B}_1$ and $\mathbf{B}_2$ make with a normal to the surface?

3.52 Above the x–y plane ($z > 0$), there exists a magnetic material with $\mu_{r_1} = 4.0$ and a field $\mathbf{H}_1 = 3.0\mathbf{a}_x + 4.0\mathbf{a}_z$ A/m. Below the plane ($z < 0$) is free space. (a) Find $\mathbf{H}_2$, assuming the boundary is free of surface current. What angle does $\mathbf{H}_2$ make with a normal to the surface? (b) Find $\mathbf{H}_2$, assuming the boundary has a surface current $\mathbf{K} = 5.0 \mathbf{a}_x$ A/m.

3.53 The x–z plane separates magnetic material with $\mu_{r_1} = 2.0$ (for $y < 0$) from magnetic material with $\mu_{r_2} = 4.0$ (for $y > 0$). In medium 1, there is a field $\mathbf{H}_1 = 2.0\mathbf{a}_x + 4.0\mathbf{a}_y + 6.0\mathbf{a}_z$ A/m. Find $\mathbf{H}_2$ assuming the boundary has a surface current $\mathbf{K} = 2.0\mathbf{a}_x - 2.0\mathbf{a}_z$ A/m.

3.54 An infinite length line of 2π-A current in the $+\mathbf{a}_z$ direction exists on the z-axis. This is surrounded by air for $\rho \le 50$ cm, at which point the magnetic medium has $\mu_{r_2} = 9.0$ for $\rho > 50$ cm. If the field in medium 2 at $\rho = 1.0$ m is $\mathbf{H} = 5.0\mathbf{a}_\phi$ A/m, find the sheet current density vector at $\rho = 50$. cm, if any.

3.9 Inductance and Magnetic Energy

3.55 Consider a long pair of straight parallel wires, each of radius a, with their centers separated by a distance d. Assuming $d \gg a$, find the inductance per unit length for this pair of wires.

3.56 In Problem 3.23 the task was to find the field at the center (radius b) of an N-turn toroid. If the radius of the toroid is large compared to the diameter of the coil (that is, if $b \gg c - a$), then the field is approximately constant from radius a to radius c. (a) Obtain an expression for the toroid's inductance. (b) Find L if there are 600 turns around a 99.8% iron core with $a = 8.0$ cm and $c = 9.0$ cm.

3.57 Consider a solid wire of radius $a = 1.0$ mm bent into a circular loop of radius 10. cm. Neglecting internal inductance of the wire, write a program to find the inductance for this loop.

MAG SOLUTIONS **3.58** Find the mutual inductance between an infinitely long wire and a rectangular wire with dimensions shown in Figure 3.58.

3.59 Consider a pair of concentric conductive loops, centered in the same plane, with radii a and b. Determine the mutual inductance between these loops if $b \gg a$.

MAG SOLUTIONS **3.60** A 4.0-cm-diameter solid nickel wire, centered on the z-axis, conducts current with a density $\mathbf{J} = 4\rho \, \mathbf{a}_z$ A/cm^2 (where ρ is in centimeters). Find the internal inductance per unit length for the wire with this current distribution.

3.10 Magnetic Circuits

MAG SOLUTIONS **3.61** Suppose 2.0 A flows through 80 turns of the toroid in Figure 3.48a that has a core cross-sectional area of 2.0 cm^2 and a mean radius of 80. cm. The core is 99.8% pure iron. (a) How much magnetic flux exists in the toroid? (b) How much energy is stored in the magnetic field contained by the toroid?

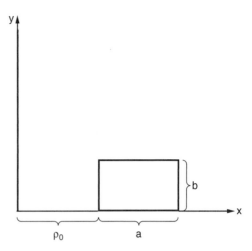

Figure 3.58 Mutual inductance schematic for Problem 3.58.

EMAG SOLUTIONS **3.62** In Figure 3.59, a 2.0-cm-diameter toroidal core with $\mu_{r_1} = 10,000$ is wrapped with a 1.0-cm-thick layer of material with $\mu_{r_2} = 3000$. The toroid has a 1.0 m mean radius. For 20. A of current driven through 50 loops of wire, find the magnetic field intensity in each material of the toroid.

3.63 Suppose the 2.0-cm-diameter core of the toroid in Figure 3.49a is characterized by the magnetization curve of

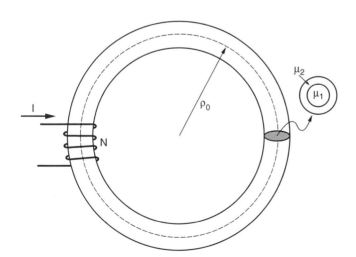

Figure 3.59 Toroid consisting of two types of magnetic material for Problem 3.62.

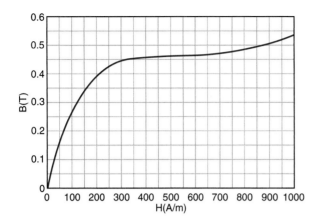

Figure 3.60 Magnetization curve for Problem 3.63.

Figure 3.60. The toroid has a mean radius of 60. cm. For 10. A of current driven through 100 loops of wire, find the magnetic field intensity in the 1.0-mm gap.

3.64 In Figure 3.52, suppose the cross-sectional area of the bar is 3.0 cm^2 and that of the electromagnet core is 2.0 cm^2. Also, the bar has a relative permittivity of 3000, whereas that of the magnetic core is 10,000. The dimensions for h and w are 12. cm and 16. cm, respectively. If the mass of the bar is 20. kg, how much current must be driven through 24 loops to hold up the bar against gravity?

3.65 Consider a 1.0-mm air gap in Figure 3.49a. The toroid mean radius and cross-sectional area are 50. cm and 2.0 cm^2, respectively. If the magnetic core has $\mu_r = 6000$ and 4.0 A is being driven through 30 loops, determine the magnitude of the force pulling the gap closed.

4

Dynamic Fields

Learning Objectives

▷ Describe charge dissipation using the current continuity equation

▷ Examine the wave equation used to describe wave propagation

▷ Define electromotive force and examine the operation of transformers and generators

▷ Define Faraday's law, showing how a time-varying magnetic field produces an electric field

▷ Define displacement current, showing how a time-varying electric field produces a magnetic field

▷ Use Maxwell's equations to demonstrate transverse electromagnetic wave propagation

▷ Introduce phasor notation to concisely describe Maxwell's equations for time-harmonic fields

We have thus far assumed no time variation in our electric and magnetic fields. This was done to build a sense of comfort in working with vectors and coordinate systems and to get a first inkling of Maxwell's equations.

Now we turn to the dynamic case, where electric and magnetic fields do change with time. First we'll consider how current is related to charge density and how rapidly charges can disperse in a material. Then, we'll review the properties of traveling waves before turning to the heart of the chapter on Faraday's law and displacement current. Finally, with the dynamic Maxwell's equations we will see the intimate linkage between electric and magnetic fields.

▷ 4.1 CURRENT CONTINUITY AND RELAXATION TIME

Consider a volume of charge Q contained within a closed surface. The only way to decrease Q within the enclosed volume is to let it flow through the surface.[4.1] This flow of charge is current, and the current must be equal to the rate of decrease in the contained charge. This can be written

$$I = \oint \mathbf{J} \cdot d\mathbf{S} = -\frac{\partial Q}{\partial t} \tag{4.1}$$

[4.1]The *principle of conservation of charge* maintains that net charge cannot be created or destroyed.

A partial derivative is used since Q can be a function of both time and position. Because a positive current out of the closed surface corresponds to a decrease in enclosed charge with time, a negative sign is appended to the derivative.

The divergence theorem (see Chapter 2, Section 2.8) can be used to rewrite the left side of (4.1) as

$$\oint \mathbf{J} \cdot d\mathbf{S} = \int (\nabla \cdot \mathbf{J}) dv \tag{4.2}$$

The right side of (4.1) can be rewritten as

$$-\frac{\partial Q}{\partial t} = -\frac{\partial}{\partial t} \int \rho_v dv \tag{4.3}$$

Now, if we fix the closed surface so that the volume containing the charge doesn't change with time, we can pull the derivative inside the integral:

$$-\frac{\partial Q}{\partial t} = -\int \frac{\partial \rho_v}{\partial t} dv \tag{4.4}$$

By comparing (4.1), (4.2), and (4.4), we see that

$$\boxed{\nabla \cdot \mathbf{J} = -\frac{\partial \rho_v}{\partial t}} \tag{4.5}$$

This is the point form of the *current continuity equation*.

In steady currents where there is no change in charge density, the continuity equation leads to *Kirchhoff's current law*, which says the currents into a junction must sum to zero.

In Chapter 2, Section 2.10 it was pointed out that free or excess charges introduced in a conductor will repel each other and race to the outside. We can use the continuity equation to determine how long it will take the charges to dissipate.

The continuity equation can be rewritten as

$$\nabla \cdot \mathbf{J} = \nabla \cdot \sigma \mathbf{E} = -\frac{\partial \rho_v}{\partial t} \tag{4.6}$$

In a homogeneous material where σ doesn't vary with position,

$$\nabla \cdot \mathbf{E} = -\frac{1}{\sigma} \frac{\partial \rho_v}{\partial t} \tag{4.7}$$

We also know by the point form of Gauss's law that

$$\nabla \cdot \mathbf{E} = \frac{\rho_v}{\varepsilon} \tag{4.8}$$

Combining (4.7) and (4.8) leaves us with the differential equation

$$\frac{\partial \rho_v}{\partial t} + \frac{\sigma \rho_v}{\varepsilon} = 0 \tag{4.9}$$

Upon separation of variables, this has the solution

$$\boxed{\rho_v = \rho_o e^{-(\sigma/\varepsilon)t}} \tag{4.10}$$

where ρ_o is the initial charge density at time $t = 0$. The charge density decreases with time, and the value reaches $1/e$ of the starting value at the *relaxation time* τ, where

$$\boxed{\tau = \frac{\varepsilon}{\sigma}}$$
(4.11)

In a good conductor, the charges are able to move rapidly and the relaxation time is very short. In contrast, in a good dielectric, it can take considerable time for a charge to dissipate.

Drill 4.1 Calculate the relaxation time for (a) copper and (b) polystyrene. (*Answer:* (a) $\tau = 1.5 \times 10^{-19}$ s, (b) $\tau = 2.6$ days)

▶ 4.2 WAVE FUNDAMENTALS

Toss a stone in a quiet pool and observe the ripples traveling radially away from the point of impact. These water waves travel, or propagate, at a particular velocity, and they carry energy with them. The medium itself (water) only bobs up and down. Other types of waves include sound waves, mechanical waves traveling as ripples in a rope, waves in a stretched slinky, and, of course, light traveling as electromagnetic waves.

In this section, we'll briefly review some of the fundamental features of waves before employing them in our study of electromagnetics. Here we will only consider continuous *time-harmonic* waves, represented by sine waves, rather than *transient* waves[4.2] (such as pulses and step functions).

Let's consider an electric field propagating in the z direction. The general solution to the wave equation, derived in the next chapter, is

$$\boxed{\mathbf{E}(z, t) = E_o e^{-\alpha z} \cos(\omega t - \beta z + \phi)\mathbf{a}_x}$$
(4.12)

The electric field in this wave expression is a function of position (z) and time (t). It is always pointing in either the plus or minus x direction, so we call this an *x-polarized* wave.[4.3] The *amplitude*, $E_o e^{-\alpha z}$, is made up of the initial amplitude at $z = 0$, E_o, and an exponential term to account for attenuation of the wave as it propagates. The *phase* inside the sine argument consists of three parts: ωt, where ω is the *angular frequency* ($\omega = 2\pi f$), βz, where β is the *phase constant* (sometimes referred to as *wave number*), and a *phase shift* ϕ.

For illustration purposes let's initially assume the phase shift ϕ is zero and look at the field versus time when $z = 0$. We then have

$$\mathbf{E}(0, t) = E_x \mathbf{a}_x = E_o \cos(\omega t)\mathbf{a}_x$$
(4.13)

which is plotted in Figure 4.1. A characteristic of a sine or cosine wave is that it repeats every 2π radians (or 360°). Put another way, we have $\cos(\omega t) = 1$ for $\omega t = n2\pi$, where $n = 0$,

---

[4.2]Transient waves are very important in the study of digital circuits, and as such they will be picked up when we study transmission lines in Chapter 6.

[4.3]Polarization will be discussed in Chapter 5.

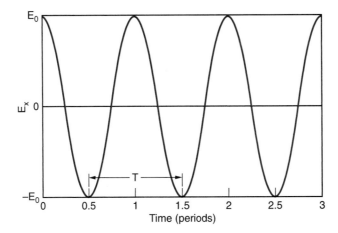

Figure 4.1 Plot of E_x versus time at $z = 0$ for the function $\mathbf{E}(0,t) = E_x\mathbf{a}_x = E_o\cos(\omega t)\mathbf{a}_x$.

1, 2…. The period T is the time elapsed for one cycle, or $\omega T = (1)2\pi$. Solving, we have the following relationship:

$$T = \frac{2\pi}{\omega} = \frac{1}{f} \tag{4.14}$$

We can reinsert the phase shift ϕ and in Figure 4.2 plot

$$\mathbf{E}(0,t) = E_o\cos(\omega t + \phi)\mathbf{a}_x \tag{4.15}$$

where we have chosen $\phi = -45°$. This trace lags the original function by $45°$.

Now let's rezero ϕ and look at the field versus position z when time $t = 0$. First let's assume the wave is in a lossless medium (such as vacuum) so that there is no attenuation. In this case, the attenuation constant α equals zero and $e^{-\alpha z} = 1$. We have

$$\mathbf{E}(z,0) = E_o\cos(-\beta z)\mathbf{a}_x \tag{4.16}$$

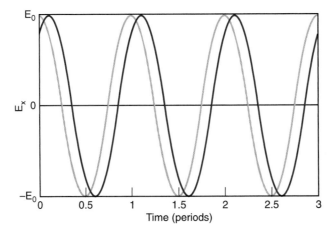

Figure 4.2 Plot of E_x versus time at $z = 0$ with $-45°$ phase shift for the function $\mathbf{E}(0,t) = E_x\mathbf{a}_x = E_o\cos(\omega t + \phi)\mathbf{a}_x$.

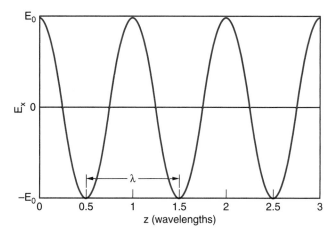

Figure 4.3 Plot of E_x versus z in a lossless medium at $t = 0$ for the function $\mathbf{E}(z,0) = E_o \cos(-\beta z)\mathbf{a}_x$.

which is plotted in Figure 4.3. We again have the 2π-radian repeat cycle, or $\cos(-\beta z) = \cos(\beta z) = 1$ when $\beta z = n2\pi$. One cycle is a wavelength long, or $\beta\lambda = (1)2\pi$. This is rearranged to give the relation between wavelength and phase constant,

$$\boxed{\beta = \frac{2\pi}{\lambda}} \tag{4.17}$$

Now let's insert attenuation:

$$\mathbf{E}(z,0) = E_o e^{-\alpha z} \cos(-\beta z)\mathbf{a}_x \tag{4.18}$$

As plotted in Figure 4.4, the amplitude is shown decreasing with increasing z with an *attenuation constant* α. The units for attenuation are in terms of nepers per meter (Np/m).

We are now ready to consider traveling waves. Let's again consider a lossless medium and we'll let $\phi = 0$ for this illustration. In Figure 4.5 we plot E_x versus position at progressive values of time using

$$\mathbf{E}(z,t) = E_o \cos(\omega t - \beta z)\mathbf{a}_x \tag{4.19}$$

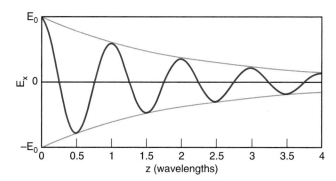

Figure 4.4 Plot of E_x versus z at $t = 0$, with attenuation included, for the function $\mathbf{E}(z,0) = E_o e^{-\alpha z} \cos(-\beta z)\mathbf{a}_x$.

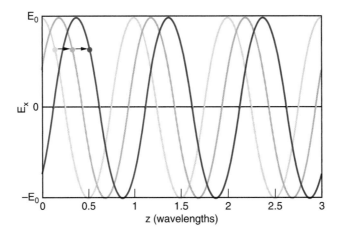

Figure 4.5 Plot of E_x versus z at progressive times showing wave travel.

For each trace a dot has been added representing a constant-phase point on the wave. We see that, as time increases, this phase point is moving in the $+z$ direction, so we call this a $+z$ traveling wave. How fast is the wave traveling? Consider the phase

$$\omega t - \beta z = C \tag{4.20}$$

where C is an arbitrary constant representing a constant-phase point such as indicated by the dots in Figure 4.5. If we take the derivative of both sides of this expression we have

$$\omega dt - \beta dz = 0 \tag{4.21}$$

which can be rearranged as

$$\boxed{u_p = \frac{dz}{dt} = \frac{\omega}{\beta} = \lambda f} \tag{4.22}$$

The *phase velocity* (also called the *propagation velocity*) of the wave is u_p. This is a property of the medium.

▶ **EXAMPLE 4.1**

Suppose we have a 1-V amplitude 100-MHz x-polarized wave in air propagating in the z direction. We want to write a wave equation like (4.12) for this case.

It is reasonable to assume that air is a lossless medium (so $\alpha = 0$). Since the frequency is 100 MHz, we know the angular frequency ω is $200\pi \times 10^6$ radians/s. Also, since the speed of light c is approximately 3×10^8 m/s, we can find the wavelength as $\lambda = c/f = 3$ m. Thus we can write

$$\mathbf{E}(z,t) = 1\cos(200\pi \times 10^6 t - (2\pi/3)z + \phi)\mathbf{a}_x \, \text{V/m}$$

To determine the phase shift ϕ, we need more information. If, for instance, we were given that $\mathbf{E}(0,0) = 1\mathbf{a}_x$ V/m, then we know $\phi = 0°$ and our wave equation becomes

$$\mathbf{E}(z,t) = 1\cos(200\pi \times 10^6 t - (2\pi/3)z) \, \mathbf{a}_x \, \text{V/m}$$

Drill 4.2 Suppose a propagating electric field is given by

$$\mathbf{E}(z,t) = 34e^{-0.002z}\cos(2\pi \times 10^9 t - 10\pi z + 45\,°)\,\text{V/m}$$

Find (a) the initial amplitude, (b) the attenuation constant, (c) the wave frequency, (d) the wavelength, and (e) the phase shift in radians. (*Answer:* (a) 34 V/m, (b) 0.002 Np/m, (c) 1 GHz, (d) 0.20 m, (e) $\pi/4$ radians)

▷ **MATLAB 4.1**

Write a program to plot the wave versus position for a fixed time. Assume the wave is in vacuum.

```
%   M-File: ML0401
%
%   Program plots wave (in vacuum) versus z-position for
%   a fixed time.
%
%   Wentworth, 7/17/02
%
%   Variables:
%   Eo      wave amplitude (V/m)
%   f       frequency (Hz)
%   omega   angular frequency (radians/s)
%   t       time snapshot (s)
%   phi     phase constant (degrees)
%   phir    phase constant (radians)
%   c       speed of light in vacuum (m/s)
%   lambda  wavelength (m)
%   B       phase constant (1/m)
%   E       electric field intensity (V/m)
%   z       position

clc         %clears the command window
clear       %clears variables

%   Initialize Variables
Eo=1;
f=1000;
t=1;
phi=0;
phir=phi*pi/180;
c=2.998e8;
lambda=c/f;
B=2*pi/lambda;
omega=2*pi*f;

%   Perform Calculation
z=0:4*lambda/100:4*lambda;
E=Eo*cos(omega*t-B*z+phir);

%   Generate the Plot
plot(z,E)
```

(continues)

```
axis('tight') %sets axes min & max data values
grid
xlabel('z(m)')
ylabel('E(V/m)')
```

Try running this program with different values for amplitude, time, phase constant, and frequency.

▷ **MATLAB 4.2**

Illustrate a traveling wave by making a movie in MATLAB.

```
%    M-File: ML0402
%
%    This program illustrates a traveling wave
%
%    Wentworth, 7/17/02
%
%    Variables:
%    Eo      wave amplitude (V/m)
%    f       frequency (Hz)
%    omega   angular frequency (radians/s)
%    t       time snapshot (s)
%    phi     phase constant (°s)
%    phir    phase constant (radians)
%    c       speed of light in vacuum (m/s)
%    lambda  wavelength (m)
%    B       phase constant (1/m)
%    E       electric field intensity
%    z       position

clc        %clears the command window
clear      %clears variables

%    Initialize Variables
Eo=1;
f=1000;
t=1;
phi=0;
phir=phi*pi/180;
c=2.998e8;
lambda=c/f;
B=2*pi/lambda;
omega=2*pi*f;

%    Perform Calculation
z=0:4*lambda/100:4*lambda;
E=Eo*cos(omega*t-B*z+phir);

%    Generate a Reference Frame
plot(z,E)
axis([0 4*lambda -2*Eo 2*Eo])
grid
xlabel('z(m)')
```

```
ylabel('E(V/m)')
pause

%   Make the Movie

t=0:1/(40*f):1/f;
for n=1:40;
    E=Eo*cos(omega*t(n)-B*z+phir);
    plot(z,E)
    axis([0 4*lambda -2*Eo 2*Eo])
    grid
    title('General Wave Equation');
    xlabel('z(m)');
    ylabel('E(V/m)');
    M(:,1)=getframe;
end
```

Run the program. After the reference frame is drawn, the program will pause and wait for you to press the return key (with the cursor in the plot window). Try changing the direction of the wave by changing the sign in front of "B*z" in the cosine equation. (You should do this for both the reference frame and the movie frame.)

This is one of several ways to perform animations in MATLAB. For more information on movies, in the command window type help movie, help moviein, and/or help getframe.

► 4.3 FARADAY'S LAW AND TRANSFORMER EMF

Following Oersted's discovery, Michael Faraday thought that if a current in a wire can produce a magnetic field, then perhaps a magnetic field can produce a current in a wire. Ten years of experiment bore out his hypothesis, which was simultaneously confirmed by Joseph Henry.[4.4] They observed that current was only induced in a circuit if the magnetic flux linking the circuit changed with time.

Figure 4.6a shows a conductive loop in a plane normal to a magnetic field that increases with time. Since the magnetic flux through the area bounded by the loop is changing, a current I_{ind} is induced in the loop as indicated by the ammeter. Notice the direction of the induced current. The flux produced by the induced current acts to oppose the *change* in flux. This statement is called Lenz's law.[4.5]

We can pull the ammeter out of the loop, leaving a pair of open terminals as shown in Figure 4.6b. Now the current induced in the loop establishes a potential difference across the terminals known as the *electromotive force*. This electromotive force, V_{emf} (or just *emf*), is related to the rate of change of flux linking a circuit by *Faraday's law*:

$$V_{emf} = -\frac{\partial \lambda}{\partial t} \qquad (4.23)$$

[4.4]But Faraday was more prompt getting his message to the publishers, so he generally gets credit for the discovery.

[4.5]Russian physicist Heinrich Lenz (1804–1865) was a contemporary of Faraday and Henry and published his law in 1834.

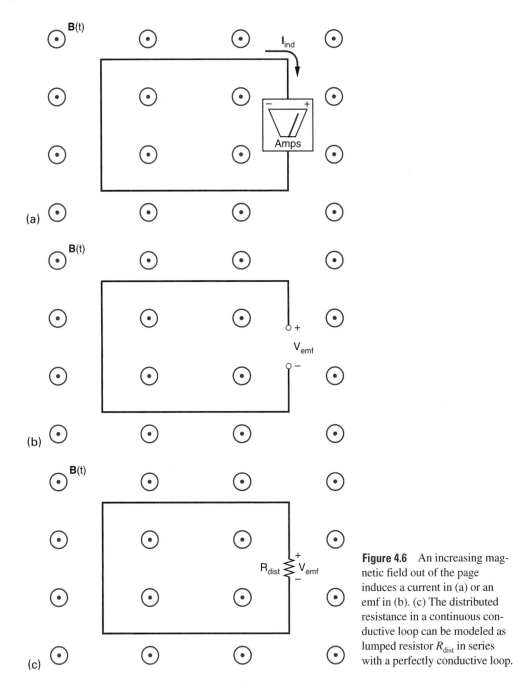

Figure 4.6 An increasing magnetic field out of the page induces a current in (a) or an emf in (b). (c) The distributed resistance in a continuous conductive loop can be modeled as lumped resistor R_{dist} in series with a perfectly conductive loop.

The negative sign in the equation is a consequence of Lenz's law. If we consider a single loop, Faraday's law can be written

$$V_{emf} = -\frac{\partial \phi}{\partial t} = -\frac{\partial}{\partial t} \int \mathbf{B} \cdot d\mathbf{S} \qquad (4.24)$$

Generating emf requires a time-varying magnetic flux linking the circuit. This occurs if the magnetic field changes with time (called *transformer emf*), or if the surface containing the flux changes with time (called *motional emf*).

The emf is measured around the closed path enclosing the area through which the flux is passing, and it can also be written

$$V_{emf} = \oint \mathbf{E} \cdot d\mathbf{L} \tag{4.25}$$

This starkly contrasts with the static field case, for which the circulation of **E** around a closed path is zero. Using (4.25), we can rewrite Faraday's law as

$$\boxed{V_{emf} = \oint \mathbf{E} \cdot d\mathbf{L} = -\frac{\partial}{\partial t} \int \mathbf{B} \cdot d\mathbf{S}} \tag{4.26}$$

In this equation, the direction of the circulation integral is related to the differential surface vector direction via the right-hand rule. For instance, in Figure 4.6b we choose $d\mathbf{S}$ pointing into the page and so the path for the circulation integral is in the same direction as that assumed for the induced current. If we are correct in our assumption, then a V_{emf} of the polarity shown will result. A negative V_{emf} will tell us the induced current is going in the other direction. It should also be noted that the surface integral on the right side of (4.26) does not have to be a planar surface bounded by the loop of the circulation integral. It can be any surface bounded by the loop. (Recall the Stoke's theorem discussion in Chapter 3 along with Figure 3.25.)

Faraday's law also applies to continuous conductive paths. There is always at least a little resistance distributed along these conductive loops. They are sometimes represented as a lumped resistance R_{dist} in series with a perfect conductor, as Figure 4.6c indicates.

Transformer EMF

Consider the case where the field is varying with time and the surface stays constant. We can pull the time derivative inside the integral in the right side of (4.26) to get

$$V_{emf} = -\int \frac{\partial \mathbf{B}}{\partial t} \cdot d\mathbf{S} \tag{4.27}$$

Partial derivatives are used inside the integral since **B** may also be a function of position. The generation of emf by a changing magnetic field is fundamental to the operation of transformers and hence is referred to as *transformer emf*.

Drill 4.3 Suppose in Figure 4.6 that the field is $\mathbf{B} = 4t\,\mathbf{a}_z$ Wb/m², where t is in seconds and $\mathbf{a}_z$ is coming out of the page. If the conductive loop has a 400 cm² area, (a) determine the emf established across the terminals in Figure 4.6b. (b) If the ammeter in Figure 4.6a is replaced by a 100-Ω resistor, determine I_{ind}. (*Answer:* (a) 160 mV, (b) 1.6 mA)

► **EXAMPLE 4.2**

Let's consider the circuit containing the pair of loops shown in Figure 4.7. Each loop has an area S. A magnetic field, normal to the plane of the loops, varies with time as

$$\mathbf{B} = B_o \sin \omega t \, \mathbf{a}_x$$

We want to calculate the voltage across the resistor, V_R.

Using the right-hand rule with the thumb pointing in the direction of $d\mathbf{S}$ (the $+\mathbf{a}_z$ direction), the fingers curl in the direction of the circulation, which in this case indicates that $V_R = V_{emf}$. Our equation for V_{emf} with N loops is

$$V_{emf} = -N \int \frac{\partial \mathbf{B}}{\partial t} \cdot d\mathbf{S}$$

where $d\mathbf{S}$ is integrated over the area of one of the loops. The result of our calculation is then

$$V_R = -2\omega B_o S \cos \omega t$$

Let's see if this answer is logical. When the magnetic field is increasing (say from $t = 0$ to $t = \pi/2$), Lenz's law implies the induced current will be directed from $-$ to $+$ using the sign convention for V_R given in the figure. So having a negative value for V_R is logical. Figure 4.8 shows the relationship between V_R and $B(t)$ for this problem.

► **EXAMPLE 4.3**

Consider the rectangular loop in Figure 4.9 moving with a velocity $\mathbf{u} = u_y \mathbf{a}_y$ in the field from an infinite length line of current I on the z-axis. From the frame of reference of the constant-area conductive loop, the magnetic field is changing with time. Assume the loop has a distributed resistance R_{dist}. Find an expression for the current in the loop (including its direction).

First we calculate the flux through the loop at an instant in time. We have

$$\mathbf{B} = \frac{\mu_o I}{2\pi\rho} \mathbf{a}_\phi = -\frac{\mu_o I}{2\pi y} \mathbf{a}$$

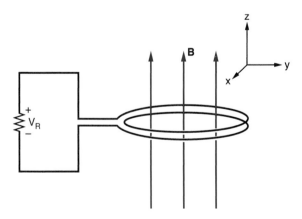

Figure 4.7   A pair of loops in a B-field increasing with time.

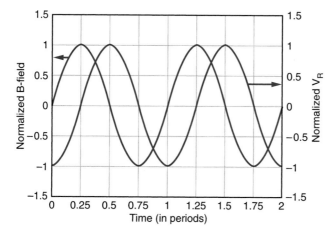

Figure 4.8 Relationship between B and V_R for Example 4.2.

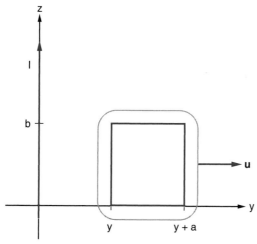

Figure 4.9 A rectangular conductive loop moves with velocity **u** away from an infinite length line of current.

and we will arbitrarily choose $d\mathbf{S}$ in the $+\mathbf{a}_x$ direction,

$$d\mathbf{S} = dydz\mathbf{a}_x$$

so the flux is easily calculated to be

$$\phi = -\frac{\mu_o I}{2\pi}\int_y^{y+a}\frac{dy}{y}\int_0^b dz = -\frac{\mu_o Ib}{2\pi}\left[\ln(y+a)-\ln(y)\right]$$

Next, we want to find how this flux changes with time, so

$$\frac{d\phi}{dt} = -\frac{\mu_o Ib}{2\pi}\frac{d}{dt}\left[\ln(y+a)-\ln(y)\right]$$

By the chain rule this yields

$$\frac{d\phi}{dt} = -\frac{\mu_o Ib}{2\pi}\left[\frac{1}{y+a}-\frac{1}{y}\right]\frac{dy}{dt}$$

Considering that $u_y = dy/dt$, and manipulating the expression inside the brackets, we arrive at

$$\frac{d\phi}{dt} = \frac{\mu_o Iabu_y}{2\pi y(y+a)}$$

Our emf is the negative of this result:

$$V_{emf} = \frac{-\mu_o Iabu_y}{2\pi y(y+a)}$$

By our choice of $d\mathbf{S}$ in the $+\mathbf{a}_x$ direction, our emf is taken from a counterclockwise circulation (looking at the loop from the $+x$-axis). Since the emf is negative, our induced current is apparently going in the clockwise direction with a value

$$I_{ind} = \frac{\mu_o Iabu_y}{2\pi y(y+a)R_{dist}}$$

Does this answer make sense? Let's see. As the loop moves away from the line of current, the flux in the loop (going in the $-\mathbf{a}_x$ direction) is decreasing. To counteract this decrease, Lenz's law says the induced current must produce a flux in the $-\mathbf{a}_x$ direction. This agrees with our calculated clockwise current result.

Drill 4.4 Referring to Example 4.2, suppose the angular frequency is 1000 radians/s, the field amplitude $B_o = 6.0$ mWb/m², and the area of one of the pair of identical loops is 144 cm². Calculate V_R at $t = 1$ ms and $t = 10$ ms. (*Answer:* –93 mV, 145 mV)

Drill 4.5 Referring to Example 4.3, plot the value of I_{ind} versus position y as y goes from 0.01 to 1 m if $a = b = 6.0$ cm, the loop velocity in the y direction is 2.0 m/s, the current on the z-axis is $I = 1.0$ A, and the distributed resistance of the rectangular loop is 10. μΩ. (*Answer:* See Figure 4.10.)

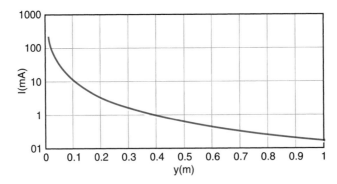

Figure 4.10 Plot of induced current versus position for Drill 4.5.

Transformers

Faraday's law is employed in transforming AC voltages and currents between a pair of windings in a magnetic circuit. Figure 4.11 illustrates a transformer consisting of *primary* and *secondary* coils wrapped around a magnetic core. The driving side of the transformer is the primary side, with number of loops N_1, and AC voltage and current v_1 and i_1, respectively. The circuit being driven is the secondary side with N_2 loops, and AC voltage and current v_2 and i_2, respectively.

Recalling the magnetic circuits discussion in Section 3.10, we know the magnetomotive force V_m is

$$V_m = NI = \mathcal{R}\phi \qquad (4.28)$$

where $\mathcal{R}$ is the magnetic circuit's reluctance, analogous to an electric circuit's resistance, and is given by

$$\mathcal{R} = \frac{\ell}{\mu A}$$

Here, ℓ and A refer to the path length around the magnetic circuit and the cross-sectional area, respectively. For the AC circuit of the figure, the magnetomotive force can be written

$$V_m = N_1 i_1 - N_2 i_2 = \mathcal{R}\phi \qquad (4.29)$$

In ideal transformers, the permeability of the core is large enough that we can consider the reluctance–flux product to be approximately zero, which implies

$$N_1 i_1 = N_2 i_2 \qquad (4.30)$$

or

$$i_2 = \frac{N_1}{N_2} i_1 \qquad (4.31)$$

From Faraday's law, we can also relate the voltage across each coil to the rate of change of flux,

$$v_1 = N_1 \frac{d\phi}{dt}, \quad v_2 = N_2 \frac{d\phi}{dt} \qquad (4.32)$$

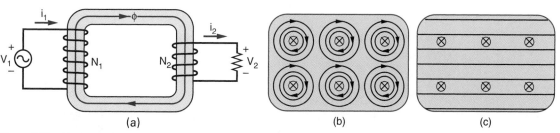

Figure 4.11 (a) A transformer. (b) Cross section of the core showing eddy currents. (c) Use of multiple layers to reduce eddy current loss.

Since the $d\phi/dt$ term is equal for both voltages, we have

$$v_2 = \frac{N_2}{N_1} v_1 \tag{4.33}$$

Transformers are routinely used in circuits to step up or step down voltages and currents, or to transform the impedance (see Problem 4.12). Power efficient circuits are desirable, and one loss mechanism that needs to be minimized is *eddy current* loss.

We have seen that when a conductive loop is in the presence of a changing magnetic field, a current is induced. We can model a conductive surface or volume as consisting of a large number of conductive loops, each one of which can have a current induced from the changing magnetic field. These induced currents are known as eddy currents. In the core of a transformer the eddy currents, such as those shown in Figure 4.11b, result in power loss in the transformer. To reduce eddy current loss, electrically insulated ferromagnetic sheets are laminated together, with the sheets continuous in the direction of the magnetic flux. This is shown in Figure 4.11c. Another way to reduce eddy current loss is to replace the ferromagnetic media with a higher electrical resistance ferrite.

> **Drill 4.6** Suppose $N_1 = 40$ turns. Assuming an ideal transformer, how many turns N_2 are required to double the voltage at the secondary? What happens to the current? (*Answer: $N_2 = 80$, current is halved*)

Point Form of Faraday's Law

Before we leave the topic of transformer emf, we can apply Stoke's theorem to the left side of (4.26) to get

$$V_{emf} = \oint \mathbf{E} \cdot d\mathbf{L} = \int (\nabla \times \mathbf{E}) \cdot d\mathbf{S} \tag{4.34}$$

Now, since the surface isn't changing with time, we can equate (4.34) and (4.27) to obtain

$$V_{emf} = \int (\nabla \times \mathbf{E}) \cdot d\mathbf{S} = -\int \frac{\partial \mathbf{B}}{\partial t} \cdot d\mathbf{S} \tag{4.35}$$

This leads us to the point or differential form of Faraday's Law,

$$\boxed{\nabla \times \mathbf{E} = -\frac{\partial \mathbf{B}}{\partial t}} \tag{4.36}$$

which is one of Maxwell's equations for dynamic fields. We'll apply this equation in Section 4.7.

▶ 4. 4 FARADAY'S LAW AND MOTIONAL EMF

Now we'll maintain a constant magnetic field and achieve a change in magnetic flux linking the circuit by changing the area of the circuit. We can modify (4.26) by pulling the differential inside the integral[4.6] as

---

[4.6]Notice that we do not need to use a partial derivative because the area doesn't change with position.

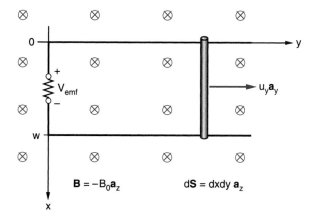

$$\mathbf{B} = -B_0\mathbf{a}_z \qquad d\mathbf{S} = dxdy\ \mathbf{a}_z$$

Figure 4.12 Conductive bar moving along a pair of parallel conductive rails.

$$V_{\text{emf}} = -\int \mathbf{B} \cdot \frac{d\mathbf{S}}{dt} \tag{4.37}$$

Let's consider a conductive bar moving with velocity **u** along a pair of conductive rails as shown in Figure 4.12. Magnetic flux density **B** is given normal to the plane of the circuit, heading into the page. As the bar moves to the right, the magnetic flux enclosed by the loop is increasing in the $-\mathbf{a}_z$ direction. So by Lenz's law, a current will be induced in the counterclockwise direction of the loop to establish a counter flux. Let's perform the circulation in that direction and therefore choose $d\mathbf{S} = dxdy\ \mathbf{a}_z$. We have

$$V_{\text{emf}} = -\int \left(-B_o\mathbf{a}_z\right) \cdot \frac{dxdy}{dt}\mathbf{a}_z$$

which can be rearranged by considering $u_y = dy/dt$ as

$$V_{\text{emf}} = B_o u_y \int_0^w dx = B_o u_y w$$

Let's examine this problem from a different point of view. In Figure 4.13, the conductive rails have been removed and we only have the bar cutting through the magnetic flux. Since the conductive bar has mobile electrons and holes available for conduction, we see that these charges are in motion in a magnetic field. We would expect, then, that acting on these charges is a Lorentz force

$$\mathbf{F}_{\text{m}} = q\mathbf{u} \times \mathbf{B} \tag{4.38}$$

Since $\mathbf{u} = u_y\mathbf{a}_y$ and $\mathbf{B} = -B_o\mathbf{a}_z$, positive charges are forced in the $\mathbf{a}_y \times (-\mathbf{a}_z) = -\mathbf{a}_x$ direction, and negative charges in the $+\mathbf{a}_x$ direction, as indicated in the figure. The charges will continue to accumulate at the ends of the bar until the coulombic attraction between the positive and negative charges is equal to the magnetic force separating them. In a good conductor with short relaxation time we quickly have no net force on a charge; that is,

$$\mathbf{F} = q\mathbf{E} + q\mathbf{u} \times \mathbf{B} = 0$$

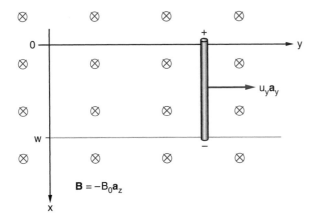

Figure 4.13 Conductive bar moving in the magnetic field in the absence of the conductive rails.

The electric force counteracting the magnetic force is therefore

$$q\mathbf{E} = -q\mathbf{u} \times \mathbf{B}$$

or

$$\mathbf{E} = -\mathbf{u} \times \mathbf{B} \tag{4.39}$$

The voltage between the ends of the bar is determined by the integration of the electric field from the negative end to the positive end along the bar length:

$$V = -\int_{-}^{+} \mathbf{E} \cdot d\mathbf{L} = \int (\mathbf{u} \times \mathbf{B}) \cdot d\mathbf{L} = \int_{w}^{0} (\mathbf{u} \times \mathbf{B}) \cdot dx \mathbf{a}_x \tag{4.40}$$

Solving for this voltage, we have

$$V = -B_o u_y \int_{w}^{0} dx = B_o u_y w \tag{4.41}$$

Now suppose we reinsert the conductive rails and the resistor as in Figure 4.12. The voltage across the bar now appears across the resistor, and we see that

$$\boxed{V_{emf} = \oint (\mathbf{u} \times \mathbf{B}) \cdot d\mathbf{L}} \tag{4.42}$$

Notice that only the portions of the close path that are moving contribute to this emf.

It should also be pointed out that the sign convention for V_{emf} is arbitrary in that it depends on the assumed direction for $d\mathbf{S}$. We arrived at (4.42) by having the direction of the circulation integral be in the same direction that we integrated to find V in (4.40).

▷ **EXAMPLE 4.4**

Consider the conductive bar moving in the $+\mathbf{a}_z$ direction at a fixed distance from an infinite length line of current on the z-axis, as shown in Figure 4.14a. Let's find the potential difference between the ends of the bar as well as the bar's polarity.

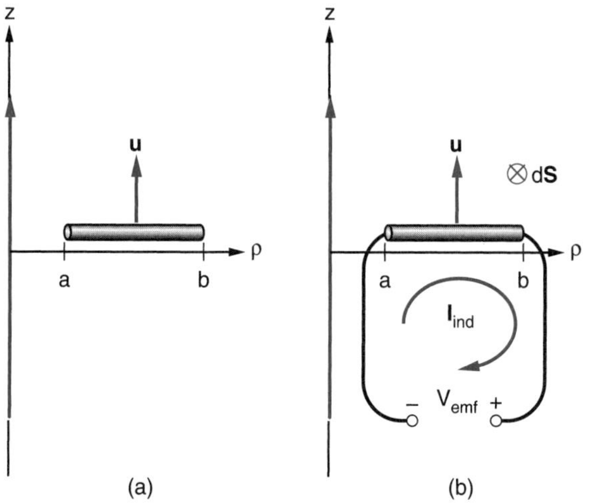

Figure 4.14   (a) Conductive bar moving in the field from a line of current. (b) A virtual loop is added for calculating a V_{emf}.

As a starting point, we create a virtual loop with a small gap in it as illustrated in Figure 4.14b. This loop will allow us to choose a direction for the integration. Here, we'll arbitrarily let dS go in the $+\mathbf{a}_\phi$ direction (the same direction as **B** from the line of current), and therefore we perform our circulation integral in the clockwise direction. If our calculated V_{emf} is positive, then the right side of the bar will be positive.

Equation (4.42), with $d\mathbf{L} = d\rho\mathbf{a}_\rho$, becomes

$$V_{emf} = \oint \left(u_z \mathbf{a}_z \times \frac{\mu_o I}{2\pi\rho} \mathbf{a}_\phi \right) \cdot d\rho\mathbf{a}_\rho = -\frac{\mu_o I u_z}{2\pi} \int_a^b \frac{d\rho}{\rho}$$

Here we only need to integrate from a to b (in the clockwise direction for our contour) since it is the only part that is moving. So we have

$$V_{emf} = -\frac{\mu_o I u_z}{2\pi} \ln\frac{b}{a}$$

Since $b > a$, the natural logarithm term is positive, and all of the other terms are positive, which means that, for our chosen loop, the V_{emf} is negative and the left end of the bar will be positive. Put another way, if we replaced the virtual portion of the loop with a stationary conductor, current would go in the counterclockwise direction.

Perhaps a simpler way to look at the bar's polarity is to consider a positive charge $+q$ initially held at the middle of the bar. If this charge is released, the Lorentz force would push it to the left of the bar; negative charges would be pushed to the right.

Drill 4.7   Find the direction and magnitude of the current induced in the circuit shown in Figure 4.15 if $B_o = 100$ $\mathbf{a}_z$ mT, $\mathbf{u} = -2.0$ $\mathbf{a}_y$ m/s, $w = 4.0$ cm, and $R = 50$ Ω. (*Answer:* = 160 μA, counterclockwise)

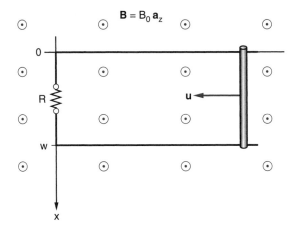

Figure 4.15 Sliding bar on conductive rails for Drill 4.7.

Generators

The electromagnetic generator converts mechanical motion to an AC electrical supply by employing Faraday's law. As illustrated in Figure 4.16, a conductive loop is turned in the presence of the magnetic field from a permanent magnet. It rotates with an angular velocity ω radians per second.

We choose to perform the circulation in the direction $1 \rightarrow 2 \rightarrow 3 \rightarrow 4 \rightarrow 1$, giving us the V_{emf} polarity as shown across the resistor. Notice that the loop sections from 2 to 3 and from 4 to 1 do not cut any magnetic flux, so we can ignore these sections in our calculations. This can be confirmed mathematically by considering that the Cartesian components of the velocity vector $\mathbf{u}$ only have $\mathbf{a}_x$ and $\mathbf{a}_y$ components, so when crossed with $B_o \mathbf{a}_y$ there is only an $\mathbf{a}_z$ component. However, for these arms of the loop, $d\mathbf{L}$ only has $\mathbf{a}_x$ and $\mathbf{a}_y$ components, which when dotted with $\mathbf{a}_z$ yields zero.

Considering the arms $1 \rightarrow 2$ and $3 \rightarrow 4$, we see that the distance traveled for a differential change in angle $d\phi$ is simply $a\,d\phi$, and $\mathbf{u} = a\,d\phi/dt\,\mathbf{a}_\phi$. Since $d\phi/dt$ is the definition of angular frequency, we have

$$\mathbf{u} = a\omega\mathbf{a}_\phi$$

Figure 4.16b shows how $\mathbf{a}_\phi$ can be decomposed into Cartesian vectors, which for the $3 \rightarrow 4$ arm is

$$(\mathbf{a}_\phi)_{3\rightarrow 4} = -\sin\phi\,\mathbf{a}_x + \cos_\phi\,\mathbf{a}_y$$

For the $1 \rightarrow 2$ arm, the Cartesian direction for $\mathbf{a}_\phi$ is

$$(\mathbf{a}_\phi)_{1\rightarrow 2} = \sin\phi\,\mathbf{a}_x - \cos\phi\,\mathbf{a}_y$$

Now we can calculate V_{emf}. The portion of V_{emf} from 1 to 2 is

$$\left(V_{emf}\right)_{1\rightarrow 2} = \int_{1}^{2}\left[a\omega\left(\sin\phi\,\mathbf{a}_x - \cos\phi\,\mathbf{a}_y\right)\times B_o\mathbf{a}_y\right]\cdot dz\mathbf{a}_z$$

This is easily integrated, going from $z = 0$ to $z = h$, to give

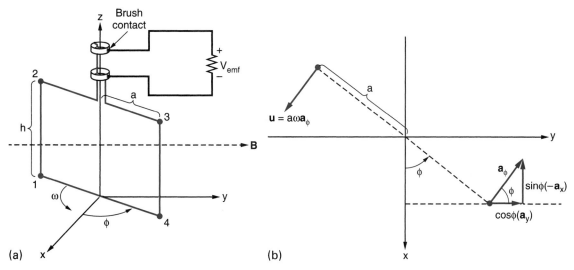

Figure 4.16 (a) An AC generator consisting of a loop rotating in the presence of a magnetic field. (b) Cross section for determination of vector elements.

$$(V_{emf})_{1\to2} = a\omega hB_o\sin\phi$$

When integrating the portion of V_{emf} from 3 to 4, the direction of **u** is opposite that of the 1 → 2 arm and the integration of dz goes from h to 0. This pair of sign changes ends up giving the same value for V_{emf}, and so the total V_{emf} is

$$V_{emf} = 2a\omega hB_o\sin\phi$$

The configuration shown in Figure 4.16 can also be used as an AC motor, if the load resistor is replaced with an AC source.

Drill 4.8 The generator of Figure 4.16 has dimensions $a = 8.0$ cm and $h = 10.$ cm, rotates at the rate of 120 revolutions per minute, and is in the presence of a 60.-mT field. Plot V_{emf} versus time t over several cycles. Assume $\phi = 0°$ at $t = 0$. (*Answer:* See Figure 4.17.)

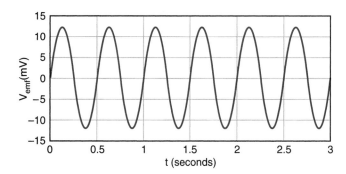

Figure 4.17 AC signal generated in Drill 4.8.

▶ 4.5 DISPLACEMENT CURRENT

Recall Ampère's circuital law for static fields from Chapter 3, rewritten as

$$\nabla \times \mathbf{H} = \mathbf{J}_c \tag{4.43}$$

Here, a "c" subscript has been added to the current density term to identify it as a conduction current density, which is related to the electric field by Ohm's law:

$$\mathbf{J}_c = \sigma \mathbf{E} \tag{4.44}$$

The current is a result of the drift of charge carriers in response to the electric field. In a vacuum where $\sigma = 0$, $\mathbf{J}_c = 0$.

A postulate of vector algebra is that the divergence of the curl of any vector field equals zero; that is,

$$\nabla \cdot (\nabla \times \mathbf{A}) = 0 \tag{4.45}$$

Let's apply this postulate to the point form of Ampère's circuital law for static magnetic fields:

$$\nabla \cdot (\nabla \times \mathbf{H}) = \nabla \cdot (\mathbf{J}_c) = 0 \tag{4.46}$$

Recalling the current continuity equation,

$$\nabla \cdot \mathbf{J}_c = -\frac{\partial \rho_v}{\partial t}$$

we see that the static form of Ampère's circuital law is clearly invalid for time-varying fields since it violates the law of current continuity.

The problem was resolved by Maxwell's introduction of an additional term to Ampère's circuital law:

$$\nabla \times \mathbf{H} = \mathbf{J}_c + \mathbf{J}_d \tag{4.47}$$

The additional term, called the *displacement current density*,[4.7] is the rate of change of the electric flux density,

$$\mathbf{J}_d = -\frac{\partial \mathbf{D}}{\partial t} \tag{4.48}$$

and thus

$$\boxed{\nabla \times \mathbf{H} = \mathbf{J}_c + \frac{\partial \mathbf{D}}{\partial t}} \tag{4.49}$$

Although the displacement current density term doesn't represent current in the conventional sense of charge flow, it does allow a time-varying electric field to be an additional source of magnetic field.

If we apply the divergence of a curl postulate to this new version of Ampère's circuital law, we have

$$\nabla \cdot (\nabla \times \mathbf{H}) = \nabla \cdot \mathbf{J}_c + \nabla \cdot \left(\frac{\partial \mathbf{D}}{\partial t} \right) = 0 \tag{4.50}$$

[4.7]Maxwell introduced this term in 1873 and it was verified experimentally by Heinrich Hertz in 1888.

Rearranging, we have

$$\nabla \cdot \mathbf{J}_c = -\nabla \cdot \left(\frac{\partial \mathbf{D}}{\partial t} \right) = -\frac{\partial}{\partial t} \nabla \cdot \mathbf{D} = -\frac{\partial \rho_v}{\partial t} \tag{4.51}$$

So we see that the additional term reconciles the vector postulate with the current continuity equation.

The addition of displacement current makes Ampère's law analogous to the point form of Faraday's law,

$$\nabla \times \mathbf{E} = -\frac{\partial \mathbf{B}}{\partial t}$$

and clearly shows the interdependency of time-varying electric and magnetic fields. Maxwell's original motivation for the additional term was to show that light is an electromagnetic wave, consisting of both electric and magnetic fields. He needed some way for magnetic field to be produced in vacuum, where conductivity (and therefore conduction current) is zero. The displacement current term fulfills this requirement.

We can integrate both sides of Ampère's circuital law over area to get

$$\int (\nabla \times \mathbf{H}) \cdot d\mathbf{S} = \int \left(\mathbf{J}_c + \frac{\partial \mathbf{D}}{\partial t} \right) \cdot d\mathbf{S} \tag{4.52}$$

then apply Stoke's theorem to get the integral form of Ampère's circuital law:

$$\boxed{\oint \mathbf{H} \cdot d\mathbf{L} = \int \mathbf{J}_c \cdot d\mathbf{S} + \frac{\partial}{\partial t} \int \mathbf{D} \cdot d\mathbf{S}} = i_c + i_d \tag{4.53}$$

where i_c and i_d represent conduction and displacement current, respectively.

To gain an understanding of displacement current, consider the simple capacitor circuit of Figure 4.18a. A sinusoidal voltage source $v(t) = V_o \sin \omega t$ is applied to the capacitor, and from circuit theory we know the voltage is related to the current $i(t)$ by the capacitance as

$$i(t) = C \frac{dv(t)}{dt} = CV_o \omega \cos \omega t$$

Now consider the loop in Figure 4.18b surrounding the plane surface S_1. By the static form of Ampère's circuital law, the circulation of **H** will be equal to the current that cuts through the surface. However, the surface doesn't have to be planar, and the same current must therefore flow through the surface S_2 that passes between the plates of the capacitor. Because no conduction current passes through an ideal capacitor ($\sigma = 0$ for an ideal dielectric), the current passing through S_2 must be entirely a displacement current.

We can calculate this current by considering the field across the capacitor is

$$\mathbf{E} = \frac{v(t)}{d} \mathbf{a}_z$$

where we let $\mathbf{a}_z$ be the direction from the positive to the negative plate. Then we have

$$\mathbf{D} = \frac{\varepsilon v(t)}{d} \mathbf{a}_z = \frac{\varepsilon V_o}{d} \sin \omega t \mathbf{a}_z$$

and the time derivative is

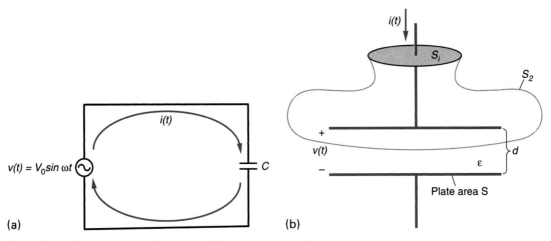

Figure 4.18 Capacitor used to demonstrate displacement current.

$$\mathbf{J}_d = \frac{\partial \mathbf{D}}{\partial t} = \frac{\omega \varepsilon V_o}{d} \cos \omega t \, \mathbf{a}_z$$

The current through the capacitor is

$$i_d = \int \mathbf{J}_d \cdot d\mathbf{S} = \frac{\omega \varepsilon S V_o}{d} \cos \omega t$$

For a parallel-plate capacitor, $C = \varepsilon S/d$, so

$$i_d = CV_o \omega \cos \omega t$$

which agrees with the circuit theory result.

The conduction current in the dielectric can be ignored if the dielectric is a low-loss (small σ) dielectric. But if that is not the case, the conduction current density is found by Ohm's law and the expression for electric field intensity between the plates as

$$\mathbf{J}_c = \frac{\sigma v(t)}{d} \mathbf{a}_z$$

For the parallel plate–capacitor example of Figure 4.18, the conduction current term is then found to be

$$i_c = \frac{\sigma V_o S}{d} \sin \omega t$$

The ratio of the conduction current magnitude to the displacement current magnitude is called the *loss tangent* and is seen to be

$$\tan \delta = \left| \frac{i_c}{i_d} \right| = \frac{\sigma}{\omega \varepsilon} \tag{4.54}$$

Loss tangent is a measure of the quality of the dielectric. A good dielectric will have a very low-loss tangent, typically less than 0.001. Loss tangent will be discussed further in Chapter 5, Section 5.3.

Drill 4.9 A pair of 100. cm^2 area plates are separated by a 1.0-mm-thick layer of lossy dielectric characterized by $\varepsilon_r = 50.$ and $\sigma = 1.0 \times 10^{-4}$ S/m. (a) Calculate the capacitance. If a voltage $v(t) = 1.0 \cos (2\pi \times 10^3 t)$ V is placed across the plates, determine (b) the conduction current, (c) the displacement current, and (d) the loss tangent. (*Answer:* (a) C = 4.4 nF, (b) $i_c = 1.0\cos(2\pi \times 10^3 t)$mA, (c) $i_d = -28 \sin (2\pi \times 10^3 t)$ µA, (d) tan δ = 36)

▶ 4.6 MAXWELL'S EQUATIONS

With his insightful inclusion of displacement current in Ampére's circuital law, Maxwell was able to unify all of the theories of electricity and magnetism into one concise set of formulas known as Maxwell's equations. With the addition of the Lorentz force equation, the constitutive relations for material media, and the current continuity equation, all of the fundamental electromagnetic equations are contained within Table 4.1.

The table contains both point and integral versions of Maxwell's equations. Both are useful in different situations. As a review, the reader is encouraged to apply the divergence theorem and Stoke's theorem to convert from the integral form to the point form, and vice versa.

Although we've covered many other equations so far in this text (and will cover many more), the equations listed in Table 4.1 are the fundamental ones from which all else follows. For example, the static form of Maxwell's equations used in Chapters 2 and 3 are found by zeroing the time derivative in the equations in Table 4.1. As another example, the relationships between field quantities at the boundary between different media are found from straightforward application of Maxwell's equations. Finally, it may be pointed out that the concept of electric potential, so useful in solving electromagnetics problems (and useful as a bridge between electromagnetics and circuits), is merely a stepping stone from the fundamental equations to the final solutions.

Table 4.1   The Fundamental Electromagnetics Equations

Maxwell's equations	Point (differential) form	Integral form
Gauss's law	$\nabla \cdot \mathbf{D} = \rho_v$	$\oint \mathbf{D} \cdot d\mathbf{S} = Q_{enc}$
Gauss's law for magnetic fields	$\nabla \cdot \mathbf{B} = 0$	$\oint \mathbf{B} \cdot d\mathbf{S} = 0$
Faraday's law	$\nabla \times \mathbf{E} = -\dfrac{\partial \mathbf{B}}{\partial t}$	$\oint \mathbf{E} \cdot d\mathbf{L} = -\dfrac{\partial}{\partial t}\int \mathbf{B} \cdot d\mathbf{S}$
Ampére's circuital law	$\nabla \times \mathbf{H} = \mathbf{J}_c + \dfrac{\partial \mathbf{D}}{\partial t}$	$\oint \mathbf{H} \cdot d\mathbf{L} = \int \mathbf{J} \cdot d\mathbf{S} + \dfrac{\partial}{\partial t}\int \mathbf{D} \cdot d\mathbf{S}$
	Lorentz force equation	$\mathbf{F} = q(\mathbf{E} + \mathbf{v} \times \mathbf{B})$
	Constitutive relations	$\begin{cases} \mathbf{D} = \varepsilon\mathbf{E} \\ \mathbf{B} = \mu\mathbf{H} \\ \mathbf{J} = \sigma\mathbf{E} \text{ (Ohm's law)} \end{cases}$
	Current continuity equation	$\nabla \cdot \mathbf{J} = -\dfrac{\partial \rho_v}{\partial t}$

A key aspect of these equations is the interdependency of the electric and magnetic fields. Because a time-varying electric field is a source of magnetic field, and vice versa, it will be shown in the next chapter how these equations led Maxwell to postulate the existence of electromagnetic waves.

▷ 4.7 LOSSLESS TEM WAVES

Let's use Maxwell's equations to study the relationship between the electric and magnetic field components of an electromagnetic wave. Consider an *x-polarized* wave propagating in the +*z* direction in some ideal medium characterized by μ and ε, with σ = 0. To say an electromagnetic wave is *x*-polarized simply means that the electric field vector is always pointing in the *x* (or −*x*) direction. We choose σ = 0 to make the medium lossless, for simplicity.

The propagating electric field is given by

$$\mathbf{E}(z,t) = E_o \cos(\omega t - \beta z)\mathbf{a}_x \tag{4.55}$$

where E_o is the wave amplitude (in volts per meter), propagating at an angular frequency of ω radians per second and having a phase constant of β radians per meter. The field is a function of its *z* position and time and is plotted versus *z* at time *t* = 0 in Figure 4.19.[4.8] Upon application of Maxwell's equations, we will find that the magnetic field also propagates in the +*z* direction, but its field vector is always normal (perpendicular) to the electric field vector. Such a wave is said to propagate in a *transverse electromagnetic wave mode*, or TEM mode for short.

We can apply Faraday's law,

$$\nabla \times \mathbf{E} = -\frac{\partial \mathbf{B}}{\partial t} = -\mu \frac{\partial \mathbf{H}}{\partial t}$$

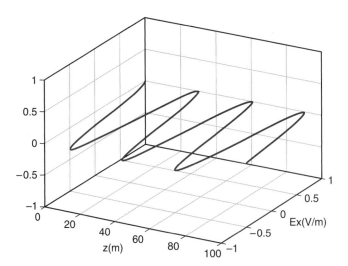

Figure 4.19 A plot of the equation $\mathbf{E}(z,0) = E_o\cos(\beta z)\mathbf{a}_x$ at 10 MHz in free space with $E_o = 1$ V/m.

[4.8]This is a MATLAB-generated figure (see MATLAB 4.3). Careful study reveals that it is indeed a right-handed coordinate system.

to (4.55) to solve for the magnetic field. Taking the curl of **E**, we have

$$\nabla \times \mathbf{E} = \begin{vmatrix} \mathbf{a}_x & \mathbf{a}_y & \mathbf{a}_z \\ \dfrac{\partial}{\partial x} & \dfrac{\partial}{\partial y} & \dfrac{\partial}{\partial z} \\ E_o \cos(\omega t - \beta z) & 0 & 0 \end{vmatrix} \tag{4.56}$$

$$= -\left(-\frac{\partial}{\partial z} E_o \cos(\omega t - \beta z)\right)\mathbf{a}_y = \beta E_o \sin(\omega t - \beta z)\mathbf{a}_y$$

This must equal the right side of Faraday's law, so

$$-\mu \frac{\partial \mathbf{H}}{\partial t} = \beta E_o \sin(\omega t - \beta z)\mathbf{a}_y \tag{4.57}$$

After dividing both sides by $-\mu$, we form the integral

$$\int d\mathbf{H} = -\frac{\beta E_o}{\mu}\int \sin(\omega t - \beta z)\mathbf{a}_y dt \tag{4.58}$$

The right side of the equation is easily integrated using $\int \sin u\, du = -\cos u$, where $u = \omega t - \beta z$ and du is ωdt. So we have

$$\mathbf{H} = \frac{\beta E_o}{\omega \mu}\cos(\omega t - \beta z)\mathbf{a}_y + C_1 \tag{4.59}$$

where C_1 is a constant of integration. Examining this problem we see that the time-varying **E** is the only source of **H**; that is, there is no conduction current given that can also generate **H**. So if we were to "turn off" **E**, **H** would also disappear. From this argument, we see that C_1 must be zero.

To the plot of **E** we can add a plot of **H** versus z at time $t = 0$ (see Figure 4.20). The amplitude of the magnetic field is given as $\beta E_o/\omega \mu$. In other words, the amplitudes of **E** and

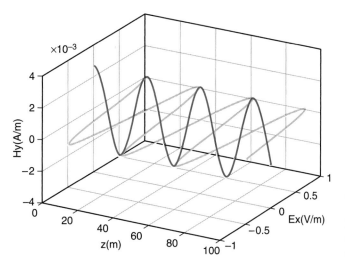

Figure 4.20 Plot of the equation $\mathbf{H}(z, 0) = (\beta E_0/\omega \mu_0)\cos(-\beta z)\mathbf{a}_y$ at 10 MHz in free space with $E_o = 1$ V/m along with the dashed plot of **E**(z,0).

H are not independent; they are related by Maxwell's equations. Also, note that both waves are traveling in the z direction, and, moreover, they are related by yet still another version of the right-hand rule! Mathematically, we can say that *the electromagnetic wave propagates in a direction given by the cross product of E and H*: Starting with fingers pointing in the *E*-field direction ($\mathbf{a}_x$), then curling them to the *H*-field direction ($\mathbf{a}_y$) means the thumb will point in the propagation direction ($\mathbf{a}_z$).

Even though we now have both fields, let's continue using Maxwell's equations on this problem. We can apply Ampère's circuital law,

$$\nabla \times \mathbf{H} = \mathbf{J}_c + \frac{\partial \mathbf{D}}{\partial t}$$

to the changing magnetic field to recalculate **E**. Since there is no conduction current ($\sigma = 0$), the $\mathbf{J}_c$ term is not present in our problem and we have

$$\nabla \times \mathbf{H} = \varepsilon \frac{\partial \mathbf{E}}{\partial t} \tag{4.60}$$

Taking the curl of **H**, we have

$$\nabla \times \mathbf{H} = \begin{vmatrix} \mathbf{a}_x & \mathbf{a}_y & \mathbf{a}_z \\ \dfrac{\partial}{\partial x} & \dfrac{\partial}{\partial y} & \dfrac{\partial}{\partial z} \\ 0 & \dfrac{\beta E_o}{\omega \mu} \cos(\omega t - \beta z) & 0 \end{vmatrix} \tag{4.61}$$

$$= -\frac{\beta E_o}{\omega \mu} \frac{\partial}{\partial z} \cos(\omega t - \beta z)\mathbf{a}_x = -\frac{\beta^2 E_o}{\omega \mu} \sin(\omega t - \beta z)\mathbf{a}_x$$

Equating this to the right side of (4.60), we have

$$\frac{\partial \mathbf{E}}{\partial t} = -\frac{\beta^2 E_o}{\omega \mu \varepsilon} \sin(\omega t - \beta z)\mathbf{a}_x \tag{4.62}$$

which we can integrate as we did for (4.58) resulting in

$$\mathbf{E} = \frac{\beta^2 E_o}{\omega^2 \mu \varepsilon} \cos(\omega t - \beta z)\mathbf{a}_x \tag{4.63}$$

For (4.63) and (4.55) to be equal, we must have

$$\beta^2 = \omega^2 \mu \varepsilon$$

or

$$\boxed{\beta = \omega\sqrt{\mu \varepsilon}} \tag{4.64}$$

Earlier in the chapter, we discovered that the propagation velocity is related to the phase constant and angular frequency by

$$u_p = \frac{\omega}{\beta}$$

or, since $\omega = 2\pi f$ and $\beta = 2\pi/\lambda$, $u_p = \lambda f$. But now we see from (4.64) that the propagation velocity is also given by

$$\boxed{u_p = \frac{1}{\sqrt{\mu\varepsilon}}} \tag{4.65}$$

This very significant result, which we found by applying Maxwell's equations to the propagating fields, relates u_p to the properties of the medium. In the absence of any media (termed *free space*), the constitutive parameters are $\mu = \mu_o$, $\varepsilon = \varepsilon_o$, and $\sigma = 0$. Plugging the by-now well-known values of μ_o and ε_o into (4.65), we are pleased to see that electromagnetic waves in free space propagate at the speed of light!

Drill 4.10   A y-polarized plane wave in air ($\mu = \mu_o$, $\varepsilon = \varepsilon_o$, and $\sigma = 0$) propagates in the x direction at 10 MHz. Write the expression for $\mathbf{E}(x,t)$ if the wave has a 10 V/m amplitude. (*Answer:* $\mathbf{E}(x,t)=10\cos(2\pi \times 10^7 t - .2\pi x/3)\mathbf{a}_y$ V/m)

► **EXAMPLE 4.5**

Suppose in a nonmagnetic medium we have an electric field

$$\mathbf{E}(x,t) = 20.\cos\left(\pi \times 10^7 t + \frac{\pi}{10}x + \frac{\pi}{6}\right)\mathbf{a}_y \text{ V/m}$$

Among other things, we want to find $\mathbf{H}(x,t)$.

By inspection, we see that the wave amplitude is $E_o = 20$ V/m. The frequency is

$$f = \frac{\omega}{2\pi} = 5 \text{ MHz}$$

With a phase constant $\beta = \pi/10$ radians/m, we have

$$u_p = \frac{\omega}{\beta} = \frac{\pi \times 10^7}{\pi/10} = 10^8 \text{ m/s}$$

Since this is in nonmagnetic medium, we have

$$u_p = \frac{c}{\sqrt{\varepsilon_r}}$$

or $\varepsilon_r = 9$.

To find $\mathbf{H}(x,t)$ we employ Faraday's law and follow the procedure from earlier in this section, arriving at

$$\mathbf{H}(x,t) = \frac{\beta E_o}{\omega\mu}\cos\left(\pi \times 10^7 t + \frac{\pi}{10}x + \frac{\pi}{6}\right)\mathbf{a}_z$$

or

$$\mathbf{H}(x,t) = 160\cos\left(\pi \times 10^7 t + \frac{\pi}{10}x + \frac{\pi}{6}\right)\mathbf{a}_z \text{ mA/m}$$

Drill 4.11 Suppose

$$\mathbf{E}(z,t) = 6.0\cos\left(2\pi \times 10^{8}t - \frac{2}{3}\pi z\right)\mathbf{a}_y \;\; \text{V/m}$$

(a) What is the wave amplitude, frequency, phase constant, wavelength, and propagation velocity? (b) Find $\mathbf{H}(z,t)$. (*Answer:* (a) 6.0 V/m, 100 MHz, $2\pi/3$ radians/m, 3m, 3×10^{8}m/s; (b) $\mathbf{H}(z,t) = -16\cos(2\pi \times 10^{8}t - 2\pi z/3)\mathbf{a}_x$ mA/m)

▶ **MATLAB 4.3**

Let's use MATLAB to create the plots in Figures 4.19 and 4.20.
First, for the E_x versus z plot we have the following:

```
% M-File: ML0403a
%
% This program generates a 3D plot of Ex versus z.
%
% Wentworth, 7/17/02
%
% Variables:
% Eo          field amplitude (V/m)
% f           frequency (Hz)
% c           speed of light in vacuum (m/s)
% lambda      wavelength (m)
% B           phase constant (1/m)
% Ex          electric field intensity (V/m)
% z           position
% null        null array

clc           %clears the command window
clear         %clears variables

% Initialize Variables
Eo=1;
f=10e6;
c=2.998e8;
lambda=c/f;
B=2*pi/lambda;

% Perform Calculation
z=0:1:100;
Ex=Eo*cos(-B*z);
null=0.*z;   %build a null array

% Generate the Plot
plot3(z,Ex,null)
grid on
view([30 30])
xlabel('z(m)')
ylabel('Ex(V/m)')
```

This is plotted in Figure 4.19. Notice the formation of a "null" array. We do this so we'll have something of the same dimension as z and E_x to plot on the three-dimensional coordinate system. (In short, it forces y to be zero for all these points.) Also, for increased clarity, the trace thickness was increased after the program was run by using the editor in the figure window.

For the E_x and H_y plots versus z, we have the following:

```
% M-File: ML0403b
%
% This program generates a 3D plot of Ex and Hy versus z.
%
% Wentworth, 7/17/02
%
% Variables:
% Eo        field amplitude (V/m)
% f         frequency (Hz)
% w         angular frequency (rad/s)
% c         speed of light in vacuum (m/s)
% uo        free space permeability (F/m)
% lambda    wavelength (m)
% B         phase constant (1/m)
% Ex        electric field intensity (V/m)
% Hy        magnetic field intensity (A/m)
% z         position
% null      null array

clc         %clears the command window
clear       %clears variables

% Initialize Variables
Eo=1;
f=10e6;
c=2.998e8;
uo=pi*4e-7;
lambda=c/f;
B=2*pi/lambda;
w=2*pi*f;

% Perform Calculation
z=0:1:100;
Ex=Eo*cos(-B*z);
Hy=((B*Eo)/(w*uo))*cos(-B*z);
null=0.*z;   %build a null array

% Generate the Plot
plot3(z,Ex,null,'-',z,null,Hy)
grid on
view([30 30])
xlabel('z(m)')
ylabel('Ex(V/m)')
zlabel('Hy(A/m)')
```

In this case, we've chosen to plot E_x versus z using a dashed line.

▶ 4.8 TIME-HARMONIC FIELDS AND PHASORS

Many, if not most, electromagnetic applications involve fields that vary sinusoidally with position and time. Such *time-harmonic* fields are encountered in a host of communications applications, and of course all AC circuitry is sinusoidal. In addition, repeated pulses of information may be treated as a Fourier series of sinusoidal waves.

A time-harmonic signal can be transformed into the frequency domain by the use of *phasors*. The utility of working in the frequency domain (also called the *phasor domain*) is that the time factor is removed from the analysis, and time derivatives and integrals become simple algebraic exercises.

Phasors are based on the use of complex numbers.[4.9] At a fixed spatial position, the value of a sine wave can be represented versus time by a polar plot of its amplitude r and phase θ. This polar plot can be superimposed onto a set of *real* (Re) and *imaginary* (Im) axes, as shown in Figure 4.21. The phasor can be written $re^{j\theta}$, or, as is evident from the figure,

$$re^{j\theta} = r \cos \theta + jr \sin \theta \qquad (4.66)$$

The relation between the polar form of the phasor ($re^{j\theta}$) and the rectangular form ($r\cos\theta + jr\sin\theta$) is given by Euler's[4.10] identity:

$$e^{j\theta} = \cos \theta + j \sin \theta \qquad (4.67)$$

It is also customary to write phasors in a shorthand polar form

$$re^{j\theta} = r \,\underline{|\theta} \qquad (4.68)$$

In Figure 4.21, corresponding points are indicated on both the time-domain sine-wave plot and the frequency-domain polar plot. The real part of the phasor is found as $\text{Re}(re^{j\theta}) = r\cos\theta$; the imaginary part is $\text{Im}(re^{j\theta}) = r \sin \theta$.

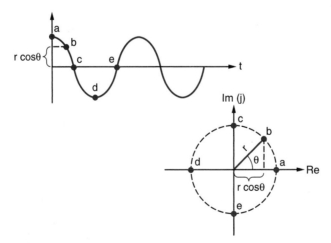

Figure 4.21 Time plot synchronized with polar plot.

[4.9]See Appendix C for a summary of complex numbers.

[4.10]The name Euler is pronounced "oiler."

▷ **MATLAB 4.4**

Show how the phasor point on a polar plot corresponds to the location on the wave plot with an animation.

```
% M-File: ML0404
%
% This program generates a time plot animation
% synchronized with a polar plot.
%
% Wentworth, 7/17/02
%
% Variables:
% Eo        field amplitude (V/m)
% E         electric field intensity (V/m)
% E1        E-field for movie
% f         frequency (Hz)
% theta     angle
% theta1    angle for movie
% T         period (1/f)
% t         time (s)
% t1        time for movie

clc            %clears the command window
clear          %clears variables

% Generate the reference frame
% Initialize Variables
Eo=1;
f=1000;
T=1/f;

% Perform Calculation
t=0:T/100:2*T;
theta=2*pi*f*t;
E=Eo*cos(theta);

% Generate the Plot
subplot(211),plot(t,E,0,Eo,'ro');
subplot(212),polar(0,Eo,'ro');
pause

%Make the Movie
t1=0:T/50:2*T;
for n=1:100
   theta1(n)=2*pi*f*t1(n);
   E1(n)=Eo*cos(theta1(n));
   subplot(211),plot(t,E,t1(n),E1(n),'ro');
   subplot(212),polar(theta1(n),Eo,'ro');
   M(:,1)=getframe;
end
```

If the animation runs too fast to follow the little red circle, try increasing the steps (from T/50 to T/100) and the upper limit on n from 100 to 200.

A general time-harmonic electric field is a function of position (x, y, z) and time (t) and can be written in *instantaneous form* as

$$\mathbf{E}(x,y,z,t) = \mathbf{E}(x,y,z)\cos(\omega t + \phi) \tag{4.69}$$

Now, using Euler's identity we can also write this as

$$\mathbf{E}(x,y,z,t) = \text{Re}\left[\mathbf{E}(x,y,z)e^{j(\omega t + \phi)}\right] \tag{4.70}$$

This can be reorganized and written as

$$\mathbf{E}(x,y,z,t) = \text{Re}\left[\mathbf{E}(x,y,z)e^{j\phi}\,e^{j\omega t}\right] = \text{Re}\left[\mathbf{E}_s e^{j\omega t}\right] \tag{4.71}$$

where the *phasor form* of the field is

$$\mathbf{E}_s = \mathbf{E}(x,y,z)e^{j\phi} \tag{4.72}$$

The phasor, written with an s subscript,[4.11] is the time-harmonic field with the time dependency stripped away.

Let's see how we would write the point form of Faraday's law in terms of phasors. We have

$$\nabla \times \mathbf{E}(x,y,z,t) = -\frac{\partial \mathbf{B}(x,y,z,t)}{\partial t}$$

Using phasors, we see that this is equivalent to

$$\nabla \times \left(\text{Re}\left[\mathbf{E}_s e^{j\omega t}\right]\right) = -\frac{\partial}{\partial t}\left(\text{Re}\left[\mathbf{B}_s e^{j\omega t}\right]\right) \tag{4.73}$$

On the left side of (4.73), the curl operator is a position derivative, and so we can pull out the Re and the $e^{j\omega t}$ terms, so

$$\nabla \times \left(\text{Re}\left[\mathbf{E}_s e^{j\omega t}\right]\right) = \text{Re}\left[(\nabla \times \mathbf{E}_s)e^{j\omega t}\right] \tag{4.74}$$

On the right side, it can be shown (see Problem 4.37) that

$$-\frac{\partial}{\partial t}\left(\text{Re}\left[\mathbf{B}_s e^{j\omega t}\right]\right) = -\text{Re}\left[\frac{\partial}{\partial t}\mathbf{B}_s e^{j\omega t}\right] \tag{4.75}$$

and since $\mathbf{B}_s$ is time independent,

$$-\text{Re}\left[\frac{\partial}{\partial t}\mathbf{B}_s e^{j\omega t}\right] = -\text{Re}\left[\mathbf{B}_s \frac{\partial}{\partial t}e^{j\omega t}\right] = -\text{Re}\left[(j\omega \mathbf{B}_s)e^{j\omega t}\right] \tag{4.76}$$

Comparing (4.74) and (4.76), we find that

$$\boxed{\nabla \times \mathbf{E}_s = -j\omega \mathbf{B}_s} \tag{4.77}$$

This is the differential phasor form of Faraday's law.

[4.11]The student may recall the s-domain in circuit analysis, where $s = j\omega$.

TABLE 4.2 Differential Phasor Form of Maxwell's Equations

$$\nabla \cdot \mathbf{D}_s = \rho_{vs}$$
$$\nabla \cdot \mathbf{B}_s = 0$$
$$\nabla \times \mathbf{E}_s = -j\omega\mathbf{B}_s$$
$$\nabla \times \mathbf{H}_s = \mathbf{J}_s + j\omega\mathbf{D}_s$$

The rest of Maxwell's equations can also be written in differential phasor form as given in Table 4.2. Derivations of Gauss's law and Ampère's circuital law are included as Problem 4.38.

The procedure for using phasors in problems is to first transform the instantaneous form of the field quantities to phasors. The problem is then solved in the phasor domain, and at the end (or at any intermediate point) the phasor form can be transformed back to an instantaneous form.

▷ **EXAMPLE 4.6**

Let's consider the problem in the previous section where we were given

$$\mathbf{E}(z,t) = E_o\cos(\omega t - \beta z)\mathbf{a}_x$$

and now want to use phasors to find $\mathbf{H}(z,t)$.

As a first step, we convert to a phasor,

$$\mathbf{E}_s = E_o e^{-j\beta z}\mathbf{a}_x$$

Next, we employ Faraday's law to find $\mathbf{B}_s$. We have

$$\nabla \times \mathbf{E}_s = -j\omega\mathbf{B}_s$$

or

$$\begin{vmatrix} \mathbf{a}_x & \mathbf{a}_y & \mathbf{a}_z \\ \dfrac{\partial}{\partial x} & \dfrac{\partial}{\partial y} & \dfrac{\partial}{\partial z} \\ E_o e^{-j\beta z} & 0 & 0 \end{vmatrix} = -j\beta E_o e^{-j\beta z}\mathbf{a}_y = -j\omega\mathbf{B}_s$$

Solving for $\mathbf{B}_s$, we find

$$\mathbf{B}_s = \frac{\beta E_o}{\omega} e^{-j\beta z}\mathbf{a}_y$$

Then, we can find $\mathbf{H}_s$ by dividing $\mathbf{B}_s$ by μ.

To find the instantaneous form, we have to reinsert $e^{j\omega t}$, employ Euler's identity, and take the real part of the result. So we have

$$\mathbf{H}(z,t) = \mathrm{Re}\left[\frac{\beta E_o}{\omega\mu} e^{-j\beta z} e^{j\omega t}\mathbf{a}_y\right] = \frac{\beta E_o}{\omega\mu}\cos(\omega t - \beta z)\mathbf{a}_y$$

▷ **EXAMPLE 4.7**

Now suppose we have a magnetic field intensity given as

$$\mathbf{H}(z,t) = H_o \sin\left(\omega t - \beta z + \frac{\pi}{4}\right)\mathbf{a}_y$$

and we want to find $\mathbf{E}(z,t)$ using phasors.

Using the relation $\sin(\alpha) = \cos(\alpha - \pi/2)$, we have

$$\sin(\omega t - \beta z + \pi/4) = \cos(\omega t - \beta z - \pi/4)$$

Converting $\mathbf{H}(z,t)$ to a phasor, we have

$$\mathbf{H}_s = H_o e^{-j(\beta z + \pi/4)}\mathbf{a}_y$$

The phasor vector $\mathbf{E}_s$ is found using Ampère's circuital law, where we can assume $\sigma = 0$ in the absence of any other information, and we have

$$\nabla \times \mathbf{H}_s = j\omega\varepsilon\mathbf{E}_s$$

Evaluating the curl, we find

$$j\beta H_o e^{-j(\beta z + \pi/4)}\mathbf{a}_x = j\omega\varepsilon\mathbf{E}_s$$

so

$$\mathbf{E}_s = \frac{\beta H_o}{\omega\varepsilon} e^{-j(\beta z + \pi/4)}\mathbf{a}_x$$

Converting to instantaneous form gives

$$\mathbf{E}(z,t) = \text{Re}\left[\frac{\beta H_o}{\omega\varepsilon} e^{-j\beta z} e^{-j\pi/4} e^{j\omega t}\mathbf{a}_x\right] = \frac{\beta H_o}{\omega\varepsilon} \cos(\omega t - \beta z - \pi/4)\mathbf{a}_x$$

Drill 4.12   Convert the following instantaneous quantities to phasors: (a) $A = 16\cos(\pi \times 10^6 t + \pi/3)$, (b) $\mathbf{A}(x, t) = A_o \sin(4\pi \times 10^8 t + 2x)\,\mathbf{a}_y$. (*Answer:* (a) $A_s = 16e^{j\pi/3}$, (b) $\mathbf{A}_s = A_o e^{j(2x - \pi/2)}\mathbf{a}_y$.)

Drill 4.13   Convert the following phasors to instantaneous quantities: (a) $A_s = 10e^{j\pi/4}$, (b) $A_s = j5e^{j3\pi/4}$, (c) $A_s = 4 + j3$. (*Answer:* (a) $A = 10\cos(\omega t + \pi/4)$, (b) $A = -5\sin(\omega t + 3\pi/4)$, (c) $A = 5\cos(\omega t + 36.9°)$)

▶ SUMMARY

- The rate of change of charge density is related to the divergence of the current density vector by the current continuity equation

$$\nabla \cdot \mathbf{J} = -\frac{\partial \rho_v}{\partial t}$$

Using this equation and Gauss's law, we can relate charge density to time as

$$\rho_v = \rho_o e^{-t/\tau}$$

where ρ_o is the initial charge density and τ is the relaxation time

$$\tau = \frac{\varepsilon}{\sigma}$$

- The general wave equation is

$$\mathbf{E}(z,t) = E_o e^{-\alpha z} \cos(\omega t - \beta z + \phi)\mathbf{a}_x$$

where E_o is the amplitude and α is the attenuation. The cosine argument is the phase, with angular frequency ω (radians/s), phase constant β (radians/m), and phase shift ϕ (radians).

- Faraday's law relates an electromotive force V_{emf} to the rate of change of flux linking a circuit by

$$V_{emf} = -\frac{\partial\lambda}{\partial t}$$

where λ is the flux linkage. For a single-loop circuit, the emf can be written

$$V_{emf} = \oint \mathbf{E}\cdot d\mathbf{L} = -\frac{\partial}{\partial t}\int \mathbf{B}\cdot d\mathbf{S}$$

- Transformer emf is for the case of a time-varying magnetic field through a fixed surface. For this case we can find the point form of Faraday's law,

$$\nabla\times\mathbf{E} = -\frac{\partial\mathbf{B}}{\partial t}$$

- Motional emf has a changing surface in a constant magnetic field. A form of Faraday's law for this case is

$$V_{emf} = \oint(\mathbf{u}\times\mathbf{B})\cdot d\mathbf{L}$$

- Displacement current density $\mathbf{J}_d$ is equal to the rate of change of the electric flux density,

$$\mathbf{J}_d = \frac{\partial\mathbf{D}}{\partial t}$$

This term is added to Ampére's circuital law, showing that a changing electric field produces a magnetic field.

- Maxwell's equations, summarized here in point (differential) form, are

$$\nabla\cdot\mathbf{D} = \rho_v$$
$$\nabla\cdot\mathbf{B} = 0$$
$$\nabla\times\mathbf{E} = -\frac{\partial\mathbf{B}}{\partial t}$$
$$\nabla\times\mathbf{H} = \mathbf{J}_c + \frac{\partial\mathbf{D}}{\partial t}$$

In integral form, Maxwell's equations are

$$\oint\mathbf{D}\cdot d\mathbf{S} = Q_{enc}$$
$$\oint\mathbf{B}\cdot d\mathbf{S} = 0$$
$$\oint\mathbf{E}\cdot d\mathbf{L} = -\frac{\partial}{\partial t}\int\mathbf{B}\cdot d\mathbf{S}$$
$$\oint\mathbf{H}\cdot d\mathbf{L} = \int\mathbf{J}\cdot d\mathbf{S} + \frac{\partial}{\partial t}\int\mathbf{D}\cdot d\mathbf{S}$$

- The direction of propagation of a transverse electromagnetic mode wave is given by the cross product $\mathbf{E}\times\mathbf{H}$, where the field vectors of both $\mathbf{E}$ and $\mathbf{H}$ are normal to the propagation direction.

- A time-harmonic field in the time domain can be represented by a phasor in the frequency domain. The transformation employs Euler's identity. For an electric field given by

$$E(x,y,z,t) = E(x,y,z)\cos(\omega t + \phi)$$

the phasor is written

$$\mathbf{E}_s = \mathbf{E}(x,y,z)\,e^{j\phi}$$

▶ PROBLEMS

4.1 Current Continuity and Relaxation Time

4.1 How long does it take for charge density to drop to 1% of its initial value in polystyrene?

4.2 At a particular point in a slab of silver, a charge density of 10^9 C/m³ is introduced. Plot ρ_v versus time for a duration of 10 relaxation times.

4.3 A current density is given by $\mathbf{J} = \rho e^{-.01t}\mathbf{a}_\rho$ A/m². Find the charge density after 10 s if it has an initial value of zero.

4.4 At $t = 0$ s, 60.0 μC is evenly distributed throughout a 2.00-cm-diameter pure silicon sphere. (a) Find the initial charge density. (b) How long does it take for the charge density to drop to 10% of its initial value? (c) What will be the final surface charge density?

4.2 Wave Fundamentals

4.5 A propagating electric field is given by

$$E(z,t) = 100.e^{-.01z}\cos\left(\pi\times10^7 t + \pi z - \frac{\pi}{4}\right)\text{V/m}$$

(a) Determine the attenuation constant, the wave frequency, the wavelength, the propagation velocity, and the phase shift.

(b) How far must the wave travel before its amplitude is reduced to 1.0 V/m?

4.6 A 10.0-MHz magnetic field travels in a fluid for which the propagation velocity is 1.0×10^8 m/s. Initially, we have $\mathbf{H}(0, 0) = 2.0\,\mathbf{a}_x$ A/m. The amplitude drops to 1.0

 EMAG SOLUTIONS

A/m after the wave travels 5.0 m in the *y* direction. Find the general expression for this wave.

4.7 Modify the simple wave program in MATLAB 4.1 to include attenuation. Generate a plot for the case where the amplitude is 4 V/m, the attenuation constant is 0.001 Np/m, and the frequency is 1 MHz. Take your snapshot in time at 0 s, and let your phase shift be 0°.

4.8 Modify the traveling wave program in MATLAB 4.2 to include attenuation. Use the parameters from Problem 4.7, except for the fixed time of course.

4.3 Faraday's Law and Transformer EMF

4.9 The magnetic flux density increases at the rate of 10 Wb/m²/s in the *z* direction. A 10×10 cm square conducting loop, centered at the origin in the *x–y* plane, has 10 Ω of distributed resistance. Determine the direction (with a sketch) and magnitude of the induced current in the conducting loop.

4.10 A bar magnet is dropped through a conductive ring. Indicate in a sketch the direction of the induced current when the falling magnet is just above the plane of the ring and when it is just below the plane of the ring, as shown in Figure 4.22.

4.11 Considering Figure 4.7, suppose the area of a single loop of the pair is 100 cm², and the magnetic flux density is constant over the area of the loops but changes with time as $\mathbf{B} = B_o e^{-\chi t} \, \mathbf{a}_z$, where $B_o = 4.0$ mWb/m² and $\chi = 0.30$ Np/s. Determine V_R at 1, 10, and 100 s.

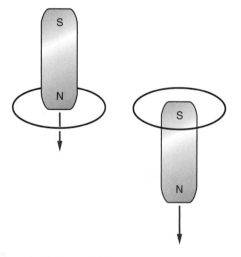

Figure 4.22 A dropped bar magnet just before and just after it passes through the plane of a conductive wire ring (for Problem 4.10).

4.12 Sometimes a transformer is used as an impedance converter, where impedance is given by *v/i*. Find an expression for the impedance Z_1 seen by the primary side of the transformer in Figure 4.11a that has a load impedance Z_2 terminating the secondary.

4.13 A 1.0-mm-diameter copper wire is shaped into a square loop of side 4.0 cm. It is placed in a plane normal to a magnetic field increasing with time as $\mathbf{B} = 1.0 \, t \, \mathbf{a}_z$ Wb/m², where *t* is in seconds. (a) Find the magnitude of the induced current and indicate its direction in a sketch. (b) Calculate the magnetic flux density at the center of the loop resulting from the induced current, and compare this with the original magnetic flux density that generated the induced current at *t* = 1.0 s.

4.14 The mean length around a nickel core of a transformer like the one shown in Figure 4.11a is 16 cm, and its cross sectional area is 1 cm². There are 30 turns on the primary side and 45 on the secondary side. If the current on the primary side is $1.0 \sin(20\pi \times 10^6 t)$ mA, (a) calculate the amplitude of the magnetic flux in the core in the absence of the output windings. (b) With the output windings in place, calculate i_2.

4.15 A triangular wire loop has its vertices at the points (2, 0, 0), (0, 3, 0), and (0, 0, 4), with dimensions in meters. A time-varying magnetic field is given by $\mathbf{B} = 4t \, \mathbf{a}_y$ Wb/m² (with *t* in seconds). If the wire has a total distributed resistance of 2 Ω, calculate the induced current and indicate its direction in a carefully drawn sketch.

4.4 Faraday's Law and Motional EMF

4.16 Referring to Figure 4.23, suppose a conductive bar of length *h* = 2.0 cm moves with velocity $\mathbf{u} = -1.0 \, \mathbf{a}_\rho$ m/s toward an infinite length line of current *I* = 4.0 A. Find an expression for the voltage from one end of the bar to the other when ρ reaches 10 cm and indicate which end is positive.

4.17 Suppose we have a conductive bar moving along a pair of conductive rails as in Figure 4.12, only now the magnetic flux density is $\mathbf{B} = 4.0 \mathbf{a}_x + 3.0 \mathbf{a}_z$ Wb/m². If *R* = 10. Ω, *w* = 20. cm, and $u_y = 3.0$ m/s, calculate the current induced and indicate its direction.

4.18 The radius *r* of a perfectly conducting metal loop in free space, situated in the *x–y* plane, increases at the rate of $(\pi r)^{-1}$ m/s. A break in the loop has a small 2.0-Ω resistor across it. Meanwhile, there exists a magnetic field $\mathbf{B} = 1.0 \, \mathbf{a}_z$ T. Determine the current induced in the loop, and show in a sketch the direction of flow.

4.19 Rederive V_{emf} for the rectangular loop of Figure 4.16 if the magnetic field is now $\mathbf{B} = B_o \mathbf{a}_z$.

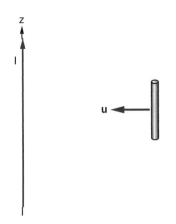

Figure 4.23 Conductive bar moving toward an infinite length line of current for Problem 4.16.

4.20 In Figure 4.16, replace the rectangular loop with a circular one of radius a and rederive V_{emf}.

4.21 A conductive rod, of length 6.0 cm, has one end fixed on a grounded origin and is free to rotate in the x–y plane. It rotates at 60 revolutions/s in a magnetic field $\mathbf{B} = 100.\ \mathbf{a}_z$ mT. Find the voltage at the end of the bar.

4.22 Consider the rotating conductor shown in Figure 4.24. The center of the $2a$-diameter bar is fixed at the origin and can rotate in the x–y plane with $\mathbf{B} = B_o \mathbf{a}_z$. The outer ends of the bar make conductive contact with a ring to make one end of the electrical contact to R; the other contact is made to the center of the bar. Given $B_o = 100.$ mWb/m^2, $a = 6.0$ cm, and $R = 50.\ \Omega$, determine I if the bar rotates at 1.0 revolution/s.

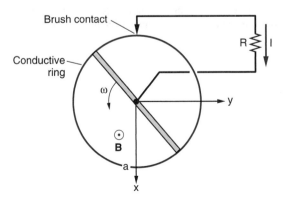

Figure 4.24 Schematic of rotating conductor for Problem 4.22.

4.23 A *Faraday disk generator* is similar to the rotating conductor of 4.22, only now the rotating element is a disk instead of a bar. Derive an expression of the V_{emf} produced by a Faraday disk generator, and using the parameters given in Problem 4.22, find I.

4.24 Consider a sliding rail problem where the conductive rails expand as they progress in the y direction as shown in Figure 4.25. If $w = 10.$ cm and the distance between the rails increases at the rate of 1.0 cm in the x direction per 1.0 cm in the y direction, and $u_y = 2.0$ m/s, find the V_{emf} across a 100.-Ω resistor at the instant when $y = 10.$ cm if the field is $B_o = 100.$ mT.

4.5 Displacement Current

4.25 Suppose a vector field is given as

$$\mathbf{A} = 3x^2 yz^3 \mathbf{a}_x$$

Verify that the divergence of the curl of this vector field is equal to zero.

4.26 Suppose a vector field is given by

$$\mathbf{A} = \rho^2 \cos\phi\ \mathbf{a}_z$$

Verify that the divergence of the curl of this vector field is equal to zero.

4.27 A pair of 60 cm^2 area plates are separated by a 2.0- mm-thick layer of ideal dielectric characterized by $\varepsilon_r = 9.0$. If a voltage $v(t) = 1.0 \sin(2\pi \times 10^3 t)$ V is placed across the plates, determine the displacement current.

4.28 Plot the loss tangent of seawater ($\sigma = 4$ S/m and $\varepsilon_r = 81$) versus log of frequency from 1 Hz to 1 GHz. At what frequency is the magnitude of the displacement current density equal to the magnitude of the conduction current density?

4.29 A 1.0-m-long coaxial cable of inner conductor diameter 2.0 mm and outer conductor diameter 6.0 mm is filled with an ideal dielectric with $\varepsilon_r = 10.2$. A voltage $v(t) = 10.\cos(6\pi \times 10^6\ t)$ mV is placed on the inner conductor and the outer conductor is grounded. Neglecting fringing fields at the ends of the coaxial cable, find the displacement current between the inner and outer conductors.

4.7 Lossless TEM Waves

4.30 Suppose in free space that $\mathbf{E}(z,\ t) = 5.0 e^{-2zt}\ \mathbf{a}_x$ V/m. Is the wave lossless? Find $\mathbf{H}(z,t)$.

4.31 An electric field propagating in a lossless nonmagnetic medium is characterized by

$$\mathbf{E}(y,t) = 100.\cos(4\pi \times 10^6 t - 0.1257y)\mathbf{a}_z \text{ V/m}$$

(a) Find the wave amplitude, frequency, propagation velocity, wavelength, and the relative permittivity of the medium.

(b) Find $\mathbf{H}(y,\ t)$.

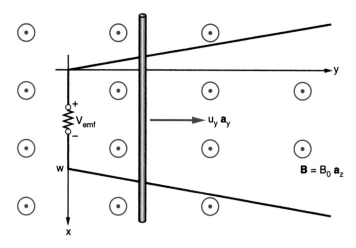

Figure 4.25 Bar sliding along a pair of widening rails for Problem 4.24.

4.32 A magnetic field propagating in free space is given by

$$\mathbf{H}(z,t) = 20.\sin(\pi \times 10^8 t + \beta z)\mathbf{a}_x \text{ A/m}$$

Find f, β, λ, and $\mathbf{E}(z, t)$.

4.33 Find the instantaneous expression for $\mathbf{E}$ for the magnetic field of Problem 4.6.

EMAG
SOLUTIONS **4.34** Given, at some point distant from a source at the origin in free space,

$$\mathbf{E}(r,t) = 8.0\cos(9\pi \times 10^6 t - \beta r)\mathbf{a}_\theta \text{ V/m}$$

find the frequency, phase constant, and $\mathbf{H}(r, t)$.

4.35 In a lossless, nonmagnetic media, the magnetic field at some point distant from a source at the origin is given by

$$\mathbf{H}(\rho,t) = 6.0\sin(3 \times 10^8 t + 10\rho)\mathbf{a}_\phi \text{ A/m}$$

Find the relative permittivity of the medium, the frequency and phase constant of the wave, and $\mathbf{E}(\rho, t)$.

4.36 Suppose, in a nonmagnetic medium of relative permittivity 3, that

$$\mathbf{E}(y,t) = 4.0\sin(\pi \times 10^7 t - \beta y)\mathbf{a}_x + 9.0\cos(\pi \times 10^7 t - \beta y)\mathbf{a}_z \text{ V/m}$$

Determine β and $\mathbf{H}(y,t)$.

4.8 Time-Harmonic Fields and Phasors

4.37 Show that

$$-\frac{\partial}{\partial t}\left(\text{Re}\left[B_s e^{j\omega t}\right]\right) = -\text{Re}\left[\left(j\omega B_s\right)e^{j\omega t}\right]$$

4.38 Derive the differential phasor form of (a) Gauss's law and (b) Ampère's circuital law. EMAG SOLUTIC

4.39 Find $\mathbf{H}(y, t)$ in Problem 4.31b using phasors.

4.40 Find $\mathbf{E}(z, t)$ in Problem 4.32 using phasors.

4.41 In free space, EMAG SOLUTI.

$$\mathbf{E}(z,t) = 10.\cos(\pi \times 10^6 t - \beta z)\mathbf{a}_x + 20.\cos(\pi \times 10^6 t - \beta z)\mathbf{a}_y \text{ V/m}$$

Find $\mathbf{H}(z, t)$.

4.42 Find $\mathbf{H}(y, t)$ in Problem 4.36 using phasors.

4.43 In MATLAB 4.4, a polar plot of the phasor corresponded to a location on a sine wave for a particular time, at a fixed position in space. You can also make a polar phasor plot for a snapshot in time, where you change position. Modify MATLAB 4.4 to provide an animation of the phasor versus sine wave as you change the position.

4.44 Repeat Problem 4.43, now accounting for attenuation. Run the program assuming an attenuation of 2×10^{-6} Np/m.

5

Plane Waves

Learning Objectives

▷ Derive the general equations for electromagnetic wave propagation

▷ Study electromagnetic wave propagation in dielectrics and define loss tangent

▷ Study electromagnetic wave propagation in conductors and define skin depth

▷ Describe electromagnetic wave power transmission using the Poynting theorem

▷ Define the polarization of an electromagnetic wave

▷ Study reflection and transmission of waves incident from one material to another

Let's recall Maxwell's equations in point form:

$$\nabla \times \mathbf{H} = \mathbf{J} + \frac{\partial \mathbf{D}}{\partial t}$$

$$\nabla \times \mathbf{E} = -\frac{\partial \mathbf{B}}{\partial t}$$

$$\nabla \cdot \mathbf{D} = \rho_v$$

$$\nabla \cdot \mathbf{B} = 0$$

In free space, the constitutive parameters are $\sigma = 0$, $\mu_r = 1$, and $\varepsilon_r = 1$, so the Ampère's circuital law and Faraday's law equations become

$$\nabla \times \mathbf{H} = \varepsilon_o \frac{\partial \mathbf{E}}{\partial t}$$

$$\nabla \times \mathbf{E} = -\mu_o \frac{\partial \mathbf{H}}{\partial t}$$

(5.1)

If we consider that at some point in this space there is a source of time-varying electric field, then by Ampère's circuital law we know that a magnetic field is induced in the surrounding region. As this magnetic field is also changing with time, by Faraday's law it in turn induces an electric field. Energy is passed back and forth between $\mathbf{E}$ and $\mathbf{H}$ fields as they radiate away from the source point at the speed of light. These fields constitute waves of electromagnetic energy radiating away from the source point.

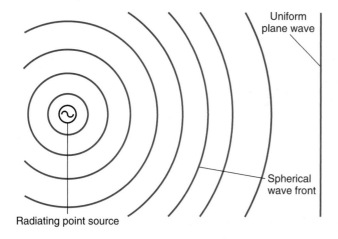

Figure 5.1 At a remote distance away from a point source, the waves appear to be planar.

The waves radiate spherically, but at a remote distance away from the source they resemble *uniform plane waves*, as depicted in Figure 5.1. In a uniform plane wave (UPW), the **E** and **H** fields are orthogonal, or transverse, to the direction of propagation. They therefore propagate in the TEM mode. The magnitude and phase of the field vectors in a UPW are equal at every point on the *wavefront*, a plane transverse to the direction of propagation. Since UPWs are handled with Cartesian coordinates, they are simpler to work with than spherical waves. This chapter therefore deals with UPWs, and spherical waves will be deferred until the study of antennas in Chapter 8.

In this chapter, we first want to show how the equations governing wave motion follow from Maxwell's equations. In particular, we'll focus on the case of sinusoidal (time-harmonic) waves. The propagation characteristics depend on the material media. We'll look at propagating media that are both ideal (lossless) and lossy. We'll also look at propagation in conductors and how current at high frequency tends to flow near the surface of the conductor. Finally, we'll examine what happens when waves pass from one medium to another.

▶ 5.1 GENERAL WAVE EQUATIONS

In this section we want to use Maxwell's equations to derive formulas governing electromagnetic wave propagation. We'll consider that the medium is free of any charge, so

$$\nabla \cdot \mathbf{D} = 0 \tag{5.2}$$

We'll also restrict our discussion to material media that are linear, isotropic, homogeneous, and time invariant—in short, *simple* media. Then Maxwell's equations may be written

$$\nabla \times \mathbf{H} = \sigma \mathbf{E} + \varepsilon \frac{\partial \mathbf{E}}{\partial t}$$

$$\nabla \times \mathbf{E} = -\mu \frac{\partial \mathbf{H}}{\partial t}$$

$$\nabla \cdot \mathbf{E} = 0 \tag{5.3}$$

$$\nabla \cdot \mathbf{H} = 0$$

To begin, let's take the curl of both sides of Faraday's law,

$$\nabla \times (\nabla \times \mathbf{E}) = \nabla \times \left(-\mu \frac{\partial \mathbf{H}}{\partial t}\right) \tag{5.4}$$

The curl on the right side of this expression is a position derivative acting on a time derivative in a homogeneous material. Equation (5.4) can therefore be written

$$\nabla \times (\nabla \times \mathbf{E}) = -\mu \frac{\partial}{\partial t}(\nabla \times \mathbf{H}) \tag{5.5}$$

Exchanging the Faraday's law equivalence for the curl of $\mathbf{H}$ gives us

$$\nabla \times (\nabla \times \mathbf{E}) = -\mu \frac{\partial}{\partial t}\left(\sigma \mathbf{E} + \varepsilon \frac{\partial \mathbf{E}}{\partial t}\right) = -\mu\sigma \frac{\partial \mathbf{E}}{\partial t} - \mu\varepsilon \frac{\partial^2 \mathbf{E}}{\partial t^2} \tag{5.6}$$

We can now manipulate the left side of the equation by invoking a vector identity, relating the curl of a curl of any vector field $\mathbf{A}$ to a divergence and a Laplacian of the field as

$$\nabla \times \nabla \times \mathbf{A} = \nabla \cdot \mathbf{A} - \nabla^2 \mathbf{A} \tag{5.7}$$

So we now have

$$\nabla \cdot \mathbf{E} - \nabla^2 \mathbf{E} = -\mu\sigma \frac{\partial \mathbf{E}}{\partial t} - \mu\varepsilon \frac{\partial^2 \mathbf{E}}{\partial t^2} \tag{5.8}$$

Because our medium is charge-free, the divergence of $\mathbf{E}$ is zero and (5.8) reduces to

$$\boxed{\nabla^2 \mathbf{E} = \mu\sigma \frac{\partial \mathbf{E}}{\partial t} + \mu\varepsilon \frac{\partial^2 \mathbf{E}}{\partial t^2}} \tag{5.9}$$

This is the *Helmholtz wave equation* for $\mathbf{E}$. A similar expression can be found for $\mathbf{H}$ (see Problem 5.1). Equation(5.9) can be broken up into three vector equations (in terms of E_x, E_y, and E_z for instance), as can the similar equation for $\mathbf{H}$. Each of the resulting six equations is a second-order differential equation that can be solved in terms of position and time. The solution is an equation defining the wave.

Drill 5.1 Write the Helmholtz wave equation for an electric field given by $\mathbf{E} = E_x(z, t)$ $\mathbf{a}_x$.

$$\left(Answer: \quad \frac{\partial^2 E_x(z,t)}{\partial z^2} = \mu\sigma \frac{\partial E_x(z,t)}{\partial t} + \mu\varepsilon \frac{\partial^2 E_x(z,t)}{\partial t^2}\right)$$

Time-Harmonic Wave Equations

Of particular interest are the Helmholtz equations for time-harmonic fields. Because the time derivative $\partial \mathbf{E}_s/\partial t$ is $j\omega \mathbf{E}_s$, (5.9) becomes

$$\nabla^2 \mathbf{E}_s = j\omega\mu(\sigma + j\omega\varepsilon)\mathbf{E}_s \tag{5.10}$$

This version of the Helmholtz wave equation is generally written in the form

$$\nabla^2 \mathbf{E}_s - \gamma^2 \mathbf{E}_s = 0 \qquad (5.11)$$

where γ (gamma) is the propagation constant, defined as

$$\gamma = \sqrt{j\omega\mu(\sigma + j\omega\varepsilon)} \qquad (5.12)$$

Since the square root of a complex number is itself complex, γ is equal to a real part (the *attenuation, α,* in nepers per meter) and an imaginary part (the *phase constant,* or β, in radians per meter). So we have

$$\gamma = \sqrt{j\omega\mu(\sigma + j\omega\varepsilon)} = \alpha + j\beta \qquad (5.13)$$

Equation (5.11) is the Helmholtz equation for time-harmonic electric fields. For time-harmonic magnetic fields we have

$$\nabla^2 \mathbf{H}_s - \gamma^2 \mathbf{H}_s = 0 \qquad (5.14)$$

To make use of these Helmholtz equations for time-harmonic fields, let's consider an *x*-polarized plane wave traveling in the *z* direction. Our electric field is therefore

$$\mathbf{E}_s(z) = E_{xs}(z)\mathbf{a}_x \qquad (5.15)$$

where we indicate that $\mathbf{E}_s$ is only a function of *z*. Recall for a UPW that the fields do not vary in the transverse direction, in this case the *x*–*y* plane, so $\mathbf{E}_s$ can only be a function of *z*. The Laplacian of $\mathbf{E}_s$ becomes a straightforward second derivative, and (5.11) becomes

$$\frac{d^2 E_{xs}}{dz^2} - \gamma^2 E_{xs} = 0 \qquad (5.16)$$

where the "function of *z*" in parentheses has been suppressed for brevity. This is a second-order, linear, homogeneous differential equation. A possible solution[5.1] for this equation is

$$E_{xs} = Ae^{\lambda z} \qquad (5.17)$$

where A and λ are arbitrary constants. It is easy to show that

$$\frac{d^2 E_{xs}}{dz^2} = \lambda^2 A e^{\lambda z}$$

and (5.16) becomes

$$\lambda^2 A e^{\lambda z} - \gamma^2 A e^{\lambda z} = 0$$

or

$$\lambda^2 - \gamma^2 = 0 \qquad (5.18)$$

[5.1]You may wish to peruse your by-now dusty and almost forgotten differential equations textbook to refresh your memory on how to solve these types of problems.

This is easily factored as

$$(\lambda + \gamma)(\lambda - \gamma) = 0 \tag{5.19}$$

The first solution of this equation is $\lambda = -\gamma$, which gives

$$E_{xs} = Ae^{-\gamma z} \tag{5.20}$$

Let's examine this solution in its instantaneous form. We substitute $\alpha + j\beta$ for γ, multiply by $e^{j\omega t}$, apply Euler's identity, and take the real part to find

$$E_x = Ae^{-\alpha z} \cos(\omega t - \beta z) \tag{5.21}$$

We can substitute for A the more informative constant E_o^+, which represents the electric field amplitude of the $+z$ traveling wave at $z = 0$. Reinserting the vector and the position and time dependencies, we can write (5.21) as

$$\mathbf{E}(z, t) = E_o^+ e^{-\alpha z} \cos(\omega t - \beta z)\mathbf{a}_x \tag{5.22}$$

Hence for this first solution of (5.19) we have a wave propagating and attenuating in the $+z$ direction with an amplitude E_o^+ at $z = 0$.

Had we chosen the second solution of (5.19), $\lambda = +\gamma$, we would have come to the solution

$$E_{xs} = E_o^- e^{+\gamma z} \tag{5.23}$$

or

$$\mathbf{E}(z,t) = E_o^- e^{\alpha z} \cos(\omega t + \beta z)\mathbf{a}_x \tag{5.24}$$

This represents a wave propagating and attenuating in the $-z$ direction with an amplitude E_o^- at $z = 0$. The general solution for E_{xs} is the linear superposition of the two solutions,

$$E_{xs} = E_o^+ e^{-\gamma z} + E_o^- e^{+\gamma z} \tag{5.25}$$

or

$$\mathbf{E}_s = (E_o^+ e^{-\gamma z} + E_o^- e^{+\gamma z})\mathbf{a}_x$$

and then the general instantaneous solution is

$$\mathbf{E}(z,t) = E_o^+ e^{-\alpha z} \cos(\omega t - \beta z)\mathbf{a}_x + E_o^- e^{\alpha z} \cos(\omega t + \beta z)\mathbf{a}_x \tag{5.26}$$

The magnetic field can be found by applying Faraday's law,

$$\nabla \times \mathbf{E}_s = -j\omega\mu\mathbf{H}_s$$

to (5.25). Evaluating the curl of $\mathbf{E}_s$ we find

$$\nabla \times \mathbf{E}_s = (-\gamma E_o^+ e^{-\gamma z} + \gamma E_o^- e^{\gamma z})\mathbf{a}_y \tag{5.27}$$

Dividing (5.27) by $-j\omega\mu$ we can solve for $\mathbf{H}_s$:

$$\mathbf{H}_s = \left(\frac{\gamma E_o^+}{j\omega\mu} e^{-\gamma z} - \frac{\gamma E_o^-}{j\omega\mu} e^{+\gamma z} \right)\mathbf{a}_y \tag{5.28}$$

Had we started with (5.14) and assumed a magnetic field of the form

$$\mathbf{H}_s(z) = H_{ys}(z)\mathbf{a}_y$$

we would have been led to the expression

$$\mathbf{H}_s = (H_o^+ e^{-\gamma z} + H_o^- e^{+\gamma z})\mathbf{a}_y \tag{5.29}$$

Comparing (5.28) and (5.29), we can find a relationship between E_o^+ and H_o^+ (and between E_o^- and H_o^-). Let's define the intrinsic impedance η (eta) of the medium as being the ratio of E_o^+ to H_o^+, that is,

$$\eta = \frac{E_o^+}{H_o^+} = \frac{j\omega\mu}{\gamma} \tag{5.30}$$

Since the units for E_o^+ and H_o^+ are volts per meter and amperes per meter, respectively, we see that the units for η are ohms. Inserting the expression for γ from (5.13), we find

$$\boxed{\eta = \sqrt{\frac{j\omega\mu}{\sigma + j\omega\varepsilon}}} \tag{5.31}$$

Further inspection of (5.28) and (5.29) reveals that

$$\eta = -\frac{E_o^-}{H_o^-} \tag{5.32}$$

The intrinsic impedance is a useful parameter for relating the electric and magnetic field amplitudes. Like the propagation constant, it is calculated from the operating frequency and the medium's constitutive parameters.

▷ **EXAMPLE 5.1**

Given a material with $\sigma = 0.100$ S/m, $\varepsilon_r = 9.00$, and $\mu_r = 1.00$ and a wave with $f = 1.00$ GHz, we want to find γ, α, β, and η.

We can first calculate $\omega = 2\pi f = 2\pi \times 10^9$ radians/s. Then we find

$$j\omega\mu = j\left(2\pi \times 10^9 \frac{\text{radians}}{\text{s}}\right)(1)\left(4\pi \times 10^{-7} \frac{\text{H}}{\text{m}}\right) = j7896 \frac{\text{H}}{(\text{m}\,\text{s})}$$

and

$$j\omega\varepsilon = j\left(2\pi \times 10^9 \frac{\text{radians}}{\text{s}}\right)(9)\left(\frac{10^{-9}}{36\pi} \frac{\text{F}}{\text{m}}\right)\left(\frac{\text{C}}{\text{F-V}}\right)\left(\frac{\text{A-s}}{\text{C}}\right)\left(\frac{\text{V}}{\text{A-}\Omega}\right) = j0.500 \frac{\text{S}}{\text{m}}$$

Here we made sure the units of $j\omega\varepsilon$ were the same as for σ.

To find γ, we employ (5.13) and find

$$\gamma = \sqrt{\left(j7896 \frac{\text{H}}{\text{m}-\text{s}}\right)(0.100 + j0.500)\left(\frac{\text{S}}{\text{m}}\right)\left(\frac{\text{V-s}}{\text{HA}}\right)\left(\frac{\text{A-}\Omega}{\text{V}}\right)} = 6.25 + j63.1 \frac{1}{\text{m}}$$

So we see that $\alpha = 6.2$ Np/m and $\beta = 63$ radians/m.

Now to find η we use (5.31),

$$\eta = \sqrt{\dfrac{j7896\,\dfrac{\text{H}}{\text{m-s}}}{\sqrt{(0.100 + j0.500)\dfrac{1}{\Omega\text{-s}}}}} = 124e^{j12.3°}\,\Omega$$

Drill 5.2   Repeat Example 5.1 at a frequency of 10 GHz. (*Answer:* $\gamma = 6.3 + j628$ /m, α = 6.3 Np/m, β = 628 radians/m, $\eta = 126e^{j0.57°}\Omega$)

Drill 5.3   Find the attenuation constant, phase constant, and intrinsic impedance of distilled water at 1.0 GHz ($\mu_r = 1$). (*Answer:* α = 0.0021 Np/m, β = 190 radians/m, $\eta = 42\Omega$)

MATLAB 5.1

Let's write a short program that prompts the user for the constitutive parameters and the frequency and then solves for γ, α, β, and η. We can use the results of Example 5.1 to test the program.

```
%  M-File: ML0501
%
%  This program prompts the user for the constitutive
%  parameters and frequency and calculates
%  attenuation, phase constant, and intrinsic impedance.
%
%  Wentworth, 7/19/02
%
%  Variables:
%  eo          free space permittivity (F/m)
%  uo          free space permeability (H/m)
%  f,w         freq (Hz) and angular freq (rad/s)
%  A,B         temporary variables
%  gamma       propagation constant (1/m)
%  alpha       attenuation (Np/m)
%  beta        phase constant (rad/m)
%  eta         intrinsic impedance (ohms)
%  meta        magnitude of eta
%  aeta        angle of eta

clc           %clears the command window
clear         %clears variables

disp('Propagation Parameter Solver')
disp(' ''
disp('Input constitutive parameters and frequency.')
disp('Program will calculate attenuation constant,')
disp('phase constant, and intrinsic impedance.')
disp(' ')
```

(continues)

```
% Initialize Variables
uo=pi*4e-7;
eo=8.854e-12;

% Prompt for Input Values
sig=input('enter sigma, in S/m,: ');
er=input('enter rel permittivity: ');
ur=input('enter rel permeability: ');
f=input('enter frequency, Hz: ');

% Perform Calculation
w=2*pi*f;  % w=angular frequency
A=i*(w*ur*uo);%one way to enter complex #
B=complex(sig,w*er*eo); %another way
gamma=sqrt(A*B);
alpha=real(gamma);
beta=imag(gamma);
eta=sqrt(A/B);
meta=abs(eta); %magnitude of eta, ohms
aeta=180*angle(eta)/pi;
%angle of eta in degrees

% Display results
disp(' ')
disp(['gamma = ' num2str(gamma) ' /m'])
disp(['alpha = ' num2str(alpha) ' Np/m'])
disp(['beta = ' num2str(beta) ' rad/m'])
disp(['eta = 'num2str(meta) '@' num2str(aeta) 'deg ohms'])
disp(' ')
```

Note that MATLAB accepts complex numbers using either "i" or "j," but reports complex numbers with an "i."

Propagating Fields Relation

By knowing that an electromagnetic wave's direction of propagation, given as a unit vector $\mathbf{a}_p$, is the same as the cross product of $\mathbf{E}_s$ and $\mathbf{H}_s$, and by knowing the relation between the amplitudes of $\mathbf{E}_s$ and $\mathbf{H}_s$, a pair of simple formulas can be derived relating $\mathbf{E}_s$, $\mathbf{H}_s$, and $\mathbf{a}_p$:

$$\boxed{\begin{aligned} \mathbf{H}_s &= \frac{1}{\eta}\mathbf{a}_p \times \mathbf{E}_s \\ \mathbf{E}_s &= -\eta\mathbf{a}_p \times \mathbf{H}_s \end{aligned}}$$

(5.33)

These formulas can greatly simplify problem solving.

▷ **EXAMPLE 5.2**

Consider the case where

$$\mathbf{E}_s = E_o^+ e^{-\gamma z}\mathbf{a}_x$$

and we want to find $\mathbf{H}_s$.

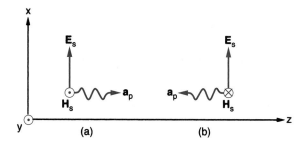

Figure 5.2 Representation of waves. In (a), the wave travels in the $\mathbf{a}_p = +\mathbf{a}_z$ direction and has $\mathbf{E}_s = E_o{}^+e^{-\gamma z}\,\mathbf{a}_x$ and $\mathbf{H}_s = (E_o{}^+/\eta)e^{-\gamma z}\,\mathbf{a}_y$. In (b), the wave travels in the $\mathbf{a}_p = -\mathbf{a}_z$ direction and has $\mathbf{E}_s = E_o{}^-e^{\gamma z}\,\mathbf{a}_x$ along with $\mathbf{H}_s = -(E_o{}^-/\eta)e^{\gamma z}\,\mathbf{a}_y$.

We know $\mathbf{a}_p = \mathbf{a}_z$ and we have from (5.33)

$$\mathbf{H}_s = \frac{1}{\eta}\mathbf{a}_z \times E_o{}^+ e^{-\gamma z}\mathbf{a}_x = \frac{E_o{}^+}{\eta}e^{-\gamma z}\mathbf{a}_y = H_o{}^+ e^{-\gamma z}\mathbf{a}_y$$

This case is shown in Figure 5.2, where a standard method of representing waves is given. This approach is somewhat simpler than rigorously performing the curl of $\mathbf{E}_s$ and dividing by $-j\omega\mu$.

Drill 5.4 Suppose in a medium with $\eta = 240\pi\ \Omega$, $\mathbf{E}(x, t) = 480\pi \cos(\omega t + 10x)\,\mathbf{a}_y$ V/m. Find $\mathbf{E}_s$, $\mathbf{a}_p$, $\mathbf{H}_s$, and $\mathbf{H}(x, t)$. (*Answer:* $\mathbf{E}_s = 480\pi\,e^{j10x}\,\mathbf{a}_y$ V/m, $\mathbf{a}_p = -\mathbf{a}_x$, $\mathbf{H}_s = -2.0e^{j10x}\,\mathbf{a}_z$ A/m, $\mathbf{H}(x, t)= -2.0\cos(\omega t + 10x)\,\mathbf{a}_z$ A/m)

▶ 5.2 PROPAGATION IN LOSSLESS, CHARGE-FREE MEDIA

We will first consider the simplest case of a time-harmonic field propagating in a charge-free ($\rho_v = 0$) medium that has zero conductivity ($\sigma = 0$). This is the case for waves traveling in vacuum, also termed *free space* since it is space free of any charges or matter. A *perfect dielectric* is also lossless since it has neither charge nor conductivity.

Evaluating our propagation constant for this case we find

$$\gamma = \sqrt{j\omega\mu(0 + j\omega\varepsilon)} = \sqrt{j^2\omega^2\mu\varepsilon} = j\omega\sqrt{\mu\varepsilon} = \alpha + j\beta \qquad (5.34)$$

Since ω, μ, and ε are nonnegative real quantities, we see that in (5.34) $\alpha = 0$ and

$$\beta = \omega\sqrt{\mu\varepsilon} \qquad (5.35)$$

Since $\alpha = 0$, the signal does not attenuate as it travels, so it is referred to as a lossless medium. Earlier we found that propagation velocity u_p was related to β and ω by

$$\boxed{u_p = \frac{\omega}{\beta} = \frac{1}{\sqrt{\mu\varepsilon}}} \qquad (5.36)$$

For free space, we calculate a propagation velocity equal to the speed of light c. In a non-magnetic lossless dielectric, we then have

$$u_{\mathrm{p}} = \frac{c}{\sqrt{\varepsilon_{\mathrm{r}}}} \tag{5.37}$$

We can also evaluate our intrinsic impedance for this lossless medium:

$$\eta = \sqrt{\frac{j\omega\mu}{(0 + j\omega\varepsilon)}} = \sqrt{\frac{\mu}{\varepsilon}} \tag{5.38}$$

We see that η is a real value. We will see in the next section that, for lossy materials, η is complex with the consequence that the **E** and **H** fields are out of phase with each other. For lossless materials, **E** and **H** are always in phase. We can rewrite (5.38) as

$$\eta = \sqrt{\frac{\mu_{\mathrm{r}}\mu_o}{\varepsilon_{\mathrm{r}}\varepsilon_o}} = \sqrt{\frac{\mu_{\mathrm{r}}}{\varepsilon_{\mathrm{r}}}}\eta_o \tag{5.39}$$

where η_o is the intrinsic impedance of free space. Evaluating η_o we find

$$\eta_o = \sqrt{\frac{\mu_o}{\varepsilon_o}} = \sqrt{\frac{4\pi \times 10^{-7}\ \mathrm{H/m}}{\left(10^{-9}\big/36\pi\right)\mathrm{F/m}}}$$

or

$$\boxed{\eta_o = 120\pi\ \Omega} \tag{5.40}$$

Considering once again the case of the x-polarized z traveling wave, for the lossless case, with $\alpha = 0$ and $\gamma = j\beta$, we have

$$\mathbf{E}_s = (E_o^+ e^{-j\beta z} + E_o^- e^{+j\beta z})\mathbf{a}_x \tag{5.41}$$

or, in instantaneous form,

$$\mathbf{E}(z,t) = (E_o^+ \cos(\omega t - \beta z) + E_o^- \cos(\omega t + \beta z))\mathbf{a}_x \tag{5.42}$$

On each wave in (5.41) we can apply our simple formulas (5.33) to find

$$\mathbf{H}_s = \left(\frac{E_o^+}{\eta}e^{-j\beta z} - \frac{E_o^-}{\eta}e^{+j\beta z}\right)\mathbf{a}_y \tag{5.43}$$

Drill 5.5   A lossless, nonmagnetic[5.2] material with $\varepsilon_{\mathrm{r}} = 36$ has $\mathbf{E}(z,\ t) = 40\pi\ \cos(\pi \times 10^7 t - \beta z)\ \mathbf{a}_x$ V/m. Find the propagation velocity, the phase constant, and the instantaneous expression for the magnetic field intensity. (*Answer:* $u_{\mathrm{p}} = 5.0 \times 10^7$ m/s, $\beta = \pi/5$ radians/m, $\mathbf{H}(z,\ t) = 2.0\ \cos(\pi \times 10^7 t - \pi z/5)\ \mathbf{a}_y$ A/m)

[5.2]"Nonmagnetic" translates to "$\mu_{\mathrm{r}} = 1$."

▶ 5.3 PROPAGATION IN DIELECTRICS

Treating a dielectric as lossless is often a good approximation, but all dielectrics are to some degree lossy. The lossy nature can be attributed to finite conductivity, polarization loss, or a combination of the two. With finite conductivity, the electric field gives rise to a conduction current density $\mathbf{J} = \sigma\mathbf{E}$. The presence of $\mathbf{E}$ and $\mathbf{J}$ results in power dissipation (as heat) via Joule's law

$$P = \int \mathbf{E} \cdot \mathbf{J}\, dv$$

as recalled from Section 2.10. This power dissipation attenuates the wave. Polarization loss comes about from the energy required of the field to flip reluctant dipoles. This loss mechanism is proportional to frequency.

A complex permittivity ε_c is written

$$\varepsilon_c = \varepsilon' - j\varepsilon'' \tag{5.44}$$

where ε' is the real part of the permittivity ($\varepsilon_r\varepsilon_o$), and ε'' is the complex part that accounts for the polarization losses. Recalling Ampère's circuital law, we now have

$$\nabla \times \mathbf{H}_s = \sigma\mathbf{E}_s + j\omega(\varepsilon' - j\varepsilon'')\mathbf{E}_s \tag{5.45}$$

which can be rearranged as

$$\nabla \times \mathbf{H}_s = [(\sigma + \omega\varepsilon'') + j\omega\varepsilon']\mathbf{E}_s \tag{5.46}$$

From (5.46) it is apparent that we can account for both conductivity and the polarization losses by an effective conductivity given by

$$\sigma_{\text{eff}} = \sigma + \omega\varepsilon'' \tag{5.47}$$

The general equations (5.13) and (5.31) continue to hold for time-harmonic fields in lossy materials. Now the propagation constant is complex, with an attenuation constant greater than zero. The intrinsic impedance is also complex, resulting in a phase difference between $\mathbf{E}$ and $\mathbf{H}$ fields.

▶ EXAMPLE 5.3

Recall the constitutive parameters from Example 5.1. With $\sigma = 0.100$ S/m (assuming $\sigma_{\text{eff}} = \sigma$ here), $\varepsilon_r = 9.00$, $\mu_r = 1.00$, and $f = 1.00$ GHz, we found $\gamma = 6.25 + j63.1$ /m and $\eta = 124e^{j5.6°}$ Ω. Suppose we have an electric field in this medium given as

$$\mathbf{E}(z, t) = 10.0e^{-6.25z}\cos(2\pi \times 10^9 t - 63.1z)\mathbf{a}_x \text{ V/m}$$

and now we want to determine $\mathbf{H}(z, t)$.

First we convert $\mathbf{E}(z, t)$ to a phasor:

$$\mathbf{E}_s = 10.0e^{-6.25z}e^{-j63.1z}\mathbf{a}_x$$

Using (5.33) with $\mathbf{a}_p = \mathbf{a}_z$, we have

$$\mathbf{H}_s = \frac{1}{\eta}\mathbf{a}_p \times \mathbf{E}_s$$

$$= \frac{1}{124e^{j5.6°}\ \Omega}\mathbf{a}_z \times 10e^{-6.25z}e^{-j63.1z}\mathbf{a}_x = 81.0e^{-j5.6°}e^{-6.25z}e^{-j63.1z}\mathbf{a}_y \text{ mA/m}$$

Converting $\mathbf{H}_s$ to instantaneous form, we get

$$\mathbf{H}(z, t) = 81.0e^{-6.25z}\cos(2\pi \times 10^9 t - 63.1z - 0.098)\mathbf{a}_y \text{ mA/m}$$

where 5.6° has been converted to 0.098 radians. Now we see that the **E** and **H** fields are out of phase by the angle of the intrinsic impedance.

Let us now find expressions for α and β in a general dielectric. From (5.12) we have

$$\gamma^2 = j\omega\mu(\sigma + j\omega\varepsilon) \tag{5.48}$$

which can be rearranged to

$$\gamma^2 = -\omega^2\mu\left(\varepsilon - j\frac{\sigma}{\omega}\right) \tag{5.49}$$

Separating the real and imaginary parts, we have

$$\gamma^2 = -\omega^2\mu\varepsilon + j\omega\mu\sigma \tag{5.50}$$

Now if we consider $\gamma = \alpha + j\beta$, we can write γ^2 as

$$\gamma^2 = (\alpha^2 - \beta^2) + j2\alpha\beta \tag{5.51}$$

The real parts of (5.50) and (5.51) must be equal, as must be the imaginary parts. After a page or two of algebra (see Problem 5.13), we can solve for α and β in terms of the material's constitutive parameters as

$$
\begin{array}{c}
\alpha = \omega\sqrt{\dfrac{\mu\varepsilon}{2}\left(\sqrt{1+\left(\dfrac{\sigma}{\omega\varepsilon}\right)^2}-1\right)} \\[2em]
\beta = \omega\sqrt{\dfrac{\mu\varepsilon}{2}\left(\sqrt{1+\left(\dfrac{\sigma}{\omega\varepsilon}\right)^2}+1\right)}
\end{array}
\tag{5.52}
$$

These equations can be used to find α and β for any material, given the constitutive parameters (see Problem 5.14).

For lossy materials with complex permittivity, σ may be replaced with σ_{eff} in these equations. Fortunately, many materials can be considered either low-loss dielectrics or good conductors over some frequency range, and (5.52) reduces to something more manageable in these cases. We'll next consider low-loss dielectrics and loss tangent; we'll save the topic of good conductors for the next section.

Low-Loss Dielectrics

A low-loss dielectric is one with a small loss tangent, that is, $(\sigma/\omega\varepsilon) \ll 1$. We can reduce (5.52) for this special case by applying a binomial series expansion to the value within the interior square root portion of the equations. The expansion is

$$(1+x)^n = 1 + nx + \frac{n(n-1)}{2!}x^2 + \cdots$$

and for $x \ll 1$ this can be approximated as

$$(1+x)^n = 1 + nx$$

So we have

$$\left(1+\left(\frac{\sigma}{\omega\varepsilon}\right)^2\right)^{1/2} \cong 1+\frac{1}{2}\left(\frac{\sigma}{\omega\varepsilon}\right)^2 \tag{5.53}$$

Inserting this approximation into (5.52) we find

$$\alpha \cong \frac{\sigma}{2}\sqrt{\frac{\mu}{\varepsilon}}, \quad \beta \cong \omega\sqrt{\mu\varepsilon} \tag{5.54}$$

As before, σ can be replaced with σ_{eff} to account for complex permittivity in these equations.

Low-loss materials have a small but definite attenuation. Otherwise, they have the same phase constant and same intrinsic impedance as lossless materials.

> **Drill 5.6** Use (5.54) to determine α and β for glass at (a) 100. Hz and (b) 1.00 MHz. Assume $\mu_r = 1.0$ and use the parameters for glass listed in Table 5.1. Compare your results with α and β calculated from (5.52). (*Answer:* (a) $\alpha = 3.4 \times 10^{-9}$ Np/m, $\beta = 6.6 \times 10^{-6}$ radians/m; (b) $\alpha = 3.3 \times 10^{-5}$ Np/m, $\beta = 0.066$ radians/m)

Loss Tangent

A standard measure of lossiness in a dielectric is given by the loss tangent, represented by Figure 5.3. Here the imaginary part of (5.46), $\omega\varepsilon'\mathbf{E}_s$, is the displacement current density. On the real axis is the real part of (5.46), $(\sigma + \omega\varepsilon'')\mathbf{E}_s$. The total current density, or $\nabla \times \mathbf{H}_s$, is the vector sum of the real and imaginary parts. We define δ as the angle by which the displacement current density leads the total current density. The tangent of this angle is

$$\tan\delta = \frac{\sigma + \omega\varepsilon''}{\omega\varepsilon'} = \frac{\sigma_{eff}}{\omega\varepsilon'} \tag{5.55}$$

called the *loss tangent*.

The loss tangent is typically applied when discussing dielectric materials, for which a small value is desirable. It is useful for classifying a material as either a good dielectric (tan $\delta \ll 1$) or a good conductor (tan $\delta \gg 1$). In a good dielectric, σ is negligible and tan $\delta = \varepsilon''/\varepsilon'$. In a good conductor, except at very high frequency, $\sigma \gg \omega\varepsilon''$ and tan $= \sigma/\omega\varepsilon'$.

As is evident from the denominator in (5.55), tanδ is a function of frequency. In Figure 5.4, the loss tangent is plotted against frequency for three materials: copper, seawater, and glass. Copper is considered a good conductor over the complete range, even though tanδ drops steadily with frequency. Glass is a good dielectric, maintaining a steady value over

TABLE 5.1 Parameters Used for MATLAB 5.2 and Figure 5.4

	σ(S/m)	ε_r'	ε_r''
Copper	5.8×10^7	1	0
Seawater	4	72	12
Glass	10^{-12}	10	0.010

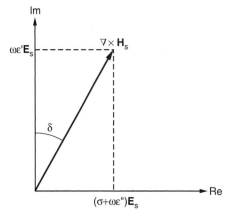

Figure 5.3 Loss tangent $\tan\delta$ is the ratio of the conduction to displacement current densities, or $\tan\delta = (\sigma + \omega\varepsilon'')/\omega\varepsilon'$.

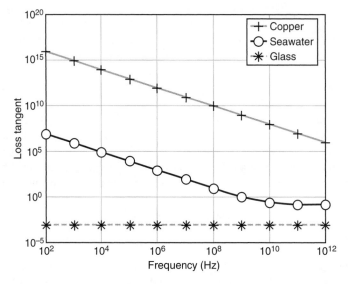

Figure 5.4 MATLAB plot of $\tan\delta$ versus frequency for three selected materials.

this frequency range. But seawater appears as a good conductor at low frequencies, turning into more of a dielectric at higher frequencies.

As a practical matter, the loss tangent of a dielectric is most often a measured value, and the individual contributions of σ and $\omega\varepsilon''$ are not readily apparent.

MATLAB 5.2

The following is the routine used to generate Figure 5.4. Equation (5.44) can also be written

$$\varepsilon_c = \varepsilon_r'\varepsilon_o - j\varepsilon_r''\varepsilon_o \qquad (5.56)$$

separating the relative permittivities from free space permittivity. Material properties are listed in Table 5.1.

```
% M-File: ML0502
%
```

```
% This program plots the loss tangent versus frequency
% for three materials: copper, seawater, and glass.
%
% Wentworth 7/22/02
%
% Variables:
% eo          free space permittivity (F/m)
% erCu1       real part of er for copper
% erCu2       imag. part of er for copper
% sigCu       conductivity of copper (S/m)
% erSe1       real part of er for seawater
% erSe2       imag. part of er for seawater
% sigSe       conductivity of seawater(S/m)
% erGl1       real part of er for glass
% erGl2       imag. part of er for glass
% sigGl       conductivity of glass (S/m)
% f           frequency (Hz)
% n           exponential factor for freq.
% w           angular frequency (rad/s)
% tndCu       loss tangent for copper
% tndSe       loss tangent for seawater
% tndGl       loss tangent for glass
clc           %clears the command window
clear         %clears variables

% Initialize Variables
eo=8.854E-12;
%copper
sigCu=5.8e7;erCu1=1;erCu2=0;
%seawater
sigSe=4;erSe1=81;erSe2=12;
%Glass
sigGl=10e-12;erGl1=10;erGl2=0.010;

% Calculations
n=2:1:12;
f=10.^n;
w=2*pi*f;
tndCu=(sigCu+w*erCu2*eo)./(w*erCu1*eo);
tndSe=(sigSe+w*erSe2*eo)./(w*erSe1*eo);
tndGl=(sigGl+w*erGl2*eo)./(w*erGl1*eo);

loglog(f,tndCu,'-+',f,tndSe,'-o',f,tndGl,'-*')
legend('copper','seawater','glass')
xlabel('frequency (Hz)')
ylabel('loss tangent')
grid on
```

▶ 5.4 PROPAGATION IN CONDUCTORS

In any decent conductor at reasonable frequencies the loss tangent, $\sigma/\omega\varepsilon$, is much greater than one.[5.3] For instance, consider stainless steel with a conductivity of 10^6 S/m. This is a

---

[5.3]For a good conductor, $\sigma/\omega\varepsilon' \gg \varepsilon''/\varepsilon'$, so the latter term is safely ignored.

relatively poor conductor compared to copper, which has $\sigma = 5.8 \times 10^7$ S/m. At 100 GHz, and assuming $\varepsilon_r = 1$, stainless steel has

$$\frac{\sigma}{\omega \varepsilon} = \frac{\left(10^6 \, \dfrac{1}{\Omega \text{-m}}\right)}{2\pi \left(100 \times 10^9 \, \dfrac{1}{\text{s}}\right)\left(\dfrac{10^{-9}}{36\pi} \, \dfrac{\text{F}}{\text{m}}\right)} = 180{,}000$$

a value considerably larger than 1. Since $\sigma / \omega \varepsilon \gg 1$ for a good conductor, the interior bracketed term of (5.52) can be written

$$\left(\sqrt{1 + \left(\frac{\sigma}{\omega \varepsilon}\right)^2} \pm 1\right) = \frac{\sigma}{\omega \varepsilon} \tag{5.57}$$

and the expressions for α and β are then easily shown to be equal:

$$\boxed{\alpha = \beta = \sqrt{\frac{\omega \mu \sigma}{2}} = \sqrt{\pi f \mu \sigma}} \tag{5.58}$$

The intrinsic impedance is approximated by

$$\eta = \sqrt{\frac{j\omega\mu}{\sigma + j\omega\varepsilon}} \cong \sqrt{\frac{j\omega\mu}{\sigma}} \tag{5.59}$$

since $\sigma \gg \omega \varepsilon$. We can rearrange this equation by considering[5.4]

$$\sqrt{j} = \frac{1 + j}{\sqrt{2}} \tag{5.60}$$

leading to

$$\boxed{\eta = \sqrt{\frac{\omega\mu}{2\sigma}}(1 + j) = \sqrt{\frac{\omega\mu}{\sigma}} e^{j45°}} \tag{5.61}$$

Note that, with a little manipulation, (5.61) can also be written

$$\eta = \sqrt{2} \frac{\alpha}{\sigma} e^{j45°} \tag{5.62}$$

So we see that, in any decent conductor, the magnetic field lags the electric field by 45°.

A consequence of the large σ is the drastic decrease in the propagation velocity and wavelength. We have

$$\boxed{u_\text{p} = \frac{\omega}{\beta} = \sqrt{\frac{2\omega}{\mu\sigma}}} \tag{5.63}$$

[5.4]Square both sides of this equation to see that $j = j$.

and since $\lambda = 2\pi/\beta$,

$$\lambda = 2\sqrt{\frac{\pi}{f\mu\sigma}} \tag{5.64}$$

Figure 5.5 represents a wave from air (essentially free space) that penetrates a good conductor.[5.5] The wave is seen to attenuate rapidly in the conductor, with a wavelength clearly much shorter than its value in air.

▷ **EXAMPLE 5.4**

Let us calculate α, β, η, and u_p for copper at 1.0 GHz and compare our results with the free space values.

With $\mu_r = 1$, we have $\alpha = \beta = 480 \times 10^3$ /m, $\eta = 2.3\ e^{j45°}$ mΩ, and $u_p = 13$ km/s. In air, $\alpha = 0$, $\beta = 6.7\pi$ /m, $\eta = 120\pi\ \Omega$, and $u_p = c$.

By way of further comparison, for a perfect conductor we have $\sigma = \alpha = \beta = \infty$ and $\eta = 0\ \Omega$. The perfect conductor appears as a "short circuit" to the wave.

A large attenuation means the fields cannot penetrate far into the conductor. The distance into a material where the field amplitude has dropped to e^{-1} (≈ 0.368) of its surface value is called the *penetration depth*. In a good conductor, the large attenuation means the penetration depth can be quite small, confining the fields near the surface, or skin, of the conductor. For good conductors it is customary to refer to the penetration depth as a *skin depth*. Solving

$$e^{-1} = e^{-\alpha\delta} \tag{5.65}$$

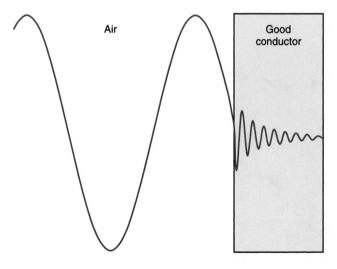

Figure 5.5 The portion of an electromagnetic wave incident from air that passes into a conductor experiences high attenuation and a decrease in wavelength.

[5.5]In such a case, most of the wave energy will be reflected. In sections 5.7 and 5.8, we'll discuss reflection and transmission of waves.

where δ is the skin depth,[5.6] we have

$$\delta = 1/\alpha = 1/\sqrt{\pi f \mu \sigma} \tag{5.66}$$

In the previous example of copper at 1 GHz, the skin depth is calculated as 2.1 μm.

> **Drill 5.7** For nickel ($\sigma = 1.45 \times 10^7$, $\mu_r = 600$) at 100 MHz, calculate α, β, η, u_p, and δ.
> (*Answer:* $\alpha = \beta = 1.85 \times 10^6$ /m, $\eta = 180\ e^{j45°}$ mΩ, $u_p = 340$ m/s, $\delta = 0.54$ μm)

Current in Conductors

At high frequency, current is confined to the outer surface or skin of a conductor. We will develop a relationship for the resistance encountered by the current in such a conductor. First, it is helpful to understand the concept of sheet resistance.

Consider the slab of conductive material in Figure 5.6. A current is driven in the $+x$ direction through this slab by the application of a voltage difference. The resistance for such a slab is

$$R = \frac{1}{\sigma}\frac{L}{wt} \tag{5.67}$$

which can be rearranged as

$$R = \frac{1}{\sigma t}\frac{L}{w} = R_{sheet}\left(\frac{\Omega}{\square}\right)\frac{L}{w}(\square) \tag{5.68}$$

where the ratio $1/\sigma t$ is represented as R_{sheet}, called the *sheet resistance* in ohms per square. Sheet resistance is useful in integrated circuit devices where a known thickness of resistive (or conductive) material is deposited on a silicon substrate, and a desired resistance is

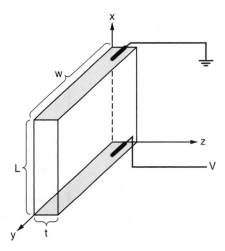

Figure 5.6 Slab of material with length L, width w, and thickness t has conductivity σ. The top and bottom faces are covered in equipotential surfaces tied to a voltage supply.

[5.6] δ is the standard symbol used to denote skin depth; it should not be confused with the angle δ used for tanδ.

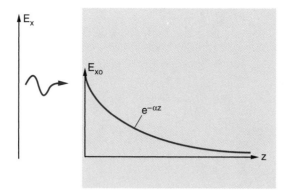

Figure 5.7 An x-polarized electric field is incident on a semiinfinite slab of σ material that occupies $z > 0$.

obtained by proper choice of the path's length to width ratio (i.e., the number of squares). We will find that sheet resistance can be adapted for high frequency to represent a *skin-effect resistance*.

Consider in Figure 5.7 the electric field incident on a semiinfinite slab of material with conductivity σ that occupies $z > 0$. Just at or below the surface of the conductor, the amplitude of the field is given as E_{xo}. As the field propagates into the slab, the amplitude decreases as

$$E_x = E_{xo}e^{-\alpha z} \qquad (5.69)$$

The corresponding current density by Ohm's law is

$$J_x = \sigma E_{xo}e^{-\alpha z} \qquad (5.70)$$

To calculate the current through a surface extending from 0 to infinity in the z direction and of width w in the y direction (out of this page toward the reader), we integrate $I = \int J_x dS$, where $dS = dy\,dz$. Then we have

$$I = \int_{z=0}^{\infty}\int_{y=0}^{w} \sigma E_{xo}e^{-\alpha z}\,dy dz = \left[\frac{-w\sigma E_{xo}}{\alpha}e^{-\alpha z}\right]_0^{\infty} = \frac{w\sigma E_{xo}}{\alpha} \qquad (5.71)$$

or, since $\delta = 1/\alpha$,

$$I = w\sigma E_{xo}\delta \qquad (5.72)$$

This current is an exponentially decaying function in the conductive slab. However, we can assume the current at the surface is constant down to a skin depth since by (5.72) this yields an equivalent total current. This is supported by Figure 5.8, where the area of a rectangle of sides E_{xo} and δ is equivalent to the area under the exponential curve.

For a distance L in the x direction, the field is related to the voltage drop by

$$V = E_{xo}L \qquad (5.73)$$

We can use this expression and the one for current to find the resistance R for a length L of slab, of width w, that extends from $z = 0$ to infinity. We have

$$R = \frac{V}{I} = \frac{E_{xo}L}{w\sigma E_{xo}\delta} = \frac{1}{\sigma\delta}\frac{L}{w} \qquad (5.74)$$

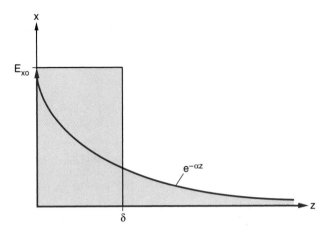

Figure 5.8 The total area under the $e^{-\alpha z}$ curve (from $z = 0$ to infinity) is equal to the product of E_{xo} and δ.

or,

$$R = R_{\text{skin}}\left(\frac{L}{w}\right) \tag{5.75}$$

where

$$R_{\text{skin}} = \frac{1}{\sigma\delta} \tag{5.76}$$

is the skin-effect resistance calculated for the field incident on a semiinfinite slab.

Skin depth plays a key role in the design of a number of high-frequency components. For instance, rectangular waveguides used for microwave power transmission are typically made of brass with a thin electroplated layer of silver. The brass is cheap but not a particularly good conductor. The silver is an excellent conductor, but it is expensive. However, since the skin depth of silver is so small, an electroplated layer is sufficient. Also, in coaxial cable designed for high frequency, the outer conductor can be very thin without a significant increase in resistance.

The skin-effect resistance assumes a semiinfinite slab of conductor. Very rarely is such a slab encountered in reality! How thick must the slab be to ensure the accuracy of our calculations? Let's find the skin-effect resistance for a conductive slab of finite thickness t. We can proceed as before, only changing the limits on our integration for the current:

$$I = \int_{z=0}^{t}\int_{y=0}^{w}\sigma E_{xo}e^{-\alpha z}\,dydz = \left[\frac{-w\sigma E_{xo}}{\alpha}e^{-\alpha z}\right]_{0}^{t} = w\sigma E_{xo}\delta\left(1 - e^{-\alpha t}\right) \tag{5.77}$$

It is then easy to show that the skin-effect resistance can be written

$$R_{\text{skin}} = \frac{1}{\sigma\delta\left(1 - e^{-t/\delta}\right)} \tag{5.78}$$

Now we can plot the ratio of the skin-effect resistance assuming a semiinfinite slab of material to the actual skin-effect resistance ((5.76)/(5.78)) against the ratio of thickness t to

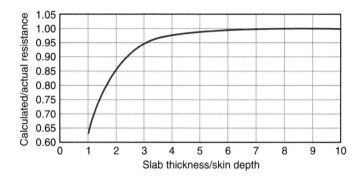

Figure 5.9 Plot showing the accuracy of the calculated skin-effect resistance as a function of slab thickness.

skin depth (Figure 5.9). At 3 skin depths, the calculated resistance is 95% that of the actual resistance. At 5 skin depths, our agreement is up to 99.3%. For most applications, 5 skin depths is sufficient to assure accuracy in our calculations.

Skin-effect resistance can also be employed for cylindrical conductors such as wire or pipe. The resistance per unit length is related to the conductor radius a by

$$\frac{R}{L} = \frac{R_{skin}}{2\pi a}$$ (5.79)

as long as $a \gg \delta$.

Drill 5.8 Calculate the sheet resistance for a 1.0-mm-thick sheet of nickel at 100 MHz. (*Answer:* 0.15 Ω per square)

► **EXAMPLE 5.5**

The concept of skin-effect resistance can be used to derive an expression for the distributed resistance of coaxial cable. As indicated in Figure 5.10, we first make the assumption that the skin depth is considerably smaller than the conductor thickness. Then, the resistance for a length L of coaxial cable is the series combination of the inner conductor resistance, R_a, and the outer conductor resistance, R_b, or $R_{tot} = R_a + R_b$. On the inner conductor

$$R_a = \frac{1}{\sigma_c} \frac{L}{2\pi a \delta}$$

where σ_c is the metal conductivity, and

$$\delta = \frac{1}{\sqrt{\pi f \mu_c \sigma_c}}$$

where μ_c is the metal permeability. On the outer conductor,

$$R_b = \frac{1}{\sigma_c} \frac{L}{2\pi b \delta}$$

So we have

$$R_{total} = \frac{L}{2\pi} \left(\frac{1}{a} + \frac{1}{b} \right) \frac{1}{\sigma_c \delta}$$

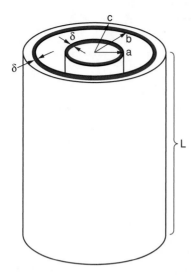

Figure 5.10 Coaxial cable indicating current is confined to within a skin-depth of the conductor surfaces.

or after rearranging the distributed resistance is

$$R' = \frac{R_{\text{total}}}{L} = \frac{1}{2\pi}\left(\frac{1}{a} + \frac{1}{b}\right)\sqrt{\frac{\pi f \mu_c}{\sigma_c}} \qquad (5.80)$$

▷ 5.5 THE POYNTING THEOREM AND POWER TRANSMISSION

If a wave is incident on a conductive surface, electrons are forced back and forth and power is dissipated as heat. The power comes from the incident electromagnetic wave.

It is relatively straightforward to derive,[5.7] from Maxwell's equations, the *Poynting theorem*[5.8] relation

$$\oint (\mathbf{E} \times \mathbf{H}) \cdot d\mathbf{S} = -\int \mathbf{J} \cdot \mathbf{E}\, dv - \frac{\partial}{\partial t}\int \frac{1}{2}\varepsilon E^2\, dv - \frac{\partial}{\partial t}\int \frac{1}{2}\mu H^2\, dv \qquad (5.81)$$

The first term on the right side of this equation is the Joule's law equation for instantaneous power dissipated in the volume (see Section 2.10). The next two terms on the right side are recognized as energy densities for the static fields from Chapters 2 and 3. In fact, derivation of (5.81) verifies the static field energy density relations that were not rigorously derived earlier. Here, with the time derivative, they are the rate of change of the energy stored in the fields.

The left side of (5.81) is a power density expression, giving the total power leaving a closed surface. The Poynting theorem is an energy conservation statement; it says the rate of decrease in energy stored in a volume's electric and magnetic fields, less the energy dissipated by heat, must be equal to the power leaving the closed surface bounding the volume.

[5.7]This form of the Poynting theorem assumes a linear, isotropic, time-invariant medium.

[5.8]The theorem was postulated by English Physicist John H. Poynting (1852–1914) in 1884.

The cross product in the left-hand side integral can be written

$$\mathbf{P} = \mathbf{E} \times \mathbf{H} \tag{5.82}$$

where $\mathbf{P}$ is known as the *instantaneous Poynting vector*. This vector represents the density and the direction of the power flow, with units of watts per meters squared. It is rather a convenient name, since $\mathbf{E} \times \mathbf{H}$ is *pointing* in the direction of power flow.

For a good illustration of the Poynting theorem consider a direct current I in a length L of wire of radius a, as indicated in Figure 5.11a. First, we recognize that DC is a static situation, so the rate of change of energy stored in the fields (the time derivative terms in (5.81)) is zero.

The current is assumed evenly distributed, so the current density is

$$\mathbf{J} = \frac{I}{\pi a^2} \mathbf{a}_z$$

By Ohm's law, the electric field is

$$\mathbf{E} = \frac{\mathbf{J}}{\sigma} = \frac{I}{\pi a^2 \sigma} \mathbf{a}_z$$

On the right-hand side of the Poynting theorem, we have

$$-\int \mathbf{J} \cdot \mathbf{E} \, dv = -\int \left(\frac{I}{\pi a^2} \mathbf{a}_z \cdot \frac{I}{\pi a^2 \sigma} \mathbf{a}_z \right) \rho \, d\rho \, d\phi \, dz$$

$$= -\frac{I^2}{\sigma \left(\pi a^2 \right)^2} \int_0^a \rho \, d\rho \int_0^{2\pi} d\phi \int_0^L dz = -I^2 \frac{1}{\sigma} \frac{L}{\pi a^2} = -I^2 R$$

On the left side of (5.81), we need $\mathbf{H}$. Using Ampère's circuital law, we can easily find that $\mathbf{H}$ at the surface ($\rho = a$) of the wire is

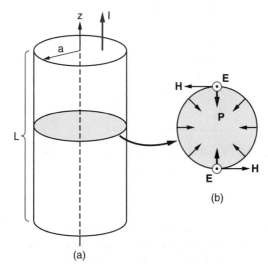

(a)

(b)

Figure 5.11 (a) A direct current I flows in the $+z$ direction in a wire of radius a. (b) Wire cross section showing the orientation of the fields and the instantaneous Poynting vector at the wire surface.

$$\mathbf{H} = \frac{I}{2\pi a}\mathbf{a}_\phi$$

The instantaneous Poynting vector is

$$\mathbf{P} = \mathbf{E} \times \mathbf{H} = \frac{I}{\pi a^2 \sigma}\mathbf{a}_z \times \frac{I}{2\pi a}\mathbf{a}_\phi$$

or

$$\mathbf{P} = -\frac{I^2}{2\pi^2 a^3 \sigma}\mathbf{a}_\rho$$

As shown in Figure 5.11b, the vector is directed radially inward on the wire. Integrating over the surface we have

$$\int \mathbf{P} \cdot d\mathbf{S} = \int \frac{-I}{2\pi^2 \sigma a^3}\mathbf{a}_\rho \cdot a\,d\phi\,dz\,\mathbf{a}_\rho$$

$$= \frac{-Ia}{2\pi^2 \sigma a^3}\int_0^{2\pi}d\phi\int_0^L dz = -I^2\frac{1}{\sigma}\frac{L}{\pi a^2} = -I^2 R$$

This side of the Poynting theorem tells us that the power flowing out of the closed surface is $-I^2R$, or, in other words, I^2R of power is flowing into the wire. To balance this influx of power, the Joule's law portion of the theorem says I^2R of power is being dissipated.

UPW Power Transmission

An expression from circuits analogous to (5.82) can be written $P = VI$ for the instantaneous power. Also from circuits, it may be recalled that for sinusoidally varying voltages and currents, an average power density can be expressed as

$$P_{ave} = \frac{1}{2}\mathrm{Re}\left[V_s I_s^*\right] \tag{5.83}$$

where the * superscript on the phasor I_s indicates a complex conjugate quantity.[5.9] Likewise, for time-harmonic electromagnetic waves the time-averaged power density in the wave can be shown[5.10] to be

$$\boxed{\mathbf{P}_{ave} = \frac{1}{2}\mathrm{Re}\left[\mathbf{E}_s \times \mathbf{H}_s^*\right]} \tag{5.84}$$

Finding the amount of power P, in watts, passing through a surface is then

$$P = \int \mathbf{P}_{ave} \cdot d\mathbf{S} \tag{5.85}$$

[5.9]If a complex number $A = \mathrm{Re} + j\mathrm{Im}$, then the complex conjugate is $A* = \mathrm{Re} - j\mathrm{Im}$.

[5.10]This is also known as the "time-average Poynting vector." Its derivation can be found in W.H. Hayt and J.A. Buck, *Engineering Electromagnetics,* 6th Ed., McGraw-Hill, 2001, pp. 365–369.

Let's consider the general case of a lossless medium containing the phasor quantity

$$\mathbf{E}_s = E_{xo}e^{-j\beta z}\mathbf{a}_x$$

We know from (5.33) that

$$\mathbf{H}_s = \frac{E_{xo}}{\eta}e^{-j\beta z}\mathbf{a}_y$$

and the complex conjugate is

$$\mathbf{H}_s^* = \frac{E_{xo}}{\eta}e^{j\beta z}\mathbf{a}_y$$

The average power density of the wave is then

$$\mathbf{P}_{ave} = \frac{1}{2}\text{Re}\left[E_{xo}e^{-j\beta z}\mathbf{a}_x \times \frac{E_{xo}}{\eta}e^{+j\beta z}\mathbf{a}_y\right] = \frac{1}{2}\frac{E_{xo}^2}{\eta}\mathbf{a}_z \qquad (5.86)$$

▷ **EXAMPLE 5.6**

Suppose we have $\mathbf{E}(z, t) = 1.0\cos(2\pi \times 10^8 t - \beta z)\mathbf{a}_x$ V/m, propagating in air. We want to find the power normally incident on a 20.-cm-diameter receiving dish.

By inspection we see that $E_{xo} = 1$ V/m, and in air $\eta = 120\pi\Omega$, so

$$\mathbf{P}_{ave} = \frac{1}{2}\frac{(1\text{V/m})^2}{120\pi\Omega}\mathbf{a}_z = 1.3\mathbf{a}_z \frac{\text{mW}}{\text{m}^2}$$

The power incident on the dish is

$$P = \int \mathbf{P}_{ave} \cdot d\mathbf{S} = \left(1.3\frac{\text{mW}}{\text{m}^2}\right)\left(\frac{\pi(0.2\text{ m})^2}{4}\right) = 42\ \mu\text{W}$$

Let us now consider

$$\mathbf{E}(z,t) = E_{xo}e^{-\alpha z}\cos(\omega t - \beta z + \phi)\mathbf{a}_x \text{ V/m}$$

the instantaneous electric field in a generally lossy media. The phasor is

$$\mathbf{E}_s = E_{xo}e^{-\alpha z}\,e^{-j\beta z}\,e^{j\phi}\mathbf{a}_x$$

The intrinsic impedance from (5.31) may be complex, and we can write this in polar form as

$$\eta = |\eta|e^{j\theta_\eta}$$

Then,

$$\mathbf{H}_s = \frac{1}{\eta}\mathbf{a}_p \times \mathbf{E}_s = \frac{E_{xo}}{|\eta|}e^{-\alpha z}e^{-j\beta z}e^{j\phi}e^{-j\theta_\eta}\mathbf{a}_y$$

This corresponds to an instantaneous magnetic field intensity

$$\mathbf{H}(z, t) = \frac{E_{xo}}{|\eta|}e^{-\alpha z}\cos\left(\omega t - \beta z + \phi - \theta_\eta\right)\mathbf{a}_y$$

which is seen to be out of phase with $\mathbf{E}(z, t)$ by the angle θ_η.

We can find the average power density from (5.84), being careful to provide the correct signs for the complex conjugate of $\mathbf{H}_s$:

$$\mathbf{P}_{ave} = \frac{1}{2} \text{Re}\left[E_{xo}e^{-\alpha z}e^{-j\beta z}e^{j\phi}\mathbf{a}_x \times \frac{E_{xo}}{|\eta|}e^{-\alpha z}e^{j\beta z}e^{-j\phi}e^{j\theta_\eta}\mathbf{a}_y \right]$$

After evaluating the cross product and taking the real part of Euler's identity, we have

$$\mathbf{P}_{ave} = \frac{1}{2}\frac{(E_{xo})^2}{|\eta|}e^{-2\alpha z}\cos\theta_\eta \mathbf{a}_z \tag{5.87}$$

This expression reduces to (5.86) for a lossless medium.

▶ **EXAMPLE 5.7**

Consider an electric field incident on a copper slab such that the field in the slab is given by

$$\mathbf{E}(z,t) = 1.0e^{-\alpha z}\cos(2\pi \times 10^7 t - \beta z)\mathbf{a}_x \text{ V/m}$$

We want to find the average power density.

Since copper is a good conductor we can use (5.58) to find

$$\alpha = \beta = \sqrt{\pi f \mu \sigma} = \sqrt{(\pi)(10^7 \text{ Hz})(4\pi \times 10^{-7} \text{ H/m})(5.8 \times 10^7 \text{ S/m})} = 47.8 \times 10^3 \text{ 1/m}$$

Then from (5.61) the intrinsic impedance is

$$\eta = \sqrt{\frac{\omega\mu}{\sigma}}e^{j45°} = 1.17e^{j45°} \text{ m}\Omega$$

Employing (5.87) we have

$$\mathbf{P}_{ave} = \frac{1}{2}\frac{(1 \text{ V/m})^2}{(1.17 \text{ m}\Omega)}e^{-2(47.8\times10^3)z}\cos 45° \, \mathbf{a}_z = 300e^{-96\times10^3 z}\mathbf{a}_z \text{ W/m}^2$$

At the surface ($z = 0$) the power density is 300 W/m². But after only one skin depth, in this case 21 μm, the wave's power density drops to e^{-2} (13.5%) of its surface value, or 41 W/m² in this case.

Drill 5.9 At the surface of a thick slab of polystyrene ($\sigma = 10^{-16}$ S/m, $\varepsilon_r = 2.6$) occupying $z > 0$, we have

$$\mathbf{E}(0, t) = 10\cos(3\pi \times 10^7 t) \, \mathbf{a}_y \text{ V/m}$$

Determine (a) $\mathbf{E}(z, t)$, (b) $\mathbf{H}(z, t)$, and (c) $\mathbf{P}_{ave}$.

(*Answer:* $\mathbf{E}(z, t) = 10e^{-\alpha z}\cos(3\pi \times 10^7 t - \beta z)\mathbf{a}_y$ V/m,
$\mathbf{H}(z, t) = -43e^{-\alpha z}\cos(3\pi \times 10^7 t - \beta z)\mathbf{a}_x$ mA/m,
$\mathbf{P}_{ave} = 210 \, e^{-2\alpha z} \, \mathbf{a}_z$ mW/m², where $\alpha = 12 \times 10^{-15}$ Np/m and $\beta = 0.51$ radians/m)

▶ 5.6 POLARIZATION

A uniform plane wave is characterized by its propagation direction and frequency. The medium's constitutive parameters determine the wave's attenuation and phase constant. To complete a UPW's description, we need to know the orientation described by its *polarization*. Formally, polarization describes the path taken by the tip of the electric field intensity vector in a fixed spatial plane orthogonal to the direction of propagation.

To better understand this definition consider a UPW characterized by the equation

$$\mathbf{E}(z,t) = E_x \mathbf{a}_x = E_o \cos(\omega t - \beta z)\mathbf{a}_x \tag{5.88}$$

Since the wave propagates in the z direction, a spatial plane orthogonal to this direction is the $z = 0$ plane. In Figure 5.12a, E_x (at $z = 0$) is plotted against time. Figure 5.12b shows the electric field intensity vector at corresponding points in time (a through e). The tip of the electric field intensity vector traces out a line segment, so (5.88) represents a *linearly polarized* wave, and in this case we say the wave is *x-polarized*.

In general, any UPW can consist of a pair of linearly polarized waves. Suppose the superposition of two such waves, an *x*-polarized wave and a *y*-polarized one, is

$$\mathbf{E}(z,t) = E_{xo}\cos(\omega t - \beta z + \phi_x)\mathbf{a}_x + E_{yo}\cos(\omega t - \beta z + \phi_y)\mathbf{a}_y \tag{5.89}$$

For simplicity, let's begin by ignoring the phase components. With $\phi_x = \phi_y = 0°$ we have

$$\mathbf{E}(z,t) = E_{xo}\cos(\omega t - \beta z)\mathbf{a}_x + E_{yo} \cos(\omega t - \beta z)\mathbf{a}_y \tag{5.90}$$

At $z = 0$, where we wish to trace the path of the total electric field intensity vector, we have

$$\mathbf{E}(0,t) = E_{xo}\cos(\omega t)\mathbf{a}_x + E_{yo} \cos(\omega t)\mathbf{a}_y \tag{5.91}$$

At $t = 0$, both linearly polarized waves have their maximum values; that is,

$$\mathbf{E}(0,0) = E_{xo}\mathbf{a}_x + E_{yo}\mathbf{a}_y \tag{5.92}$$

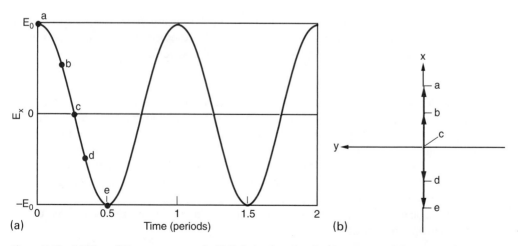

Figure 5.12 (a) Plot of E_x versus t at $z = 0$. (b) Polarization plot showing a trace of the electric field vector on the plane $z = 0$ at various times.

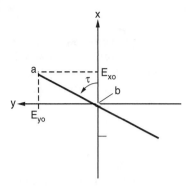

Figure 5.13 The vector from Eq. (5.91) traces out a linearly polarized wave.

which is shown as point *a* in Figure 5.13. At $t = T/4$ (one fourth of a period, where $\omega t = \pi/2$), both waves are minimum and

$$\mathbf{E}(0,T/4) = 0 \tag{5.93}$$

which is shown as point *b* in the figure. If we plot the tip of the total vector over a complete cycle, we get the line segment shown. As before, the tip still traces out a line, so again the wave is linearly polarized. The *tilt angle* τ (tau) is the angle this line makes with the *x*-axis.

Linear polarization occurs when the two linearly polarized waves are in phase ($\phi_y - \phi_x = 0°$), or when they are 180° out of phase ($\phi_y - \phi_x = \pm 180°$). In general, polarization depends on the relative phase difference ($\phi_y - \phi_x$) between the two waves. If we choose the *x*-polarized wave as our reference ($\phi_x = 0°$), then the entire phase difference is accounted for by ϕ_y. If $\phi_x = 0°$ and $\phi_y = 45°$ ($\pi/4$ radians) we have

$$\mathbf{E}(0,t) = E_{xo}\cos(\omega t)\mathbf{a}_x + E_{yo}\cos(\omega t + \pi/4)\mathbf{a}_y \tag{5.94}$$

In this case, the tip traces out an ellipse on a polarization plot, as shown in Figure 5.14, and we say the wave is *elliptically polarized*. The tilt angle in this case is the angle the long axis of the ellipse makes with the *x*-axis. Another commonly used term is *axial ratio*: the ratio of the long axis of an ellipse to the short axis.

The most general type of polarization is elliptical. Linear polarization is a special case of elliptical polarization that has an infinite axial ratio. Another special case often of interest is when the amplitudes E_{xo} and E_{yo} are equal and the waves are out of phase by $\pi/2$ radians ($\phi_y - \phi_x = \pm 90°$). This case, shown in Figure 5.15, is known as *circular polarization*. It has a unity axial ratio.

A final descriptor for a polarized wave is its *handedness*. For instance, a wave is right-hand circular polarized (RHCP) if the thumb of the right hand points in the direction of propagation and the fingers curl in the direction that the vector traces with time on the polarization plot. An RHCP wave has $\phi_y - \phi_x = -90°$. If the trace goes in the other direction, or $\phi_y - \phi_x = +90°$, the wave is left-hand circular polarized (LHCP). Handedness also applies to elliptical polarization where you may have left-hand elliptical polarization (LHEP) or right-hand elliptical polarization (RHEP).

It can be useful to use phasors in representing polarized waves. Equation (5.89) can be written in phasor form as

$$\mathbf{E}_s(z = 0) = E_{xo}e^{j\phi_x}\mathbf{a}_x + E_{yo}e^{j\phi_y}\mathbf{a}_y \tag{5.95}$$

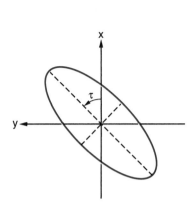

Figure 5.14 Polarization plot for elliptical polarization. The axial ratio is the major axis length divided by the minor axis length.

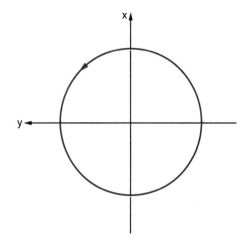

Figure 5.15 Polarization plot for circular polarization. The arrow indicates the direction of the trace as a function of time in this right-handed circular polarized wave.

For an LHCP wave, where we let $\phi_x = 0°$ and $\phi_y = +90°$, we then have

$$\mathbf{E}_s (z = 0) = E_{xo}(\mathbf{a}_x + e^{j\pi/2}\, \mathbf{a}_y) \tag{5.96}$$

By applying Euler's identity this becomes

$$\mathbf{E}_s (z = 0) = E_{xo}(\mathbf{a}_x + j\mathbf{a}_y) \tag{5.97}$$

Likewise, the RHCP wave would be

$$\mathbf{E}_s (z = 0) = E_{xo} (\mathbf{a}_x + e^{-j\pi/2}\, \mathbf{a}_y) = E_{xo} (\mathbf{a}_x - j\mathbf{a}_y) \tag{5.98}$$

This way of representing polarization will come in handy when discussing polarization efficiency in Chapter 8.

Wave polarization is of practical importance for radio and television broadcasts. Amplitude modulation (AM) radio is broadcast with polarization vertical to the earth's surface. As such, dipole-type antennas (see Chapter 8) are oriented vertically. Frequency modulation (FM) broadcasts are generally circularly polarized and reception is quite forgiving of antenna orientation.

It is important enough to mention again that any elliptically polarized wave may be represented as the superposition of two linearly polarized waves. In problems involving elliptical waves, it is customary to decompose the wave into its linear parts since they are easier to manipulate.

MATLAB 5.3

This program traces the polarization ellipse and determines handedness. Representing the phases ϕ_x and ϕ_y by fx and fy, respectively, we have

$$\mathbf{E}(0, t) = Exo\cos(\omega t + fx)\mathbf{a}_x + Eyo\cos(\omega t + fy)\mathbf{a}_y$$

(continues)

The program prompts for *Exo*, *fx*, *Eyo*, and *fy*. The trace direction (and hence the handedness) is indicated by moving from the "o" mark (at $\omega t = 0$) to the "+" mark (at $\omega t = \pi/4$) on the resulting plot (see, for example, Figure 5.16).

```
%    M-File: ML0503
%
%    This program traces polarization ellipses, given the
%    amplitude and phase of a pair of linearly polarized
%    waves.
%
%    Wentworth 7/22/02
%
%    Variables:
%    Exo,Eyo    amplitudes for pair of waves
%    fxd,fyd    phase angle for each wave
%    fx,fy      phase (radians) for each wave
%    wtd        ang freq * time, in degrees
%    wtr        ang freq * time, in radians
%    x,y        superposed position
%    x0,y0      position at wtd=0 degrees
%    x45,y45    position at wtd=45 degrees
%
clc              %clears the command window
clear            %clears variables

%    Prompt for input values
disp('Polarization Plot')
disp(' ')
Exo=input('enter x-amplitude: ');
fxd=input('enter x-phase angle (deg): ');
fx=fxd*pi/180;
Eyo=input('enter y-amplitude: ');
fyd=input('enter y-phase angle (deg): ');
fy=fyd*pi/180;
disp(' ')
disp('To determine direction of polarization, move')
disp('from o to + along the plot.')
disp(' ')

%    Perform calculations
wtd=0:360; %wt in degrees
wtr=wtd*pi/180;
x=Exo*cos(wtr+fx);
y=Eyo*cos(wtr+fy);
x0=Exo*cos(fx);
y0=Eyo*cos(fy);
x45=Exo*cos(fx+pi/4);
y45=Eyo*cos(fy+pi/4);

%    Make the plot
plot(x,y,x0,y0,'ok',x45,y45,'+k')
xlabel('x')
ylabel('y')
title('Polarization Plot')
axis('equal')
```

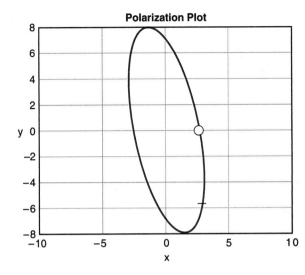

Figure 5.16 Polarization plot of the LHEP wave from Drill 5.11 using ML0503.

Drill 5.10 Suppose $E(z,t) = 3.0\cos(\omega t - \beta z)a_x + 4.0\cos(\omega t - \beta z)a_y$ V/m. What is the wave polarization and tilt angle? (*Answer:* linear, 53°)

Drill 5.11 $E(z,t) = 3.0\cos(\omega t - \beta z - 30°)a_x + 8.0\cos(\omega t - \beta z + 90°)a_y$ V/m. Determine the polarization of this wave. (*Answer:* LHEP; see Figure 5.16 plotted using ML0503)

Practical Application: Liquid Crystal Displays

Laptop computers, calculators, and microwave ovens are but a few of the devices that may contain liquid crystal displays (LCDs). Such displays are much more energy efficient than cathode ray tubes (CRTs) and also are much lighter and more compact. These features are of obvious use to laptop computers, which have limited battery life.

The *liquid crystals* used in LCDs are transparent rod-shaped organic molecules. Like a liquid, the molecules are free to move around. But like a crystal, they tend to align themselves with one another (think of a school of fish). Two features make liquid crystals useful for displays. First, the molecular alignment can be set by placement on a surface containing appropriate sized grooves, but the alignment can be changed by application of an electric field. Second, the molecules can guide the orientation of polarized light.

Basic operation of an LCD can be understood by considering Figure 5.17a. Here, the liquid crystal is squeezed between a pair of polarized, grooved plates. The grooves and polarization of the top plate are at a right angle to those of the bottom plate. The liquid crystal molecules are aligned with the grooves of the top plate and gradually twist around to be aligned with the bottom plate. Light passing through the top plate will be polarized. This polarized light will be guided along the twist of the liquid crystal until exiting through the other polarized plate. However, when an electric field is applied, as shown in Figure 5.17b, the liquid crystals align with the field and the polarized light doesn't twist around to match the bottom plate polarization. Transmission of light is therefore blocked.

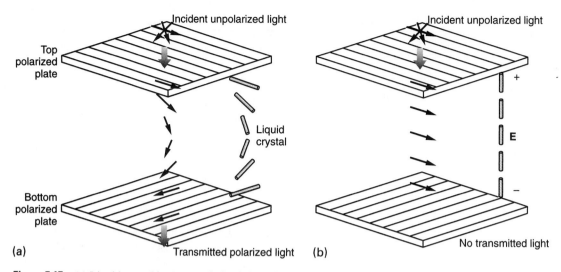

Figure 5.17 (a) Liquid crystal between polarized plates in the absence of a field. (b) Applied field changes the alignment of the liquid crystal molecules.

An electric field can be applied by coating the top and bottom plates with a thin-film transparent electrode material such as indium-tin-oxide. A typical color LCD display will be backlit by fluorescent light and will consist of an array of red, green, and blue pixels formed by adding an appropriate color filter to each LCD sandwich.

▷ 5.7 REFLECTION AND TRANSMISSION AT NORMAL INCIDENCE

In practice, plane waves are always encountering obstacles in their path. This section and the next discuss what happens when a wave is incident on a different medium. A good example is what happens when a light wave is incident on a mirror; most of the light gets reflected and a portion gets transmitted (with the transmitted part rapidly attenuating in the silver backing of the mirror). How much of a plane wave is transmitted or reflected depends on the constitutive parameters of the two involved media.

First we'll look at plane waves that are normally incident from one medium to another. By "normally incident" we mean the planar boundary separating the two media is perpendicular to the wave's propagation direction. We'll start with the general lossy case, and then see what happens in lossless and good conductor situations. Section 5.8 will deal with waves that are *obliquely* incident, where the propagation direction is no longer perpendicular to the plane separating the media.

General Case

To begin, let's consider a time-harmonic x-polarized electric field incident from medium 1 (with constitutive parameters μ_{r_1}, ε_{r_1}, and σ_1) to medium 2 (with constitutive parameters μ_{r_2}, ε_{r_2}, and σ_2). We'll indicate the incident fields with an i superscript, and we have

$$\mathbf{E}^i(z,t) = E_o^i e^{-\alpha_1 z} \cos(\omega t - \beta_1 z)\mathbf{a}_x \tag{5.99}$$

where E_o^i is the amplitude of the incident electric field intensity at $z = 0$, the location of the planar boundary separating the two media. It is far easier to carry out the upcoming calculations using phasors, and we will assume the reflected and transmitted waves maintain x-polarization. The magnetic fields are found using (5.33), and reflected and transmitted field values are indicated with r and t superscripts, respectively. We have the following set of equations:

Incident Fields

$$\mathbf{E}_s^i = E_o^i e^{-\alpha_1 z} e^{-j\beta_1 z} \mathbf{a}_x$$

$$\mathbf{H}_s^i = \frac{E_o^i}{\eta_1} e^{-\alpha_1 z} e^{-j\beta_1 z} \mathbf{a}_y \qquad (5.100)$$

Reflected Fields

$$\mathbf{E}_s^r = E_o^r e^{\alpha_1 z} e^{j\beta_1 z} \mathbf{a}_x$$

$$\mathbf{H}_s^r = -\frac{E_o^r}{\eta_1} e^{\alpha_1 z} e^{j\beta_1 z} \mathbf{a}_y \qquad (5.101)$$

Transmitted Fields

$$\mathbf{E}_s^t = E_o^t e^{-\alpha_2 z} e^{-j\beta_2 z} \mathbf{a}_x$$

$$\mathbf{H}_s^t = \frac{E_o^t}{\eta_2} e^{-\alpha_2 z} e^{-j\beta_2 z} \mathbf{a}_y \qquad (5.102)$$

where E_o^r and E_o^t represent the amplitudes for the electric field intensities at $z = 0$. In Figure 5.18, the general propagation directions are given by $\mathbf{a}_i$, $\mathbf{a}_r$, and $\mathbf{a}_t$.

Now let's find a relationship among E_o^i, E_o^r and E_o^t. We note that, for a UPW normally incident on a planar surface, only tangential fields are present. The boundary conditions relating to tangential fields are from (2.102)

$$\mathbf{E}_{t_1} = \mathbf{E}_{t_2} \qquad (5.103)$$

and from (3.76)

$$\mathbf{a}_{21} \times (\mathbf{H}_1 - \mathbf{H}_2) = \mathbf{K} \qquad (5.104)$$

In the absence of a surface current at the interface, (5.104) becomes

$$\mathbf{H}_{t_1} = \mathbf{H}_{t_2} \qquad (5.105)$$

Applying (5.103), we know that at the $z = 0$ boundary the total field in medium 1, $\mathbf{E}_s^i + \mathbf{E}_s^r$, must be equal to the field in medium 2, $\mathbf{E}_s^t$. So we have

$$E_o^i e^{-\alpha_1(0)} e^{-j\beta_1(0)} \mathbf{a}_x + E_o^r e^{\alpha_1(0)} e^{j\beta_1(0)} \mathbf{a}_x = E_o^t e^{-\alpha_2(0)} e^{-j\beta_2(0)} \mathbf{a}_x \qquad (5.106)$$

or

$$E_o^i + E_o^r = E_o^t \qquad (5.107)$$

The tangential **H** fields, by (5.105), are also equal, so

$$\frac{E_o^i}{\eta_1} - \frac{E_o^r}{\eta_1} = \frac{E_o^t}{\eta_2} \qquad (5.108)$$

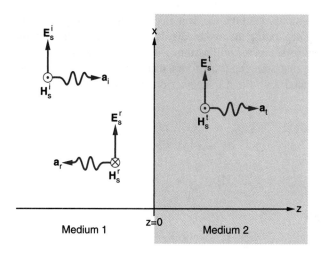

Figure 5.18 Plane wave incident from medium 1 to medium 2 results in a reflected wave and a transmitted wave.

or

$$E_o^i - E_o^r = \frac{\eta_1}{\eta_2} E_o^t \tag{5.109}$$

Using (5.107) and (5.109), we can arrive at equations relating the reflected and transmitted amplitudes to the incident amplitude. We find

$$\boxed{E_o^r = \frac{\eta_2 - \eta_1}{\eta_2 + \eta_1} E_o^i = \Gamma E_o^i} \tag{5.110}$$

and

$$\boxed{E_o^t = \frac{2\eta_2}{\eta_2 + \eta_1} E_o^i = \tau E_o^i} \tag{5.111}$$

Here we define a *reflection coefficient* Γ (gamma), and a *transmission coefficient* τ (tau), where

$$\Gamma = \frac{\eta_2 - \eta_1}{\eta_2 + \eta_1} = \frac{E_o^r}{E_o^i} \tag{5.112}$$

and

$$\tau = \frac{2\eta_2}{\eta_2 + \eta_1} = \frac{E_o^t}{E_o^i} \tag{5.113}$$

by comparing (5.112) and (5.113), it is easy to show

$$\boxed{\tau = 1 + \Gamma} \tag{5.114}$$

By knowing the constitutive parameters, we can solve for Γ and τ. Then, if we know any of the field quantities (incident, reflected, or transmitted), we can solve for all the other fields.

▶ **EXAMPLE 5.8**

Let's consider that the plane $z = 0$ separates two lossless, nonmagnetic media. Medium 1 ($z < 0$) has $\varepsilon_r = 4.0$ and medium 2 ($z > 0$) is air. We are given the incident field

$$\mathbf{E}^i = 1.0\cos(\omega t - \beta_1 z)\mathbf{a}_x \text{ V/m}$$

and we want to find all the incident, reflected, and transmitted field quantities involved, as well as their average power densities.

We can calculate the intrinsic impedances as $\eta_1 = 60\pi \, \Omega$ and $\eta_2 = 120\pi \, \Omega$. Then, we see that the phasors for the incident field are

$$\mathbf{E}_s^i = 1e^{-j\beta_1 z}\mathbf{a}_x$$

and

$$\mathbf{H}_s^i = \frac{1}{60\pi}e^{-j\beta_1 z}\mathbf{a}_y$$

where again we used (5.33). We can calculate the time-averaged power density from (5.84) as

$$\mathbf{P}_{ave}^i = \frac{1}{2}\text{Re}\left[\mathbf{E}_s^i \times \mathbf{H}_s^{i*}\right] = \frac{1}{2}\left(1\frac{V}{m}\right)\left(\frac{1}{60\pi}\frac{A}{m}\right)\mathbf{a}_z = 2.7 \, \mathbf{a}_z \, \text{mW/m}^2$$

To find the reflected wave, we need to first find Γ. From (5.112),

$$\Gamma = \frac{120\pi - 60\pi}{120\pi + 60\pi} = \frac{1}{3}$$

Then, using (5.110) and (5.101) we have

$$\mathbf{E}_s^r = \frac{1}{3}e^{j\beta_1 z}\mathbf{a}_x \text{ V/m}$$

$$\mathbf{H}_s^r = -\frac{1}{180\pi}e^{j\beta_1 z}\mathbf{a}_y \text{ A/m}$$

Just as with the incident field case, we find the average power density is

$$\mathbf{P}_{ave}^r = 0.3(-\mathbf{a}_z) \, \text{mW/m}^2$$

This is 0.3 mW/m² propagating in the $-\mathbf{a}_z$ direction. The transmission coefficient from (5.113) or (5.114) is 4/3, so from (5.111) and (5.102) we have

$$\mathbf{E}_s^t = \frac{4}{3}e^{-j\beta_2 z}\mathbf{a}_x \text{ V/m}$$

$$\mathbf{H}_s^t = \frac{4}{360\pi}e^{-j\beta_2 z}\mathbf{a}_y \text{ A/m}$$

and

$$\mathbf{P}_{ave}^t = 2.4\mathbf{a}_z \, \text{mW/m}^2$$

Unlike the electric or magnetic fields, *the power must be conserved across the boundary.* In other words, the power incident at the boundary must be accounted for by reflection and transmission. The incident 2.7 mW/m² is divided into a 0.30 mW/m² reflected part and a 2.4 mW/m² transmitted part.

Drill 5.12 A wave is normally incident from air onto a material with $\mu_r = 36$ and $\varepsilon_r = 4$. Calculate (a) the reflection coefficient and (b) the transmission coefficient. (*Answer:* (a) 1/2, (b) 3/2)

Drill 5.13 Suppose for the situation in Drill 5.12 that the transmitted electric field is found to be $\mathbf{E}^t = 15\cos(\omega t - \beta_2 z)\mathbf{a}_x$ mV/m. Determine the incident and reflected electric fields. (*Answer:* $\mathbf{E}^i = 10\cos(\omega t - \beta_1 z)\mathbf{a}_x$ mV/m, $\mathbf{E}^r = 5\cos(\omega t + \beta_1 z)\mathbf{a}_x$ mV/m)

▶ **EXAMPLE 5.9**

Suppose the field

$$\mathbf{E}^i = 1.0\cos(\omega t - \beta_1 z)\mathbf{a}_x \text{ V/m}$$

is now incident from air (medium 1, $z < 0$) onto a good conductor (medium 2, $z > 0$). We recall from Section 5.4 that good conductors ($\sigma \gg \omega\varepsilon$) have small impedances. With $\eta_1 \gg \eta_2$, it is easy to see from (5.112) and (5.113) that $\Gamma \cong -1$ and $\tau \cong 0$. Therefore, the reflected electric field intensity is

$$\mathbf{E}^r = -1.0\cos(\omega t + \beta_1 z)\mathbf{a}_x \text{ V/m}$$

Since τ is not exactly zero, there will be some small amount of transmitted field that rapidly attenuates in the good conductor. However, if the conductor is perfect, $\Gamma = -1$ and $\tau = 0$ and there will be no power transmitted.

Drill 5.14 What is the expression for $\mathbf{H}^r$ in the good conductor example? (*Answer:* $\mathbf{H}^r = (1/120\pi)\cos(\omega t + \beta_1 z)\mathbf{a}_y$ A/m).

▶ **EXAMPLE 5.10**

Let's apply our general equations for finding the reflected and transmitted fields to the case of our wave

$$\mathbf{E}^i = 1.0\cos(\omega t - \beta_1 z)\mathbf{a}_x \text{ V/m}$$

incident from air ($z < 0$) onto a lossy media ($z > 0$). The constitutive parameters for this lossy media are $\sigma = 0.010$ S/m, $\mu_r = 1.0$, and $\varepsilon_r = 2.0$. In the lossless and good conductor cases, we could get by without knowing the frequency. Here, though, the frequency is an important part of the problem. Let's work this problem at 100 MHz. At this frequency in air, $\beta_1 = 2\pi/3$ radians/m.

We calculate $\sigma/\omega\varepsilon = 0.90$. This is a quasi-conductor; it really cannot be considered a good conductor or a low-loss material. We can employ (5.52) and (5.31) to find our propagation parameters α_2, β_2, and η_2, or we can just use our program ML0501. In either case, we find $\alpha_2 = 1.2$ Np/m, $\beta_2 = 3.2$ radians/m, and $\eta_2 = 230\,e^{j21°}\ \Omega$.

Next we calculate the reflection and transmission coefficients from (5.112) and (5.113):

$$\Gamma = \frac{\eta_2 - \eta_1}{\eta_2 + \eta_1} = \frac{230e^{j21°} - 120\pi}{230e^{j21°} + 120\pi} = 0.30e^{j145°}$$

and

$$\tau = 1 + \Gamma = 0.77e^{j13°}$$

Now, since $E_o^r = \Gamma E_o^i$, we have

$$\mathbf{E}_s^r = 0.30e^{j145°}\, e^{j\beta_1 z}\, \mathbf{a}_x$$

or

$$E^r = 0.30\cos\left(2\pi \times 10^8 t + \frac{2\pi}{3}z + 145°\right)\mathbf{a}_x \text{ V/m}$$

Likewise, since $E_o^t = \tau E_o^i$, we have

$$\mathbf{E}_s^t = 0.77e^{j13°}\, e^{-1.2z}\, e^{-j3.2z}\, \mathbf{a}_x$$

or

$$\mathbf{E}^t = 0.77e^{-1.2z} \cos(2\pi \times 10^8 t - 3.2z + 13°)\mathbf{a}_x \text{ V/m}$$

The associated magnetic fields and average power densities are found in the usual way. For instance, using (5.33) we find

$$\mathbf{H}^t = 3.4e^{-1.2z} \cos(2\pi \times 10^8 t - 3.2z - 8°)\mathbf{a}_y \text{ mA/m}$$

And applying (5.83). we find

$$\mathbf{P}^t_{ave} = 1.2e^{-2.5z}\, \mathbf{a}_z \text{ mW/m}^2$$

Drill 5.15 Find the expression for $\mathbf{H}^r$ and $\mathbf{P}^r_{ave}$ in the lossy media example 5.10.

(*Answer:* $\mathbf{H}^r = -0.81 \cos(2\pi \times 10^8 t + (2\pi/3)z + 145°)\mathbf{a}_y$ mA/m, $\mathbf{P}^r_{ave} = 120 \ \mu\text{W/m}^2 \ (-\mathbf{a}_z)$)

Standing Waves

Because of reflection at the boundary, the medium containing the incident wave also contains a reflected wave. The superposition of the two waves can set up a *standing wave* pattern as exhibited in Figure 5.19.

Let's suppose wave $\mathbf{E}^i = 1\cos(\omega t - \beta z)\, \mathbf{a}_x$ V/m is incident on a boundary at $z = 0$ that presents a $\Gamma = 0.5$ reflection coefficient. By combining incident and reflected waves, our total instantaneous wave in medium 1 is

$$E = \cos(\omega t - \beta z) + 0.5\cos(\omega t + \beta z)$$

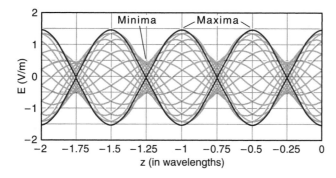

Figure 5.19 Standing wave pattern for an incident wave in a lossless medium reflecting off a second medium at $z = 0$ where $\Gamma = 0.5$.

A series of plots is shown in Figure 5.19. Each is a plot of E versus z over two wavelengths. They are plotted at 20° increments for ωt ranging from 0° to 360°.

We can make a couple of observations. First, we note that a standing wave pattern is set up with adjacent maxima separated by a half wavelength and adjacent minima also separated by a half wavelength. Second, the maxima are seen to be related to the reflection coefficient by

$$E_{max} = 1 + |\Gamma| \tag{5.115}$$

in this case reaching a value of $1 + 0.5 = 1.5$. Similarly, the minima are

$$E_{min} = 1 - |\Gamma| \tag{5.116}$$

in this case dropping to 0.5. The ratio of the maximum amplitude of the standing wave to the minimum is known as the *standing wave ratio*,[5.11]

$$SWR = \frac{E_{max}}{E_{min}} = \frac{1+|\Gamma|}{1-|\Gamma|} \tag{5.117}$$

One special case for the *SWR* is a wave normally incident on a good conductor. In such a case, $\Gamma = -1$ and the *SWR* is infinite. This happens because $E_{min} = 0$ for $(N\lambda/2)$, where $N = 0, 1, 2, 3 \ldots$.

As a practical matter, it is reasonably straightforward to measure *SWR* by probing the electric field strength at particular points away from the second medium. In this way we can determine the degree to which an incident wave is being reflected. This can be especially important in radome applications where a minimal reflection is required; such circumstances may necessitate design of an impedance-matching section.

We will see more of Γ, *SWR*, and impedance matching in the next chapter on transmission lines.

► 5.8 REFLECTION AND TRANSMISSION AT OBLIQUE INCIDENCE

Normal incidence of waves from one medium to another constitutes a special circumstance of the more general case where the waves are obliquely incident. For normal incidence, the fields are tangential to the boundary. This is not always the case for oblique incidence, and hence the treatment becomes a bit more complicated.

Let's first look at Figure 5.20 and define some of our terminology. The propagation directions are $\mathbf{a}_i$, $\mathbf{a}_r$, and $\mathbf{a}_t$ for the incident, reflected, and transmitted waves, respectively. We define the *plane of incidence* as the plane containing both a normal to the boundary and the incident wave's propagation direction. In the figure, the propagation direction is $\mathbf{a}_i$ and the normal is $\mathbf{a}_z$, so the plane of incidence is the x–z plane. The *angle of incidence* θ_i is the angle the incident field makes with a normal to the boundary. Likewise, θ_r is the *angle of reflection* and θ_t is the *angle of transmission*.

[5.11]The standing wave ratio is quite often referred to as *voltage standing wave ratio*, or *VSWR*. It is also often abbreviated by s. We'll use *SWR* in this text when referring to unguided waves (this chapter), and *VSWR* when referring to guided waves (Chapters 6 and 7).

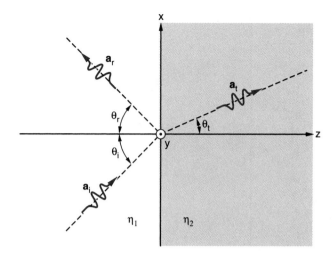

Figure 5.20 A UPW traveling in the $\mathbf{a}_i$ direction is obliquely incident from medium 1 onto medium 2.

Any UPW obliquely incident on the boundary can be decomposed into a pair of polarizations. In one of these, the electric field is perpendicular, or transverse, to the plane of incidence. This is therefore called *perpendicular polarization*, or more commonly, *transverse electric (TE) polarization*. The other polarization has the electric field parallel to the plane of incidence and is termed *parallel polarization*. In this second case, the magnetic field is transverse to the plane of incidence, so this is called *transverse magnetic (TM) polarization*.

In oblique incidence problems, the approach is to decompose the UPW into its TE and TM components and solve each one separately. After all the reflected and transmitted fields for each polarization are determined, they can be recombined for a final answer.

As we did for the normal incidence case, we want to find the relationship among the amplitudes of the incident, reflected, and transmitted waves. These will be related in terms of reflection and transmission coefficients, specific for each polarization. In the oblique case, we also want to find a relationship among the angles θ_i, θ_r, and θ_t.

To slightly simplify our treatment, we'll consider lossless media characterized by intrinsic impedances η_1 (for $z < 0$) and η_2 (for $z > 0$). The phasor equations developed will hold for general media by replacing $j\beta$ with γ.

TE Polarization

We first consider TE polarization, as depicted in Figure 5.21 where the electric field intensity vector is directed out of the page, or transverse to the plane of incidence.

Ignoring the second medium for a moment, consider the incident wave propagating as shown in Figure 5.22a. We superimpose an artificial pair of axes, x' and z', such that the incident wave's electric field is

$$\mathbf{E}_s^i = E_o^i e^{-j\beta_1 z'} \, \mathbf{a}_y \tag{5.118}$$

We can find the wave's magnetic field using (5.33),

$$\mathbf{H}_s^i = \frac{E_o^i}{\eta_1} e^{-j\beta_1 z'} \left(-\mathbf{a}_{x'}\right) \tag{5.119}$$

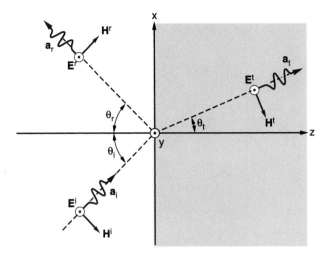

Figure 5.21 TE polarization.

Now, by trigonometry arguments, Figure 5.22a shows we can relate z' to our original coordinate system, and we obtain for the electric field

$$\mathbf{E}_s^i = E_o^i e^{-j\beta_1(x\sin\theta_i + z\cos\theta_i)}\mathbf{a}_y \qquad (5.120)$$

For the magnetic field, we find $-\mathbf{a}_{x'}$ in terms of our original coordinate system using Figure 5.22b, resulting in

$$\mathbf{H}_s^i = \frac{E_o^i}{\eta_1} e^{-j\beta_1(x\sin\theta_i + z\cos\theta_i)}\left(-\cos\theta_i\mathbf{a}_x + \sin\theta_i\mathbf{a}_z\right) \qquad (5.121)$$

The reflected and transmitted fields are found in a similar manner.

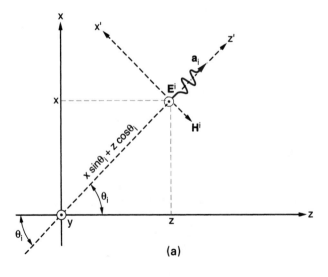

(a)

Figure 5.22 A pair of axes (x' and z') superimposed on our coordinate system to find an equivalence for (a) z'.

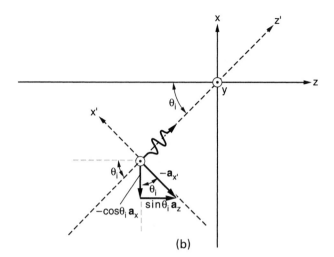

Figure 5.22 A pair of axes (x' and z') superimposed on our coordinate system to find an equivalence for (b) $-\mathbf{a}_{x'}$.

For TE polarization, the fields are summarized as follows:

Incident Fields

$$
\mathbf{E}_s^i = E_o^i e^{-j\beta_1\left(x\sin\theta_i + z\cos\theta_i\right)}\mathbf{a}_y
$$

$$
\mathbf{H}_s^i = \frac{E_o^i}{\eta_1} e^{-j\beta_1\left(x\sin\theta_i + z\cos\theta_i\right)}\left(-\cos\theta_i\mathbf{a}_x + \sin\theta_i\mathbf{a}_z\right)
$$

(5.122)

Reflected Fields

$$
\mathbf{E}_s^r = E_o^r e^{-j\beta_1\left(x\sin\theta_r - z\cos\theta_r\right)}\mathbf{a}_y
$$

$$
\mathbf{H}_s^r = \frac{E_o^r}{\eta_1} e^{-j\beta_1\left(x\sin\theta_r - z\cos\theta_r\right)}\left(\cos\theta_r\mathbf{a}_x + \sin\theta_r\mathbf{a}_z\right)
$$

(5.123)

Transmitted Fields

$$
\mathbf{E}_s^t = E_o^t e^{-j\beta_2\left(x\sin\theta_t + z\cos\theta_t\right)}\mathbf{a}_y
$$

$$
\mathbf{H}_s^t = \frac{E_o^t}{\eta_2} e^{-j\beta_2\left(x\sin\theta_t + z\cos\theta_t\right)}\left(-\cos\theta_t\mathbf{a}_x + \sin\theta_t\mathbf{a}_z\right)
$$

(5.124)

Now we need to relate the amplitudes for the three waves, and to do so we use the tangential boundary conditions (as we did for the normal incidence case). With TE polarization, all of the electric field is tangential to the surface. At $z = 0$ we have

$$
\mathbf{E}_o^i e^{-j\beta_1 x\sin\theta_i}\mathbf{a}_y + \mathbf{E}_o^r e^{-j\beta_1 x\sin\theta_r}\mathbf{a}_y = \mathbf{E}_o^t e^{-j\beta_2 x\sin\theta_t}\mathbf{a}_y
$$

(5.125)

For this equality to hold, the phases must match:

$$\beta_1 x \sin\theta_i = \beta_1 x \sin\theta_r = \beta_2 x \sin\theta_t \qquad (5.126)$$

From this we have Snell's law of reflection,

$$\boxed{\theta_i = \theta_r} \qquad (5.127)$$

In addition, we see that

$$\boxed{\frac{\beta_1}{\beta_2} = \frac{\sin\theta_t}{\sin\theta_i}} \qquad (5.128)$$

which is a version of Snell's law of refraction.

Equation (5.128) can be solved for the angle of transmission to give us

$$\theta_t = \sin^{-1}\left[\frac{\beta_1}{\beta_2}\sin\theta_i\right] \qquad (5.129)$$

Something interesting happens when $\beta_1 > \beta_2$. As θ_i increases from normal incidence ($0°$), θ_t increases more rapidly until at some *critical angle* for θ_i, θ_t has reached $90°$. This occurs at

$$\left(\theta_i\right)_{\text{critical}} = \sin^{-1}\left(\frac{\beta_2}{\beta_1}\right) \qquad (5.130)$$

For θ_i greater than the critical angle, there is total reflection of the wave and no transmission of power into medium 2. This total reflection is key to operation of fiber optic transmission lines, to be studied in Chapter 7.

It should be noted that, for the case of total reflection, fields do extend into the second medium, where they decay exponentially with z. However, the transmitted electric and magnetic fields are $90°$ out of phase, so no power is transmitted. These fields are termed *evanescent* waves when referring to their presence in waveguides.

Returning to (5.125), with the phases matched we have

$$E_o^i + E_o^r = E_o^t \qquad (5.131)$$

our first equation relating the amplitudes.

We need a second equation, found from considering the equivalence of the tangential magnetic field across the boundary (assuming that it carries no surface current). Here, only the x component of **H** is tangential, and we have

$$\frac{E_o^i}{\eta_1}e^{-j\beta_1 x \sin\theta_i}\left(-\cos\theta_i \mathbf{a}_x\right) + \frac{E_o^r}{\eta_1}e^{-j\beta_1 x \sin\theta_r}\left(\cos\theta_r \mathbf{a}_x\right)$$

$$= \frac{E_o^t}{\eta_2}e^{-j\beta_2 x \sin\theta_t}\left(-\cos\theta_t \mathbf{a}_x\right) \qquad (5.132)$$

Since the phases are equal, and since $\theta_i = \theta_r$, we can express this as

$$\frac{E_o^i - E_o^r}{\eta_1} \cos \theta_i = \frac{E_o^t}{\eta_2} \cos \theta_t \tag{5.133}$$

or

$$E_o^i - E_o^r = E_o^t \frac{\eta_1}{\eta_2} \frac{\cos \theta_t}{\cos \theta_i} \tag{5.134}$$

This is our second equation relating the amplitudes, and we can solve (5.131) and (5.134) simultaneously to find

$$E_o^r = \frac{\eta_2 \cos \theta_i - \eta_1 \cos \theta_t}{\eta_2 \cos \theta_i + \eta_1 \cos \theta_t} E_o^i = \Gamma_{TE} E_o^i \tag{5.135}$$

and

$$E_o^t = \frac{2\eta_2 \cos \theta_i}{\eta_1 \cos \theta_t + \eta_2 \cos \theta_i} E_o^i = \tau_{TE} E_o^i \tag{5.136}$$

A TE subscript is added to the reflection and transmission coefficients, as these values are quite different from ones we'll find for the parallel polarization (TM) case. For TE polarization, it can be seen that

$$\tau_{TE} = 1 + \Gamma_{TE} \tag{5.137}$$

It should be noted that in terms of power conservation, we only consider power directed normal to the boundary. For the TE polarization case of Figure 5.22, we have

$$P^i_{ave,z} = P^r_{ave,z} + P^t_{ave,z} \tag{5.138}$$

leading to

$$\frac{\left(E_o^i\right)^2}{2\eta_1} \cos \theta_i = \frac{\left(E_o^r\right)^2}{2\eta_1} \cos \theta_i + \frac{\left(E_o^t\right)^2}{2\eta_2} \cos \theta_t \tag{5.139}$$

► **EXAMPLE 5.11**

Consider a 100.-MHz wave with amplitude 6.00 V/m obliquely incident from air onto a slab of lossless, nonmagnetic material with $\varepsilon_r = 9.00$. The angle of incidence is 60.0° and the wave is TE polarized. We want to find the incident, reflected, and transmitted fields.

In air ($\eta_1 = 120\pi \ \Omega$) at 100 MHz the wavelength is 3 m, so $\beta_1 = (2\pi/3)$ radians/m. Then, since sin 60° = 0.866 and cos 60° = 0.500, we can write the incident fields from (5.122) as

$$\mathbf{E}_s^i = 6e^{-j(1.814x \, + \, 1.047z)}\mathbf{a}_y \ \text{V/m}$$

and

$$\mathbf{H}_s^i = \frac{6}{120\pi} e^{-j(1.814x + 1.047z)} \left(-0.5\mathbf{a}_x + 0.866\mathbf{a}_z\right) \ \text{A/m}$$

To get the reflected and transmitted amplitudes, we need Γ_{TE} and τ_{TE}, which we can evaluate from (5.135) and (5.136) once β_2, η_2, and θ_t are known. For $\varepsilon_r = 9.00$ in medium 2, we have

$$\beta_2 = \frac{\omega\sqrt{\varepsilon_r}}{c} = 2\pi \text{ radians/m}$$

and

$$\eta_2 = \sqrt{\frac{\mu}{\varepsilon}} = \frac{120\pi \ \Omega}{\sqrt{9}} = 40\pi \ \Omega$$

The angle of transmission is calculated from (5.129) to be 16.8°.
With these values in hand, we finally arrive at

$$\Gamma_{TE} = -0.613$$

and

$$\tau_{TE} = 0.387$$

Then, since $\theta_i = \theta_r$, we can write (5.123) as

$$\mathbf{E}_s^r = -3.68e^{-j(1.814x - 1.047z)}\mathbf{a}_y \text{ V/m}$$

and

$$\mathbf{H}_s^r = -9.76e^{-j(1.814x - 1.047z)}(-0.5\mathbf{a}_x + 0.866\mathbf{a}_z) \text{ mA/m}$$

The transmitted fields are then calculated from (5.124) as follows:

$$\mathbf{E}_s^t = 2.32e^{-j(1.82x + 6.02z)}\mathbf{a}_y \text{ V/m}$$

and

$$\mathbf{H}_s^t = 18.5e^{-j(1.82x + 6.02z)}(-0.96\mathbf{a}_x + 0.29\mathbf{a}_z) \text{ mA/m}$$

Drill 5.16 A 1.0-GHz TE wave is incident at a 30° angle of incidence from air onto a thick slab of nonmagnetic, lossless dielectric with $\varepsilon_r = 16$. Find Γ_{TE} and τ_{TE}.

(*Answer:* $\Gamma_{TE} = -0.64$, $\tau_{TE} = 0.36$)

TM Polarization

Oblique incidence of a TM polarized wave is indicated in Figure 5.23. Now all of the magnetic fields are tangential to the boundary, but only the x component of the electric field is tangential. Analysis reveals that the reflection and transmission angle relations are the same as for the TE case.

By similar geometric arguments as before, we arrive at the following fields:

Incident Fields

$$\boxed{\begin{aligned} \mathbf{E}_s^i &= E_o^i e^{-j\beta_1(x\sin\theta_i + z\cos\theta_i)}(\cos\theta_i\mathbf{a}_x - \sin\theta_i\mathbf{a}_z) \\ \mathbf{H}_s^i &= \frac{E_o^i}{\eta_1}e^{-j\beta_1(x\sin\theta_i + z\cos\theta_i)}\mathbf{a}_y \end{aligned}}$$

(5.140)

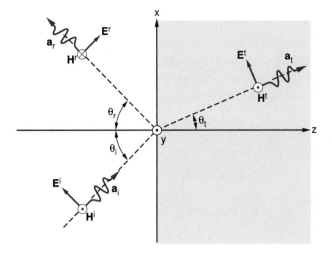

Figure 5.23 TM polarization.

Reflected Fields

$$\mathbf{E}_s^r = E_o^r e^{-j\beta_1 \left(x \sin\theta_r - z \cos\theta_r \right)} \left(\cos\theta_r \mathbf{a}_x + \sin\theta_r \mathbf{a}_z \right)$$

$$\mathbf{H}_s^r = -\frac{E_o^r}{\eta_1} e^{-j\beta_1 \left(x \sin\theta_r - z \cos\theta_r \right)} \mathbf{a}_y$$

(5.141)

Transmitted Fields

$$\mathbf{E}_s^t = E_o^t e^{-j\beta_2 \left(x \sin\theta_t + z \cos\theta_t \right)} \left(\cos\theta_t \mathbf{a}_x - \sin\theta_t \mathbf{a}_z \right)$$

$$\mathbf{H}_s^t = \frac{E_o^t}{\eta_2} e^{-j\beta_2 \left(x \sin\theta_t + z \cos\theta_t \right)} \mathbf{a}_y$$

(5.142)

Employing the boundary conditions, we find the following expressions relating the field amplitudes:

$$E_o^r = \frac{\eta_2 \cos\theta_t - \eta_1 \cos\theta_i}{\eta_2 \cos\theta_t + \eta_1 \cos\theta_i} E_o^i = \Gamma_{TM} E_o^i$$

(5.143)

and

$$E_o^t = \frac{2\eta_2 \cos\theta_i}{\eta_1 \cos\theta_i + \eta_2 \cos\theta_t} E_o^i = \tau_{TM} E_o^i$$

(5.144)

A TM subscript has been added to the reflection and transmission coefficients to represent parallel (TM) polarization. The relationship between the reflection and transmission coefficients for the TM case is

$$\tau_{TM} = \left(1 + \Gamma_{TM}\right)\frac{\cos\theta_i}{\cos\theta_t} \tag{5.145}$$

For TM polarizations, there exists an incidence angle at which all of the wave is transmitted into the second medium. This is known as the *Brewster angle*, $\theta_i = \theta_{BA}$, and it can be found by first setting the numerator of the reflection coefficient in (5.143) equal to zero; that is,

$$\eta_2\cos\theta_t = \eta_1\cos\theta_{BA} \tag{5.146}$$

If we square both sides of this equation, we have

$$\eta_1^2\cos^2\theta_{BA} = \eta_2^2\cos^2\theta_t$$

or

$$\eta_1^2(1 - \sin^2\theta_{BA}) = \eta_2^2(1 - \sin^2\theta_t) \tag{5.147}$$

Using Snell's law of refraction, we can replace the $\sin\theta_t$ term in (5.147) and do some manipulating to obtain

$$\sin\theta_{BA} = \sqrt{\frac{\beta_2^2\left(\eta_2^2 - \eta_1^2\right)}{\eta_2^2\beta_1^2 - \eta_1^2\beta_2^2}} \tag{5.148}$$

This rather cumbersome expression is greatly simplified when considering lossless, nonmagnetic media characterized only by ε_{r_1} and ε_{r_2}. In this case we have

$$\sin\theta_{BA} = \sqrt{\frac{1}{1 + \dfrac{\varepsilon_{r_1}}{\varepsilon_{r_2}}}} \tag{5.149}$$

When a randomly polarized wave (such as light) is incident on a material at the Brewster's angle, the TM polarized portion is totally transmitted but a TE component is partially reflected. This principle is employed in gas lasers, where quartz windows at each end of the laser tube are set at the Brewster's angle to produce linearly polarized laser output.

Drill 5.17 A 100.-MHz TM wave is incident at the Brewster's angle from air onto a thick slab of lossless, nonmagnetic material with $\varepsilon_{r_2} = 2.0$. Calculate the angle of transmission in medium 2. (*Answer:* 35°)

▶ SUMMARY

- The Helmholtz equations governing propagation of electromagnetic waves are derived directly from Maxwell's equations. For the electric field intensity, the equation is

$$\nabla^2 E = \mu\sigma\frac{\partial E}{\partial t} + \mu\varepsilon\frac{\partial^2 E}{\partial t^2}$$

which for time-harmonic fields reduces to

$$\nabla^2 E_s - \gamma^2 E_s = 0$$

where the propagation constant γ is related to the wave attenuation constant α and phase constant β as

$$\gamma = \sqrt{j\omega\mu(\sigma + j\omega\varepsilon)} = \alpha + j\beta$$

Solution of the Helmholtz equation for the case of a general x-polarized field yields

$$\mathbf{E}(z, t) = E_o^+ e^{-\alpha z}\cos(\omega t - \beta z)\mathbf{a}_x + E_o^- e^{\alpha z}\cos(\omega t + \beta z)\mathbf{a}_x$$

- The propagating wave relationship for phasors, where $\mathbf{a}_p$ is the direction of propagation, is

$$\mathbf{H}_s = \frac{1}{\eta}\mathbf{a}_P \times \mathbf{E}_s$$

$$\mathbf{E}_s = -\eta\mathbf{a}_P \times \mathbf{H}_s$$

Here η is the intrinsic impedance of the media given by

$$\eta = \sqrt{\frac{j\omega\mu}{\sigma + j\omega\varepsilon}}$$

The intrinsic impedance of free space, η_o, is $120\pi\ \Omega$.

- The parameters α and β are functions of a material's constitutive parameters as follows:

$$\alpha = \omega\sqrt{\frac{\mu\varepsilon}{2}\left(\sqrt{1 + \left(\frac{\sigma}{\omega\varepsilon}\right)^2} - 1\right)}$$

and

$$\beta = \omega\sqrt{\frac{\mu\varepsilon}{2}\left(\sqrt{1 + \left(\frac{\sigma}{\omega\varepsilon}\right)^2} + 1\right)}$$

- In low-loss dielectrics, characterized by $(\sigma/\omega\varepsilon) \ll 1$,

$$\alpha \cong \frac{\sigma}{2}\sqrt{\frac{\mu}{\varepsilon}}, \quad \beta \cong \omega\sqrt{\mu\varepsilon}, \quad \eta \cong \sqrt{\frac{\mu}{\varepsilon}}$$

- In good conductors, characterized by $(\sigma/\omega\varepsilon) \gg 1$,

$$\alpha \cong \beta \cong \sqrt{\pi f \mu\sigma}$$

and

$$\eta \cong \sqrt{\frac{\omega\mu}{\sigma}}e^{j45°} \cong \sqrt{2}\frac{\alpha}{\sigma}e^{j45°}$$

- Fields attenuate rapidly as they propagate in a good conductor. The skin depth indicates where the field amplitude has dropped to e^{-1} of its value at the surface and is given by

$$\delta = 1/\sqrt{\pi f \mu\sigma}$$

The skin-effect resistance R_{skin} in a conductor of width t is

$$R_{skin} = \frac{1}{\sigma\delta\left(1 - e^{-t/\delta}\right)}$$

- The time-averaged power density in a wave is given by the phasor Poynting theorem:

$$\mathbf{P}_{ave} = \frac{1}{2}\mathrm{Re}\left[\mathbf{E}_s \times \mathbf{H}_s^*\right]$$

For x-polarized propagation in the z direction within general media this can be expressed as

$$\mathbf{P}_{ave} = \frac{1}{2}\frac{(E_{xo})^2}{|\eta|}e^{-2\alpha z}\cos\theta_\eta\mathbf{a}_z$$

The amount of power P, in watts, passing through a surface is

$$P = \int \mathbf{P}_{ave} \cdot d\mathbf{S}$$

- Polarization describes the path taken by the tip of the electric field intensity vector in a fixed spatial plane orthogonal to the direction of propagation. Any UPW can be decomposed into a pair of linearly polarized waves. The amplitudes and phases of the two waves determine the polarization type, the most general of which is elliptical. Elliptical polarization reduces to linear polarization if the two waves are in phase. It reduces to circular polarization if the waves are out of phase by 90° and have equal amplitudes.

- When a UPW is incident normally from one medium to another, some of it may be reflected and some transmitted. The amplitudes of the electric fields for these waves are related by the reflection coefficient Γ and the transmission coefficient τ given by

$$\Gamma = \frac{\eta_2 - \eta_1}{\eta_2 + \eta_1} = \frac{E_o^r}{E_o^i}$$

and

$$\tau = \frac{2\eta_2}{\eta_2 + \eta_1} = \frac{E_o^t}{E_o^i}$$

These are related by

$$\tau = 1 + \Gamma$$

- The superposition of an incident wave and a reflected wave can lead to a standing wave that has locations of maximum and minimum amplitudes. The standing wave ratio SWR is given by

$$SWR = \frac{E_{max}}{E_{min}} = \frac{1 + |\Gamma|}{1 - |\Gamma|}$$

- Waves obliquely incident on a surface can be decomposed into TE polarized and TM polarized waves. For TE waves, the reflected and transmitted wave amplitudes are related to the incident wave amplitude by the reflection and transmission coefficients:

$$E_o^r = \frac{\eta_2 \cos\theta_i - \eta_1 \cos\theta_t}{\eta_2 \cos\theta_i + \eta_1 \cos\theta_t} E_o^i = \Gamma_{TE} E_o^i$$

and

$$E_o^t = \frac{2\eta_2 \cos\theta_i}{\eta_1 \cos\theta_t + \eta_2 \cos\theta_i} E_o^i = \tau_{TE} E_o^i$$

For TM waves,

$$E_o^r = \frac{\eta_2 \cos\theta_t - \eta_1 \cos\theta_i}{\eta_2 \cos\theta_t + \eta_1 \cos\theta_i} E_o^i = \Gamma_{TM} E_o^i$$

and

$$E_o^t = \frac{2\eta_2 \cos\theta_i}{\eta_1 \cos\theta_i + \eta_2 \cos\theta_t} E_o^i = \tau_{TM} E_o^i$$

- Snell's laws of reflection and refraction are, respectively,

$$\theta_i = \theta_r$$

and

$$\frac{\beta_1}{\beta_2} = \frac{\sin\theta_t}{\sin\theta_i}$$

- At the Brewster's angle, for TM polarization, all of the incident wave is transmitted. This angle occurs at

$$\sin\theta_{BA} = \sqrt{\frac{\beta_2^2 (\eta_2^2 - \eta_1^2)}{\eta_2^2 \beta_1^2 - \eta_1^2 \beta_2^2}}$$

▶ PROBLEMS

5.1 General Wave Equations

5.1 Starting with Maxwell's equations for simple, charge-free media, derive the Helmholtz equation for **H**.

5.2 Derive (5.10) by starting with the phasor point form of Maxwell's equations for simple, charge-free media.

5.3 A wave with $\lambda = 6.0$ cm in air is incident on a non-magnetic, lossless liquid medium. In the liquid, the wavelength is measured as 1.0 cm. What is the wave's frequency (a) in air and (b) in the liquid? (c) What is the liquid's relative permittivity?

5.4 Suppose $\mathbf{H}_s(z) = H_{ys}(z)\mathbf{a}_y$. Start with (5.14) and derive (5.29).

5.5 Given $\sigma = 1.0 \times 10^{-5}$ S/m, $\varepsilon_r = 2.0$, $\mu_r = 50.$, and $f = 10.$ MHz, find γ, α, β, and η.

5.6 In some material, the constitutive parameters are constant over a large frequency range and are given as $\sigma = 0.10$ S/m , $\varepsilon_r = 4.0$, and $\mu_r = 600$. Write a MATLAB routine that will plot α, β, and η (magnitude and phase) versus the log of frequency from 1 Hz up to 100 GHz.

5.7 Suppose $\mathbf{E}(x, y, t) = 5.0\cos(\pi \times 10^6 t - 3.0x + 2.0y)\,\mathbf{a}_z$ V/m. Find the direction of propagation, $\mathbf{a}_p$, and $\mathbf{H}(x, y, t)$.

5.8 Suppose in free space, $\mathbf{H}(x, t) = 100.\cos(2\pi \times 10^7 t - \beta x + \pi/4)\,\mathbf{a}_z$ mA/m. Find $\mathbf{E}(x, t)$.

5.2 Propagation in Lossless, Charge-Free Media

5.9 Start with the Helmholtz equation (5.11), and using $\gamma = j\beta$, derive the traveling wave equation (5.41).

5.10 A 100-MHz wave in free space propagates in the y direction with an amplitude of 1 V/m. If the electric field vector for this wave has only an $\mathbf{a}_z$ component, find the instantaneous expression for the electric and magnetic fields.

5.11 In a lossless, nonmagnetic material with $\varepsilon_r = 16$, $\mathbf{H} = 100 \cos(\omega t - 10y)\,\mathbf{a}_z$ mA/m. Determine the propagation velocity, the angular frequency, and the instantaneous expression for the electric field intensity.

5.12 Given $\mathbf{E} = 120\pi \cos(6\pi \times 10^6 t - 0.080\pi y)\,\mathbf{a}_z$ V/m and $\mathbf{H} = 2.00\cos(6\pi \times 10^6 t - 0.080\pi y)\,\mathbf{a}_x$ A/m, find μ_r and ε_r.

5.3 Propagation in Dielectrics

5.13 Work through the algebra to derive the α and β equations (5.52) from Eqs. (5.50) and (5.51).

5.14 Write a routine to prompt the user for a material's constitutive parameters and an operating frequency, and calculate the α and β from (5.52). Verify the program by running Drill 5.6.

5.15 Given a material with $\sigma = 1.0 \times 10^{-3}$ S/m, $\mu_r = 1.0$, and $\varepsilon_r' = 3.0$, $\varepsilon_r'' = 0.015$, compare a plot of α versus frequency from 1 Hz to 1 GHz using (5.52) to a similar plot using (5.54). At what frequency does the percentage error exceed 2%?

5.16 In a medium with properties $\sigma = 0.00964$ S/m , $\varepsilon_r = 1.0$, and $\mu_r = 100.$, a 1.0 mA/m amplitude magnetic field travels in the $+x$ direction at 100. MHz with its field vector in the z direction. Find the instantaneous form of the related electric field intensity.

5.17 Make a pair of plots similar to Figure 5.4 for the three materials of Table 5.1. Instead of plotting loss tangent, plot the magnitude of η on one plot and the phase of η on the other.

5.4 Propagation in Conductors

5.18 Starting with (5.13), show that $\alpha = \beta$ for a good conductor.

5.19 In seawater, a propagating electric field is given by $\mathbf{E}(z, t) = 20.e^{-\alpha z} \cos(2\pi \times 10^6 t - \beta z + 0.5)\,\mathbf{a}_y$ V/m. Assuming $\varepsilon'' = 0$, find (a) α and β and (b) the instantaneous form of $\mathbf{H}$.

5.20 Calculate the skin depth at 1.00 GHz for (a) copper, (b) silver, (c) gold, and (d) nickel.

5.21 For nickel ($\sigma = 1.45 \times 10^7$, $\mu_r = 600$), make a table of α, β, η, u_p, and δ for 1 Hz, 1 kHz, 1 MHz, and 1 GHz.

5.22 A semiinfinite slab exists for $z > 0$ with $\sigma = 300$ S/m, $\varepsilon_r = 10.2$, and $\mu_r = 1.0$. At the surface ($z = 0$),

$$\mathbf{E}(0, t) = 1.0\cos(\pi \times 10^6 t)\,\mathbf{a}_x \text{ V/m}$$

Find the instantaneous expressions for $\mathbf{E}$ and $\mathbf{H}$ anywhere in the slab.

5.23 In a nonmagnetic material,

$$\mathbf{E}(z, t) = 10.e^{-200z} \cos(2\pi \times 10^9 t - 200z)\,\mathbf{a}_x \text{ mV/m}$$

Find $\mathbf{H}(z,t)$.

5.24 A 0.1-μm layer of copper is deposited atop a very thick slab of nickel. For a field incident on the copper surface, (a) calculate R_{skin} at 1.0 GHz. Compare this with R_{skin} at 1.0 GHz for (b) a semiinfinite slab of copper and (c) a 0.1-μm thickness of copper by itself.

5.25 Calculate the DC resistance per meter length of a 4.0-mm-diameter copper wire. Now find the resistance at 1.0 GHz.

5.5 The Poynting Theorem and Power Transmission

5.26 In air, $\mathbf{H}(z,t) = 12.\cos(\pi \times 10^6 t - \beta z + \pi/6)\,\mathbf{a}_x$ A/m. Determine the power density passing through a 1.0 m² surface that is normal to the direction of propagation.

5.27 A 600-MHz uniform plane wave incident in the z direction on a thick slab of Teflon ($\varepsilon_r = 2.1$, $\mu_r = 1.0$) imparts a 1.0 V/m amplitude y-polarized electric field intensity at the surface. Assuming $\sigma = 0$ for Teflon, find in the Teflon (a) $\mathbf{E}(z,t)$, (b) $\mathbf{H}(z,t)$, and (c) $\mathbf{P}_{ave}$.

5.28 Assume distilled water ($\sigma = 10^{-4}$ S/m, $\varepsilon_r = 81$, $\mu_r = 1.0$) fills the region $z > 0$. At the surface, we have $\mathbf{E}(0, t) = 8.0\cos(2\pi \times 10^8 t)\,\mathbf{a}_x$ V/m. Determine (a) $\mathbf{E}(z, t)$, (b) $\mathbf{H}(z, t)$, (c) $\mathbf{P}_{ave}$, and (d) the power passing through a 10 m² surface located at $z = 1.0$ m.

5.29 The density of solar radiation is approximately 150 W/m² at some locations on the earth's surface. How much solar power is incident on a typical "100-W" solar panel (0.6×1.6 m² area) if the panel is normal to the radiation propagation direction? How much power is incident if the panel is tilted 45° to the radiation propagation direction?

5.30 A 200-MHz uniform plane wave incident on a thick copper slab imparts a 1.0 mV/m amplitude at the surface. How much power passes through a square meter at the surface? How much power passes through a square meter area 10. μm beneath the surface?

5.6 Wave Polarization

5.31 Suppose $\mathbf{E}(z,t) = 10.\cos(\omega t - \beta z)\mathbf{a}_x + 5.0\cos(\omega t - \beta z)\mathbf{a}_y$ V/m. What is the wave polarization and tilt angle?

5.32 Given $\mathbf{E}(z,t) = 10.\cos(\omega t - \beta z)\mathbf{a}_x - 20.\cos(\omega t - \beta z - 45°)\mathbf{a}_y$ V/m, find the polarization and handedness.

5.33 Given $\mathbf{H}(z,t) = 2.0\cos(\omega t - \beta z)\mathbf{a}_x + 6.0\cos(\omega t - \beta z - 120°)\mathbf{a}_y$ A/m, find the polarization and handedness.

5.34 Given

$$\mathbf{E}(z, t) = E_{xo} \cos(\omega t - \beta z)\mathbf{a}_x + E_{yo}\cos(\omega t - \beta z + \phi)\mathbf{a}_y$$

we say that E_y leads E_x for $0° < \phi < 180°$ and that E_y lags E_x when $-180° < \phi < 0°$. Determine the handedness for each of these two cases.

5.35 For a general elliptical polarization represented by $\mathbf{E}(z,t) = E_{xo} \cos(\omega t - \beta z)\mathbf{a}_x + E_{yo} \cos(\omega t - \beta z + \phi)\mathbf{a}_y$, the axial ratio and tilt angle can be found from the following formulas (from K. R. Demarest, *Engineering Electromagnetics*, Prentice-Hall, 1998, pp. 451–453):

$$a = |E_{xo}|, \quad b = |E_{yo}|$$

MAJ = length of majority-axis

MIN = length of minority-axis

$$MAJ = 2\sqrt{\frac{1}{2}\left[a^2 + b^2 + \sqrt{a^4 + b^4 + 2a^2b^2 \cos 2\phi}\right]}$$

$$MIN = 2\sqrt{\frac{1}{2}\left[a^2 + b^2 - \sqrt{a^4 + b^4 + 2a^2b^2 \cos 2\phi}\right]}$$

$$\text{axial ratio} = \frac{MAJ}{MIN}$$

$$\tau = \frac{1}{2}\tan^{-1}\left[\frac{2ab}{a^2 - b^2}\cos\phi\right]$$

Compose a program that not only draws a polarization plot like MATLAB 5.3 but also calculates the axial ratio and tilt angle. Run the program for Drill 5.11.

5.7 Reflection and Transmission at Normal Incidence

5.36 Starting with (5.107) and (5.109), derive (5.110) and (5.111)

5.37 A UPW is normally incident from medium 1 ($z < 0$, $\sigma = 0$, $\mu_r = 1.0$, $\varepsilon_r = 4.0$) to medium 2 ($z > 0$, $\sigma = 0$, $\mu_r = 8.0$, $\varepsilon_r = 2.0$). Calculate the reflection and transmission coefficients seen by this wave.

5.38 Suppose medium 1 ($z < 0$) is air and medium 2 ($z > 0$) has $\varepsilon_r = 16$. The transmitted magnetic field intensity is known to be $\mathbf{H}^t = 12\cos(\omega t - \beta_2 z)\mathbf{a}_y$ mA/m. (a) Determine the instantaneous value of the incident electric field. (b) Find the reflected time-averaged power density.

 5.39 Suppose a UPW in air carrying an average power density of 100 mW/m² is normally incident on a nonmagnetic material with $\varepsilon_r = 11$. What are the time-averaged power densities of the reflected and transmitted waves?

5.40 A UPW in a lossless nonmagnetic $\varepsilon_r = 16$ medium (for $z < 0$) is given by

$$\mathbf{E}(z,t) = 10.\cos(\omega t - \beta_1 z)\mathbf{a}_x + 20.\cos(\omega t - \beta_1 z + \pi/3)\mathbf{a}_y \text{ V/m}.$$

This is incident on a lossless medium characterized by $\mu_r = 12$ and $\varepsilon_r = 6.0$ (for $z > 0$). Find the instantaneous expressions for the reflected and transmitted electric field intensities.

5.41 The wave $\mathbf{E}^i = 100\cos(\pi \times 10^6 t - \beta_1 z + \pi/4)\,\mathbf{a}_x$ V/m is incident from air onto a perfect conductor. Find $\mathbf{E}^r$ and $\mathbf{E}^t$.

5.42 A UPW given by $\mathbf{E}(z,t) = 10.\cos(\omega t - \beta_1 z)\mathbf{a}_x + 20.\cos(\omega t - \beta_1 z + \pi/3)\mathbf{a}_y$ V/m is incident from air (for $z < 0$) onto a perfect conductor (for $z > 0$). Find the instantaneous expression for the reflected electric field intensity and the SWR.

5.43 The wave $\mathbf{E}^i = 10.\cos(2\pi \times 10^8 t - \beta_1 z)\,\mathbf{a}_x$ V/m is incident from air onto a copper conductor. Find $\mathbf{E}^r$, $\mathbf{E}^t$, and the time-averaged power density transmitted at the surface.

 5.44 Given a UPW incident from medium 1 ($\sigma = 0$, $\mu_r = 1.0$, $\varepsilon_r = 25.$) to medium 2 ($\sigma = 0.0080$, $\mu_r = 1.0$, $\varepsilon_r = 81.$), calculate Γ, SWR, and τ at 1 kHz, 1 MHz, and 1 GHz.

5.45 Write a program that prompts the user for the constitutive parameters in medium 1 and medium 2 separated by a planar surface. You are to assume a wave is normally incident from medium 1 to medium 2. The program is to plot the magnitudes of Γ and τ versus a frequency range supplied by the user. Plot the values from 100 Hz to 10 GHz for the pair of media specified in the previous problem.

5.46 A wave specified by $\mathbf{E}^i = 100.\cos(\pi \times 10^7 t - \beta_1 z)\mathbf{a}_x$ V/m is incident from air (at $z < 0$) to a nonmagnetic media ($z > 0$, $\sigma = 0.050$ S/m, $\varepsilon_r = 9.0$). Find $\mathbf{E}^r$, $\mathbf{E}^t$, and SWR. Also find the time-averaged power densities for the incident, reflected and transmitted waves.

5.47 A wave specified by $\mathbf{E}^i = 12\pi\cos(2\pi \times 10^7 t - \beta_1 z + \pi/4)\mathbf{a}_x$ V/m is incident from a nonmagnetic, lossless, $\varepsilon_r = 9.0$ medium (at $z < 0$) to a medium ($z > 0$) with $\sigma = 0.020$ S/m, $\mu_r = 2.0$, and $\varepsilon_r = 16$. Find $\mathbf{H}^i$, $\mathbf{E}^r$, $\mathbf{H}^r$, $\mathbf{E}^t$, $\mathbf{H}^t$, and the time-averaged power densities for the incident, reflected, and transmitted waves.

5.8 Reflection and Transmission at Oblique Incidence

5.48 A 100-MHz TE polarized wave with amplitude 1.0 V/m is obliquely incident from air ($z < 0$) onto a slab of lossless, nonmagnetic material with $\varepsilon_r = 25$ ($z > 0$). The angle of incidence is 40°. Calculate (a) the angle of transmission, (b) the reflection and transmission coefficients, and (c) the incident, reflected, and transmitted fields.

5.49 A 100-MHz TM polarized wave with amplitude 1.0 V/m is obliquely incident from air ($z < 0$) onto a slab of lossless, nonmagnetic material with $\varepsilon_r = 25$ ($z > 0$). The angle of incidence is 40°. Calculate (a) the angle of transmission, (b) the reflection and transmission coefficients, and (c) the incident, reflected, and transmitted fields.

5.50 A randomly polarized UPW at 200 MHz is incident at the Brewster's angle from air ($z < 0$) onto a thick slab of lossless, nonmagnetic material with $\varepsilon_r = 16$ ($z > 0$). The wave can be decomposed into equal TE and TM parts, each with an incident electric field amplitude of 10. V/m. Find expressions for the instantaneous values of the incident, reflected, and transmitted electric fields.

PART II

Applied Electromagnetics

CHAPTER 6

Transmission Lines

Learning Objectives

▷ Develop equations for wave propagation on a transmission line and define characteristic impedance and propagation constant

▷ Investigate wave reflection from terminated transmission lines and define input impedance and standing wave ratio

▷ Introduce the Smith chart, a graphical tool for the study of transmission lines, and use it to develop impedance matching networks

▷ Introduce design and analysis equations for microstrip transmission lines

▷ Study the behavior of transient signals on a terminated transmission line

▷ Investigate the dispersion of a signal pulse as it travels along a transmission line

The first application of electromagnetic theory to be studied is the *transmission line,* or *T-line* for short. Power lines, telephone lines, and cable TV lines are all examples. Wires used to breadboard a typical circuit can also be treated as T-lines, and in fact must be if the operating frequency is high enough. T-lines are characterized by their ability to guide propagation of electromagnetic energy and their length, which is comparable to or larger than a wavelength.

In the conventional study of electronic circuits, power supplies and circuit elements (resistors, capacitors, etc.) are connected together by pieces of ideal wire (see Figure 6.1a). These wires are considered lossless lines of negligible length. The signal phase at the load is the same as at the source, as represented by Figure 6.1b. By contrast, the length of a transmission line is very important. In Figure 6.1c, a quarter-wavelength long transmission line (represented by the pair of thick lines) is inserted between the supply and the resistor. As Figure 6.1d shows, the signal is delayed in traveling from the source to the load, and some shift in the phase is introduced. Some interesting things happen on a T-line as a result of this phase difference from one end to the other. For instance, a voltage wave traveling along the T-line may be partially reflected when it encounters the load, a decidedly different result from the ideal case of Figure 6.1a.

With the ultimate goal of learning how waves behave on a T-line, this chapter begins with an explanation of *distributed parameters* for a dual-conductor transmission line. For the special case of a time-harmonic (sinusoidal) source, these parameters are used to find

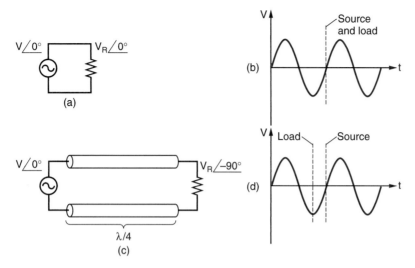

Figure 6.1 In (a), a sinusoidal voltage is dropped across a resistor. The supply and resistor are connected by an ideal (negligible length) conductor, and these are shown in (b) to be in phase. In (c) a quarter-wavelength long transmission line is added between the supply and the resistor and the voltage at the resistor in (d) is 90° out of phase with the supply voltage.

equations governing wave behavior and to determine what happens on terminated lines. Finding the equations will be accomplished using conventional circuit theory, and it will be observed that the equations are very similar to those found for uniform plane waves. This chapter is restricted to TEM mode propagation, where the electric and magnetic fields are always transverse to the direction of propagation. In the next chapter we will consider other modes of propagation.

Understanding transmission lines is of extreme importance to anyone working with modern high-speed integrated circuits and circuit boards. It is a prerequisite topic for understanding the electromagnetic interference problems encountered in Chapter 9 and the microwave and RF circuits studied in Chapter 10.

▶ 6.1 DISTRIBUTED-PARAMETER MODEL

Three of the most common dual-conductor transmission line types are shown in Figure 6.2. The *twin-lead* T-line may be familiar to some students as the line attached from a television to an aerial antenna. It otherwise finds rather limited use. The *coaxial* (or *coax*) T-line finds heavy use in connecting high-frequency equipment together, and it will be the focus of much of this chapter. It consists of an inner conductor of radius a surrounded by a dielectric sheath out to a radius b, then another conductive layer. *Microstrip* T-line is most applicable at the circuit board level and will be discussed in Section 6.6. In Figure 6.2, the microstrip line consists of copper that has been electroplated on an alumina (Al_2O_3) substrate. The entire backside of the substrate (not shown) also has plated copper.

All three of the T-lines shown in Figure 6.2 can be modeled as a simple twin-lead configuration. Figure 6.3 shows that a differential segment of the line can be modeled using the series distributed elements R' (resistance/meter) and L' (inductance/meter) and the shunt distributed elements G' (conductance/meter) and C' (capacitance/meter). The primes indicate these are "per unit length" or distributed values. When these parameters are multiplied by the length of the differential segment, Δz in meters, "pure" element values ($R, L, G,$ and C) result. The parameters are considered to be evenly distributed along the length of the T-line.

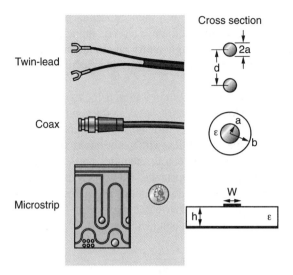

Figure 6.2 Transmission line examples along with schematic cross sections. A quarter is shown for scale.

Although this distributed-parameter model will hold for all three of the structures in Figure 6.2, the distributed parameters themselves will be different.

When a signal travels along a conventional wire, it encounters resistance. This series resistance is very small for good conductors, and in fact it can vanish altogether by using superconductors. However, some printed circuit board manufacturers are now using conductive inks that can be screen-printed on the circuit board (similar to how a tee shirt may be screen printed) and cured at high temperature to drive out the solvents. The material left behind is a thin conductor that is far from ideal and may in fact have considerable resistance.

The two wires in the dual-conductor T-line are separated by some dielectric material that will ideally be a perfect insulator. Real dielectrics do in fact conduct a small amount of shunt current. The parameter used to quantify this is the *conductance* (the inverse of resistance). Note that this conductance is a property of the dielectric and has nothing to do with the series resistance of the conductor.

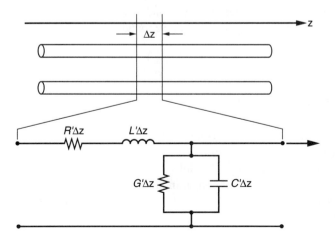

Figure 6.3 The distributed parameters for a differential segment of transmission line.

Recalling Chapter 2, we know there is a shunt capacitance between the two conductor lines. And from Chapter 3 we recall that there is a series inductance associated with signal propagation along the line. For a given geometry and material composition, formulas for each of the distributed parameters can be derived using electromagnetics concepts developed earlier.

Let's consider the coaxial cable shown in Figure 6.2. Coax is a good transmission line in that the fields are confined to the dielectric between the conductors. Very little escapes as noise outside the cable, and likewise very little noise from outside can get in. There are literally hundreds of coaxial cable assemblies, with variations in dimensions, conductor metal and type (solid or stranded), type of dielectric, and outer sleeve material. Some of the more common have an "RG" designator. For instance, "RG-6/U coaxial cable" has a 75-Ω *characteristic impedance* (discussed in the next section) and is routinely employed for use in home cable systems. Formulas for the distributed parameters follow from earlier sections. From Section 2.10, the distributed shunt conductance G' is

$$G' = \frac{2\pi\sigma_d}{\ln(b/a)} \tag{6.1}$$

where σ_d is the conductance of the dielectric. From Section 2.13, the distributed shunt capacitance is

$$C' = \frac{2\pi\varepsilon}{\ln(b/a)} \tag{6.2}$$

From Section 3.9, the distributed series inductance is

$$L' = \frac{\mu}{2\pi}\ln(b/a) \tag{6.3}$$

This inductance does not include internal inductance in the conductors. We are assuming this internal inductance component is negligible for high frequencies where the skin depth is small and current runs on the outer skin of the inner conductor and on the inner skin of the outer conductor.

From Section 5.4, considering that the distributed resistance will be a series combination of the resistance in the inner and outer conductors, and assuming the skin depth is much smaller than the metal thickness involved, we found a distributed series resistance of

$$R' = \frac{1}{2\pi}\left(\frac{1}{a} + \frac{1}{b}\right)\sqrt{\frac{\pi f \mu}{\sigma_c}} \tag{6.4}$$

where σ_c is the conductor conductivity.

Drill 6.1   Find the distributed parameters for RG-58/U cable, at 1.0 GHz, if the radius of the inner conductor is 0.45 mm, and the outer conductor goes from a radius of 1.47 mm to 2.4 mm. Polyethylene is the dielectric (assume $\sigma_d = 0$) and copper is the conductor. Note that the skin depth for copper in this problem is ~2 μm, much smaller than the metal dimensions employed. (*Answer:* $R' = 3.8$ Ω/m, $L' = 240$ nH/m, $G' = 0$ S/m, $C' = 110$ pF/m)

▶ **MATLAB 6.1**

The following program calculates the distributed parameters for coaxial cable after the user inputs the inner and outer radii and the material properties. It assumes only non-magnetic materials are used. This program is very similar to MATLAB 5.1.

```
%   M-File: ML0601
%
%This program calculates the distributed
%parameters for coaxial cable given the
%dimensions and material properties.
%
%   Wentworth, 7/30/02
%
%   Variables:
%   a          coaxial inner radius (mm)
%   b          coaxial outer radius (mm)
%   er         diel. rel. permittivity (F/m)
%   sigd       dielectric conductivity (S/m)
%   sigc       conductor conductivity (S/m)
%   f          operating frequency (Hz)
%   muo        free space permeability
%   eo         free space permittivity
%   G          distrib. conductance (S/m)
%   C          distrib. capacitance (F/m)
%   L          distrib. inductance (H/m)
%   R          distrib. resistance (ohms/m)
%   Rs         conductor sheet res. (ohms/m)

clc          %clears the command window
clear        %clears variables

%   Initialize variables
eo=8.854e-12;
muo=pi*4e-7;
%   Prompt for input values
disp('Calc Dist. Parameters for Coax')
disp(' ')
a=input('inner radius, in mm, = ');
b=input('outer radius, in mm, = ');
er=input('relative permittivity, er= ');
sigd=input('diel. conductivity,S/m, = ');
sigc=input('cond. conductivity,S/m, = ');
f=input('frequency, in Hz, = ');
disp(' ')

%   Perform calculations
G=2*pi*sigd/log(b/a);
C=2*pi*er*eo/log(b/a);
L=muo*log(b/a)/(2*pi);
Rs=sqrt(pi*f*muo/sigc);
R=(1000*((1/a)+(1/b))*Rs)/(2*pi);

%   Display results
disp(['G = ' num2str(G) ' S/m'])
```

(continues)

```
disp(['C = ' num2str(C) ' F/m'])
disp(['L = ' num2str(L) ' H/m'])
disp(['R = ' num2str(R) ' ohm/m'])
```

Executing the program for the coax parameters of Drill 6.1, we get the following:

```
Calc Coax Distributed Parameters
inner radius, in mm, = .45
outer radius, in mm, = 1.47
relative permittivity, er= 2.26
diel. conductivity, in S/m, = 0
cond. conductivity, in S/m, = 5.8e7
frequency, in Hz, = 1e9
G = 0 S/m
C = 1.0606e-010 F/m
L = 2.3675e-007 H/m
R = 3.8112 ohm/m
>>
```

Telegraphist's Equations

We can develop the equations for wave propagation on transmission line beginning with simple circuit theory applied to the differential segment of Figure 6.3, redrawn in Figure 6.4 with instantaneous voltages and currents at each end of the segment. We notice the voltage on the left side of the segment is $v(z,t)$, indicating it is a function of both position z and time t. On the right side of the circuit model, the voltage is at a position Δz further along the line and is therefore $v(z + \Delta z, t)$. Similar comments can be made about the current entering the model, $i(z,t)$, and that leaving the model, $i(z+\Delta z, t)$.

Applying Kirchhoff's voltage law, and recalling from circuit theory that the voltage v across an inductor is related to the rate of change in current by $v = L\, di/dt$, we obtain

$$v(z,t) - v(z + \Delta z, t) = i(z,t)R'\Delta z + L'\Delta z \frac{\partial i(z,t)}{\partial t} \tag{6.5}$$

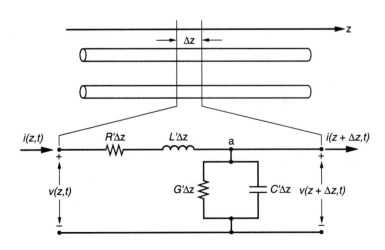

Figure 6.4 The distributed-parameter model including instantaneous voltage and current.

Now, we divide both sides by Δz, and take the limit as Δz approaches zero, to get

$$\lim_{\Delta z \to 0} \frac{v(z,t) - v(z + \Delta z, t)}{\Delta z} = i(z,t)R' + L' \frac{\partial i(z,t)}{\partial t} \tag{6.6}$$

The limit on the left is the definition of a derivative,[6.1] so we obtain

$$-\frac{\partial v(z,t)}{\partial z} = i(z,t)R' + L' \frac{\partial i(z,t)}{\partial t} \tag{6.7}$$

A similar expression can be found by applying Kirchhoff's current law at node a and recalling $i = C \, dv/dt$ for a capacitor:

$$i(z,t) - i(z + \Delta z, t) = v(z + \Delta z, t)G' \Delta z + C' \Delta z \frac{\partial v(z + \Delta z, t)}{\partial t} \tag{6.8}$$

Once again dividing by Δz and taking the limit, we obtain

$$-\frac{\partial i(z,t)}{\partial z} = v(z,t)G' + C' \frac{\partial v(z,t)}{\partial t} \tag{6.9}$$

Equations (6.7) and (6.9) are together the general transmission line equations, also known as the *telegraphist's equations*. They are quite general, being applicable to problems involving transients (such as step changes in voltage) or time-harmonics (sine waves). Note that the transient solution finds application in high-speed digital electronics, where voltages are routinely switched between high and low states. However, for simplified mathematics and to gain an understanding of the functioning of a transmission line, we will first turn our attention to the time-harmonic case. The transient case will be deferred to Section 6.7.

6.2 TIME-HARMONIC WAVES ON TRANSMISSION LINES

Referring once again to Figure 6.4, if the voltage is a sinusoidal function of time, it can be represented at any position and time along the line by

$$v(z, t) = V(z)\cos(\omega t + \phi) \tag{6.10}$$

where $V(z)$ depends only on position along the line. We can write

$$v(z, t) = \text{Re}\left[V(z)e^{j(\omega t + \phi)}\right] \tag{6.11}$$

where application of Euler's identity and taking only the real part recovers (6.10). This can also be expressed as

$$v(z, t) = \text{Re}\left[V_s(z)e^{j\omega t}\right] \tag{6.12}$$

[6.1]For a function $f(x)$, its derivative is defined as $-\dfrac{df(x)}{dx} = \lim_{\Delta x \to 0} \dfrac{f(x + \Delta x) - f(x)}{\Delta x}$.

where the phasor

$$V_s(z) = V(z)e^{j\phi} \tag{6.13}$$

For simplicity, we'll let the phase shift $\phi = 0$ for the rest of our derivation.

Likewise, the current $i(z,t)$ can be written in phasor form as

$$i(z,t) = \mathrm{Re}\left[I_s(z)e^{j\omega t}\right] \tag{6.14}$$

The utility of using phasors is that the time derivatives in (6.7) and (6.9) can be replaced by $j\omega$. For instance,

$$\frac{\partial v(z,t)}{\partial t} = j\omega V_s(z)$$

So, by employing phasors we can rewrite the telegraphist's equations as

$$\boxed{\frac{dV_s(z)}{dz} = -(R' + j\omega L')I_s(z)} \tag{6.15}$$

and

$$\boxed{\frac{dI_s(z)}{dz} = -(G' + j\omega C')V_s(z)} \tag{6.16}$$

Notice that we no longer need partial derivatives since the phasors are only a function of position. The task is now to solve for the two unknowns, $V_s(z)$ and $I_s(z)$, in these two equations.

Taking the position derivative of both sides of (6.15), we have

$$\begin{aligned}
\frac{d}{dz}\frac{dV_s(z)}{dz} &= -\frac{d}{dz}(R' + j\omega L')I_s(z) \\
&= -(R' + j\omega L')\frac{dI_s(z)}{dz}
\end{aligned} \tag{6.17}$$

Here we were able to pull $R' + j\omega L'$ out of the position derivative since the distributed properties are not a function of position. Replacing the derivative of $I_s(z)$ with (6.16), we have

$$\frac{d^2V_s(z)}{dz^2} = (R' + j\omega L')(G' + j\omega C')V_s(z) \tag{6.18}$$

Upon rearranging, we have

$$\frac{d^2V_s(z)}{dz^2} - \gamma^2 V_s(z) = 0 \tag{6.19}$$

where γ is the propagation constant defined by

$$\boxed{\gamma = \sqrt{(R' + j\omega L')(G' + j\omega C')} = \alpha + j\beta} \tag{6.20}$$

This is quite familiar to us from our study of the propagation of uniform plane waves. By knowing the distributed parameters of the T-line, we are able to find the propagation constant that consists of both an attenuation constant α and a phase constant β. Solving the homogeneous second-order differential equation follows the same procedure we used in Section 5.1. We finally arrive at a general solution

$$V_s(z) = V_o^+ e^{-\gamma z} + V_o^- e^{+\gamma z} \tag{6.21}$$

Here, V_o^+ and V_o^- represent the amplitudes at $z = 0$ of the waves traveling in the $+z$ and $-z$ directions, respectively. The instantaneous form, found by reinserting $e^{j\omega t}$ and taking the real part using Euler's identity, is then

$$\boxed{v(z,t) = V_o^+ e^{-\alpha z} \cos(\omega t - \beta z) + V_o^- e^{+\alpha z} \cos(\omega t + \beta z)} \tag{6.22}$$

Had we started by taking the position derivative of both sides of (6.16), the eventual results would have been

$$I_s(z) = I_o^+ e^{-\gamma z} + I_o^- e^{+\gamma z} \tag{6.23}$$

and

$$\boxed{i(z,t) = I_o^+ e^{-\alpha z} \cos(\omega t - \beta z) + I_o^- e^{+\alpha z} \cos(\omega t + \beta z)} \tag{6.24}$$

Equations (6.21) and (6.23) (or (6.22) and (6.24)) are the *traveling wave equations* for the transmission line.

Characteristic Impedance

An extremely useful transmission line parameter is the *characteristic impedance* Z_o, defined as the ratio of the positive traveling voltage wave amplitude to the positive traveling current wave amplitude,

$$Z_o = \frac{V_o^+}{I_o^+} \tag{6.25}$$

We can relate Z_o to the distributed parameters by inserting our wave equations (6.21) and (6.23) into one of the phasor-form telegraphist's equations. Inserting into (6.15) gives us

$$\frac{d}{dz}\left(V_o^+ e^{-\gamma z} + V_o^- e^{+\gamma z}\right) = -\left(R' + j\omega L'\right)\left(I_o^+ e^{-\gamma z} + I_o^- e^{+\gamma z}\right) \tag{6.26}$$

Evaluating the derivative, we have

$$-\gamma V_o^+ e^{-\gamma z} + \gamma V_o^- e^{+\gamma z} = -(R' + j\omega L')(I_o^+ e^{-\gamma z} + I_o^- e^{+\gamma z}) \tag{6.27}$$

The $e^{-\gamma z}$ components on each side of (6.27) must be equal, so equating these components and rearranging we arrive at

$$Z_o = \frac{V_o^+}{I_o^+} = \frac{R' + j\omega L'}{\gamma} \tag{6.28}$$

or

$$Z_o = \sqrt{\frac{R' + j\omega L'}{G' + j\omega C'}} \tag{6.29}$$

Equating the $e^{+\gamma z}$ components of (6.27), we find that the characteristic impedance is also related to the negative traveling wave amplitudes as

$$Z_o = \frac{-V_o^-}{I_o^-} \tag{6.30}$$

> **Drill 6.2** Find γ, α, β, and Z_o for the T-line characterized by the distributed parameters of Drill 6.1 at 1 GHz. (*Answer:* $\gamma = 0.04 + j31$ m^{-1}, $\alpha = 0.04$ Np/m, $\beta = 31$ radians/m, $Z_o = 47 - j.06$ Ω)

Lossless Line

Commercially available transmission lines are made with good conductors, like copper, such that R' tends to be small. They are also made with good dielectrics, like Teflon or poly-ethylene, such that G' is small. If $R' \ll \omega L'$ and $G' \ll \omega C'$, we can often assume $R' = G' = 0$ and consider the transmission line to be *lossless*. Evaluating the propagation constant (Eq. 6.20) for this case, we have

$$\gamma = j\omega\sqrt{L'C'} = \alpha + j\beta \tag{6.31}$$

There is no attenuation ($\alpha = 0$), as we would expect for a lossless line. The phase constant is

$$\beta = \omega\sqrt{L'C'} \tag{6.32}$$

From this equation we can find the propagation velocity

$$u_p = \frac{\omega}{\beta} = \frac{1}{\sqrt{L'C'}} \tag{6.33}$$

The characteristic impedance from (6.29) is

$$Z_o = \sqrt{\frac{L'}{C'}} \tag{6.34}$$

If we consider the formulas for coaxial cable, we can see from (6.2) and (6.3) that

$$L'C' = \mu\varepsilon \tag{6.35}$$

Although we will not prove it here, this relation holds for all transmission lines, not just coaxial cable. So we have

$$u_p = \frac{1}{\sqrt{\mu \varepsilon}} \qquad (6.36)$$

and in most cases a nonmagnetic material is used for the dielectric, so

$$u_p = \frac{c}{\sqrt{\varepsilon_r}} \qquad (6.37)$$

Also, from (6.2) and (6.3) we see that

$$\frac{L'}{C'} = \frac{1}{(2\pi)^2} \frac{\mu}{\varepsilon} \ln\left(\frac{2b}{a}\right) \qquad (6.38)$$

so the characteristic impedance is simplified to

$$Z_o = \frac{60}{\sqrt{\varepsilon_r}} \ln\left(\frac{b}{a}\right) \Omega \qquad (6.39)$$

▶ **EXAMPLE 6.1**

A 1.0-mm-diameter copper wire is surrounded by a 1.0-mm thickness of Teflon, then jacketed by copper. Assuming this coaxial cable is lossless, we want to find the propagation velocity u_p and the characteristic impedance Z_o.

The propagation velocity in this case can be found by straightforward application of (6.37). For Teflon, $\varepsilon_r = 2.1$ from Appendix E, so we have

$$u_p = \frac{c}{\sqrt{\varepsilon_r}} = \frac{3 \times 10^8 \text{ m/s}}{\sqrt{2.1}} = 2.1 \times 10^8 \text{ m/s}$$

To calculate Z_o, we use (6.39) where $a = 0.50$ mm and $b = 1.5$ mm. So we have

$$Z_o = \frac{60}{\sqrt{\varepsilon_r}} \ln\left(\frac{b}{a}\right) = \frac{60}{\sqrt{2.1}} \ln\left(\frac{1.5}{0.5}\right) = 46 \ \Omega$$

Drill 6.3 What outer radius of Teflon dielectric is required in Example 6.1 to give the line a 50-Ω characteristic impedance? (*Answer:* 1.7 mm)

Power Transmission

The lossless assumption is convenient for working many problems involving moderate length cable. However, significant runs of cable can experience large losses in power even if they have very low attenuation.

The instantaneous power $P_i^+(z,t)$ in the $+z$ traveling wave at any point along the T-line is simply

$$P_i^+(z,t) = v(z,t)i(z,t) = \frac{V_o^{+2}}{Z_o}e^{-2\alpha z}\cos^2(\omega t - \beta z) \tag{6.40}$$

Because we are most often interested in the time-averaged power $P_{ave}^+(z)$, we can integrate (6.40) over one cycle:

$$P_{ave}^+(z) = \frac{1}{T}\int_0^T P_i^+(z,t)\,dt = \frac{V_o^{+2}}{Z_o}e^{-2\alpha z}\frac{1}{T}\int_0^T \cos^2(\omega t - \beta z)\,dt \tag{6.41}$$

The integral can be solved by using the half-angle cosine formula,[6.2] resulting in

$$\boxed{P_{ave}^+(z) = \frac{V_o^{+2}}{2Z_o}e^{-2\alpha z}} \tag{6.42}$$

▶ **EXAMPLE 6.2**

Let us consider a coaxial cable of dimensions $a = 1.0$ mm and $b = 3.0$ mm, filled with a nonmagnetic dielectric with $\varepsilon_r = 5.0$ and $\tan\delta = 0.00010$ measured at 2.0 GHz. Copper metal is used, and the outer conductor is assumed thick enough that we can use Eq. (6.4) to find R'. Our task is to find how much power is lost in a meter length of this cable as represented by Figure 6.5.

Referring to Section 5.3 of Chapter 5, we can use the loss tangent to calculate an effective conductivity σ_{eff} of the dielectric,

$$\sigma_{eff} = \omega\varepsilon\tan\delta = 55.6\times 10^{-6}\text{S/m}$$

This conductivity can be used in (6.1) to find G'. Using (6.1)–(6.4), or MATLAB 6.1, we find the distributed parameters and then take the real part of (6.20) to find $\alpha = 0.047$ Np/m for this line.

In measuring power loss, the power at some position z can be ratioed to the power at $z = 0$, that is,

$$\frac{P_{ave}^+(z)}{P_{ave}^+(0)} = e^{-2\alpha z} \tag{6.43}$$

This is convenient since we don't have to know the voltage amplitude or even the impedance. For a 1-m length of our cable, the ratio is 0.91.

It is convenient and customary to measure power ratios on a logarithmic scale, called the *decibel*[6.3] scale. The power ratio can be expressed as a *gain* G(dB), where

[6.2]$\cos^2\theta = (1 + \cos 2\theta)/2$.

[6.3]Originally used to express power ratios in telephone lines, the decibel is named in honor of Alexander Graham Bell.

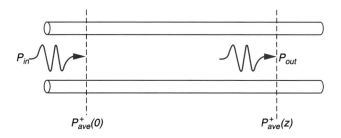

Figure 6.5 Section of T-line for attenuation calculations.

$$G(\text{dB}) = 10 \log\left(\frac{P_{\text{out}}}{P_{\text{in}}}\right)$$

(6.44)

In Example 6.2, for instance, $G(\text{dB}) = -0.4$ dB.[6.4]

The convenience of decibels is twofold. First, using decibels reduces the size of numbers needed to express very large or very small values. For instance, if $P_{\text{out}}/P_{\text{in}} = 10^{10}$, then $G(\text{dB}) = 100$ dB. Second, it is very easy to multiply power ratios by just adding their decibels. Microwave engineers can rapidly determine the overall gain of a multiple block microwave circuit by adding the individual gains. For instance, suppose a filter circuit with a -1.5-dB gain ($P_{\text{out}}/P_{\text{in}} = 0.707$) is in series with an amplifier that has a 9-dB gain ($P_{\text{out}}/P_{\text{in}} = 7.94$). It is easy to see the overall gain is 7.5 dB. Multiplying the ratio products is a bit harder.

Although the decibel scale expresses a power ratio, it is sometimes convenient to express an absolute power. In this case, a power reference is needed. One of the most common ways to represent absolute power levels is the dB_{m} scale, where the reference is chosen as 1 mW. Then

$$G(\text{dB}_{\text{m}}) = 10 \log\left(\frac{P}{1 \text{ mW}}\right)$$

(6.45)

As an example, a 1-MW power level is 90 dB_{m}.

Decibels are also related to nepers. If we consider, for instance, a 10-W input at $z = 0$ of a T-line and a 1-W output at $z = 1$ m, then attenuation(dB) = 10 dB. However, we know from (6.43) that

$$\frac{1}{10} = e^{-2\alpha(1\text{m})}$$

or

$$-2\alpha = \ln\left(\frac{1}{10}\right)$$

and

$$\alpha = -\frac{1}{2}\ln\left(\frac{1}{10}\right) = 1.151 \text{ Np}$$

[6.4]A negative gain corresponds to a positive attenuation, sometimes written "attenuation(dB) = 10 log ($P_{\text{in}}/P_{\text{out}}$)."

Therefore we see that 1.151 Np is equal to 10 dB, or

$$\boxed{1 \text{ Np} = 8.686 \text{ dB}}$$ (6.46)

> **Drill 6.4** The output of a 10-dB amplifier is measured at 10 mW. How much input power was applied? (*Answer:* 1 mW)

> **Drill 6.5** Express the input and output power from Drill 6.4 in dB_m. (*Answer:* 0 dB_m, 10 dB_m)

> **Drill 6.6** A 12-dB amplifier is in series with a 4-dB attenuator. What is the overall gain of the circuit? (*Answer:* 8 dB)

▶ 6.3 TERMINATED T-LINES

Most of the practical problems involving T-lines relate to what happens when the line is terminated. Figure 6.6 shows such a terminated line, where the load is located at $z = 0$. The load itself is considered a *lumped element* in that it is small compared to a wavelength, and the wires connecting the T-line to the load are considered to be negligibly short. The load impedance is simply the ratio of the voltage to the current at the load. Applying the wave equations (6.21) and (6.23) at $z = 0$ we have

$$Z_L = \frac{V_s(z=0)}{I_s(z=0)} = \frac{V_o^+ e^{-\gamma(0)} + V_o^- e^{+\gamma(0)}}{I_o^+ e^{-\gamma(0)} + I_o^- e^{+\gamma(0)}}$$ (6.47)

or

$$Z_L = \frac{V_o^+ + V_o^-}{I_o^+ + I_o^-}$$ (6.48)

By applying our Z_o relations from (6.28) and (6.30), we can manipulate (6.48) to get

$$Z_L = Z_o \frac{V_o^+ + V_o^-}{V_o^+ - V_o^-}$$ (6.49)

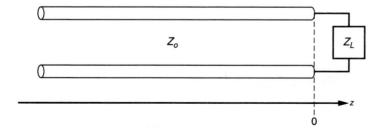

Figure 6.6 A T-line terminated with load impedance Z_L.

which can be rearranged as

$$V_o^- = \frac{Z_L - Z_o}{Z_L + Z_o} V_o^+ \qquad (6.50)$$

It is useful to imagine launching a positive traveling wave along the terminated transmission line as shown in Figure 6.7. We will consider this to be the *incident wave*, and for the moment we will ignore whatever is to the left of $z = -\ell$. Equation (6.50) tells us that if the load is unequal to the characteristic impedance of the line, then a wave must be reflected from the load. The degree of *impedance mismatch* is represented by the *reflection coefficient* at the load, given by

$$\boxed{\Gamma_L = \frac{V_o^-}{V_o^+} = \frac{Z_L - Z_o}{Z_L + Z_o}} \qquad (6.51)$$

It is easy to see that the reflection coefficient for a shorted load ($Z_L = 0$), a matched load ($Z_L = Z_o$), and an open load ($Z_L = \infty$) are -1, 0, and $+1$, respectively. Thus, the reflection coefficient magnitude ranges from 0 to 1.

In general, the reflection coefficient at any point along the T-line is given by the ratio of the reflected wave to the incident wave, that is,

$$\Gamma = \frac{V_o^- e^{+\gamma z}}{V_o^+ e^{-\gamma z}} = \Gamma_L e^{+2\gamma z} \qquad (6.52)$$

The reflection coefficient at $z = -\ell$, for instance, would be

$$\Gamma = \Gamma_L e^{-2\gamma\ell}$$

Superposition of the incident and reflected waves creates a standing wave pattern. The voltage standing wave ratio, *VSWR*, which is the ratio of the maximum to the minimum voltage amplitudes, is related to the reflection coefficient by

$$\boxed{VSWR = \frac{1 + |\Gamma_L|}{1 - |\Gamma_L|}} \qquad (6.53)$$

Since the reflection coefficient can range in magnitude from 0 to 1, the *VSWR* can range from 1 to infinity.

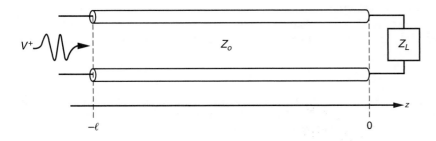

Figure 6.7 Voltage wave V^+ incident at $z = -\ell$ of a terminated T-line.

The reflection coefficient Γ and the *VSWR* for terminated T-lines are similar to the expressions found for uniform plane waves normally incident from one medium, specified as having an intrinsic impedance η_1, to another medium with η_2. Recall that the reflection coefficient for this case was

$$\Gamma = \frac{\eta_2 - \eta_1}{\eta_2 + \eta_1}$$

and the expression for *SWR* was the same as (6.53).

> **Drill 6.7** A 50-Ω line is terminated in a 150-Ω load. Find (a) Γ_L and (b) *VSWR*. (*Answer:* $\frac{1}{2}$, 3)

Input Impedance

At any point along the T-line, we can find the ratio of the total voltage to the total current. This ratio is known as *input impedance*.[6.5] Looking into the line at $z = -\ell$ in Figure 6.8, the input impedance Z_{in} is

$$Z_{in} = \frac{V_s(z = -\ell)}{I_s(z = -\ell)} = \frac{V_o^+ e^{+\gamma\ell} + V_o^- e^{-\gamma\ell}}{V_o^+ e^{+\gamma\ell} - V_o^- e^{-\gamma\ell}} Z_o \qquad (6.54)$$

This can be manipulated (see Problem 6.11) by using (6.50), Euler's identity, and definitions for the hyperbolic functions sinh, cosh, and tanh (see Appendix D) to give

$$\boxed{Z_{in} = Z_o \frac{Z_L + Z_o \tanh(\gamma\ell)}{Z_o + Z_L \tanh(\gamma\ell)}} \qquad (6.55)$$

For the special lossless case, (6.55) becomes

$$\boxed{Z_{in} = Z_o \frac{Z_L + jZ_o \tan(\beta\ell)}{Z_o + jZ_L \tan(\beta\ell)}} \qquad (6.56)$$

The utility of this concept is that the T-line beyond wherever the input impedance is determined can be replaced by a lumped-element impedance Z_{in}, as indicated in the pair of equivalent circuits shown in Figure 6.8.

> **Drill 6.8** Suppose the terminated T-line for Drill 6.7 is lossless. Find Z_{in} for a length of line equal to (a) $\lambda/8$ and (b) $\lambda/4$. (*Note: Make sure your calculator is set to radians rather than degrees for calculations involving (6.55) and (6.56).*)(*Answer:* $30 - j40 \ \Omega$, $16.7 \ \Omega$)

[6.5]The difference between the input impedance and the characteristic impedance should be emphasized. Whereas Z_{in} and Z_L refer to ratios of the total voltages and currents, Z_o is related to the ratio of the voltage to the current for only one of the waves, the incident one or the reflected one.

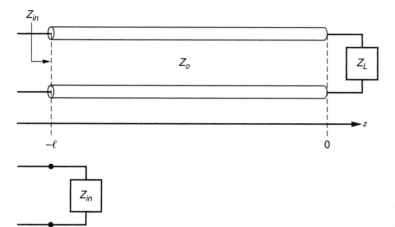

Figure 6.8 The terminated T-line can be replaced by an equivalent lumped-element input impedance.

► **MATLAB 6.2**

Let's create a function that will allow input of the variables in (6.55) and will calculate input impedance. Hyperbolic functions are included in MATLAB, and you can find out how they are handled by seeking help in the command line window. For instance,

```
>> help tanh
```

```
TANH  Hyperbolic tangent.
 TANH(X) is the hyperbolic tangent of the elements of X.
```

For the function, we have the following:

```
function Zin=Zinput(Zo,ZL,G,L)
%enter Zinput(Zo, ZL, G, L) where
%  Zo = complex char imp of line, ohms
%  ZL = complex load imp, ohms
%  G = propagation constant (gamma), 1/m
%  L = T-line length, m
%returns Zin, the input impedance
num=ZL+Zo*tanh(G*L);
den=Zo+ZL*tanh(G*L);
Zin=Zo*num/den;
```

Let's test this function for a 0.60-m-long line of characteristic impedance $75 + j25\ \Omega$ having a propagation constant $0.01 + j0.1\ \mathrm{m^{-1}}$ and terminated in a load $25 - j40\ \Omega$.

```
>> Zo=75+i*25;
>> G=0.01+i*0.1;
>> ZL=25-i*40;
>> L=.6;
>> Zinput(Zo,ZL,G,L)
```

```
ans =
 22.9622 -34.2355i
```

So our result is $Z_{in} = 23 - j34\ \Omega$.

Complex Loads

Input impedances or loads exhibiting complex impedance may be modeled using simple resistor, inductor, and capacitor lumped elements. Figure 6.9 indicates the elements along with their s-domain (phasor) values.

For instance, consider a load $Z_L = 100 + j200\ \Omega$. We can model this as a 100-Ω resistor in series with an inductor. The value of inductance will depend on the frequency, so if we specify 1 GHz then

$$j\omega L = j200\ \Omega$$

or

$$L = \frac{200\ \Omega}{2\pi\left(1\times 10^9\ \text{Hz}\right)}\left(\frac{\text{Hz-s}}{1}\right)\left(\frac{\text{H-A}}{\text{V-s}}\right)\left(\frac{\text{V}}{\Omega\text{-A}}\right) = 32\ \text{nH}$$

One case of interest is a lossless line terminated in a purely reactive load. If we consider $Z_o = R_o$, representing the all-real characteristic impedance for a lossless T-line, and $Z_L = jX_L$, representing a purely reactive load, then the reflection coefficient is

$$\Gamma_L = \frac{jX_L - R_o}{jX_L + R_o}$$

This clearly has a unity magnitude. This is as expected, since no energy can be dissipated in a purely reactive load. The wave is completely reflected. There is, however, a phase shift associated with the reactive load.

> **Drill 6.9**   A $\lambda/6$ line with $Z_o = 50\ \Omega$ is terminated in a 25-Ω resistance. Find the input impedance along with its equivalent lumped-element circuit at 1.0 GHz. (*Answer:* $Z_{in} = 57 + j37\ \Omega$; this resembles a 57-$\Omega$ resistor in series with a 5.9-nH inductor)

The Complete Circuit

Now let's add a source, the phasor voltage V_{ss}, and source impedance Z_s as indicated in Figure 6.10. The voltage at any point on the T-line requires that we know V_o^+. We can solve for this in terms of V_{ss} by considering that voltage division will give us the voltage across the input impedance,

$$V_{in} = V_{ss}\frac{Z_{in}}{Z_S + Z_{in}} = V_s(z = -\ell) \tag{6.57}$$

Then, at any point on the line the voltage is given by

$$V_s(z) = V_o^+ e^{-\gamma z} + V_o^- e^{+\gamma z} = V_o^+ (e^{-\gamma z} + \Gamma_L e^{+\gamma z}) \tag{6.58}$$

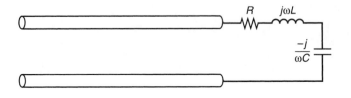

Figure 6.9   The s-domain impedance values of R, L, and C.

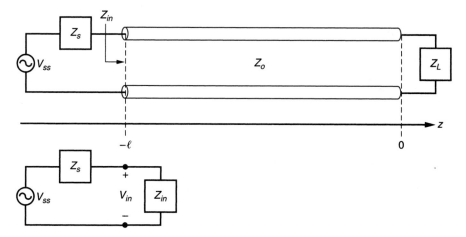

Figure 6.10   The circuit after adding a source, and the equivalent circuit.

Evaluating (6.58) at $z = -\ell$, and inserting this into (6.57), we can solve for V_o^+ as

$$V_o^+ = \frac{V_{in}}{e^{+\gamma\ell} + \Gamma_L e^{-\gamma\ell}} \tag{6.59}$$

Then, the voltage across the load can be calculated by evaluating (6.58) at $z = 0$, that is,

$$V_L = V_s(z = 0) = V_o^+ (1 + \Gamma_L) \tag{6.60}$$

▷ **EXAMPLE 6.3**

Consider the lossless T-line circuit of Figure 6.11. We want to find the voltage across the 100-Ω load. To begin, the source voltage is converted to its phasor,

$$V_{SS} = 10e^{j30°} V$$

To find V_L using (6.60), we need Γ_L and V_o^+. We have

$$\Gamma_L = \frac{Z_L - Z_o}{Z_L + Z_o} = \frac{100\ \Omega - 50\ \Omega}{100\ \Omega + 50\ \Omega} = \frac{1}{3}$$

Finding V_o^+ requires that we know Z_{in}, which we can find from (6.56) for a lossless T-line,

$$Z_{in} = Z_o \frac{Z_L + jZ_o \tan(\beta\ell)}{Z_o + jZ_L \tan(\beta\ell)}$$

Since in this example

$$\beta\ell = \frac{2\pi}{\lambda} \frac{\lambda}{4} = \frac{\pi}{2}$$

we have

$$\tan \beta\ell = \tan \frac{\pi}{2} = \infty$$

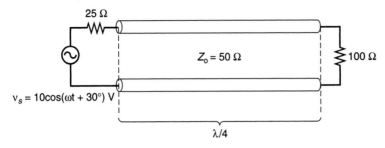

Figure 6.11   Circuit for Example 6.3.

The equation for Z_{in} reduces to

$$Z_{in} = \frac{Z_o^2}{Z_L} = 25 \ \Omega$$

Now we have for the input voltage

$$V_{in} = V_{ss} \frac{Z_{in}}{Z_s + Z_{in}} = 10e^{j30°} \frac{25}{25 + 25} = 5e^{j30°} \ V$$

To evaluate (6.59) for V_o^+ we also need $e^{\gamma \ell}$ and $e^{-\gamma \ell}$. Euler's equation can be used to convert $e^{\gamma \ell}$ and $e^{-\gamma \ell}$ into complex numbers in rectangular coordinates if so desired. For the lossless case, $\gamma = j\beta$ and $\gamma \ell = j\beta \ell = j\pi/2$. So

$$V_o^+ = \frac{V_{in}}{e^{+\gamma \ell} + \Gamma_L e^{-\gamma \ell}} = \frac{5e^{j30°}}{e^{j90°} + e^{-j90°}/3} = 7.5e^{-j60°} \ V$$

Finally, we can apply (6.60) to get

$$V_L = 7.5e^{-j60°} \left(1 + \tfrac{1}{3}\right) = 10e^{-j60°} \ V$$

Converting this phasor to its instantaneous form gives us the voltage across the load,

$$v_L = 10 \cos(\omega t - 60°) V$$

Drill 6.10   Rework Example 6.3 after interchanging the source and load resistances, and find the phasor voltage across the load V_L. (*Answer*: $2.5e^{-j60°}$ V)

▷ 6.4   THE SMITH CHART

The Smith Chart, shown in Figure 6.12, is a graphical tool for use with transmission line circuits and microwave circuit elements. It was created in the 1930s by Philip H. Smith, an engineer at Bell Telephone Labs, and was heavily used by engineers developing microwave systems in World War II. It was initially a very useful way of circumventing the complex-number arithmetic required in transmission line problems, only requiring a straight edge and a compass to operate. It has grown to be a very handy tool for understanding the behavior of microwave circuit elements and continues to be used even though computers are quite capable of handling the complex math. In fact, it is common for Microwave CAD packages to display their solutions on a Smith Chart.

For hand calculations, the Smith Chart is most useful assuming the T-line is lossless. Although lossy lines can be modeled with the Smith Chart, it is not often practical to do so.

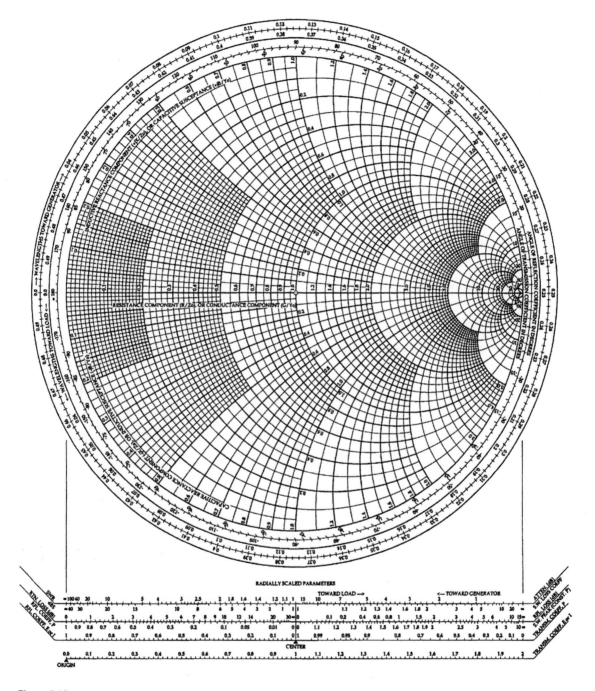

Figure 6.12 The *Transmission Line Calculator*, commonly referred to as the *Smith[a] Chart*.

[a]SMITH is a registered trademark of Analog Instrument Co., Box 950, New Providence , NJ 07974.

A printable version of the Smith chart along with other Smith chart resources is available at http://www.sss-mag.com/smith.html.

Therefore, only lossless T-Lines will be considered in the following discussion. The Smith Chart can be thought of as two graphs in one. First, it plots the normalized impedance at any point along a T-line. Second, it plots the reflection coefficient at any point along the line. Let's first look at how the chart is derived.

Smith Chart Derivation

As depicted in Figure 6.13a, the complex reflection coefficient at a load is related to the load and line impedance as

$$\Gamma_L = \frac{Z_L - Z_o}{Z_L + Z_o} \tag{6.61}$$

We can normalize the load impedance to the characteristic impedance, writing

$$z_L = \frac{Z_L}{Z_o}$$

and then

$$\Gamma_L = \frac{z_L - 1}{z_L + 1} \tag{6.62}$$

Now, as Figure 6.13b suggests, we can replace the load along with any arbitrary length of T-line by an input impedance, and the reflection coefficient at this new load can be written

$$\Gamma = \Gamma_L e^{j2\beta z} = \frac{z - 1}{z + 1} \tag{6.63}$$

Here the reflection coefficient has the same magnitude as it does at the load but is phase shifted by $2\beta z$, where z is the distance along the line from the load to the input impedance point.

The reflection coefficient and normalized load terms are complex and can be expanded into real and imaginary parts as follows:

$$\Gamma = \Gamma_{Re} + j\Gamma_{Im}, z = r + jx \tag{6.64}$$

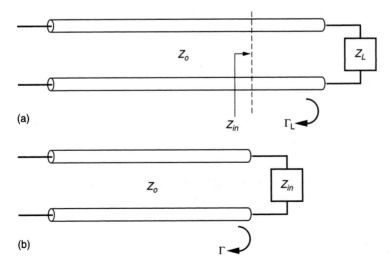

Figure 6.13 (a) T-line terminated in a load Z_L is characterized as having a reflection coefficient Γ_L. (b) A section of line containing the load is replaced by input impedance Z_{in}.

Rearranging (6.64) to solve for z in terms of Γ, we get

$$z = \frac{1+\Gamma}{1-\Gamma} \qquad (6.65)$$

or

$$r + jx = \frac{1+\Gamma_{Re}+j\Gamma_{Im}}{1-\Gamma_{Re}-j\Gamma_{Im}} \qquad (6.66)$$

Manipulating the right side of (6.66) into its real and imaginary parts, we find

$$r = \frac{1-\Gamma_{Re}{}^2-\Gamma_{Im}{}^2}{\left(1-\Gamma_{Re}\right)^2+\Gamma_{Im}{}^2} \qquad (6.67)$$

and

$$jx = \frac{j\Gamma_{Im}}{\left(1-\Gamma_{Re}\right)^2+\Gamma_{Im}{}^2} \qquad (6.68)$$

Now, the general equation for a circle of radius a, centered at $x = m$ and $y = n$, is

$$(x-m)^2 + (y-n)^2 = a^2 \qquad (6.69)$$

Equations (6.67) and (6.68) can be rearranged into forms that give circular functions of r and x:

$$\left(\Gamma_{Re}-\frac{r}{r+1}\right)^2+\Gamma_{Im}{}^2 = \left(\frac{1}{r+1}\right)^2 \qquad (6.70)$$

and

$$\left(\Gamma_{Re}-1\right)^2+\left(\Gamma_{Im}-\frac{1}{x}\right)^2 = \left(\frac{1}{x}\right)^2 \qquad (6.71)$$

These circles will be plotted on the Γ_{Im} versus Γ_{Re} axes shown in Figure 6.14a.

Equation (6.70) can be used to plot normalized *resistance circles*. Consider a normalized resistance $r = 1$. Then we have

$$\left(\Gamma_{Re}-\frac{1}{2}\right)^2+\Gamma_{Im}{}^2 = \frac{1}{4}$$

This is the equation for a circle centered at $\Gamma_{Re} = \frac{1}{2}$ and $\Gamma_{Im} = 0$ with a radius $\frac{1}{2}$ as shown in Figure 6.14b. Also shown in this figure is the $r = 0$ circle, with a radius 1 centered at $\Gamma_{Re} = \Gamma_{Im} = 0$. The area within this $r = 0$ circle represents all the possible points for Γ, which must have a magnitude less than or equal to one. Similar circles can be drawn within this allowed space for any other value of r in the range $0 \le r \le \infty$.

Using (6.71), we can plot normalized *reactance circles*. Consider a normalized reactance $x = 1$. Then we have

$$\left(\Gamma_{Re}-1\right)^2+\left(\Gamma_{Im}-1\right)^2 = 1$$

This is a circle of radius 1 centered at $\Gamma_{Re} = \Gamma_{Im} = 1$. But notice that we only draw the circle for that part of the chart where $|\Gamma| \le 1$. If the reactance is $x = -1$, we have

$$\left(\Gamma_{Re}-1\right)^2+\left(\Gamma_{Im}+1\right)^2 = 1$$

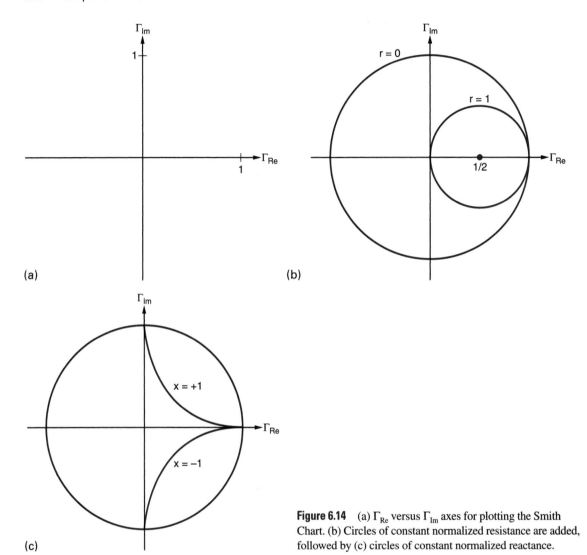

Figure 6.14 (a) Γ_{Re} versus Γ_{Im} axes for plotting the Smith Chart. (b) Circles of constant normalized resistance are added, followed by (c) circles of constant normalized reactance.

or a radius 1 circle centered at $\Gamma_{Re} = 1$ and $\Gamma_{Im} = -1$. The allowed portions of these two circles are shown in Figure 6.14c. More circles can be drawn for reactance, where it is seen that the upper half of the Smith Chart represents positive reactance (appearing inductive) and the bottom half represents negative reactance (appearing capacitive).

> **MATLAB 6.3**
>
> The following routine draws a Smith Chart. It calls a pair of functions, "realcirc" and "imcirc" to draw the lines. You can easily customize your own chart by modifying the "now add real circles" and "now add +/– x circles" portions of the program.

This routine, inspired by Soeren Laursen's "A Smith Chart Toolbox" in 1995, contains a number of useful Smith Chart utilities written in MATLAB. Most recently this toolbox was found at http://filebox.vt.edu/users/brindlec/smith.html.

```
%   M-File: ML0603
%
%   This program plots a simple Smith Chart. It calls
%   on the functions realcirc, imcirc, and z2gamma.
%
%   Wentworth, 8/3/02
%
%   Variables:
%   a        circle radius
%   m,n      x,y center of circle
%   theta    angle (degrees)
%   z        a complex location
%   rvalues  values to plot for the real circles
%   xvalues  values to plot for the imag circles
%   xpos     x location of text
%   ypos     y location of text

clc         %clears the command window
clear       %clears variables

%first plot real = 0 circle
theta=linspace(-pi,pi,180);
a=1;
m=0;n=0;
Re=a*cos(theta)+m;
Im=a*sin(theta)+n;
z=Re+i*Im;
plot(z,'k')
axis('equal')
axis('off')
hold on

%add the x = 0 line
plot([-1 1],[0,0],'k');

%now add real circles
rvalues=[0.5 1 2 4];
for r=rvalues
    realcirc(r);
    xpos=z2gamma(r);
    h=text(xpos,0,num2str(r));
set(h,'VerticalAlignment','top','HorizontalAlignment'
'right');
end

%now add +/-x circles
xvalues=[0.2 0.5 1 2];
for x=xvalues
    imcirc(x);
    imcirc(-x);
    xpos=real(z2gamma(i*x));
```

(continues)

```
      ypos=imag(z2gamma(i*x));
      h=text([xpos xpos], [ypos -ypos], [' j' num2str(x);'-j' num2str(x)]);
      set(h(1),'VerticalAlignment','bottom');
      set(h(2),'VerticalAlignment','top');
      if xpos==0
      set(h,'HorizontalAlignment','center');
      elseif xpos<0
      set(h,'HorizontalAlignment','right');
      end
end

h=text(-1,0,'0');
set(h,'VerticalAlignment','middle','HorizontalAlignment','right');
end
```

The next three functions (z2gamma(r), realcirc(r), and imcirc(r)) were entered and saved separately.

```
function [result] = z2gamma(z)
%Z2GAMMA Convert impedance to reflection coefficient
%GAMMA = Z2GAMMA(Z)
%converts the impedances in matrix Z to reflection
%coefficients in GAMMA.

result = (z-1)./(z+1);

      _____

function [h]=realcirc(r)

%REALCIRC(r) draws circle of constant real with
%normalized r;
phi=1:1:360;
theta=phi*pi/180;
a=1/(1+r);
m=r/(r+1);n=0;
Re=a*cos(theta)+m;
Im=a*sin(theta)+n;
z=Re+i*Im;
h=plot(z,'k');
axis('equal')
axis('off')

      _____
function [h]=imcirc(x)
%IMCIRC(x) draws a circle of constant imag with
%normalized x;
a=abs(1/x);
m=1;
n=1/x;
k=1;
for t=1:1:360
   angle(t)=t*pi/180;
   Re(t)=a*cos(angle(t))+m;
   Im(t)=a*sin(angle(t))+n;
   z(t)=Re(t)+i*Im(t);
   if abs(z(t)) <= 1
```

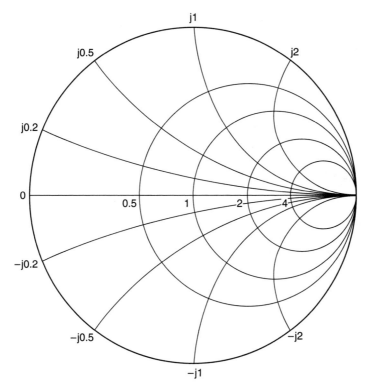

Figure 6.15 Smith Chart generated using ML0603.

```
        zz(k)=z(t);
        k=k+1;
    end
end
h=plot(zz,'k');
axis('equal')
axis('off')
```

A Smith Chart generated using this routine is displayed in Figure 6.15.

Using the Smith Chart

The Smith Chart is a plot of normalized impedance. For instance, suppose you have a $Z_o = 50\ \Omega$ line terminated in a load $Z_L = 50 + j100\ \Omega$ as shown in Figure 6.16a. To locate this point on the Smith Chart, you would first normalize the load impedance to obtain $z_L = 1 + j2$ (Figure 6.16 b). Then, as shown in Figure 6.16c, the normalized impedance is located at the intersection of the $r = 1$ circle and the $x = +2$ circle.

To see how the Smith Chart is also a plot of the reflection coefficient, we recall that the reflection coefficient at any point z along a lossless T-line is

$$\Gamma = \Gamma_{Re} + j\Gamma_{Im} = \Gamma_L e^{j2\beta z} = |\Gamma_L| e^{j\theta_\Gamma} \tag{6.72}$$

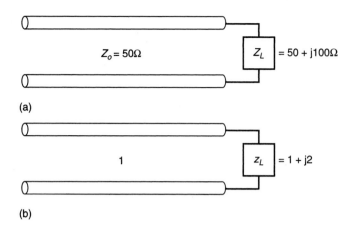

(a)

(b)

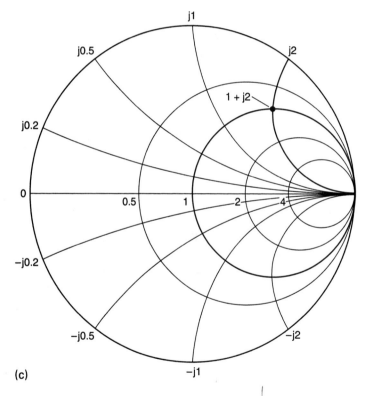

(c)

Figure 6.16 A T-line terminated in a load (a) shown with values normalized to Z_o in (b). (c) The location of the normalized load impedance is found on the Smith Chart.

In this equation, we note that the reflection coefficient has a magnitude $|\Gamma_L|$ and an angle θ_Γ equal to its angle at the load plus $2\beta z$. The actual value of $|\Gamma_L|$ is found by taking the distance from the center of the chart to the point divided by the distance from the center of the chart to the periphery ($|\Gamma_L| = 1$). To avoid this calculation, a scale for magnitude of reflection coefficient is provided below the Smith Chart, as seen in Figure 6.12. The angle of the reflection coefficient θ_Γ is indicated on the *angle of reflection coefficient* scale, shown just outside the $|\Gamma_L| = 1$ circle on the chart.

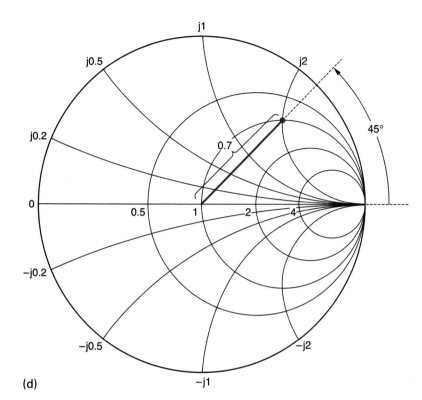

(d)

Figure 6.16 (d) Reflection coefficient and angle.

Continuing our example, we read from the Smith Chart (see Figure 6.16d) that the normalized load $z_L = 1 + j2$ corresponds to $|\Gamma_L| = 0.7$ and $\theta_\Gamma = 45°$. This can be verified by calculation:

$$\Gamma_L = \frac{Z_L - Z_o}{Z_L + Z_o} = \frac{50 + j100 - 50}{50 + j100 + 50} = 0.5 + j0.5 = 0.707e^{j45°}$$

The values of normalized impedance and reflection coefficient are functions of position along the T-line (see Figure 6.17). After locating the normalized impedance point, it is useful to draw the *constant-$|\Gamma_L|$ circle*. Since

$$\Gamma = |\Gamma_L|e^{j\theta_\Gamma}$$

we let θ_Γ change as we hold $|\Gamma_L|$ constant, and this traces out a circle of constant $|\Gamma_L|$ on the chart. Now, moving along this constant-$|\Gamma_L|$ circle is akin to moving along the T-line. Recalling that $\theta_\Gamma = 2\beta z + \phi$, we see that the input impedance at some point $z = -\ell$ along the T-line corresponds to $\theta_\Gamma = -2\beta\ell + \phi$. Thus, moving away from the load corresponds to moving in the clockwise direction on the Smith Chart.

Since the function $e^{j\theta_\Gamma}$ is sinusoidal, it repeats for

$$2\beta z = N2\pi$$

(where $N = 1, 2, 3...$). So, since $\beta = 2\pi/\lambda$, we see that at values of z where

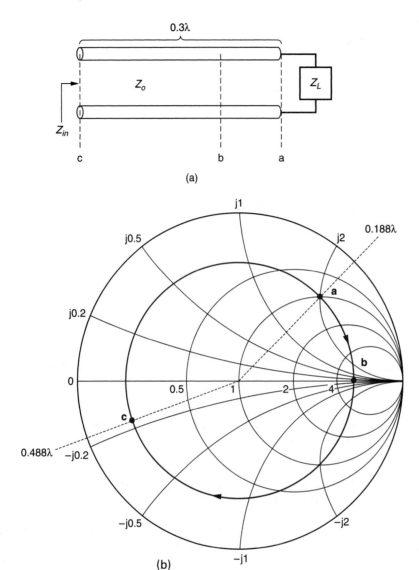

(a)

(b)

Figure 6.17 Movement along a T-line corresponds to movement along a constant-$|\Gamma_L|$ circle on the Smith Chart.

$$z = N\frac{\lambda}{2}$$

Γ must repeat, and this distance corresponds to travel of an integral multiple of half wavelengths along the line.

Continuing our example, we suppose the line is 0.3λ in length. Drawing a line from the center of the chart to the outside *Wavelengths Toward Generator* (WTG) scale,[6.6] as shown in Figure 6.17b, we observe a starting point at 0.188λ. Adding 0.3λ moves us along the

---

[6.6]The WTG scale is used when moving clockwise, or away from the load. The *Wavelengths Toward Load* (WTL) scale is used when moving counterclockwise on the chart.

constant-$\left|\Gamma_L\right|$ circle to 0.488λ on the WTG scale, corresponding to a normalized input impedance of $z_{in} = 0.175 - j0.08$. Denormalizing, we find an input impedance of

$$Z_{in} = z_{in}Z_o = 8.75 - j4 \; \Omega$$

The voltage standing wave ratio can be determined by reading the value of r at the $\theta_\Gamma = 0°$ crossing for the constant-$\left|\Gamma_L\right|$ circle (point b in Figure 6.17). This can be seen by first recalling that

$$z = r + jx = \frac{1+\Gamma}{1-\Gamma}$$

Notice that, at $x = 0$ and $r > 1$, $\left|\Gamma_L\right| = \Gamma$, so

$$z = r = \frac{1+\left|\Gamma\right|}{1-\left|\Gamma\right|} = VSWR$$

In the continuing example, we find $VSWR = 5.9$ by reading from the Smith Chart. More exact calculation gives 5.83.

The point where $VSWR$ is taken is also the location of the maximum value of r along the constant-$\left|\Gamma_L\right|$ circle. The minimum value of r occurs a distance $\lambda/4$ away (at $\theta_\Gamma = 180°$). These two points for maximum and minimum values of r correspond to the locations of relative voltage maximum and minimum on the T-line.

Drill 6.11   Locate the following load impedances terminating a 50-Ω T-line: (a) $Z_L = 0$ (a short circuit), (b) $Z_L = \infty$ (an open circuit), (c) $Z_L = 100 + j100 \; \Omega$, (d) $Z_L = 100 - j100$ Ω, and (e) $Z_L = 50 \; \Omega$. (*Answer:* See Figure 6.18)

Drill 6.12   A 0.334λ-long $Z_o = 50 \; \Omega$ T-line is terminated in a load $Z_L = 100 - j100 \; \Omega$. Use the Smith Chart to find (a) Γ_L, (b) $VSWR$, (c) Z_{in}, and (d) the distance from the load to the first voltage minimum. (*Answer:* (a) $0.61e^{-j30°}$, (b) 4.3, (c) $Z_{in} = 22.5 + j45 \; \Omega$, and (d) 0.208λ)

Impedance Measurement

Measuring a device's input impedance requires connecting it to a measuring instrument. However, the presence of the connection cable between the device and the measurement apparatus can significantly influence the measured impedance, especially at high frequency. Hence, several schemes for accurately measuring impedance have been devised.

One technique for measuring input impedance employs a slotted coaxial air line, as shown in Figure 6.19a. A probe can be slid along the slotted line to measure electric field amplitude. A scale is affixed to the line for accurate location of the voltage maxima and minima. The ratio of the voltage maxima to the minima is the $VSWR$. To determine the input impedance, the field magnitude is plotted with the load in place and compared to the field plotted with the load replaced with a short circuit. An example details this procedure.

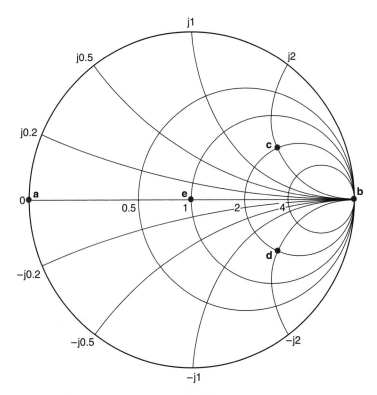

Figure 6.18 Solution for Drill 6.11.

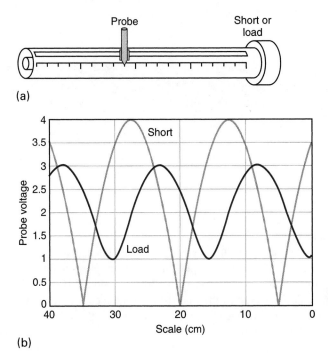

(a)

(b)

Figure 6.19 (a) Slotted coaxial air line. (b) Field magnitude versus position along the slotted line.

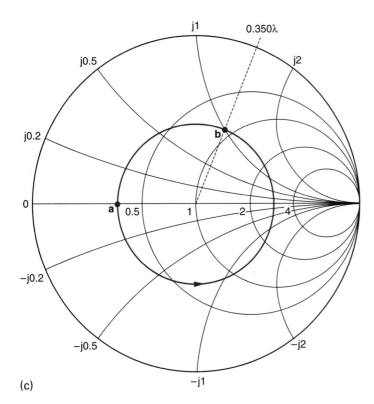

(c)

Figure 6.19 (c) Smith Chart solution to Example 6.4.

▶ **EXAMPLE 6.4**

An unknown load is attached to a 50-Ω impedance slotted coaxial air line. The load is determined with the following procedure:

1. Terminate the line in a short circuit. Determine the guide wavelength and the location of the voltage minima. Choose one of these minima as the reference location of the load.

   The light blue line in Figure 6.19b shows the field amplitude plotted against the scale position for a short-circuit termination. The "0 cm" end of the scale is close to the load, but not necessarily right at the load. Voltage minima are located at 5, 20, and 35 cm, and we'll arbitrarily choose 5 cm to be the reference location for our load. From our understanding of the Smith Chart we know the difference in two minima (15 cm) corresponds to $\lambda/2$. So $\lambda = 30$ cm and because we are dealing with an air line ($u_p = c$) we have $f = c/\lambda = 1$ GHz.

2. Terminate the line in the load. Determine the $VSWR$ and draw a constant-$|\Gamma|$ circle. Also determine the location of voltage minima.

   The heavy blue line in Figure 6.19b represents the field plot with the load in place that has a voltage maxima of 3 and minima of 1, so we have a $VSWR = 3$. This is used to draw the constant-$|\Gamma|$ circle on the Smith Chart of Figure 6.19c. Finally, from Figure 6.19b, we see that the voltage minima with the load in place are located at 0.5, 15.5, and 30.5 cm.

3. Move from one of the load minima to the reference location of the load.

   On the Smith Chart, we can begin at the 15.5-cm voltage minimum (point a). If we move to our 5-cm reference location, then we must move 10.5 cm or

$$\frac{10.5 \text{ cm}}{30 \text{ cm} / \lambda} = 0.350\lambda$$

toward the load. This is point b on our constant-$|\Gamma|$ circle, corresponding to $z = 0.8 + j1.0$, or

$$Z_L = 40 + j50 \ \Omega$$

Had we instead begun at the 0.5-cm minimum, then we would have moved 4.5 cm (0.150λ) toward the generator and reached the same point b.

Drill 6.13 Suppose in Example 6.4 that the 50-Ω coaxial air line extends all the way from the 0 cm scale location to the location of the termination. What is the shortest length this extension can be? (*Answer:* 10 cm)

Rather than using a slotted line, it is much more common to measure the input impedance of a device or network using a *network analyzer* (see Chapter 10). The effect of the connection cable can be *calibrated out* by using a set of standard terminations. To get an idea of how this is accomplished, the following example uses a calibration short circuit to determine the impedance of a load connected to the end of a section of transmission line.

▷ **EXAMPLE 6.5**

The calibration short circuit is used to determine the electrical length of the test connection to the location of the device to be tested. Suppose when we attach a short circuit to a 50-Ω T-line, as indicated in Figure 6.20a, that we measure an input impedance of $Z_{insc} = +j128 \ \Omega$. Normalizing this to 50 Ω, we find $z = j2.56$, or point a on the chart of Figure 6.20c. We know that the actual normalized impedance of a short is $z = 0 + j0$ (point b on the chart). Moving from the measured short at point a toward the actual load at point b, we see that the electrical length of the line is 0.191λ.

Now suppose we attach the unknown load, as indicated in Figure 6.20b, and measure an impedance of $Z_{inL} = 30 - j40 \ \Omega$. We normalize and locate this as point c ($z = 0.6 - j0.8$). This point is located at 0.125λ on the WTL scale. We then move toward the load a distance of 0.191λ, or to 0.316λ on the WTL scale (point d). This corresponds to $z_L = 1.3 + j1.4$, or $Z_L = 65 + j65 \ \Omega$.

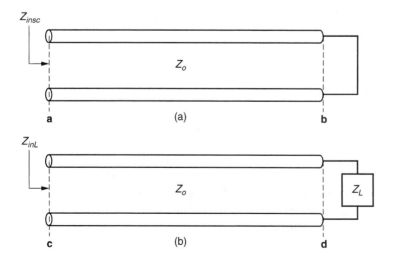

Figure 6.20 The short-circuited T-line of (a) determines the electrical length of the line that is then used in (b) to find the unknown load with the aid of a Smith Chart (c).

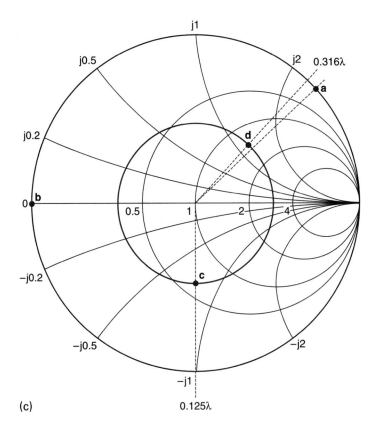

(c)

Figure 6.20   The short-circuited T-line of (a) determines the electrical length of the line that is then used in (b) to find the unknown load with the aid of a Smith Chart (c).

The Smith Chart is used in a variety of microwave engineering applications, in particular impedance matching and amplifier design. Impedance matching is the topic of the next section, and it will be revisited in Chapter 10 prior to the topic of amplifier design.

▷ 6.5   IMPEDANCE MATCHING

We often desire that all of the power propagating along a T-line be dropped across the terminating load impedance. But as we've seen, an impedance mismatch can result in the reflection of much of this power. An impedance *matching network*, as indicated in Figure 6.21, provides a solution. The impedance looking into the network is matched to the line impedance. If the network itself consists only of reactive elements, then it will dissipate no power and consequently all the power will be dropped across the load.

A number of techniques are available for constructing matching networks. Practicality is one concern, since the network should be of fairly simple design and be easy to implement. The most practical designs include quarter-wave transformers, single-stub tuners, lumped-element tuners, and multisection transformers. With the exception of multisection transformers, all of these matching networks tend to operate over a fairly narrow bandwidth. One desirable trait for these networks is to have tuning capability in the event that the load impedance changes. Using variable reactive elements in a lumped-element tuner or using an adjustable-length T-line stub in a stub tuner can provide some tuning capability.

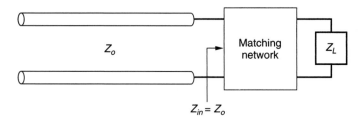

Figure 6.21 Adding an impedance-matching network ensures that all power will make it to the load.

In this section we will first investigate the simple quarter-wave transformer. Then we'll see how the Smith Chart may be used to construct a variety of stub matching networks. Lumped-element networks will be deferred until Chapter 10. Multisection tuning networks are beyond the scope of this text but can be studied in several of the references cited at the end of Chapter 10.

Quarter-Wave Transformer

If the load impedance is all real (no reactive component), then a quarter-wave matching network can be constructed as shown in Figure 6.22. We can apply (6.56) to find the impedance looking into the quarter-wave-long section of lossless Z_s impedance line terminated in a resistive load R_L. We have

$$Z_{in} = Z_S \frac{R_L + jZ_S \tan \beta \ell}{Z_S + jR_L \tan \beta \ell}$$

For a quarter-wave line length, $\beta \ell = \pi/2$ and $\tan(\beta \ell) = \infty$. This leads to

$$Z_{in} = \frac{Z_S^2}{R_L} = Z_o$$

for an impedance-matched line, or

$$Z_S = \sqrt{Z_o R_L} \tag{6.73}$$

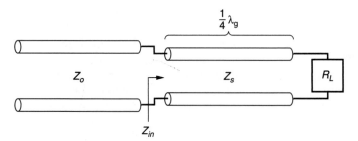

Figure 6.22 Quarter-wave transformer.

> **Drill 6.14** Suppose a 50-Ω T-line is terminated in a 100-Ω load. Determine the required impedance of a quarter-wave matching section of T-line. (*Answer:* 70.7 Ω)

Matching with the Smith Chart

Consider a normalized load impedance z_L located at an arbitrarily selected point on the Smith Chart of Figure 6.23. Since, as shown in Figure 6.21, we want $Z_{in} = Z_o$, the normalized input impedance will be located at the center of the Smith Chart, where $|\Gamma| = 0$. Our matching network must therefore move us from z_L to the center of the chart.

Moving away from the load along the Z_o line generates the constant-$|\Gamma|$ circle shown in the figure. Of special interest are the two points where this circle crosses the $1 \pm jx$ circle. At either of these points, inserting the appropriate reactive element (that is, adding $\mp jx$) will then move us to the matched condition at the center of the chart.

▷ **EXAMPLE 6.6**

Let us construct a simple matching network by adding a reactive element at a suitable location along a 50-Ω T-line terminated in an $11 + j25$ Ω load. This starting condition is shown in Figure 6.24a.

We first locate the normalized load impedance z_L on the Smith Chart, and then we draw the constant-$|\Gamma|$ circle. This is shown in Figure 6.23, where $z_L = 0.22 + j0.5$. Then, we move from z_L clockwise (toward the generator) a distance d along the constant-$|\Gamma|$ circle until the point $1 + j2.0$ is

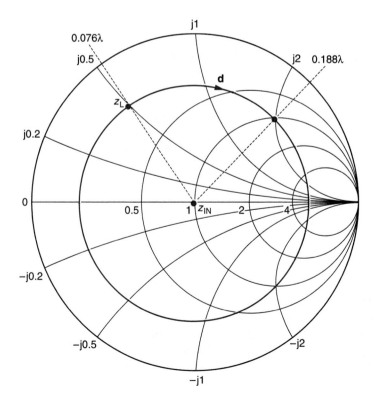

Figure 6.23   The objective of a matching network is to move to the center of the Smith Chart. The values shown are for Example 6.6.

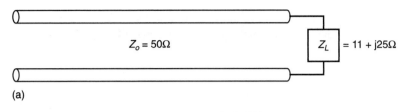

(a)

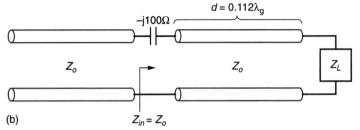

(b)

Figure 6.24 (a) A 50-Ω T-line terminated in an $11+ j25$ Ω load for Example 6.6. (b) T-line with tuning capacitor added at the appropriate distance from the load.

reached. Using the WTG scale, we are moving from $0.076\lambda_g$ to $0.188\lambda_g$, a distance of $0.112\lambda_g$. We cut the T-line at this point and insert a series capacitive element of normalized reactance $-j2.0$, corresponding to $-j100$ Ω. The normalized input impedance is then

$$z_{in} = 1 + j2.0 - j2.0 = 1 + j0$$

corresponding to the center of the Smith Chart. The result is shown in Figure 6.24b. The value of capacitance required depends on frequency, that is,

$$-j100 \ \Omega = \frac{-j}{\omega C}$$

Note that we could have proceeded along Z_o until the second intersection was reached (at $1 - j2.0$), in which case a series inductor could have been added ($+j100$ Ω) to provide the impedance match.

Drill 6.15 Suppose, for Example 6.6, that the 50-Ω line is a coaxial cable made with a Teflon dielectric, and it must operate at 800 MHz. Determine (a) the length of the coaxial line between the load and the capacitor and (b) the value of the series capacitor added to provide an impedance match. (*Answer: d* = 2.9 cm, (b) *C* = 2.0 pF)

Before going on to shunt-stub matching, we first need to discuss the admittance of open-ended and shorted T-line stubs.

Admittance of Shunt Stubs

Sometimes, as in the case of T-line stubs, we find it much more convenient to add shunt elements rather than series elements. With shunt elements, it is much easier to work in terms of admittances. Admittance is the inverse of impedance, allowing direct addition of parallel elements. Figure 6.25a shows the relationship between admittances and impedances. Here we see that the characteristic admittance Y_o is simply $1/Z_o$, and the load admittance Y_L is $1/Z_L$. The convenience of admittances is that shunt values may be added, for instance as shown in Figure 6.25b.

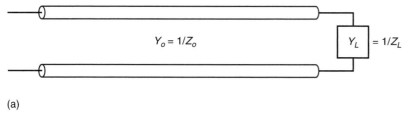

(a)

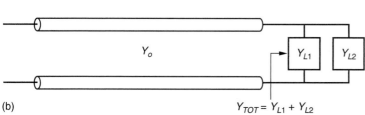

(b)

Figure 6.25 (a) Admittance relationship to impedance. (b) Adding shunt elements using admittances.

It may be noted that the Smith Chart is also a chart of normalized admittance. The normalized load admittance is

$$y_L = \frac{Y_L}{Y_o} = \frac{1}{z_L} \tag{6.74}$$

It is customary to represent the normalized admittance as $y = g + jb$.

Consider the normalized impedance $z_L = 2 + j1$ as shown in Figure 6.26. We can calculate the normalized admittance as $y_L = 1/z_L = 0.40 - j0.20$. With the Smith Chart it is easy to find the normalized admittance. We simply move to a point on the opposite side of the constant-$|\Gamma|$ circle, as depicted in the figure.

> **Drill 6.16** A 50-Ω line is terminated in a pair of parallel load impedances of $50 + j100$ Ω and $50 - j100$ Ω. Determine the total load admittance and impedance seen by the line. (*Answer:* 8 mS, 125 Ω)

Lossless T-line stubs that are terminated in either a short or an open end can be used as purely reactive tuning elements. Consider a shorted T-line as shown in Figure 6.27a. We see that the constant-$|\Gamma|$ circle follows the periphery of the Smith Chart ($z = 0 \pm jx$). Proper selection of the T-line length d allows us to choose any value of reactance that we want, whether it's capacitive or inductive. This is verified from (6.56), where it is seen that the input impedance at some distance d from the short is

$$Z_{in} = jZ_o \tan(\beta d) \tag{6.75}$$

or

$$z_{in} = j \tan(\beta d) \tag{6.76}$$

For instance, if we travel from the short to a distance $0.125\lambda_g$ on the WTG scale, the Smith Chart gives us an input impedance $Z_{in} = +j50$ Ω, corresponding to an inductance. This is confirmed by using (6.75), where

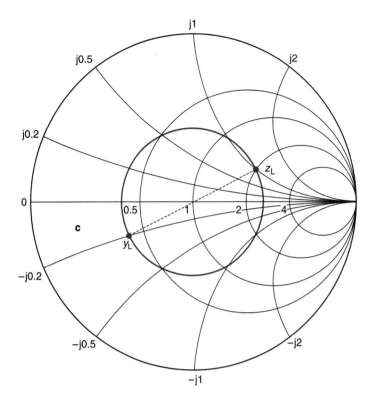

Figure 6.26 Finding normalized admittance with a Smith Chart.

$$\beta d = \left(\frac{2\pi}{\lambda_g}\right)\left(\frac{\lambda_g}{8}\right) = \frac{\pi}{4}$$

results in $\tan(\pi/4) = 1.0$, leading to $Z_{in} = j50 \ \Omega$.

The stub admittance is found by starting from the point on the right side of the chart ($y_{short} = \infty + j\infty$), and traveling clockwise to the point $y_{in} = 0 - j1.0$. Mathematically we have

$$y_{in} = -j \cot(\beta d) \tag{6.77}$$

An open-ended stub will be very similar. In Figure 6.27b, the z_{in} and y_{in} curves would be interchanged for an open stub.

Shunt-Stub Matching

In Smith Chart terms, the objective of a shunt-stub matching network as shown in Figure 6.28a is to move to the center of the chart. Since a shunt stub will be added, we will work in the admittance chart. From the normalized load admittance, a section of through line is traversed to arrive at the $1 \pm jb$ circle. At this point we add a shunt stub of normalized admittance $0 \mp jb$. The sum of these admittances takes us to the center of the chart where $|\Gamma| = 0$, $Y_{in} = Y_o$, and $Z_{in} = Z_o$, and matching is complete.

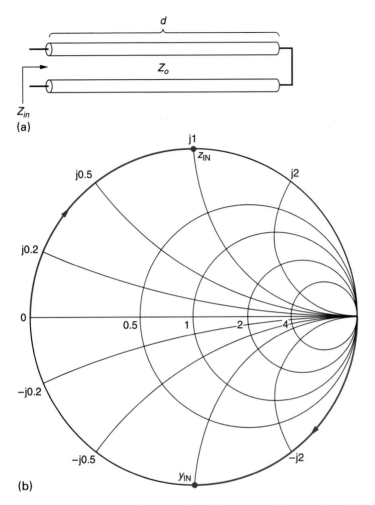

(a)

(b)

Figure 6.27 (a) A shorted T-line stub. (b) Smith Chart view of z_{in} and y_{in} for $d = \lambda/8$.

The procedure for constructing a stub matching network is as follows:

1. Locate z_L, the normalized load impedance.

2. Draw the constant-$|\Gamma|$ circle and use it to locate y_L.

3. From y_L, move clockwise along the constant-$|\Gamma|$ circle to an intersection with the $1 \pm jb$ circle, at which point the impedance looking into the through line is $y_d = 1 \pm jb$. The distance moved is found using the WTG scale and represents the through-line length d.

4. If a shorted shunt stub is employed, consider that its normalized admittance is located on the periphery of the chart at $0.250\lambda_g$ on the WTG scale, or at $\infty + j\infty$ on the admittance chart. Move clockwise along the periphery of the chart to $0 \mp jb$. The distance traveled is the length of the stub, ℓ, and the normalized admittance looking into the stub is therefore $y_\ell = \mp jb$.

5. The total admittance (Figure 6.28b) is $y_{tot} = y_d + y_\ell = 1 + j0$, and the matching network is complete.

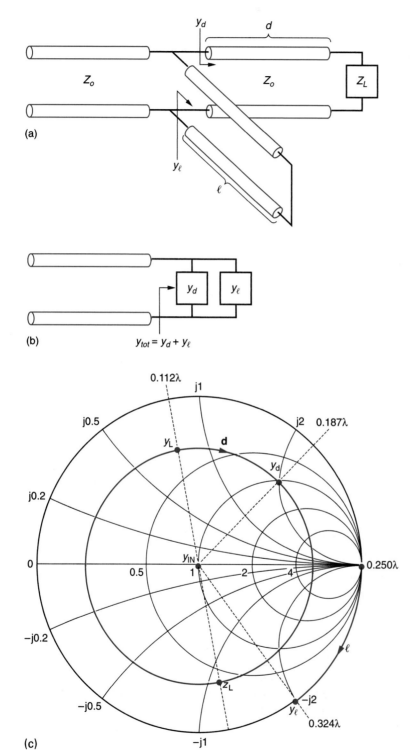

(a)

(b)

$y_{tot} = y_d + y_\ell$

(c)

Figure 6.28 (a) The generic layout of the shorted shunt-stub matching network. (b) Adding shunt admittances. (c) Using the Smith Chart to find through line and stub lengths. Values on the chart apply to Example 6.7.

▶ **EXAMPLE 6.7**

Construct the shorted shunt-stub matching network for a 50-Ω line terminated in a load $Z_L = 20 - j55$ Ω.

  The points indicated in the Smith Chart of Figure 6.28c apply to this problem. First, we locate the normalized load impedance, $z_L = Z_L/Z_o = 0.4 - j1.1$, and draw the constant-$|\Gamma|$ circle. Then we locate $\mathbf{y_L}$. Moving to the first intersection with the $1 \pm jb$ circle (in this case, at $1 + j2.0$), we travel from $0.112\lambda_g$ to $0.187\lambda_g$ on the WTG scale, so our through-line length d is $0.075\lambda_g$.

  Next we insert the shorted shunt stub. On the admittance chart, the location of the short is on the right side of the chart at WTG = $0.250\lambda_g$. We must move clockwise (toward the generator) until we reach the point $0 - j2.0$, located at WTG = $0.324\lambda_g$. This gives us a stub length of $0.324\lambda_g - 0.250\lambda_g = 0.074\lambda_g$.

▶ **EXAMPLE 6.8**

Now we want to construct an open-ended shunt-stub matching network for a 50-Ω line terminated in a load $Z_L = 150 + j100$ Ω.

  Referring to Figure 6.29b, we first locate $z_L = Z_L/Z_o = 3.0 + j2.0$ and draw the constant-$|\Gamma|$ circle. Then we locate $\mathbf{y_L}$. Moving to the first intersection with the $1 \pm jb$ circle (in this case, at $1 + j1.6$), we travel from WTG = $0.474\lambda_g$ to WTG = $0.178\lambda_g$. We add a half a wavelength to the end point, so the length of our through line is $d = (0.500\lambda_g) + 0.178\lambda_g - 0.474\lambda_g = 0.204\lambda_g$.

  Next we insert the open-ended shunt stub. On the admittance chart, the location of the open end is on the left side of the chart at WTG = $0.000\lambda_g$. We must move clockwise (toward the generator) until we reach the point $0 - j1.6$ located at WTG = $0.339\lambda_g$, for a stub length $\ell = 0.339\lambda_g$.

Drill 6.17 In Example 6.7, we chose the first intersection with the $1 \pm jb$ circle (at $1 + j2.0$) in designing our matching network. We could also have continued on to the second intersection, occurring at $1 - j2.0$. Determine the through-line length d, and the stub length ℓ for the matching network using this second intersection. (*Answer:* $d = 0.200\lambda_g$, $\ell = 0.426\lambda_g$)

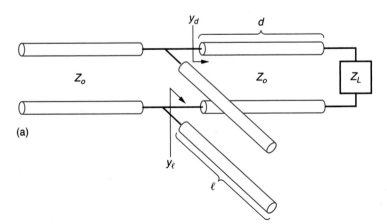

(a)

Figure 6.29 (a) The generic layout of the open-ended shunt-stub matching network.

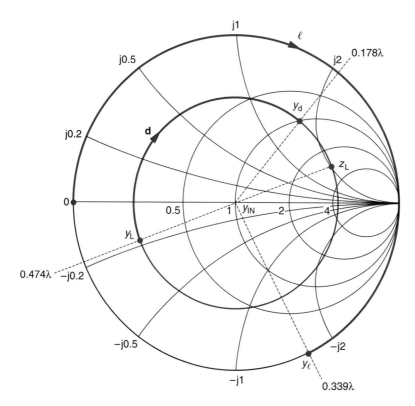

Figure 6.29 (b) Smith Chart solution to Example 6.8.

> **Drill 6.18** Determine the through-line length d and the stub length ℓ for the open-ended shunt matching network of Example 6.8 if the other intersection with the $1 \pm jb$ circle is used. (*Answer:* $d = 0.348\lambda_g$, $\ell = 0.161\lambda_g$)

▶ 6.6 MICROSTRIP

High-frequency circuits are very often constructed on small, flat boards using *microstrip* T-line interconnects. A cross section of microstrip is shown in Figure 6.30. On the bottom of the board (or *substrate*) is a continuous sheet of metal termed the ground plane. On top is a narrow ribbon of metal termed the signal line. The combination of ground plane, signal line, and dielectric make up the microstrip.

The convenience of microstrip is that circuit elements such as transistors and capacitors are easily mounted and supported atop the substrate. The microstrip impedance is a function primarily of the signal line width, dielectric thickness, and dielectric relative permittivity. Fabrication typically employs a low-loss-tangent dielectric substrate that is pre-covered with metal on both sides. A photoresist pattern is applied to the top side, and immersion in an acid bath removes, or *etches*, unwanted metal. In addition to making controlled impedance T-lines, such simple processing can also make many high-frequency circuit components such as filters and couplers (see Chapter 10).

A typical field pattern is shown for a cross section of microstrip in Figure 6.31a. Notice that although most of the field lines are in the dielectric, some are in air. The inhomoge-

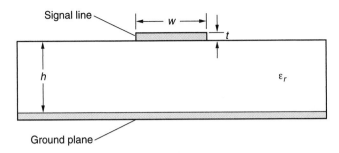

Figure 6.30 Cross section of microstrip T-line.

neous dielectric means waves do not propagate in a pure TEM mode; there are some field components in the direction of propagation. However, most field components are TEM, and it is customary to model microstrip as a signal line buried in a continuous dielectric of effective relative permittivity ε_{eff}, as shown in Figure 6.31b. Propagation in such a model is said to be *quasi-TEM* mode, indicating that we are assuming the propagation is TEM for simplicity. In such a case the propagation velocity u_p is related to the speed of light by

$$u_p = \frac{c}{\sqrt{\varepsilon_{eff}}}$$ (6.78)

where we are assuming a nonmagnetic dielectric, and the phase constant along the line is

$$\beta = \frac{2\pi f}{u_p}$$ (6.79)

The physical length of one wavelength at a particular frequency along the T-line is called the *guide wavelength*, given by

$$\lambda_g = \frac{u_p}{f} = \frac{\lambda_o}{\sqrt{\varepsilon_{eff}}}$$ (6.80)

where $\lambda_o = c/f$.

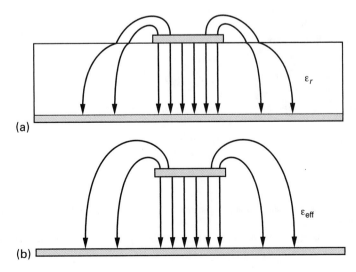

(a)

(b)

Figure 6.31 (a) Typical electric field lines in a cross section of microstrip. (b) Field lines where the air and dielectric have been replaced by a homogeneous medium of effective relative permittivity ε_{eff}.

An exact solution for microstrip is extremely difficult, and *semiempirical*[6.7] equations have been developed. The effective relative permittivity can be written

$$\varepsilon_{eff} = \frac{\varepsilon_r + 1}{2} + \frac{\varepsilon_r - 1}{2\sqrt{1 + 12\,h/w}} \tag{6.81}$$

The characteristic impedance is broken into two parts, depending on the value of the ratio w/h. For $w/h \leq 1$,

$$Z_o = \frac{60}{\sqrt{\varepsilon_{eff}}} \ln\left(\frac{8h}{w} + \frac{w}{4h}\right) \Omega \tag{6.82}$$

and for $w/h > 1$,

$$Z_o = \frac{1}{\sqrt{\varepsilon_{eff}}} \frac{120\pi\ \Omega}{\dfrac{w}{h} + 1.393 + 0.667 \ln\left(\dfrac{w}{h} + 1.444\right)} \tag{6.83}$$

This particular microstrip model doesn't take into account the thickness t of the metal layers, nor does it consider the propagation dependence on frequency. But for most purposes it is sufficiently accurate. The analysis equations for this model are incorporated into ML0604.

Microstrip's maximum operating frequency is limited by loss, dispersion, and the excitation of non-TEM propagation modes. For $w < 2h$, a useful approximation for microstrip's maximum frequency limit is

$$f_{max} = \frac{c}{4h\sqrt{\varepsilon_r}} \tag{6.84}$$

▷ **EXAMPLE 6.9**

Suppose a typical microwave substrate is 40. mils thick[6.8] alumina (Al_2O_3), with $\varepsilon_r = 9.90$. Copper forms the ground plane and an 8.0-mil-wide signal line. We want to find the impedance and propagation velocity for this microstrip.

Using (6.81), we have

$$\varepsilon_{eff} = \frac{9.90 + 1}{2} + \frac{9.90 - 1}{2\sqrt{1 + 12(40)/(8)}} = 6.02$$

Inserting this value into (6.78) gives us a propagation velocity

$$u_p = \frac{c}{\sqrt{\varepsilon_{eff}}} = 1.22 \times 10^8 \ \text{m/s}$$

[6.7]A semiempirical expression is based on a mathematically derived model, to which correction factors are added so that the equation matches reality.

[6.8]English units, in particular *mils*, are still in common use by American circuit board manufacturers. There are 1000 mils to the inch, or 25.4 μm per mil.

Since $w/h < 1$, we use (6.82) and find

$$Z_o = \frac{60}{\sqrt{6.02}} \ln\left(\frac{8(40)}{8} + \frac{8}{4(40)} \right) \Omega = 90 \ \Omega$$

Drill 6.19   What is the maximum frequency of operation for the microstrip of Example 6.9? (*Answer:* 23 GHz)

▷ **MATLAB 6.4**

Here we'll put the microstrip analysis equations into a MATLAB program, save it as ML0604, and run it for the values given in Example 6.9.

```
%   M-File: ML0604
%
%   Microstrip Analysis
%
%   Given the physical dimensions and er, this will
%   calculate eeff, Zo, and up for microstrip.
%
%   Wentworth, 8/3/02
%
%   Variables:
%   w           line width
%   h           substrate thickness
%   er          substrate relative permittivity
%   eeff        effective relative permittivity
%   up          propagation velocity (m/s)
%   Zo          characteristic impedance (ohms)

clc             %clears the command window
clear           %clears variables

disp('Microstrip Analysis')
disp('enter width & thickness in the same units')
disp(' ')

%   Prompt for input values
w=input('enter line width: ');
h=input('enter substrate thickness: ');
er=input('enter substrate er: ');

%   Perform Calculations
eeff=((er+1)/2)+(er-1)/(2*sqrt(1+12*h/w));
up=2.998e8/sqrt(eeff);
if w/h<=1
Zo=60*log((8*h/w)+(w/(4*h)))/sqrt(eeff);
else if w/h>1
Zo=120*pi/(sqrt(eeff)*((w/h)+1.393+0.667*log((w/h)+1.444)));
    end
end
```

(*continues*)

```
%   Display results
disp(['eeff = ' num2str(eeff) ])
disp(['up = ' num2str(up) 'm/s'])
disp(['Zo = ' num2str(Zo) 'ohms'])
```

Now, to run the program for the values of Example 6.9, we have the following:

```
>> ML0604
Microstrip Analysis
enter width & thickness in the same units
enter line width: 8
enter substrate thickness: 40
enter substrate er: 9.90
eeff = 6.0198
up = 122191751.9462m/s
Zo = 90.2408ohms
```

Drill 6.20 Suppose a 20.-mil-thick alumina substrate is used, and the signal line remains 8.0 mils wide. Recalculate u_p and Z_o. What is the maximum operating frequency for this microstrip? (*Answer:* $u_p = 1.20 \times 10^8$ m/s, $Z_o = 72\ \Omega$, $f_{max} = 47$ GHz)

Rather than analyzing a board to find Z_o, it is more often the case that a particular Z_o is sought for microstrip on a known dielectric substrate. Then, it is necessary to choose the width that corresponds to the particular Z_o. The design equations are as follows, broken into two sets based on the ratio w/h:

For $w/h \leq 2$,

$$\frac{w}{h} = \frac{8e^A}{e^{2A} - 2} \tag{6.85}$$

and for $w/h > 2$,

$$\frac{w}{h} = \frac{2}{\pi}\left[B - 1 - \ln(2B-1) + \frac{\varepsilon_r - 1}{2\varepsilon_r}\left(\ln(B-1) + 0.39 - \frac{0.61}{\varepsilon_r}\right)\right] \tag{6.86}$$

The variables A and B in these equations are given by

$$A = \frac{Z_o}{60}\sqrt{\frac{\varepsilon_r + 1}{2}} + \frac{\varepsilon_r - 1}{\varepsilon_r + 1}\left(0.23 + \frac{0.11}{\varepsilon_r}\right) \tag{6.87}$$

and

$$B = \frac{377\pi}{2Z_o\sqrt{\varepsilon_r}} \tag{6.88}$$

> **MATLAB 6.5**

Now we'll put the microstrip design equations into MATLAB, save it as ML0605, and run it for Drill 6.20.

```
%   M-File: ML0605
%
%   Microstrip Design
%
%   Given the desired Zo, substrate thickness, and er,
%   this program will calculate w, eeff, and up.
%
%   Wentworth, 8/3/02
%
%   Variables:
%   w              line width
%   h              substrate thickness
%   er             substrate relative permittivity
%   eeff           effective relative permittivity
%   up             propagation velocity (m/s)
%   Zo             characteristic impedance (ohms)
%   A,B            calculation variables
%   smallratio     calc variable
%   bigratio       calc variable

clc                    %clears the command window
clear                  %clears variables%MstripDesign

disp('Microstrip Design')
disp('width & thickness will be in the same units')
disp(' ')

%   Prompt for input values
Zo=input('enter desired impedance: ');
h=input('enter substrate thickness: ');
er=input('enter substrate er: ');

%Perform Calculations
A=(Zo/60)*sqrt((er+1)/2)+((er-1)/(er+1))*(0.23+0.11/er);
B=377*pi/(2*Zo*sqrt(er));
smallratio=8*exp(A)/(exp(2*A)-2);
bigratio=(2/pi)*(B-1-log(2*B-1)+((er-1)/(2*er))*(log(B-1)+0.39-0.61/er));
if smallratio<=2
    w=smallratio*h;
end
if bigratio>=2
    w=bigratio*h;
end

eeff=((er+1)/2)+(er-1)/(2*sqrt(1+12*h/w));
up=2.998e8/sqrt(eeff);
```

(continues)

```
%Display results
disp(['w = ' num2str(w) ])
disp(['eeff = ' num2str(eeff) ])
disp(['up = ' num2str(up) 'm/s'])
```

Now running the program for Drill 6.21, we obtain:

```
>> ML0605
Microstrip Design
width & thickness will be in the same units

enter desired impedance: 50
enter substrate thickness: 40
enter substrate er: 9.9
w = 38.6273
eeff = 6.6644
up = 116131354.2486m/s
>>
```

> **Drill 6.21** Design a 50-Ω impedance microstrip line on a 40-mil-thick alumina substrate. (*Answer:* w = 38.6 mils)

Attenuation

Attenuation of a signal propagating in microstrip can arise from conductor loss, dielectric loss, and radiation loss. Unintentional loss to radiation can be minimized by avoiding sharp angles or discontinuities in the microstrip line. Most attenuation therefore arises from conductor and dielectric loss, and the total attenuation α_{tot} is the sum

$$\alpha_{tot} = \alpha_c + \alpha_d \tag{6.89}$$

where α_c and α_d are conductor and dielectric attenuation, respectively.

A simple approximation for the conductor loss is given by

$$\alpha_c = \frac{R_{skin}}{Z_o w}\left(\frac{Np}{m}\right)$$
$$\alpha_c = 8.686\frac{R_{skin}}{Z_o w}\left(\frac{dB}{m}\right) \tag{6.90}$$

where w must be in meters and R_s is the conductor skin-effect resistance. The resistance R_s is ideally given as

$$R_{skin} = \frac{1}{\sigma\delta} \tag{6.91}$$

where δ is the well-known skin depth.[6.9] However, if the conductor is thin, then

$$R_{skin} = \frac{1}{\sigma\delta\left(1 - e^{-t/\delta}\right)} \tag{6.92}$$

[6.9] Recall from Chapter 5 that $\delta = 1/\sqrt{\pi f \mu\sigma}$.

is more accurate. Neither of these formulas take into account surface roughness where the conductor meets the dielectric, but it can be gleaned from (6.90) that a wide conductor on a smooth dielectric will minimize conductor loss.

Dielectric loss is approximated by

$$\alpha_d = \frac{2\pi f}{c} \frac{\varepsilon_r(\varepsilon_{eff}-1)}{2\sqrt{\varepsilon_{eff}}(\varepsilon_r-1)} \tan \delta \left(\frac{Np}{m} \right)$$

(6.93)

where the loss tangent is the most critical variable.[6.10]

The formulas for attenuation are approximate and do not, for instance, take into account surface roughness or environmental conditions (temperature, humidity, etc.). Therefore, it is common to measure the attenuation over the frequency range of interest for a particular microstrip configuration.

> **Drill 6.22** Calculate the conductor and dielectric attenuations (both in dB/m) for the microstrip of Drill 6.21 at 1.0 GHz if 6.0-μm-thick copper conductor is used. (*Answer:* $\alpha_c = 1.6$ dB/m, $\alpha_d = .022$ dB/m)

Other Planar T-Lines

Of the planar T-line structures used for high-frequency circuits, microstrip is the most common. A disadvantage of microstrip is that it tends to be *dispersive*, meaning different frequency components travel at different velocities along the line. Dispersion is the topic of Section 6.8. Also, for a component atop the substrate to make ground contact, a metallized hole (called a *via*) must be drilled through the board and filled with metal. Finally, for a given substrate thickness, the T-line width must be fixed if you want to maintain a constant line impedance.

Two other planar T-line structures are shown in Figure 6.32. *Stripline* has the advantage of being very well shielded since it has ground planes on both top and bottom. Also, since

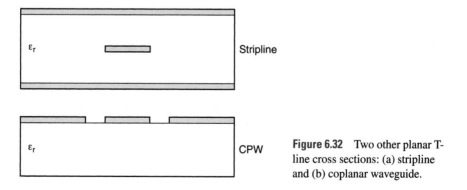

Figure 6.32 Two other planar T-line cross sections: (a) stripline and (b) coplanar waveguide.

[6.10]Note that tanδ is frequency dependent. It is common practice to use a provided value of tanδ, such as those given in Appendix E, even when the value is not cited at your particular frequency. For improved accuracy it may be necessary to conduct your own measurements.

the electric field encounters a homogeneous dielectric in stripline, it doesn't suffer from dispersion as much as microstrip.[6.11] A disadvantage of stripline is the difficulty in contacting discrete components like transistors.

Coplanar waveguide (CPW)[6.12] has both signal line and ground lines on the same side of the substrate and so it is the easiest of the planar structures upon which to place discrete components. The impedance is controlled by the ratio of the center line width w to the gap space s. This means very narrow lines can fan out to wider lines all the while maintaining a constant impedance. A downside of CPW is the requirement for a relatively large amount of substrate area to build a circuit. A variation on CPW is conductor-backed coplanar waveguide T-line, or CBCPW, which as the name implies has a backside ground plane in addition to the two topside ground planes. This has better shielding than CPW and has some of the characteristics of both microstrip and CPW.

▶ 6.7 TRANSIENTS

Our study of transmission lines has so far concerned propagation of steady-state sinusoidal signals at a single frequency. We are now interested in the *transient* situation, where a sudden change in voltage or current is instigated at one end of a line.

The most common example of transients on T-lines is the propagation of signals along the interconnects between digital circuits. Information is carried as ones and zeros, typically corresponding to voltage levels of 6 and 0 V. Switching from 0 to 6 V, for instance, involves a step change in voltage that propagates along the interconnect.

Launching a voltage step function on a T-line can be accomplished as shown in Figure 6.33a by closing the switch at $t = 0$. For simplicity, we'll assume this T-line is lossless.

The voltage launched at the T-line is determined based on voltage division between the source impedance Z_s and the impedance seen by the signal looking into the T-line. However, unlike the steady-state case, the transient signal has no knowledge of how the T-line is terminated. It only sees Z_o. So a voltage

$$V_o = V_s \frac{Z_o}{Z_o + Z_s} \tag{6.94}$$

is initially launched on the line. Since the voltage in general will depend on the location along the line as well as time, it will be represented as $V(z,t)$. The initial launched voltage can therefore be written

$$V(-\ell, 0) = V_o \tag{6.95}$$

The *transit time* t_ℓ, or time for the signal to traverse the length ℓ of the line, is related to propagation velocity u_p by

$$t_\ell = \frac{\ell}{u_p} \tag{6.96}$$

[6.11]If ε_r varies with frequency, dispersion will result in both microstrip and stripline.

[6.12]It is no coincidence that CPW are also the initials for the inventor of coplanar waveguide, C. P. Wen, an engineer at Texas Instruments, Inc.

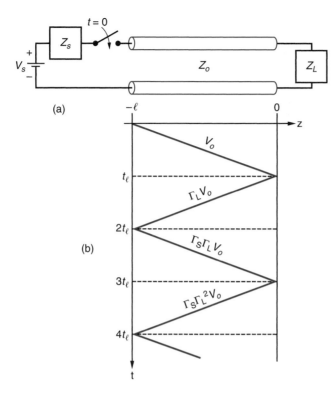

(a)

(b)

Figure 6.33 (a) A voltage step function is realized by closing the switch at $t = 0$. (b) Bounce diagram for transient voltage on the T-line.

Upon reaching the load end at time t_ℓ, part of the signal gets reflected. The reflection coefficient is the same as before,

$$\Gamma_L = \frac{Z_L - Z_o}{Z_L + Z_o}$$

and a wave $\Gamma_L V_o$ is reflected. Therefore, the total voltage at the load end, just after the incident voltage has reached the load, is

$$V(0,t_\ell) = V_o(1 + \Gamma_L) \tag{6.97}$$

The reflected signal travels back toward the source. An additional t_ℓ time is required to reach the source end, where it sees a reflection coefficient

$$\Gamma_s = \frac{Z_s - Z_o}{Z_s + Z_o} \tag{6.98}$$

and a portion of $\Gamma_L V_o$ is reflected back toward the load. We therefore have

$$V(-\ell, 2t_\ell) = V_o(1 + \Gamma_L + \Gamma_L \Gamma_s) \tag{6.99}$$

With all the reflections and rereflections, keeping track of $V(z,t)$ can be difficult. To ease this difficulty we can employ a *bounce diagram*, also referred to as a reflection diagram, as shown in Figure 6.33b. The position along the T-line is shown as the horizontal axis, here aligned for convenience with the T-line of Figure 6.33a. Time is on the vertical axis. The diagonals represent the location of the leading edge of a wave. The top-most diagonal, traveling from time $t = 0$ to $t = t_\ell$, shows the location of the front of the incident wave V_o. The second diagonal shows the front of the first reflected wave $\Gamma_L V_o$, and so on. A vertical line drawn on the bounce diagram shows the voltage at that location as a function of time. At a particular time on the vertical line, the total voltage will consist of all the waves crossing the vertical above that point.

▷ **EXAMPLE 6.10**

An example will show how a step change in voltage propagates along a section of T-line and how the bounce diagram works. Figure 6.34a shows the circuit, where for simplicity we are using only resistive values for the source and load impedances. We want to plot the voltage at the middle of the 6.0-cm-long T-line as a function of time to 8.0 ns.

Our first step is to calculate the reflection coefficients. At the load, we have

$$\Gamma_L = \frac{125\Omega - 75\Omega}{125\Omega + 75\Omega} = \frac{1}{4}$$

At the source end, we have

$$\Gamma_s = \frac{25\Omega - 75\Omega}{25\Omega + 75\Omega} = -\frac{1}{2}$$

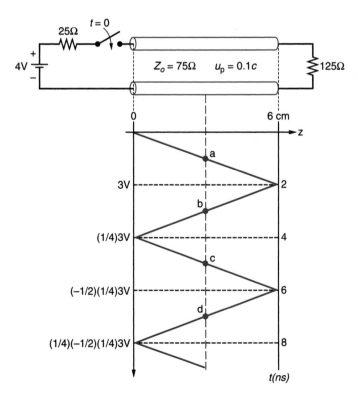

Figure 6.34 (a) T-line circuit and bounce diagram for a voltage step change.

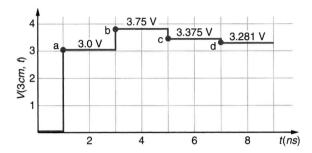

Figure 6.34 (b) Plot of the voltage at the middle of the T-line.

We can also calculate the transit time for the 6.0-cm line, given a propagation velocity of 0.1c:

$$t_\ell = \frac{6 \text{ cm}}{(0.1)\left(3 \times 10^8 \text{ m/s}\right)}\left(\frac{1 \text{ m}}{100 \text{ cm}}\right)\left(\frac{10^9 \text{ ns}}{\text{s}}\right) = 2 \text{ ns}$$

Next, we indicate on the bounce diagram the values for each wave, shown to the left of the diagram in Figure 6.34a. We tally these up at the middle of the T-line and plot the result in Figure 6.34b. Initially, there is no voltage seen in the middle of the line. But after 1 ns has elapsed, the incident wave V_o is present, as indicated by point a on the two figures. This voltage remains constant until the reflected wave arrives, 2 ns later (point b).

After a long enough time, the reflections will settle down and the voltage at every point along of the line will equal

$$V = V_o \frac{Z_L}{Z_L + Z_s} = 4 \text{ } V \frac{125 \text{ } \Omega}{125 \text{ } \Omega + 25 \text{ } \Omega} = 3.33 \text{ } V$$

Drill 6.23 Swap the source and the load resistors for Figure 6.34a and generate a plot of the voltage at the source end of the line to 8 ns. (*Answer:* See Figure 6.35)

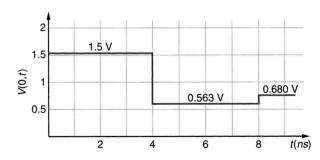

Figure 6.35 Plot of voltage at the source end for Drill 6.23.

Pulse Response

Instead of a voltage step, suppose we launch the square pulse shown in Figure 6.36a onto a T-line circuit. This can be represented by adding a second switch that shorts the supply at $t = T$, as shown in Figure 6.36b.

We can model this situation by placing a pair of voltage step changes on the bounce diagram of Figure 6.37a. A V_o incident wave is launched at $t = 0$, and a $-V_o$ incident wave is launched at $t = T$. Finding the voltage at any point along the T-line at a particular time follows the usual bounce-diagram approach. For instance, the hypothetical pulse response at the middle of the T-line is shown in Figure 6.37b.

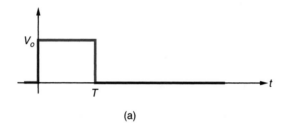

(a)

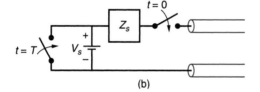

(b)

Figure 6.36   A pulse input as shown in (a) can be realized by the circuit in (b).

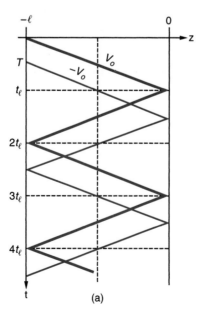

(a)

Figure 6.37   (a) Bounce diagram at the middle of the line for a pulse input.

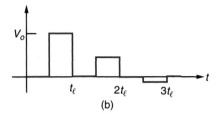

Figure 6.37 (b) Voltage at the middle of the line for a pulse input.

▷ **EXAMPLE 6.11**

To better illustrate the pulse response procedure, suppose in Example 6.10 the 4-V step voltage is replaced with a 4-V pulse of 3-ns duration.

The bounce diagram is shown in Figure 6.38a with the values of the waves as indicated. The voltage at the middle of the line is plotted in Figure 6.38b.

Drill 6.24 For the 3-ns pulse example, plot the voltage at the load out to 8 ns. (*Answer:* See Figure 6.39)

The bounce diagram is a handy tool for step changes and for square pulses. However, realistic digital signals have significant rise and fall times, resulting in pulses that have sloped edges. The bounce diagram becomes very difficult to use in such cases. The

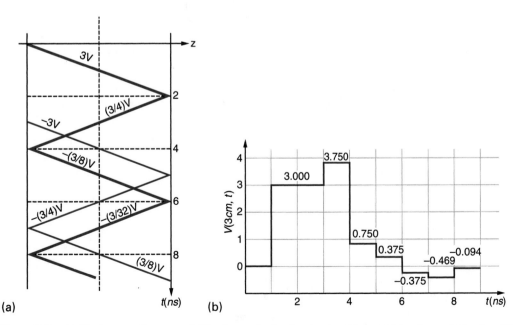

(a)

(b)

Figure 6.38 (a) The bounce diagram and (b) voltage plot for the circuit of Figure 6.34a when a 4-V pulse of 3-ns duration is applied.

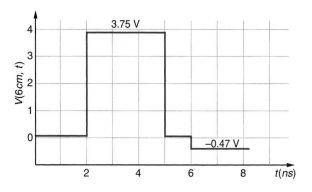

Figure 6.39 Plot for Drill 6.24.

following MATLAB routine plots *V(z,t)* resulting from a triangular pulse source. With modification, the program can be used for pulses with other shapes, including stepped and square pulses.

MATLAB 6.6

The following routine traces the voltage at an arbitrary point along a T-line resulting from a triangular pulse. The pulse is to ramp up linearly from 0 to 10 V over 1.5 ns, then abruptly drop to zero, as indicated in Figure 6.40. We'll also consider the Example 6.10 situation of 6-cm line length and 0.1*c* propagation velocity. However, we'll short the load and want to find *v(z,t)* at *z* = 4.5 cm.

To keep the routine simple, we choose to impedance match the source so that we only have a single positive traveling wave and, perhaps, a negative traveling wave. The program consists of two main sections. First, the triangular function is defined and placed into a time-discretized array. Second, the positive and the negative wave data are generated and combined.

```
%   M-File: ML0606
%   Analysis of a triangular pulse (matched source
%   impedance) traveling down a T-line and reflecting
%   off a resistive load. We want to be able to trace
%   the voltage at an arbitrary point along the line.
%
%   Wentworth, 4/25/03

%   Variables
%   Vo    pulse height (V)
%   t1    pulse start (ns)
%   t2    pulse end (ns)
%   L     line length (cm)
%   T     transit time (ns)
%   z     location to find pulse (cm)
%   tau   time "location" to find pulse (ns)
%   up    propagation velocity (m/s)

%   Zo,ZL  line,load impedance (ohms)
%   N     number of points
```

```
% GL       load reflection coefficient
clc
clear

%enter variables
Vo=10;
t1=0;
t2=1.5;
L=6;
z=4.5;
up=3e7;
Zo=50;
ZL=0;

T=1e9*(L/up)/100;
tau=1e9*(z/up)/100;

N=500;
GL=(ZL-Zo)/(ZL+Zo);
%initialize array
for i=1:N+1
  v(i)=0;
end

dt=2*T/N;
%enter triangular pulse function
m=0.5*Vo/(t2-t1);   %slope
for i=1:N+1
  t(i)=i*dt;
  if t(i)<t1
    vo(i)=0;
  end
  if and(t(i)>t1,t(i)<t2)
    vo(i)=m*(t(i)-t1);
  end
  if t(i)>t2
    vo(i)=0;
  end
end

%Generate + wave data
for i=1:N+1
  ta=i*dt;
  if ta>tau
    j=ceil((ta-tau)/dt);
    vplus(i)=vo(j);
  end
end

%Generate - wave data
for i=1:N+1
  ta=i*dt;
  tb=2*T-tau;
  if ta>tb
    j=ceil((ta-tb)/dt);
```

(*continues*)

```
      vmin(i)=GL*vo(j);
   end
end

%Sum the data
for i=1:N+1
   v(i)=vplus(i)+vmin(i);
end

plot(t,v)
xlabel('time (ns)')
ylabel('voltage')
AXIS([0 2*T -Vo Vo])
grid on
```

Note use of the function "ceil." Whereas the related function "round" will go to the nearest integer, the function "ceil" will round up, to keep $j > 0$.

Suppose we want to use the routine to model the pulse function of Example 6.11 (except with a matched source—see Problem 6.51). We create the unit step function:

```
function U=step(t,T)
%Unit step function. When time t exceeds a time T,
%value of the function is 1.
U=t>=T;
```

Then we modify the program, replacing the "enter triangular pulse function" with "enter rectangular pulse function":

```
%enter rectangular pulse function
t=0:dt:2*T;
vo=0.5*Vo*(step(t,t1)-step(t,t2));
```

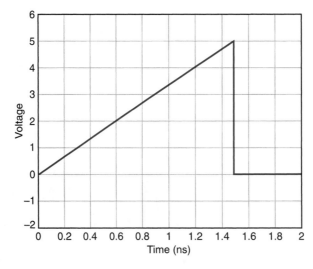

Figure 6.40 Triangular pulse source used in MATLAB 6.6.

Practical Application: Schottky-Diode Terminations

We have seen that signal reflection can be eliminated by terminating the end of the line in a matched load. But the load impedance in a digital circuit may be unknown, or it may vary depending on its logic state. Even if the load impedance is known, and a resistive termination is possible, for many digital circuits the presence of matching resistors will result in continual unwanted power consumption.

Circuit designers have used a number of termination strategies to improve high-frequency performance in digital interconnects. A popular technique employs a pair of Schottky diodes placed at the load end of a transmission line as shown in Figure 6.41. The diode D_1 is said to *clamp* to the supply voltage V_{cc} whereas D_2 clamps to ground. Consider the case where a pulse of value V_{cc} is incident from the transmission line onto an open circuit ($Z_L = \infty$). Without D_1, the reflected pulse would result in $2V_{cc}$ at the load end for the duration of the pulse. But with D_1 in place, this *overshoot* is shunted directly to V_{cc}. Likewise, an *undershoot*, a negative voltage at the load resulting, say, from a shorted load ($Z_L = 0$), would be shunted to ground by D_2. Voltages are therefore bounded by V_{cc} and ground, plus and minus, respectively, the forward bias drop in the Schottky diode. We therefore see that signal reflections are reduced by the clamping action of the diodes, rather than by an impedance match. Also, although a resistive termination would draw continual power, the Schottky diodes only conduct when needed.

The diode of choice for this application is the Schottky, since it has a very small forward bias drop, is relatively fast, and is easily integrated with digital logic.

Reactive Loads

Transient analysis is a bit more difficult when the load is reactive. Consider the inductive load shown in Figure 6.42. At the load end of this circuit we can write the total voltage as

$$v_L(t) = (V_o^i + V_o^r(t))U(\tau) \tag{6.100}$$

where, V_o^i and $V_o^r(t)$ represent the incident and reflected waves, respectively. The unit step function $U(\tau)$ is defined as

$$U(\tau) = \begin{cases} 0 \text{ for } \tau < 0 \\ 1 \text{ for } \tau \geq 0 \end{cases} \tag{6.101}$$

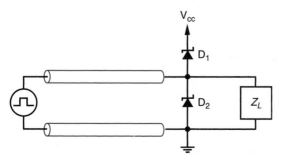

Figure 6.41 Schottky-diode termination.

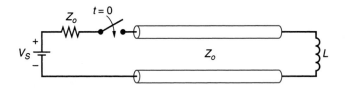

Figure 6.42 A voltage step incident on a T-line with an inductive termination.

Since the incident wave doesn't arrive at the load end until $t = t_\ell$, τ for (6.100) is defined as

$$\tau = t - t_\ell$$

Rather than carry the unit step function through the calculations, it will be inserted again where needed in the final equations.

The reflected wave $V_o^r(t)$ is a function of time and is no longer related to the incident wave by a simple reflection coefficient. Instead, for transient analysis we must consider the inductance relation

$$v_L(t) = L\frac{di_L(t)}{dt} \tag{6.102}$$

We can write an expression similar to (6.100) for the current at the load:

$$i_L(t) = I_o^i + I_o^r(t) \tag{6.103}$$

Then since we know

$$Z_o = \frac{V_o^i}{I_o^i} = -\frac{V_o^r(t)}{I_o^r(t)}$$

we can rewrite (6.103) as

$$i_L(t) = \frac{1}{Z_o}\left(V_o^i - V_o^r(t)\right) \tag{6.104}$$

Adding (6.100) and (6.104) we can eliminate $V_o^r(t)$ and find

$$L\frac{di_L(t)}{dt} + Z_o i_L(t) = 2V_o^i \tag{6.105}$$

This is a linear, first-order differential equation that can be solved[6.13] for $i_L(t)$ as

$$i_L(t) = \frac{2V_o^i}{Z_o}\left(1 - e^{-Z_o \tau/L}\right)U(\tau) \tag{6.106}$$

The unit step equation has been reinserted, and τ replaces t in the exponential, where $\tau = t - t_\ell$ at the load end.

Inserting (6.106) into (6.102), we evaluate the derivative and find

$$v_L(t) = 2V_o^i e^{-Z_o \tau/L}U(\tau) \tag{6.107}$$

[6.13]You should consult your differential equations book. LaPlace or Fourier transforms may also be used.

We can plot $v_L(t)$ as shown in Figure 6.43a.
Finally, the reflected voltage is

$$V_o^r(t) = v_L(t) - V_o^i = V_o^i(2e^{-Z_o\tau/L} - 1)U(\tau) \tag{6.108}$$

We've simplified this example by matching the source impedance to the load impedance so there is no reflection at the source end. We can plot the voltage at the source end, $v_s(t)$, by considering that $\tau = t - 2t_\ell$. We have

$$v_s(t) = V_o^i + V_o^r(t) = V_o^i + V_o^i(2e^{-Z_o\tau/L} - 1)\,U(\tau) \tag{6.109}$$

This is plotted in Figure 6.43b. The inductor appears as a short circuit after sufficient time has elapsed.
A capacitive termination can be solved using the same approach (see Problem 6.53). In Figure 6.44, the voltage at the load end is

$$v_L(t) = 2V_o^i\big[1 - e^{-\tau/Z_oC}\big]U(\tau) \tag{6.110}$$

where $\tau = t - t_\ell$. After the capacitor charges, it resembles an open circuit.

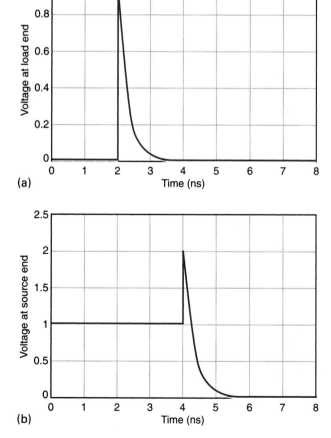

(a)

(b)

Figure 6.43 The 6-cm-long, 75-Ω, $u_p = 0.1c$ T-line is terminated in a 20-nH inductor. (a) The voltage at the load end, and (b) the voltage at the source end.

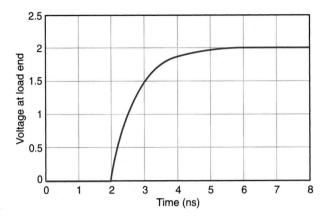

Figure 6.44 The T-line of Figure 6.42 is now terminated in a capacitor. Plot shows the voltage at the load end.

Drill 6.25 Estimate the value of the terminating capacitor using Figure 6.44. (*Answer:* 10 pF)

Time-Domain Reflectometry

Finding the location of discontinuities in T-lines can be accomplished using *time-domain reflectometry*, or *TDR* for short. Here, the response to a step change in voltage is observed at the sending end of the T-line. A wave is reflected from the discontinuity, and the time it takes to receive the reflected signal gives the discontinuity location. The shape of the reflected signal indicates the discontinuity type. Figure 6.45 shows the TDR response for a number of terminations.

If the propagation velocity for the T-line is known, then the measured round-trip travel time can be used to find the discontinuity location by

$$\ell = u_p \frac{1}{2}\left(2t_\ell\right)$$

(6.111)

As might be expected, locating faults in buried cable using TDR can minimize digging. But TDR can also be a useful tool in high-frequency circuit analysis.

▶ **EXAMPLE 6.12**

Let's analyze the TDR response shown for a 50-Ω T-line with $u_p = 0.6c$ in Figure 6.46a.
We can determine the distance to the discontinuity as

$$\ell = (0.6)(3 \times 10^8 \text{ m/s})(1/2)(24 \times 10^{-9} \text{ s}) = 2.2 \text{ m}$$

From a study of the Figure 6.45 plots, the discontinuity appears resistive. We have

$$\Gamma = \frac{V^-}{V^+} = \frac{V_{\text{tot}} - V^+}{V^+} = \frac{1.5 - 1}{1} = 0.5$$

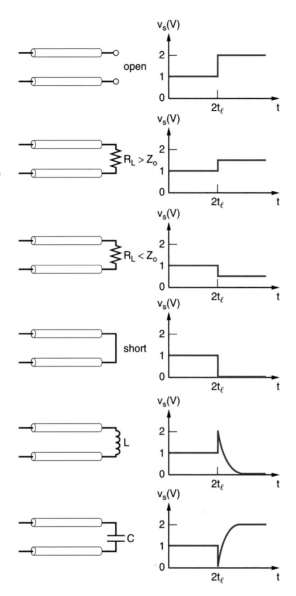

Figure 6.45 TDR plots for Z_o T-line with various terminations. A 1-V incident step change in voltage is assumed.

and since

$$\Gamma = \frac{R_L - Z_o}{R_L + Z_o}$$

then

$$R_L = Z_o \frac{1+\Gamma}{1-\Gamma} = 50 \ \Omega \frac{1+0.5}{1-0.5} = 150 \ \Omega$$

Figure 6.46b shows a TDR plot based on the reflection coefficient. This is a common way to represent TDR plots by highly sophisticated instruments such as vector network analyzers.

▷

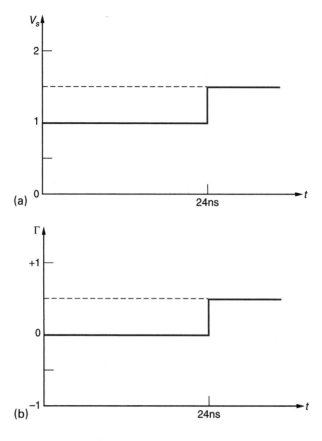

Figure 6.46 (a) Example TDR plot. (b) A reflection coefficient TDR plot.

6.8 DISPERSION

Digital signals on T-lines are often sent as pulses. These pulses can be decomposed into a number of sinusoidal frequency components using a Fourier series. In realistic media, the propagation velocity is a function of frequency. The different frequency components that make up the pulse therefore travel at different velocities, resulting in a spreading out of the pulse as it propagates. This spreading due to change in velocity with frequency is known as *dispersion*.

Let's first look at an example of an ideal pulse, shown in Figure 6.47. This 6-V pulse occurs over 4 ns and repeats every 20 ns. We want to decompose this pulse into its Fourier series.

A very brief review of Fourier series is in order. Any periodic function $f(t)$ may be expressed as a summation of sinusoids,

$$f(t) = a_0 + \sum_{n=1}^{N} \left[a_n \cos(n\omega_o t) + b_n \sin(n\omega_o t) \right] \tag{6.112}$$

where a_0 is the average value of the waveform, a_n and b_n are the Fourier series coefficients, ω_o is the fundamental angular frequency, and $N = \infty$ for an exact representation of the function. The angular frequency is related to the pulse period T_o (20 ns in the example) by

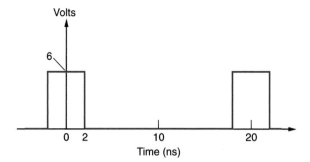

Figure 6.47 An ideal pulse.

$$\omega_o = \frac{2\pi}{T_o} \tag{6.113}$$

For the special case of *even-function symmetry*, that is, symmetry about the vertical axis as in our example, the Fourier series coefficients are

$$
\begin{aligned}
a_o &= \frac{2}{T_o} \int_0^{T_o/2} f(t)\, dt \\
a_n &= \frac{4}{T_o} \int_0^{T_o/2} f(t) \cos(n\omega_o t)\, dt \\
b_n &= 0
\end{aligned}
\tag{6.114}
$$

In our example, it is straightforward to show that $a_0 = 1.2$ V, and

$$a_n = \frac{12}{n\pi} \sin\left(\frac{n\pi}{5}\right) \text{V}$$

Using the coefficients, we plot the Fourier series representation of the pulse for $N = 10$, 100, and 1000 in Figure 6.48. The values of n correspond to harmonics of the fundamental frequency, where

$$f(n) = n f_o \tag{6.115}$$

In our example, $f_o = 1/T_o = 50$ MHz. For $N = 10$, frequency components up to 500 MHz are required. As evidenced from Figure 6.48, there is considerable ripple in the Fourier series representation. Increasing N to 100 increases the highest frequency component to 5 GHz. Here, the ripple is quite small, and the edge of the pulse is almost vertical. Increasing N to 1000 (or frequency up to 50 GHz), we see a very good approximation to the ideal pulse.

In actual digital signals, the rise time and fall time for a pulse are not instantaneous, and it doesn't take quite so many harmonics in the Fourier series to provide an accurate representation. It may be noted, however, that microprocessors touted as operating at 1 GHz actually carry frequency components many times higher!

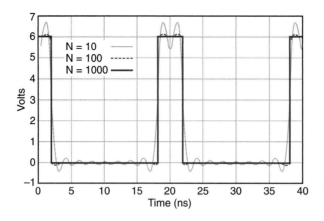

Figure 6.48 The ideal pulse of Figure 6.47 modeled with different values of *N*.

▷ **MATLAB 6.7**

Find the Fourier series coefficients and generate a plot for the example pulse of Figure 6.48.

```
%    M-File: ML0607
%
%    This program assembles a pulse using Fourier series.
%
%    Wentworth, 8/3/02
%
%    Variables:
%    N          number of Fourier coefficients
%    a0         avg value of the waveform (V)
%    T          period (s)
%    fo         fundamental frequency (Hz)
%    wo         fund angular freq (rad/s)
%    t          time (sec)
%    ftot       fourier sum at a particular time(volts)

clc              %clears the command window
clear            %clears variables

%    Initialize variables
clear
N=100;
a0=1.2;
T=20e-9;
fo=1/T;
wo=2*pi*fo;

%    Evaluate Fourier Series Coefficients
for n=1:N
    a(n)=(12/(pi*n))*sin(n*pi/5);
end

%    Generate data and plot
for i=1:180
    t(i)=i*T/90;
```

```
    for n=1:N
    f(n)=a(n)*cos(n*wo*t(i));
    end
    ftot(i)=a0+sum(f);
end

plot(t,ftot)
xlabel('time(s)')
ylabel('volts')
grid on
```

Data for each value of N were inserted into Microsoft Excel™ to generate Figure 6.48. To save the data for $N = 100$, for instance, add the following line:

```
Save 'Fourier100' t ftot –ascii
```

This saves the data in ASCII format in the file Fourier100 in your MATLAB work folder.

Each harmonic of the pulse propagates along the T-line by

$$v_n(z,t) = a_n \cos(\omega_n t - \beta_n z) \qquad (6.116)$$

where $\omega_n = n\omega_o$, and the propagation velocity is

$$\left(u_p\right)_n = \frac{\omega_n}{\beta_n} \qquad (6.117)$$

It is customary to plot the ω versus β for a T-line. If the dielectric is lossless with constant relative permittivity over all the harmonics, the plot is a straight line indicating constant u_p. In this case, the pulses do not disperse as they propagate.

Let's consider what happens when ε_r is a function of frequency. Suppose we have

$$\varepsilon_r = 6 - \left(\frac{f}{5 \times 10^{10}} \right)$$

valid up to 50 GHz. This is not a drastic change, since ε_r will only decrease from a value of 6 to 5 over this large frequency range. We can generate an ω–β diagram, as shown in Figure 6.49. This plot shows little deviation from linear behavior.[6.14] To show how the small change in ε_r affects a pulse on the line, we can modify MATLAB 6.7 to calculate the value at each harmonic of the pulse at some distance along the T-line. At $z = 10$ m, we calculate each component using (6.116) and add them together to arrive at Figure 6.50. There is clearly significant degradation in the signal caused by dispersion.

[6.14]It is easier to see small curvature in the lines if you hold the page such that you are looking from one end of the line.

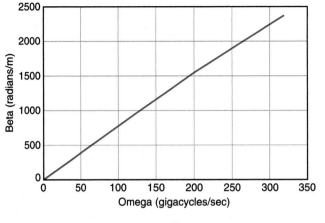

Figure 6.49 ω–β example plot.

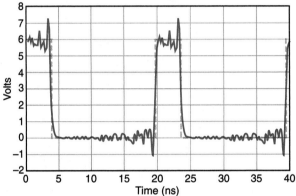

Figure 6.50 Dispersion example using $N = 1000$ after a 10-m distance is traveled along the T-line. The dashed line indicates the original signal.

▷ **MATLAB 6.8**

Let us see how to modify MATLAB 6.7 to generate the pulse shown in Figure 6.50.

```
%    M-File: ML0608
%
%    Pulse with Dispersion
%
%    This program assembles a pulse using Fourier series,
%    where er is a function of frequency. The pulse is
%    initiated at z=0 and is inspected at a location z.
%
%    Wentworth, 8/3/02
%
%    Variables:
%    N          number of Fourier coefficients
%    a0         avg value of the waveform (V)
%    T          period (s)
%    fo         fundamental frequency (Hz)
%    wo         fund angular freq (rad/s)
%    t          time (sec)
%    Vtot       fourier sum at a particular time(volts)
```

```
%   er          relative permittivity
%   beta        phase constant(rad/m)
%   z           location to look at pulse

clc             %clears the command window
clear           %clears variables%MstripDesign

%   Initialize variables
clear
N=1000;
a0=1.2;
T=20e-9;
fo=1/T;
wo=2*pi*fo;
z=10;

%evaluate Fourier series coefficients
for n=1:N
  a(n)=(12/(pi*n))*sin(n*pi/5);
end

%Generate data
for i=1:180
    t(i)=i*T/90;
    for n=1:N
       f(n)=n*50e6;
       %er(n)=6-f(n)/5e10;
       er(n)=6;
       beta(n)=2*pi*f(n)*sqrt(er(n))/3e8;
       V(n)=a(n)*cos(n*wo*t(i)-beta(n)*z);
    end
    Vtot(i)=a0+sum(V);
end

plot(t,Vtot)
xlabel('time(ns)')
ylabel('Volts')
grid on

save 'dispoff' t Vtot -ascii
```

▶ SUMMARY

- T-lines may be represented by distributed parameter models consisting of $R'(\Omega/\text{m})$, $L'(\text{H/m})$, $G'(\text{S/m})$, and $C'(\text{F/m})$. The values of these parameters vary according to the particular T-line. For coaxial cable, we have

$$R' = \frac{1}{2\pi}\left(\frac{1}{a}+\frac{1}{b}\right)\sqrt{\frac{\pi f \mu}{\sigma_c}}$$

$$L' = \frac{\mu}{2\pi}\ln\left(\frac{b}{a}\right)$$

$$G' = \frac{2\pi\sigma_d}{\ln(b/a)}$$

and

$$C' = \frac{2\pi\varepsilon}{\ln(b/a)}$$

- The general transmission line equations, or telegraphist's equations, are

$$-\frac{\partial v(z,t)}{\partial z} = i(z,t)R' + L'\frac{\partial i(z,t)}{\partial t}$$

and

$$-\frac{\partial i(z,t)}{\partial z} = v(z,t)G' + C'\frac{\partial v(z,t)}{\partial t}$$

For time-harmonic waves on T-lines these equations are written

$$\frac{\partial V_s(z)}{\partial z} = -(R' + j\omega L')I_s(z)$$

and

$$\frac{\partial I_s(z)}{\partial z} = -(G' + j\omega C')V_s(z)$$

- The propagation constant γ (m^{-1}), attenuation α (Np/m), and phase constant β (radians/m) are related to the distributed parameters by

$$\gamma = \sqrt{(R' + j\omega L')(G' + j\omega C')} = \alpha + j\beta$$

- The instantaneous form of the time-harmonic traveling wave equations are

$$v(z,t) = V_o^+ e^{-\alpha z} \cos(\omega t - \beta z) + V_o^- e^{+\alpha z} \cos(\omega t + \beta z)$$

and

$$i(z,t) = I_o^+ e^{-\alpha z} \cos(\omega t - \beta z) + I_o^- e^{+\alpha z} \cos(\omega t + \beta z)$$

- Characteristic impedance $Z_o (\Omega)$ of a T-line is related to the wave amplitudes and the distributed parameters by

$$Z_o = \frac{V_o^+}{I_o^+} = \frac{-V_o^-}{I_o^-} = \sqrt{\frac{R' + j\omega L'}{G' + j\omega C'}}$$

- The average power $P_{ave}^+(z)$ traveling in the $+z$ direction along a T-line is

$$P_{ave}^+(z) = \frac{\left(V_0^+\right)^2}{2Z_o} e^{-2\alpha z}$$

- Power ratios are often expressed in terms of decibels. The power gain, or ratio of the output power to the input power of a device, can be expressed as

$$G(dB) = 10 \log\left(\frac{P_{out}}{P_{in}}\right)$$

Decibels are related to nepers by

$$1 \text{ Np} = 8.868 \text{ dB}$$

- For a Z_o impedance line terminated in a load Z_L, a portion of the incident wave (V_o^+) is reflected (V_o^-), with the amount given by the reflection coefficient at the load Γ_L,

$$V_o^- = \frac{Z_L - Z_o}{Z_L + Z_o} V_o^+ = \Gamma_L V_o^+$$

The superposition of the reflected and incident waves establishes a standing wave pattern on the line. The ratio of the maximum to minimum amplitudes is the voltage standing wave ratio, given by

$$VSWR = \frac{1 + |\Gamma_L|}{1 - |\Gamma_L|}$$

- A length of T-line can be replaced in a circuit by an equivalent input impedance, which is given in general as

$$Z_{in} = Z_o \frac{Z_L + Z_o \tanh(\gamma \ell)}{Z_o + Z_L \tanh(\gamma \ell)}$$

For lossless T-line, the input impedance expression becomes

$$Z_{in} = Z_o \frac{Z_L + jZ_o \tan(\beta \ell)}{Z_o + jZ_L \tan(\beta \ell)}$$

- The Smith Chart is a very useful graphical tool for working a variety of T-line problems. It is a plot of normalized impedance or admittance at any point along a section of T-line and is also a plot of the reflection coefficient along the line.

- The impedance of a complex load can be matched to the impedance of a T-line with matching networks designed using the Smith Chart. The object is to move from the normalized load impedance (or admittance) toward the center of the chart where $|\Gamma| = 0$.

- A T-line shunt stub terminated in either a short or an open circuit can be a useful reactive element for a matching network. To construct a shunt-stub matching network, you locate the load admittance, move along the constant circle until reaching the $1 \pm jb$ circle, and then add the appropriate length shunt stub to add reactance $0 \mp jb$.

- Microstrip is the most heavily used of the various planar T-lines. For calculation and design purposes, it is considered to operate in a quasi-TEM mode, neglecting any field components in the direction of propagation. A number of semiempirical expressions have been developed for microstrip analysis and design.

- Problems involving sudden signal changes on a T-line, called transients, are often dealt with using bounce diagrams. The response to a step change in voltage by a variety of terminations is the operating principle of time-domain reflectometry.

- The spreading or distortion of a signal caused by the variation in propagation velocity with frequency is known as dispersion. Pulses may be modeled by their Fourier series components, each at a different frequency and propagating at a different speed on a T-line.

▶ PROBLEMS

6.1 Distributed Parameters Model

6.1 RG-223/U coax has an inner conductor radius $a = 0.47$ mm and inner radius of the outer conductor $b = 1.435$ mm. The conductor is copper, and polyethylene comprises the dielectric. Calculate the distributed parameters at 800 MHz.

6.2 Modify MATLAB 6.1 to account for a magnetic conductive material. Apply this program to Problem 6.1 with the copper conductor replaced by nickel.

6.3 Modify (6.3) to include internal inductance of the conductors. To simplify the calculation, assume current is evenly distributed across the conductors. Find the new value of L' for the coax of Drill 6.1.

6.2 Time-Harmonic Waves on Transmission Lines

6.4 Modify MATLAB 6.1 to also calculate γ, α, β, and Z_o. Confirm the program using Drill 6.2.

6.5 The impedance and propagation constant at 100 MHz for a T-line are determined to be $Z_o = 18.6 - j0.253\ \Omega$ and $\gamma = 0.0638 + j4.68\ \mathrm{m}^{-1}$. Calculate the distributed parameters.

6.6 The specifications for RG-214 coaxial cable are as follows:

- 2.21-mm diameter copper inner conductor,
- 7.24-mm inner diameter of outer conductor,
- 9.14-mm outer diameter of outer conductor, and
- Teflon dielectric ($\varepsilon_r = 2.10$).

Calculate the characteristic impedance and the propagation velocity for this cable.

6.7 For the RG-214 coax of Problem 6.6 operating at 1.0 GHz, how long is this T-line in terms of wavelengths if its physical length is 50. cm?

6.8 If 1.0 W of power is inserted into a coaxial cable, and 1.0 μW of power is measured 100 m down the line, what is the line's attenuation in dB/m?

6.9 Starting with a 1.0-mm-diameter solid copper wire, you are to design a 75-Ω coaxial T-line using mica as the dielectric. Determine (a) the inner diameter of the outer

copper conductor, (b) the propagation velocity on the line, and (c) the approximate attenuation, in dB/m, at 1 MHz.

6.10 A coaxial cable has a solid copper inner conductor of radius $a = 1$ mm and a copper outer conductor of inner radius b. The outer conductor is much thicker than a skin depth. The dielectric has $\varepsilon_r = 2.26$ and $\sigma_{\mathrm{eff}} = 0.0002$ at 1 GHz. Letting the ratio b/a vary from 1.5 to 10, generate a plot of the attenuation (in dB/m) versus the line impedance. Use the lossless assumption to calculate impedance.

6.3 Terminated T-Lines

6.11 Start with (6.54) and derive (6.55).

6.12 Derive (6.56) from (6.55) for a lossless line.

6.13 A 2.4-GHz signal is launched on a 1.5-m length of T-line terminated in a matched load. It takes 6.25 ns to reach the load and suffers 1.2 dB of loss. Find the propagation constant.

6.14 A source with 50-Ω source impedance drives a 50-Ω T-line that is 1/8 of a wavelength long, terminated in a load $Z_L = 50 - j25\ \Omega$. Calculate Γ_L, *VSWR*, and the input impedance seen by the source.

6.15 A 1-m-long T-line has the following distributed parameters: $R' = 0.10\ \Omega/\mathrm{m}$, $L' = 1.0\ \mu\mathrm{H/m}$, $G' = 10.0\ \mu\mathrm{S/m}$, and $C' = 1.0\ \mathrm{nF/m}$. If the line is terminated in a 25-Ω resistor in series with a 1.0-nH inductor, calculate, at 200 MHz, Γ_L and Z_{in}.

6.16 The reflection coefficient at the load for a 50-Ω line is measured as $\Gamma_L = 0.516e^{j8.2°}$ at $f = 1.0$ GHz. Find the equivalent circuit for Z_L.

6.17 The input impedance for a 30.-cm length of lossless 100-Ω impedance T-line operating at 2.0 GHz is $Z_{\mathrm{in}} = 92.3 - j67.5\ \Omega$. The propagation velocity is $0.70c$. Determine load impedance.

6.18 For the lossless T-line circuit shown in Figure 6.51, determine the input impedance Z_{in} and v_L, the instantaneous voltage at the load end.

6.19 Referring to Figure 6.10, consider a lossless 75-Ω

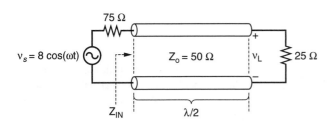

Figure 6.51 Complete circuit for Problem 6.18.

T-line with $u_p = 0.8c$ that is 30. cm long. The supply voltage is $v_s = 6.0 \cos(\omega t)$V with $Z_s = 75\ \Omega$. If $Z_L = 100 + j125\ \Omega$ at 600 MHz, find (a) Z_{in}, (b) the voltage at the load end of the T-line, and (c) the voltage at the sending end of the T-line.

6.20 Suppose the T-line for Figure 6.10 is characterized by the following distributed parameters at 100 MHz: $R' = 5.0\ \Omega/m$, $L' = 0.010\ \mu H/m$, $G' = 0.010\ S/m$, and $C' = 0.020\ nF/m$. If $Z_L = 50 - j25\ \Omega$, $v_s = 10 \cos(\omega t)$ V, $Z_s = 50\ \Omega$, and the line length is 1.0 m, find the voltage at each end of the T-line.

6.4. The Smith Chart

6.21 Locate on a Smith Chart the following load impedances terminating a 50-Ω impedance T-line. (a) $Z_L = 200\ \Omega$, (b) $Z_L = j25\ \Omega$, (c) $Z_L = 50 + j\,50\ \Omega$, and (d) $Z_L = 25 - j200\ \Omega$.

6.22 Repeat Problem 6.14 using the Smith Chart.

6.23 A 0.690λ-long lossless $Z_o = 75\ \Omega$ T-line is terminated in a load $Z_L = 15 + j67\ \Omega$. Use the Smith Chart to find (a) Γ_L, (b) $VSWR$, (c) Z_{in}, and (d) the distance between the input end

of the line and the first voltage maximum from the input end.

6.24 A 0.269λ-long lossless $Z_o = 100\ \Omega$ T-line is terminated in a load $Z_L = 60 + j40\ \Omega$. Use the Smith Chart to find (a) Γ_L, (b) $VSWR$, (c) Z_{in}, and (d) the distance from the load to the first voltage maximum.

6.25 The input impedance for a 100-Ω lossless T-line of length 1.162λ is measured as $12 + j42\ \Omega$. Determine the load impedance.

6.26 On a 50-Ω lossless T-line, the $VSWR$ is measured as 3.4. A voltage maximum is located 0.079λ away from the load. Determine the load.

6.27 Figure 6.52 is generated for a 50-Ω slotted coaxial air line terminated in a short circuit and then in an unknown load. Determine (a) the measurement frequency, (b) the $VSWR$ when the load is attached, and (c) the load impedance.

6.28 Figure 6.53 is generated for a 50-Ω slotted coaxial air line terminated in a short circuit and then in an unknown load. Determine (a) the measurement frequency, (b) the $VSWR$ when the load is attached, and (c) the load impedance.

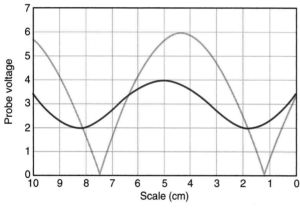

Figure 6.52 Field pattern with air line terminated in a short (light blue line) and in an unknown load (dark blue line) for Problem 6.27.

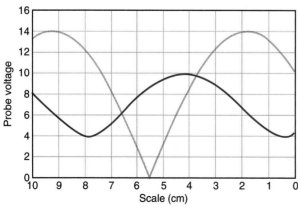

Figure 6.53 Field pattern with air line terminated in a short (light blue line) and in an unknown load (dark blue line) for Problem 6.28.

6.29 Referring to Figure 6.20, suppose we measure Z_{insc} = +$j25$ Ω and Z_{inL} = 35 + $j85$ Ω. What is the actual load impedance? Assume Z_o = 50 Ω.

6.30 Modify MATLAB 6.3 to draw the normalized load point and the constant-$|\Gamma_L|$ circle, given Z_o and Z_L. Demonstrate your program with the values from Drill 6.11.

6.5 Impedance Matching

6.31 A matching network, using a reactive element in series with a length d of T-line, is to be used to match a 35 − $j50$ Ω load to a 100-Ω T-line at 1.0 GHz. Find the through-line length d and the value of the reactive element if (a) a series capacitor is used and (b) a series inductor is used.

6.32 A matching network consists of a length of T-line in series with a capacitor. Determine the length (in wavelengths) required of the T-line section and the capacitor value needed (at 1.0 GHz) to match a 10 − $j35$ Ω load impedance to the 50-Ω line.

6.33 You would like to match a 170-Ω load to a 50-Ω T-line. (a) Determine the characteristic impedance required for a quarter-wave transformer. (b) What through-line length and stub length are required for a shorted shunt-stub matching network?

6.34 A load impedance Z_L = 200 + $j160$ Ω is to be matched to a 100-Ω line using a shorted shunt-stub tuner. Find the solution that minimizes the length of the shorted stub.

6.35 Repeat Problem 6.34 for an open-ended shunt-stub tuner.

6.36 A load impedance Z_L = 25 + $j90$ Ω is to be matched to a 50-Ω line using a shorted shunt-stub tuner. Find the solution that minimizes the length of the shorted stub.

6.37 Repeat Problem 6.36 for an open-ended shunt-stub tuner.

6.38 (a) Design an open-ended shunt-stub matching network to match a load Z_L = 70 + $j110$ Ω to a 50-Ω impedance T-line. Choose the solution that minimizes the length of the through line. (b) Now suppose the load turns out to be Z_L = 40 + $j100$ Ω. Determine the reflection coefficient seen looking into the matching network.

6.6 Microstrip

6.39 A 6.00-cm-long microstrip transmission line is terminated in a 100.-Ω resistive load. The signal line is 0.692 mm wide atop a 0.500-mm-thick polyethylene substrate. What is the input impedance of this line at 1.0 GHz? What is the maximum frequency at which this microstrip can operate?

6.40 A 75-Ω impedance microstrip line is to be designed on a 2.0-mm-thick Teflon substrate using copper metallization. What is the maximum operating frequency for this microstrip? Now determine w and the physical length of a quarter-wave section of line at 800. MHz.

6.41 Analysis of a 2.56-cm-long microstrip line reveals that it has a 50-Ω characteristic impedance and an effective relative permittivity of 5.49. It is terminated in a 60-Ω resistor in series with a 1.42-pF capacitor. Determine the input impedance looking into this terminated line at 1.60 GHz.

6.42 A 100-Ω impedance microstrip line is to be designed using copper metallization on a 0.127-cm-thick dielectric of relative permittivity 3.8. Determine (a) w, (b) f_{max}, and at 2.0 GHz find (c) u_p and λ_G.

6.43 Modify MATLAB 6.4 to calculate attenuation. Try out your program using the parameters of Drills 6.21 and 6.22.

6.44 A 50-Ω impedance microstrip line is desired for operation at 2.4 GHz. It is to be built on a 20-mil-thick mica substrate using a 10-μm-thick copper conductor. Calculate (a) w, (b) α_c, (c) α_d, and (d) α_{tot} at this frequency.

6.45 One type of board routinely used to build microwave circuits is 50-mil-thick Rogers Corporation RT/Duroid, with ε_r = 10.8 and $\tan\delta$ = 0.0028. It is coated on both sides by "1/4 oz copper." This translates to a 0.35-mil thickness of copper. Find w and u_p for a 50-Ω line. Then determine the α_c, α_d, and α_{tot} at three frequencies: 1, 10, and 20 GHz. What is the maximum frequency of operation for this microstrip?

6.46 A 1.5-in length of microstrip line of width 48.86 mils sits atop a 50-mil-thick substrate with dielectric constant 4. Determine the impedance looking into this circuit at 2 GHz if it is terminated in a 300-Ω resistor. Assume ideal conductors and lossless dielectric.

6.47 The top-down view of a microstrip circuit is shown in Figure 6.54. If the microstrip is supported by a 40-mil-thick alumina substrate, (a) determine the line width required to achieve a 50-Ω impedance line. (b) What is the guide wavelength on this microstrip line at 2.0 GHz? (c) Suppose at this frequency the load impedance is Z_L = 150 − $j100$ Ω. Determine the length of the stubs (d_{thru} and ℓ_{stub}) required to impedance match the load to the line.

6.48 Suppose the microstrip circuit shown in Figure 6.54 is realized atop the RT/Duroid board of Problem 6.45. Assuming the board material is lossless, (a) determine the line width required to achieve a 75-Ω impedance line. (b) Now suppose at 1.0 GHz the load impedance is Z_L =

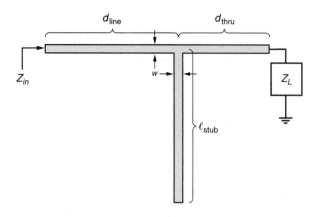

Figure 6.54 The top-down view of an open-ended microstrip stub-matching circuit for Problems 6.47 and 6.48.

$150 + j150$ Ω. Find the length of the stubs (d_{thru} and ℓ_{stub}) required to impedance match the load to the line.

6.7 Transients

6.49 Consider Figure 6.33 with the following values: $V_s = 10$ V, $Z_s = 30$ Ω, $Z_o = 50$ Ω, $u_p = 0.666c$, $Z_L = 150$ Ω, and $\ell = 10$ cm. Plot, out to 2 ns, (a) the voltage at the source end, (b) the voltage at the middle, and (c) the voltage at the load end of the T-line.

6.50 Repeat Problem 6.49 for a 10-V pulse of duration 0.4 ns.

6.51 Consider a 12-cm-long 50-Ω transmission line terminated in a 25-Ω load and having a matched source impedance ($Z_s = 50$ Ω). The propagation velocity on the T-line is $0.67c$. The source is a 0.4-ns square pulse of amplitude 6 V. Modify MATLAB 6.6 to plot $v(z, t)$ at two points: $z = 2$ cm and $z = 10$ cm.

6.52 Modify MATLAB 6.6 to plot $v(z, t)$ at $z = 4.5$ cm if the source pulse is as indicated in Figure 6.55.

6.53 The expressions for $i_L(t)$ and $v_L(t)$ of (6.106) and (6.107) were derived for a T-line terminated in an inductor. Find similar expressions for a T-line terminated in a capacitor.

6.54 For Figure 6.42, $Z_o = 100$ Ω and $u_p = 0.1c$. Estimate L if the v_L versus t plot is given in Figure 6.56.

6.55 A 50-Ω T-line with $u_p = 0.5c$ is terminated in some resistive load such that the TDR plot is given by Figure 6.57. Determine the location and the value of the load.

6.56 The TDR plot for a 75-Ω T-line with $u_p = 0.2c$ is given in Figure 6.58. What type of components terminate the line? Estimate the component values.

6.8 Dispersion

6.57 Use Fourier series to construct a 5-V pulse of duration 5 ns that repeats every 10 ns.

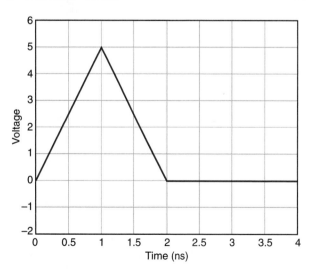

Figure 6.55 Triangular pulse source for Problem 6.52.

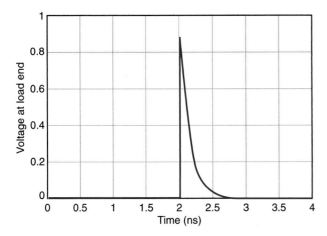

Figure 6.56 Voltage versus time plot for Problem 6.54.

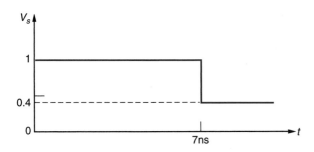

Figure 6.57 TDR plot for Problem 6.55.

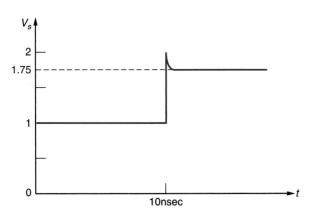

Figure 6.58 TDR plot for Problem 6.56.

6.58 Actual pulses have some slope to the leading and trailing edges. Suppose a symmetrical pulse is 5 V from −2 ns to +2 ns and has a linear slope to 0 V on each edge of duration 0.2 ns. The pulse repeats every 20 ns. Construct this pulse using Fourier series for $N = 10$, 100, and 1000. Comment on how this pulse compares to the one of Figure 6.48.

6.59 A material has a constant $\varepsilon_r = 4$ from DC up to 20 GHz. Then

$$\varepsilon_r = 4\cos\left(\frac{f = 20 \times 10^9}{10 \times 10^9}\right)$$

for 20 GHz $< f <$ 50 GHz. Show the pulse from Problem 6.58 after it has traveled 10 meters along a coaxial T-line with this dielectric.

CHAPTER 7

Waveguide

Learning Objectives

▷ Develop equations governing wave propagation in rectangular waveguide

▷ Describe propagation modes, cutoff frequency, impedance, and wave propagation in rectangular waveguide

▷ Discuss propagation modes and field distribution in dielectric waveguide

▷ Describe the components of an optical fiber communications system

▷ Use power and rise-time budgets to design optical fiber systems

In the previous chapter, we saw how a pair of conductors was used to guide electromagnetic wave propagation. This propagation was via the TEM mode, meaning both the electric and magnetic field components were transverse, or perpendicular, to the direction of propagation. In this chapter we investigate wave-guiding structures that support propagation in non-TEM modes, namely in the TE and TM modes.

The generic term *waveguide* can mean any structure that supports propagation of a wave. Although T-lines are technically a subset of waveguides, in general usage the term waveguide refers to constructs that only support non-TEM mode propagation. Such constructs share an important trait: They are unable to support wave propagation below a certain frequency, termed the *cutoff frequency*. The most common waveguide types are shown in Figure 7.1.

Figures 7.1a and 7.1b are referred to as *metallic waveguide* structures. The first of these, *rectangular waveguide* (Figure 7.1a), is often employed in high-power microwave applications. It is relatively simple to fabricate and can have much less attenuation than coaxial T-line. It is rather limited in frequency range, however, and also suffers from dispersion. We will examine rectangular waveguide in the first two sections of this chapter.

Circular waveguide, shown in Figure 7.1b, has even higher power handling capability than rectangular waveguide. Analysis requires use of Bessel functions, a task beyond the scope of this text.

The *dielectric waveguide* shown in Figures 7.1c and 7.1d can have much smaller loss than metallic waveguide at high frequencies. Optical fiber (Figure 7.1d) also has a tremendous bandwidth advantage over metallic waveguide, and despite the absence of metallic

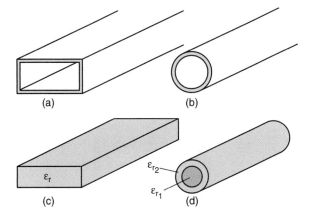

Figure 7.1 Non-TEM mode waveguide structures include (a) rectangular waveguide, (b) circular waveguide, (c) dielectric slab waveguide, and (d) fiber optic waveguide.

boundaries, there is extremely good signal isolation between adjacent fibers. The advantages of optical fiber over other waveguide and T-line types make it a workhorse in the communications industry. The last three sections of this chapter concern optical fiber.

7.1 RECTANGULAR WAVEGUIDE FUNDAMENTALS

In this section, we present the basic equations and operating principles necessary for working with rectangular waveguide. We will take these largely on faith until we reach the next section where field equations are derived starting with Maxwell's equations.

A cross section of rectangular waveguide is shown in Figure 7.2. Propagation is in the $+z$ direction, or out of the page. The conducting walls are typically brass, copper, or aluminum. They are chosen thick enough to provide mechanical rigidity (1- to 3-mm thick) and are several skin depths thick over the frequency range of interest. The inside wall is smoothly polished to reduce loss. The inside may also be electroplated with silver or gold to improve performance.

The interior dimensions are $a \times b$, where it is customary to label the longer side a. Choice of the a dimension determines the frequency range of the dominant, or lowest order, mode of propagation. Higher order modes have higher attenuation and can be difficult to extract from the guide, so it is usually desirable to operate rectangular waveguide in the lowest propagating mode. The b dimension affects attenuation; smaller b has higher attenuation. It also sets the maximum power capacity of the guide, determining at what level voltage breakdown occurs. However, if the b dimension is increased beyond $a/2$, the next mode will be excited earlier, thus decreasing the useful frequency range. In practice, the b dimension is often chosen to be half that of the a dimension.

Waveguide can support TE and TM modes. In TE modes, the electric field is transverse to the direction of propagation. Some magnetic field component must be in the direction of propagation; otherwise, the mode would be TEM, which as we will see in a moment is unsupported in hollow waveguide. For TM modes, it is the magnetic field that is transverse and an electric field component that must be in the propagation direction.

Why is TEM not supported by rectangular guide? The TEM mode requires at least a pair of conductors to propagate and is therefore not supported by hollow guide like rectangular

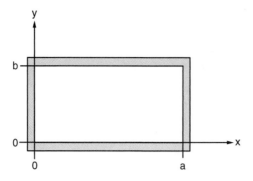

Figure 7.2 Cross section of rectangular waveguide.

waveguide. To see why this is so, let us suppose that a hollow guide does support the TEM mode. By definition, the magnetic field must be entirely in the transverse plane, and from Gauss's law for magnetic fields, $\nabla \cdot \mathbf{B} = 0$, these field lines must form closed loops (see Section 3.5). Now, by Ampère's circuital law,

$$\oint \mathbf{H} \cdot d\mathbf{L} = \int \mathbf{J_c} \cdot d\mathbf{S} + \frac{\partial}{\partial t} \int \mathbf{D} \cdot d\mathbf{S}$$

Since no conductive element can be enclosed in the hollow waveguide, the conduction-current term must be zero. The displacement-current term requires a component of $\mathbf{D}$, and therefore $\mathbf{E}$, in the direction of propagation, that is, normal to the transverse plane. But for TEM mode propagation, the $\mathbf{E}$ must be entirely transverse. Therefore, the TEM mode cannot be supported by hollow waveguide.

The *order* of the mode refers to the field configuration in the guide and is given by m and n integer subscripts, as TE_{mn} and TM_{mn}. The m subscript corresponds to the number of half-wave variations of the field in the x direction, and the n subscript is the number of half-wave variations in the y direction.

In conjunction with the guide dimensions, m and n determine the cutoff frequency[7.1] for a particular mode. We have

$$f_{c_{mn}} = \frac{1}{2\sqrt{\mu\varepsilon}} \sqrt{\left(\frac{m}{a}\right)^2 + \left(\frac{n}{b}\right)^2} \tag{7.1}$$

For conventional rectangular waveguide filled with air, where $a = 2b$, the dominant or lowest order mode is TE_{10} with a cutoff frequency $f_{c_{10}} = c/2a$.

The relative cutoff frequencies for the first 12 modes of this waveguide are shown in Figure 7.3. These cutoff frequencies are ratioed to the dominant mode. Notice that at some frequencies there may exist more than one mode. For instance, owing to this particular waveguide's condition $a = 2b$, the TE_{20} and TE_{01} modes have the same cutoff frequency. Also, the TE_{11} and TM_{11} modes share a cutoff frequency. Notice also that there are no modes where both m and n are zero and also no TM modes with either m or n equal to zero. As will be seen in the next section, the fields for such modes are not supported in waveguide.

[7.1]A particular mode is only supported above its cutoff frequency. Although this would argue for using the term "cut on frequency" instead of cutoff frequency, we will yield to the standard terminology.

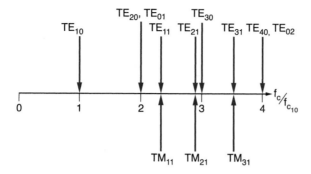

Figure 7.3 Location of modes relative to the dominant TE_{10} mode in standard rectangular waveguide where $a = 2b$.

Table 7.1 lists some of the more common commercially available types of waveguide. The dimensions are given in inches, since the a dimension corresponds to the waveguide's WR (waveguide rectangular) designation number. Notice that the cited useful frequency range starts somewhat above the value of $f_{c_{10}}$, and ends somewhat below the value of the next mode.

The fields within the waveguide will be determined in the next section. But we can look at the field pattern for two modes, TE_{10} and TE_{20}, shown in Figure 7.4. In both cases, **E** only varies in the x direction; since $n = 0$, the field is constant in the y direction. For TE_{10}, the electric field has a half sine wave pattern, whereas for TE_{20} a full sine wave pattern is observed.

▶ **EXAMPLE 7.1**

Let us calculate the cutoff frequency for the first four modes of WR284 waveguide.

From Table 7.1 the guide dimensions are $a = 2.840$ inches and $b = 1.340$ inches. Converting to metric units we have $a = 7.214$ cm and $b = 3.404$ cm.

For $f_{c_{10}}$, (7.1) reduces to

$$f_{c_{10}} = \frac{c}{2a}$$

TABLE 7.1   Some Standard Rectangular Waveguides

Waveguide designation	a (in)	b (in)	t (in)	$f_{c_{10}}$ (GHz)	Frequency range (GHz)
WR975	9.750	4.875	0.125	0.605	0.75–1.12
WR650	6.500	3.250	0.080	0.908	1.12–1.70
WR430	4.300	2.150	0.080	1.375	1.70–2.60
WR284	2.840	1.340	0.080	2.08	2.60–3.95
WR187	1.872	0.872	0.064	3.16	3.95–5.85
WR137	1.372	0.622	0.064	4.29	5.85–8.20
WR90	0.900	0.450	0.050	6.56	8.2–12.4
WR62	0.622	0.311	0.040	9.49	12.4–18

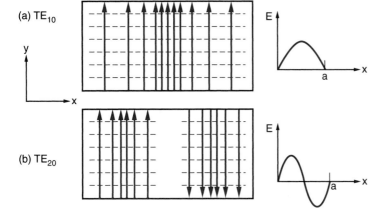

(a) TE$_{10}$

y

x

(b) TE$_{20}$

E

a

x

E

a

x

Figure 7.4 The field patterns and associated field intensities in a cross section of rectangular waveguide for (a) TE$_{10}$ and (b) TE$_{20}$. Solid lines indicate electric field; dashed lines are the magnetic field.

and we have

$$f_{c_{10}} = \frac{3 \times 10^8 \, \text{m/s}}{2(7.214 \, \text{cm})} \frac{100 \, \text{cm}}{1 \, \text{m}} = 2.08 \, \text{GHz}$$

This agrees with the cutoff frequency cited in Table 7.1. Next, we have

$$f_{c_{01}} = \frac{c}{2b} = \frac{3 \times 10^8 \, \text{m/s}}{2(3.404 \, \text{cm})} \frac{100 \, \text{cm}}{1 \, \text{m}} = 4.41 \, \text{GHz}$$

Then,

$$f_{c_{20}} = \frac{c}{a} = 4.16 \, \text{GHz}$$

We notice that $f_{c_{01}}$ is not equal to $f_{c_{20}}$ in this example since $a \neq 2b$.

The fourth mode cutoff frequency is $f_{c_{11}}$. We apply (7.1) and have

$$f_{c_{11}} = \frac{3 \times 10^8 \, \text{m/s}}{2} \sqrt{\left(\frac{1}{7.214 \, \text{cm}}\right)^2 + \left(\frac{1}{3.404 \, \text{cm}}\right)^2} \frac{100 \, \text{cm}}{1 \, \text{m}} = 4.87 \, \text{GHz}$$

Drill 7.1   Calculate the cutoff frequency for the first four modes of WR90 waveguide. (*Answer:* $f_{c_{10}} = 6.56 \, \text{GHz}, f_{c_{20}} = 13.12 \, \text{GHz}, f_{c_{01}} = 13.12 \, \text{GHz}, f_{c_{11}} = 14.67 \, \text{GHz}$)

Wave Propagation

We can achieve a qualitative understanding of wave propagation in waveguide by considering the wave to be a superposition of a pair of TEM waves. Figure 7.5a shows a TEM wave propagating in the z direction. Figure 7.5b shows the *wavefronts*, with bold lines indicating constant phase at the maximum value of the field (+E_o), and lighter lines indicating constant phase at the minimum value (–E_o). The waves propagate at a velocity u_u, where the u subscript indicates media unbounded by guide walls. In air, $u_u = c$.

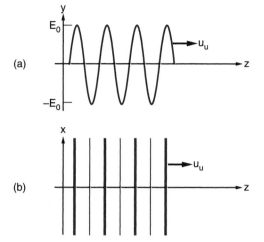

Figure 7.5   (a) A y-polarized TEM plane wave propagates in the $+z$ direction. (b) Wavefront view of the propagating wave.

Now consider a pair of identical TEM waves, labeled as $u+$ and $u-$ in Figure 7.6a. The $u+$ wave is propagating at a $+\theta$ angle to the z axis, whereas the $u-$ wave propagates at a $-\theta$ angle. These waves are combined in Figure 7.6b. Notice that horizontal lines can be drawn on the superposed waves that correspond to zero total field. Along these lines the $u+$ wave is always 180° out of phase with the $u-$ wave.

Since we know $\mathbf{E} = 0$ on a perfect conductor, we can replace the horizontal lines of zero field with perfect conducting walls, as shown in Figure 7.7a. Now, $u+$ and $u-$ are reflected off the walls as they propagate along the guide. The field pattern in this region is identical to what we had in Figure 7.6b.

The distance separating adjacent zero-field lines in Figure 7.6b, or separating the conducting walls in Figure 7.7a, is given as the dimension a in Figure 7.7b. This a distance is

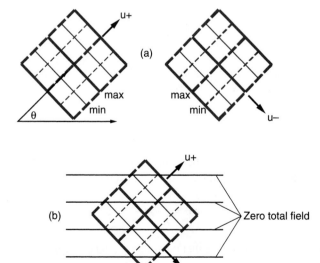

Figure 7.6   We take two identical y-polarized TEM waves, rotate one by $+\theta$ and the other by $-\theta$ as shown in (a), and combine them in (b).

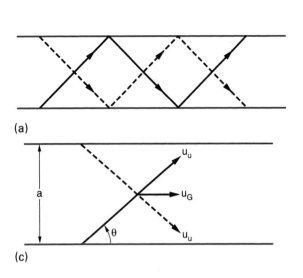

(a)

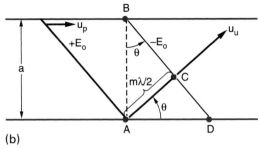

(b)

(c)

Figure 7.7 (a) Replacing adjacent zero field lines with conducting walls, we get an identical field pattern inside. (b) The $u+$ wavefronts for a supported propagation mode are shown for an arbitrary angle θ. (c) The velocity of the superposed fields, or group velocity, is u_G.

determined by the angle θ and by the distance between wavefront peaks, or the wavelength λ. For a given wave velocity u_u, the frequency is $f = u_u/\lambda$.

If we fix the wall separation at a, and change the frequency, we must then also change the angle θ if we are to maintain a propagating wave. Figure 7.7b shows wavefronts for the $u+$ wave. The edge of a $+E_o$ wavefront (point A) will line up with the edge of a $-E_o$ wavefront (point B), and the two fronts must be $\lambda/2$ apart for the $m = 1$ mode. For any value of m, we can write by simple trigonometry

$$\sin\theta = \frac{m\lambda/2}{a} \tag{7.2}$$

or

$$\lambda = \frac{2a}{m}\sin\theta = \frac{u_u}{f} \tag{7.3}$$

The waveguide can support propagation as long as the wavelength is smaller than a critical value that occurs at $\theta = 90°$, given by

$$\lambda_c = \frac{2a}{m} = \frac{u_u}{f_c} \tag{7.4}$$

where f_c is the cutoff frequency for the propagating mode. Combining (7.3) and (7.4) we can relate the angle θ to the operating frequency and the cutoff frequency by

$$\sin\theta = \frac{\lambda}{\lambda_c} = \frac{f_c}{f} \tag{7.5}$$

Examining Figure 7.7b, we see that the time t_{AC} it takes for the wavefront to move from A to C (a distance l_{AC}) is

$$t_{AC} = \frac{l_{AC}}{u_u} = \frac{m\lambda/2}{u_u} \tag{7.6}$$

Meanwhile, a constant phase point moves along the wall from A to D. Calling this phase velocity u_p, and given the distance

$$l_{AD} = \frac{m\lambda/2}{\cos\theta} \tag{7.7}$$

we have the time t_{AD} to travel from A to D of

$$t_{AD} = \frac{l_{AD}}{u_p} = \frac{m\lambda/2}{\cos\theta\, u_p} \tag{7.8}$$

Since the times t_{AD} and t_{AC} must be equal, we have

$$u_p = \frac{u_u}{\cos\theta} \tag{7.9}$$

We can use our θ relation from (7.5) to arrive[7.2] at

$$\boxed{u_p = \frac{u_u}{\sqrt{1-\left(f_c/f\right)^2}}} \tag{7.10}$$

This very interesting result says that the phase velocity can be considerably faster than the velocity of the wave in unbounded media, tending toward infinity as f approaches f_c. Note that nothing is physically moving at this velocity. A good analogy is to consider an ocean wave striking the beach at some slight angle off normal. The point of contact with the beach moves along much faster than the waves move.

The phase constant associated with this phase velocity is

$$\beta = \beta_u \sqrt{1-\left(f_c/f\right)^2} \tag{7.11}$$

where β_u is the phase constant in unbounded media. The wavelength in the guide is related to this phase velocity by $\lambda = 2\pi/\beta$, or

$$\boxed{\lambda = \frac{\lambda_u}{\sqrt{1-\left(f_c/f\right)^2}}} \tag{7.12}$$

The propagation velocity of the superposed wave is given by the *group velocity* u_G. From Figure 7.7c it is apparent that

$$u_G = u_u \cos\theta \tag{7.13}$$

[7.2] $\cos\theta = \sqrt{\cos^2\theta} = \sqrt{1-\sin^2\theta} = \sqrt{1-(f_c/f)^2}$

or

$$u_G = u_u \sqrt{1 - \left(\frac{f_c}{f}\right)^2}$$

(7.14)

This group velocity is slower than that of an unguided wave, which is to be expected since the guided wave propagates in a zig-zag path, bouncing off the waveguide walls.

Drill 7.2 Suppose WR284 is filled with Teflon ($\varepsilon_r = 2.10$). At an operating frequency of 2.00 GHz, find (a) u_u, (b) u_p, and (c) u_G. (*Answer:* (a) 2.07×10^8 m/s, (b) 2.97×10^8 m/s, and (c) 1.44×10^8 m/s)

Waveguide Impedance

The ratio of the transverse electric field to the transverse magnetic field for a propagating mode at a particular frequency is the *waveguide impedance*, also referred to as the *transverse wave impedance*. This can be a useful term for problems involving, for instance, reflection from loads.

For a particular TE mode, the wave impedance is

$$Z_{mn}^{TE} = \frac{\eta_u}{\sqrt{1 - \left(\frac{f_c}{f}\right)^2}}$$

(7.15)

where η_u is the intrinsic impedance of the propagating media. In air, $\eta_u = \eta_o = 120\pi \ \Omega$.

For a TM mode,

$$Z_{mn}^{TM} = \eta_u \sqrt{1 - \left(\frac{f_c}{f}\right)^2}$$

(7.16)

MATLAB 7.1

We want to plot the wave impedance for the TE_{11} and TM_{11} modes of WR90 waveguide versus frequency from 15 to 25 GHz.

```
%   M-File: ML0701
%
%   Waveguide Impedance Plot
%   Plots the impedance for TE11 and TM11 vs freq. for
%   air-filled waveguide
%
%   Wentworth, 11/26/02
%
%   Variables
% Zo                characteristic impedance of air
% ainches,binches   guide dimensions in inches
% a,b               guide dimensions in meters
% fc                TE11 mode cutoff frequency (Hz)
% f                 frequency (Hz)
```

```
% fghz            frequency (GHz)
% Factor          the factor sqrt(1-(fc/f)^2)
% Zof             array filled with Zo
% ZTE,ZT          MTE and TM mode impedance

clc               %clears the command window
clear             %clears variables

%   Initialize variables
c=2.998e8;  %speed of light
Zo=120*pi;
ainches=0.900;binches=0.450;

%   convert to metric
a=ainches*0.0254;b=binches*0.0254;

%   calc fc11
fc=c*sqrt((1/a)^2+(1/b)^2)/2;

%   Perform calculations
f=15e9:.1e9:25e9;
fghz=f/1e9;
Factor=sqrt(1-(fc./f).^2);
Zof=Zo.*f./f;
ZTE=Zo.*Factor;
ZTM=Zo./Factor;

%   Display results
plot(fghz,ZTE,'-.k',fghz,ZTM,'--k',fghz,Zof,'-k')
legend('ZTE11','ZTM11','Zo')
xlabel('frequency, (GHz)')
ylabel('impedance (ohms)')
grid on
```

The plot is shown in Figure 7.8 (after employing the editing function to make the fonts bigger and the lines bolder). Notice that both TE and TM impedances tend toward free-space impedance as the frequency increases.

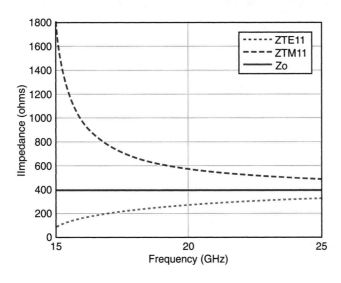

Figure 7.8 Waveguide impedance of the TE_{11} and TM_{11} modes versus frequency for WR90.

▷ **EXAMPLE 7.2**

Let's determine the TE mode impedance looking into a 20.-cm-long section of shorted WR90 waveguide operating at 10. GHz.

At 10 GHz, only the TE_{10} mode is supported. From (7.15) we have

$$Z_{10}^{TE} = \frac{120\pi \ \Omega}{\sqrt{1 - \left(\dfrac{6.56 \ \text{GHz}}{10 \ \text{GHz}}\right)^2}} = 500 \ \Omega$$

We find Z_{IN} using (6.56),

$$Z_{IN} = jZ_o \tan(\beta\ell)$$

for a shorted line. Now, β is found from (7.11) as

$$\beta = \beta_u \sqrt{1 - \left(\frac{f_c}{f}\right)^2} = \frac{2\pi f}{c}\sqrt{1 - \left(\frac{f_c}{f}\right)^2}$$

$$= \frac{2\pi\left(10 \times 10^9 \ \text{Hz}\right)}{3 \times 10^8 \ \text{m}/_s}\sqrt{1 - \left(\frac{6.56 \ \text{GHz}}{10 \ \text{GHz}}\right)^2} = 158 \ \text{radians}/\text{m}$$

so $\beta\ell$ in the Z_{IN} equation is

$$\beta\ell = \left(158\frac{\text{rad}}{\text{m}}\right)(0.2 \ \text{m}) = 31.6 \ rad$$

and Z_{IN} is then calculated as

$$Z_{IN} = j(500 \ \Omega)\tan(31.6) = j100 \ \Omega$$

Drill 7.3 Repeat Example 7.2 if the 20-cm line is terminated in a 50-Ω load instead of a short. (*Answer:* $Z_{IN} = 52 + j100 \ \Omega$).

Practical Application: Microwave Ovens

The ubiquitous microwave oven owes its existence to the invention of the cavity magnetron as a compact source for radar in World War II. The magnetron converts DC power to 2.45-GHz microwave radiation that penetrates food and attempts to make the water molecules flip at this frequency, thus delivering the radiation as heat. Because of radiation leakage generating noise to the spectrum, 2.45 GHz is set aside specifically for microwave oven use. It is a convenient frequency that is high enough to be absorbed by the food, yet low enough that the radiation passes right through glass and plastic and penetrates well into the food. There is a misconception that "2.45 GHz" corresponds to a natural resonance frequency of the water molecule. It does not.

The oven itself (Figure 7.9) is a metal box that contains the microwave radiation, with the food typically supported by a rotating glass carousel for even heating. The small holes in the door of the oven are too small to permit significant leakage of microwave power.

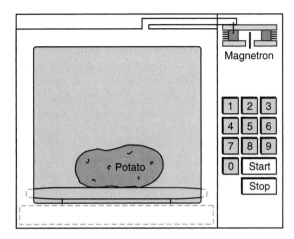

Figure 7.9 Depiction of a microwave oven.

Microwave energy from the magnetron is tapped by a small probe that feeds a short section of waveguide. The radiation then passes into the oven.

A magnetron is depicted in Figure 7.10a. A 3-kV potential difference establishes a strong electric field between the cathode and anode. The hot cathode emits electrons that accelerate toward the anode. The paths of these electrons are curved by the presence of a magnetic field supplied by permanent magnets (recall the Lorentz force equation). As shown in Figure 7.10b, six to eight tuned cavities surround the main chamber and are connected to it by narrow slots. Alternate segments between cavities are connected with metal straps. The strapped-together cavities form a parallel combination of resonators with an overall 2.45-GHz resonant frequency, set by the dimensions of the cavities and slots. The

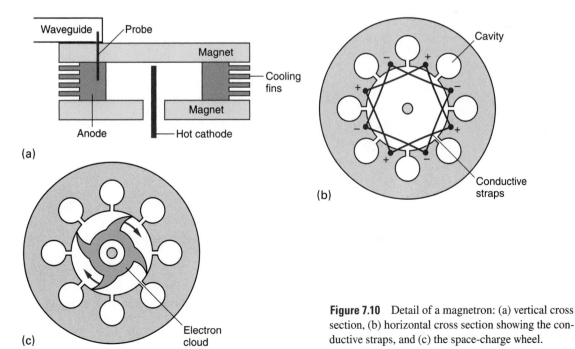

Figure 7.10 Detail of a magnetron: (a) vertical cross section, (b) horizontal cross section showing the conductive straps, and (c) the space-charge wheel.

DC power is converted to microwave energy that rotates within the magnetron as a *space-charge wheel*, the electron cloud depicted in Figure 7.10c.

It is commonly believed that metal cannot be placed within a microwave oven. Thin metals (aluminum foil) or metals with sharp corners (forks) will spark as the induced currents have no place to go, but smooth objects like spoons are okay. Finally, it is a bad idea to operate the oven without a dissipative load (i.e., food) present. The power cannot dissipate and will make its way back to the magnetron, possibly causing damage.

▶ 7.2 WAVEGUIDE FIELD EQUATIONS

Beginning with Maxwell's equations, we want to develop the time-harmonic field equations for rectangular waveguide. For simplicity, we will consider the guide to be filled with a lossless, charge-free media and the walls to be perfect conductors. The phasor form of Maxwell's equations becomes

$$\nabla \cdot \mathbf{E}_s = 0$$

$$\nabla \cdot \mathbf{H}_s = 0$$

$$\nabla \times \mathbf{E}_s = -j\omega\mu\mathbf{H}_s$$

$$\nabla \times \mathbf{H}_s = j\omega\varepsilon\mathbf{E}_s$$

(7.17)

For the waveguide cross section of Figure 7.2, the field components in Cartesian coordinates are

$$\mathbf{E}_s = E_{xs}\mathbf{a}_x + E_{ys}\mathbf{a}_y + E_{zs}\mathbf{a}_z$$

$$\mathbf{H}_s = H_{xs}\mathbf{a}_x + H_{ys}\mathbf{a}_y + H_{zs}\mathbf{a}_z$$

(7.18)

Inserting (7.18), we can expand (7.17) to a set of eight equations. The four of these we will need are

$$\frac{\partial E_{zs}}{\partial y} - \frac{\partial E_{ys}}{\partial z} = -j\omega\mu H_{xs}$$

(7.19)

$$\frac{\partial E_{xs}}{\partial z} - \frac{\partial E_{zs}}{\partial x} = -j\omega\mu H_{ys}$$

(7.20)

$$\frac{\partial H_{zs}}{\partial y} - \frac{\partial H_{ys}}{\partial z} = j\omega\varepsilon E_{xs}$$

(7.21)

$$\frac{\partial H_{xs}}{\partial z} - \frac{\partial H_{zs}}{\partial x} = j\omega\varepsilon E_{ys}$$

(7.22)

Now consider that the fields only propagate in the +z direction with velocity u_G and associated phase constant β. We have, for instance,

$$E_{xs} = E_x e^{-j\beta z}$$

It may be noted that although the phasor E_{xs} is a function of position (x, y, z), E_x is a phasor that is only a function of x and y. To indicate the difference, and for brevity, we drop the s subscript. The partial derivative of E_{xs} with respect to z is

$$\frac{\partial E_{xs}}{\partial z} = -j\beta E_x e^{-j\beta z}$$

The other five field components and their partial derivatives with respect to z can be written in a similar way. Since the $e^{-j\beta z}$ term will be present for each component, it can be eliminated from the equations, which become

$$\frac{\partial E_z}{\partial y} + j\beta E_y = -j\omega\mu H_x \qquad (7.23)$$

$$-j\beta E_x - \frac{\partial E_z}{\partial x} = -j\omega\mu H_y \qquad (7.24)$$

$$\frac{\partial H_z}{\partial y} + j\beta H_y = j\omega\varepsilon E_x \qquad (7.25)$$

$$-j\beta H_x - \frac{\partial H_z}{\partial x} = j\omega\varepsilon E_y \qquad (7.26)$$

Using these equations, we can find expressions for the four transverse components (E_x, E_y, H_x, and H_y) in terms of the z-directed components (E_z and H_z). For instance, solving (7.23) for H_x we find

$$H_x = \frac{j}{\omega\mu}\frac{\partial E_z}{\partial y} - \frac{\beta}{\omega\mu}E_y$$

Inserting this value of H_x into (7.26), we can solve for E_y as

$$E_y = \frac{j\omega\mu\dfrac{\partial H_z}{\partial x} - j\beta\dfrac{\partial E_z}{\partial y}}{\beta_u^2 - \beta^2} \qquad (7.27)$$

where, for lossless propagating media, we have

$$\beta_u = \omega\sqrt{\mu\varepsilon}$$

If we solve (7.26) for E_y and insert it into (7.23), we have

$$H_x = \frac{j\omega\varepsilon\dfrac{\partial E_z}{\partial y} - j\beta\dfrac{\partial H_z}{\partial x}}{\beta_u^2 - \beta^2} \qquad (7.28)$$

In a similar manner, using (7.24) and (7.25) we would find

$$E_x = \frac{-j\omega\mu\dfrac{\partial H_z}{\partial y} - j\beta\dfrac{\partial E_z}{\partial x}}{\beta_u^2 - \beta^2} \qquad (7.29)$$

and

$$H_y = \frac{-j\omega\varepsilon\dfrac{\partial E_z}{\partial x} - j\beta\dfrac{\partial H_z}{\partial y}}{\beta_u^2 - \beta^2} \tag{7.30}$$

Now, if we consider a TM mode, then $H_z = 0$. We would solve for E_z, then use (7.27) through (7.30) to find the transverse components. Likewise, for the TE mode, we would solve for H_z before finding the transverse components.

TM Mode

We'll first look at the TM mode, where $H_z = 0$, and find an expression for E_z. The Helmholtz equation (5.11) for propagation of the electric field in a lossless medium can be written

$$\nabla^2 \mathbf{E}_s + \beta_u^2 \mathbf{E}_s = 0 \tag{7.31}$$

Expanding this equation for our z-propagating fields we have

$$\frac{\partial^2 E_z}{\partial x^2} + \frac{\partial^2 E_z}{\partial y^2} + \left(\beta_u^2 - \beta^2\right)E_z = 0 \tag{7.32}$$

To solve this equation, we employ the method of *separation of variables*, by assuming

$$E_z(x, y) = XY \tag{7.33}$$

Here, E_z can be expressed as the product of a function X, which only depends on x, and a function Y, which only depends on y. Using (7.33) to expand (7.32), we find

$$Y\frac{d^2 X}{dx^2} + X\frac{d^2 Y}{dy^2} + \left(\beta_u^2 - \beta^2\right)XY = 0 \tag{7.34}$$

Dividing both sides of (7.34) by XY and rearranging, we obtain

$$\beta^2 = \beta_u^2 + \frac{1}{X}\frac{d^2 X}{dx^2} + \frac{1}{Y}\frac{d^2 Y}{dy^2} \tag{7.35}$$

Notice that the second term on the right side of (7.35) only depends on x, and the third term only depends on y. For this equation to be true for all values of x and y, each of these terms must be a constant. It will be convenient to express the constants as

$$\beta_x^2 = -\frac{1}{X}\frac{d^2 X}{dx^2} \tag{7.36}$$

and

$$\beta_y^2 = -\frac{1}{Y}\frac{d^2 Y}{dy^2} \tag{7.37}$$

and (7.35) becomes

$$\beta = \sqrt{\beta_u^2 - \beta_x^2 - \beta_y^2} \tag{7.38}$$

We can use (7.36) to solve for X and then employ our boundary conditions at the $x = 0$ and $x = a$ walls of the waveguide to find β_x. Likewise, we can solve for Y from (7.37) and use the boundary conditions at $y = 0$ and $y = b$ to solve for β_y.

First, we have

$$\frac{d^2 X}{dx^2} + \beta_x^2 X = 0$$

This differential equation has the general solution

$$X = c_1 \cos(\beta_x x) + c_2 \sin(\beta_x x)$$

where c_1 and c_2 are constants.

Now, we know that the tangential electric fields at the walls of the waveguide must be zero. This means the function X must equal zero for $x = 0$ and $x = a$. Applying $X = 0$ at $x = 0$, we immediately see that $c_1 = 0$. Also, since $X = 0$ at $x = a$, we have

$$0 = c_2 \sin(\beta_x a)$$

which is true whenever

$$\beta_x a = m\pi$$

($m = 0, 1, 2, 3,...$). Therefore, we have

$$\beta_x = \frac{m\pi}{a} \tag{7.39}$$

Likewise,

$$\frac{d^2 Y}{dy^2} + \beta_y^2 Y = 0$$

has the general solution

$$Y = c_3 \cos(\beta_y y) + c_4 \sin(\beta_y y)$$

And since $Y = 0$ at $y = 0$ and $y = b$, we realize $c_3 = 0$ and

$$\beta_y = \frac{n\pi}{b} \tag{7.40}$$

($n = 0, 1, 2, 3,...$). The waveguide phase constant is then seen to be

$$\boxed{\beta = \sqrt{\beta_u^2 - \left(\frac{m\pi}{a}\right)^2 - \left(\frac{n\pi}{b}\right)^2}} \tag{7.41}$$

It may be noted that as long as the argument inside the square root of (7.41) is positive, then propagation will proceed in the z direction. Manipulation of (7.41) directly results in (7.1) (see Problem 7.11).

The general solution for the z-directed electric field for TM mode propagation is therefore

$$\boxed{E_{zs} = E_o \sin\left(\frac{m\pi x}{a}\right) \sin\left(\frac{n\pi y}{b}\right) e^{-j\beta z}} \tag{7.42}$$

where E_o is the product of the c_2 and c_4 constants.

We can now find the transverse field components by using (7.27) to (7.30) and reinserting the $e^{-j\beta z}$ term. Evaluating the derivative of (7.42) with respect to y, we find

$$\frac{\partial E_{zs}}{\partial y} = \left(\frac{n\pi}{b}\right) E_o \sin\left(\frac{m\pi x}{a}\right)\cos\left(\frac{n\pi y}{b}\right)e^{-j\beta z}$$

Inserting this into (7.27) and (7.28) (with $H_z = 0$), we have

$$E_{ys} = \frac{-j\beta}{\beta_u^2 - \beta^2}\left(\frac{n\pi}{b}\right) E_o \sin\left(\frac{m\pi x}{a}\right)\cos\left(\frac{n\pi y}{b}\right)e^{-j\beta z} \tag{7.43}$$

and

$$H_{xs} = \frac{j\omega\varepsilon}{\beta_u^2 - \beta^2}\left(\frac{n\pi}{b}\right) E_o \sin\left(\frac{m\pi x}{a}\right)\cos\left(\frac{n\pi y}{b}\right)e^{-j\beta z} \tag{7.44}$$

The derivative with respect to x is

$$\frac{\partial E_{zs}}{\partial x} = \left(\frac{m\pi}{a}\right) E_o \cos\left(\frac{m\pi x}{a}\right)\sin\left(\frac{n\pi y}{b}\right)e^{-j\beta z}$$

Inserting this derivative into (7.29) and (7.30), we find

$$E_{xs} = \frac{-j\beta}{\beta_u^2 - \beta^2}\left(\frac{m\pi}{a}\right) E_o \cos\left(\frac{m\pi x}{a}\right)\sin\left(\frac{n\pi y}{b}\right)e^{-j\beta z} \tag{7.45}$$

and

$$H_{ys} = \frac{-j\omega\varepsilon}{\beta_u^2 - \beta^2}\left(\frac{m\pi}{a}\right) E_o \cos\left(\frac{m\pi x}{a}\right)\sin\left(\frac{n\pi y}{b}\right)e^{-j\beta z} \tag{7.46}$$

Inspection of these TM field components shows that if either m or n is equal to zero, then all the fields will be zero as well. Therefore, the TM_{11} mode is the first viable TM mode.

Let's find the instantaneous expressions for the TM_{11} mode for an air-filled waveguide. The first component is found by applying

$$E_z(x, y, z, t) = \text{Re}[E_{zs}e^{j\omega t}] \tag{7.47}$$

to (7.42). We find

$$E_z(x, y, z, t) = E_o \sin\left(\frac{\pi x}{a}\right)\sin\left(\frac{\pi y}{b}\right)\cos(\omega t - \beta z) \tag{7.48}$$

To find $E_y(x, y, z, t)$, we first consider that $-j$ can be written as $e^{-j90°}$. After reinserting $e^{j\omega t}$ and taking the real part, we then have the term $\cos(\omega t - \beta z - 90°)$, which is equal to $\sin(\omega t - \beta z)$. So we have

$$E_y(x,y,z,t) = \frac{\beta}{\beta_u^2 - \beta^2}\frac{\pi}{b}E_o\sin\left(\frac{\pi x}{a}\right)\cos\left(\frac{\pi y}{b}\right)\sin(\omega t - \beta z) \qquad (7.49)$$

The other components are found similarly to be

$$H_x(x,y,z,t) = \frac{-\omega\varepsilon}{\beta_u^2 - \beta^2}\frac{\pi}{b}E_o\sin\left(\frac{\pi x}{a}\right)\cos\left(\frac{\pi y}{b}\right)\sin(\omega t - \beta z) \qquad (7.50)$$

$$E_x(x,y,z,t) = \frac{\beta}{\beta_u^2 - \beta^2}\frac{\pi}{a}E_o\cos\left(\frac{\pi x}{a}\right)\sin\left(\frac{\pi y}{b}\right)\sin(\omega t - \beta z) \qquad (7.51)$$

$$H_y(x,y,z,t) = \frac{\omega\varepsilon}{\beta_u^2 - \beta^2}\frac{\pi}{a}E_o\cos\left(\frac{\pi x}{a}\right)\sin\left(\frac{\pi y}{b}\right)\sin(\omega t - \beta z) \qquad (7.52)$$

The TM$_{11}$ field patterns are shown in Figure 7.11.

▶ **MATLAB 7.2**

This program displays the TM$_{11}$ E_z field pattern inside the waveguide. The results are for a generic rectangular waveguide and are normalized. This routine shows how to make a contour plot as well as a three-dimensional surface plot. The results are given in Figure 7.12. A modified version of this routine was employed to generate Figure 7.11.

```
%   M-File: ML0702
%
%   TM11 Ez Field Pattern
%   Generates contour and surface plots
%
%   Wentworth, 11/26/02
%
%   Variables
%   m,n          mode indicators
%   a,b          unitless guide dimensions
%   betax        x component of phase constant
%   betay        y component of phase constant
%   Ez           Ez for contour plot
%   Ezc          Ez for conventional plot
%   Ezs          Ez for surface plot

clc            %clears the command window
clear          %clears variables

%   Initialize variables
m=1;n=1;
a=40;b=20;
betax=m*pi/a;
betay=n*pi/b;

%   Generate data for contour plot
for i=1:a/40:a
    x(i)=(i/40)*a;
```

(continues)

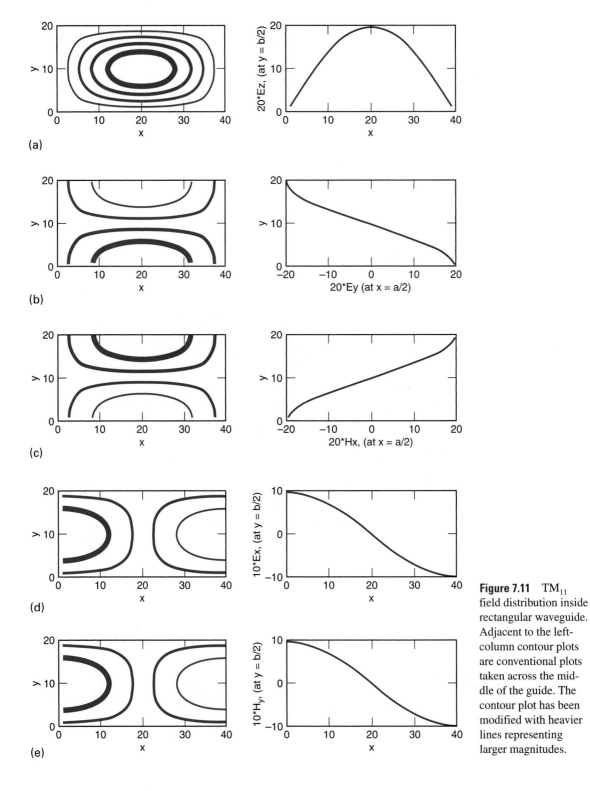

Figure 7.11 TM$_{11}$ field distribution inside rectangular waveguide. Adjacent to the left-column contour plots are conventional plots taken across the middle of the guide. The contour plot has been modified with heavier lines representing larger magnitudes.

```
      for j=1:b/20:b
      y(j)=(j/20)*b;
      Ez(j,i)=sin(betax*x(i))*sin(betay*y(j));
      end
end

%   Generate data for conventional plot at
%   y=b/2
yc=b/2;
xc=1:a/40:a;
Ezc=20*sin(betax*xc)*sin(betay*yc);

%   Generate data for surface plot
[X,Y]=meshgrid(0:a,0:b);
Ezs=10*sin(betax.*X).*sin(betay.*Y);

subplot(3,1,1)
contour(x,y,Ez,4)
title('Ez')
ylabel('y')
axis('equal')
axis([0 a 0 b])

subplot(3,1,2)
plot(xc,Ezc)
xlabel('x')
ylabel('20*Ez, (at y=b/2)')
axis('equal')
axis([0 a 0 b])subplot(3,1,3)
surf(Ezs)
axis([0 40 0 20 0 10])
axis('equal')
xlabel('x')
ylabel('y')
zlabel('Ezs*10')
```

It may be noted that multiplicative factors were inserted to make the plots scale properly.

TE Mode

Solution of the TE case proceeds exactly as the TM case up to the point where boundary conditions are applied. We begin with the Helmholtz wave equation

$$\nabla^2\mathbf{H}_s + \beta_u^2\mathbf{H}_s = 0 \tag{7.53}$$

and eventually reach the expression

$$H_{zs} = XYe^{-j\beta z} \tag{7.54}$$

where

$$X = c_1\cos(\beta_x x) + c_2\sin(\beta_x x) \tag{7.55}$$

and

$$Y = c_3\cos(\beta_y y) + c_4\sin(\beta_y y) \tag{7.56}$$

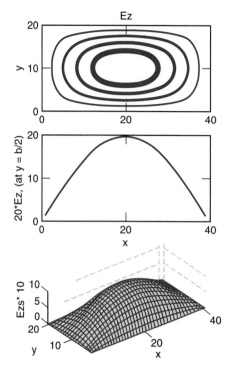

Figure 7.12 The TM$_{11}$ E_z plots of MATLAB 7.2. This is a black and white rendition of plots that will appear in color when you run the program. The contour plot has been modified with heavier lines representing larger magnitudes.

We again apply the boundary condition that tangential E must be zero at the conductive walls. This means that for $x = 0$ and $x = a$, $E_y = 0$. Since $E_y = 0$ and $E_z = 0$ for TE modes, it is apparent from (7.27) that at these two boundaries for x we must have

$$\frac{\partial H_z}{\partial x} = 0 \tag{7.57}$$

Since the only portion of H_{zs} that varies with x in (7.54) is the X part, we can apply our $x = 0$ boundary condition to (7.55) to get

$$\left[\frac{dX}{dx}\right]_{x=0} = -\beta_x c_1 \sin(\beta_x x) + \beta_x c_2 \cos(\beta_x x) = 0 \tag{7.58}$$

This is true only for $c_2 = 0$. Applying the $x = a$ boundary condition to (7.55), we have

$$\left[\frac{dX}{dx}\right]_{x=a} = -\beta_x c_1 \sin(\beta_x a) = 0 \tag{7.59}$$

which is true for $\beta_x a = m\pi$ ($m = 0, 1, 2, 3,...$). This gives the same value for β_x that we have for the TM case.

Since $c_2 = 0$, we have

$$\frac{dX}{dx} = -\beta_x c_1 \sin(\beta_x x) \tag{7.60}$$

After integrating,[7.3] we have

$$x = c_1 \cos(\beta_x x) \tag{7.61}$$

At $y = 0$ and $y = b$, the boundary conditions are that $E_x = 0$. Using this in conjunction with (7.29) we find at these boundaries that

$$\frac{\partial H_z}{\partial y} = 0 \tag{7.62}$$

Since the only portion of H_z that varies with y is Y, from (7.56) we find for the first boundary condition

$$\left[\frac{dY}{dy} \right]_{y=0} = -\beta_y c_3 \sin(\beta_y y) + \beta_y c_4 \cos(\beta_y y) = 0 \tag{7.63}$$

From this we see that $c_4 = 0$. Finally,

$$\left[\frac{dY}{dy} \right]_{y=b} = -\beta_y c_3 \sin(\beta_y y) = 0 \tag{7.64}$$

means that $\beta_y b = n\pi$ ($n = 0, 1, 2, 3,\ldots$). This gives the same value for β_y found for the TM case.

So for

$$\frac{dY}{dy} = -\beta_y c_3 \sin(\beta_y y) \tag{7.65}$$

we integrate to get

$$Y = c_3 \cos(\beta_y y) \tag{7.66}$$

The z-directed magnetic field is therefore

$$H_{zs} = H_o \cos\left(\frac{m\pi x}{a} \right) \cos\left(\frac{n\pi y}{b} \right) e^{-j\beta z} \tag{7.67}$$

where H_o is the product of c_1 and c_3.

The other field components, from (7.27) to (7.30), are as follows:

$$E_{ys} = \frac{-j\omega\mu}{\beta_u^2 - \beta^2} \frac{m\pi}{a} H_o \sin\left(\frac{m\pi x}{a} \right) \cos\left(\frac{n\pi y}{b} \right) e^{-j\beta z} \tag{7.68}$$

$$H_{xs} = \frac{-j\beta}{\beta_u^2 - \beta^2} \frac{m\pi}{a} H_o \sin\left(\frac{m\pi x}{a} \right) \cos\left(\frac{n\pi y}{b} \right) e^{-j\beta z} \tag{7.69}$$

[7.3]We can safely ignore the constant of integration.

$$E_{xs} = \frac{j\omega\mu}{\beta_u^2 - \beta^2} \frac{n\pi}{b} H_o \cos\left(\frac{m\pi x}{a}\right) \sin\left(\frac{n\pi y}{b}\right) e^{-j\beta z} \tag{7.70}$$

$$H_{ys} = \frac{j\beta}{\beta_u^2 - \beta^2} \frac{n\pi}{b} H_o \cos\left(\frac{m\pi x}{a}\right) \sin\left(\frac{n\pi y}{b}\right) e^{-j\beta z} \tag{7.71}$$

As with the TM case, we see that if both m and n are zero then all the fields disappear. But we do have fields if only one of m or n is zero. For instance, for the TE_{10} mode, since $n = 0$ there will be no E_{xs} or H_{ys}. The instantaneous expressions for the rest of the fields for the TE_{10} mode are as follows:

$$H_z(x, y, z, t) = H_o \cos\left(\frac{\pi x}{a}\right) \cos(\omega t - \beta z) \tag{7.72}$$

$$E_y(x, y, z, t) = \frac{\omega\mu}{\beta_u^2 - \beta^2} \frac{\pi}{a} H_o \sin\left(\frac{\pi x}{a}\right) \sin(\omega t - \beta z) \tag{7.73}$$

$$H_x(x, y, z, t) = \frac{\beta}{\beta_u^2 - \beta^2} \frac{\pi}{a} H_o \sin\left(\frac{\pi x}{a}\right) \sin(\omega t - \beta z) \tag{7.74}$$

Normalized values of these fields in a cross section of guide are shown in Figure 7.13.

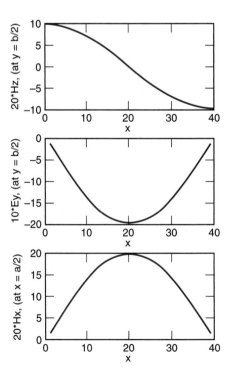

Figure 7.13 TE_{10} field plots are constant in the y direction.

Drill 7.4 What fields are present for the (a) TE$_{01}$ mode, (b) TM$_{10}$ mode, and (c) TM$_{11}$ mode? (*Answer:* (a) H_z, E_x, and H_y, (b) none, (c) E_z, E_y, H_x, E_x, and H_y)

▷ 7.3 DIELECTRIC WAVEGUIDE

Metallic rectangular waveguide finds practical application at microwave frequencies typically between 1 and 40 GHz. However, as we have seen, the dimensions of rectangular waveguide operating in the TE$_{10}$ mode must be on the same order as the wavelength, and at higher frequencies it becomes impractical to fabricate such small guide. For instance, at an optical frequency of 300 THz, the wavelength is only 1 μm. Making a metallic waveguide at such dimensions would be a serious challenge even for MEMS technology. It is unlikely that such waveguide would be desirable even if it could be built, because ohmic losses in the diminishing conductive skin of the walls would be excessive.

Dielectric waveguide overcomes these problems. Before turning our attention to fiber optic waveguide, we will first consider the simpler case of a rectangular slab of waveguide as shown in Figure 7.1c. Such structures are the basis for the planar lightguides used in integrated optical circuits. Also, a background in how such guides operate will aid us in understanding the more complicated propagation characteristics of optical fiber.

Wave propagation in rectangular dielectric waveguide can be analyzed via Maxwell's equations and application of suitable boundary conditions. One difficulty is that, unlike metallic rectangular waveguide, fields also exist outside of the dielectric guide. The mathematical treatment is therefore somewhat more complicated than that for the rectangular waveguide field equations we derived in Section 7.2. Some of the dielectric waveguide field equations will be given (with no derivation) at the end of this section, but now we will analyze the guide another way.

In Chapter 5 we studied the reflection and transmission of electromagnetic waves incident at an oblique angle from one dielectric medium to another. The incident, reflected, and transmitted angles, θ_i, θ_r, and θ_t, are related by Snell's laws as

$$\theta_i = \theta_r \qquad \text{(Snell's law of reflection)}$$

$$\frac{\beta_1}{\beta_2} = \frac{\sin \theta_t}{\sin \theta_i} \quad \text{(Snell's law of refraction)}$$

where

$$\beta_1 = \omega\sqrt{\mu_1 \varepsilon_1}, \quad \beta_2 = \omega\sqrt{\mu_2 \varepsilon_2}$$

If we consider nonmagnetic media where the wave is incident from the higher to lower permittivity medium, then the transmitted wave bends more sharply than the incident one (see Figure 7.14a), that is, $\theta_t > \theta_i$. If the angle of incidence is increased sufficiently, a *critical angle* is reached whereby the incident wave is completely reflected (Figure 7.14b). The transmitted wave rapidly attenuates in the second medium.

The critical angle, in terms of the relative permittivity of each medium, can be written

$$\left(\theta_i\right)_{\text{critical}} = \sin^{-1}\left(\frac{\sqrt{\varepsilon_{r_2}}}{\sqrt{\varepsilon_{r_1}}}\right) \tag{7.75}$$

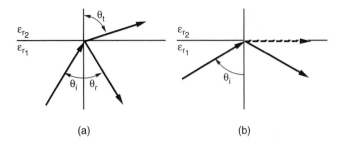

Figure 7.14 (a) A wave incident at an angle θ_i from ε_{r_1} material to ε_{r_2} material ($\varepsilon_{r_1} > \varepsilon_{r_2}$). (b) A critical angle for θ_i is reached where the entire wave is reflected.

(a)   (b)

It is convenient and customary to cite a dielectric's *index of refraction n*, especially when dealing with optical problems. This index is the ratio of the speed of light in vacuum to the speed of light in the unbounded medium:

$$n = \frac{c}{u_u} \tag{7.76}$$

In nonmagnetic material, this can be written

$$n = \sqrt{\varepsilon_r} \tag{7.77}$$

so we have

$$\boxed{(\theta_i)_{critical} = \sin^{-1}\left(\frac{n_2}{n_1}\right)} \tag{7.78}$$

Only incident angles larger than this critical angle will result in propagating modes. Snell's law of refraction for this nonmagnetic media case can be rewritten

$$\boxed{\frac{n_1}{n_2} = \frac{\sin \theta_t}{\sin \theta_i}} \tag{7.79}$$

Drill 7.5 A slab of dielectric with index of refraction 3.00 is suspended in air. What is the relative permittivity of the dielectric? At what angle from a normal to the boundary will light be totally reflected within the dielectric? (*Answer:* 9, 19.5°)

As with the metallic waveguide case, constructive addition of reflected and rereflected waves is required for propagation. But now, at the guide wall, the electric field is not fixed at zero and the reflected wave encounters a phase shift that is a function of θ_i.

Consider Figure 7.15a, where a wavefront is shown just before striking the wall of the dielectric guide at point A. It will be reflected at A, rereflected at B, and when it reaches point C its phase must be some integral multiple of 2π radians from what it was just prior to striking point A. The phase at point C must match the phase at point A. If we call the wave just before it strikes the wall at point A

$$E_{A-} = E_o$$

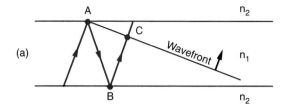

(a)

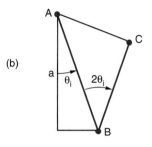

(b)

Figure 7.15 (a) The wavefront for a
supported propagation mode must have
the same phase at points A and C.
(b) An expanded view of the problem's
geometry.

then just after striking the wall at A we have

$$E_{A+} = \Gamma E_{A-} = |\Gamma| E_o e^{j\phi}$$

The particular reflection coefficient and its phase are functions of the type of wave we are considering. We'll return to this in a moment.

Just prior to striking the wall at point B, the wave has had to travel a length l_{AB}, so we have

$$E_{B-} = E_{A+} e^{-j\beta_1 l_{AB}}$$

After striking this second wall,

$$E_{B+} = \Gamma E_{B-} = |\Gamma|^2 E_o e^{-j\beta_1 l_{AB}} e^{j2\phi}$$

Finally, at point C we have

$$E_C = |\Gamma|^2 E_o e^{-j\beta_1(l_{AB} + l_{BC})} e^{j2\phi}$$

The phase of E_C must be equal to zero or an integral multiple of 2π radians to satisfy the phase-matching restriction, so we have

$$\beta_1(l_{AB} + l_{BC}) - 2\phi = 2\pi m \qquad (7.80)$$

(where $m = 0, 1, 2, 3,...$).

Figure 7.15b shows segments l_{AB} and l_{BC} in more detail. Here we see that

$$l_{AB} = \frac{a}{\cos\theta_i}$$

and

$$l_{BC} = l_{AB} \cos 2\theta_i$$

Adding these two lengths, and invoking the half-angle formula,[7.4] we find

$$l_{AB} + l_{BC} = 2a \cos\theta_i$$

[7.4]$\cos^2\theta = (1/2)(\cos 2\theta + 1)$

Our phase relationship becomes

$$\beta_1 2a \cos\theta_i - 2\phi = 2\pi m \tag{7.81}$$

The phase shift upon reflection at a wall will depend on the angle of incidence and on the type of wave (TE or TM).

TE Mode

Considering first the TE wave, we have from Chapter 5

$$\Gamma = \frac{\eta_2 \cos\theta_i - \eta_1 \cos\theta_t}{\eta_2 \cos\theta_i + \eta_1 \cos\theta_t} \tag{7.82}$$

For nonmagnetic media this can be rewritten as

$$\Gamma = \frac{\eta_1 \cos\theta_i - \eta_2 \cos\theta_t}{\eta_1 \cos\theta_i + \eta_2 \cos\theta_t} \tag{7.83}$$

Using Snell's law of refraction in this equation, we can show that

$$\Gamma_{TE} = \frac{\cos\theta_i + j\sqrt{\sin^2\theta_i - (n_2/n_1)^2}}{\cos\theta_i - j\sqrt{\sin^2\theta_i - (n_2/n_1)^2}} \tag{7.84}$$

The magnitude of Γ_{TE} is unity and

$$\phi_{TE} = 2\tan^{-1}\left(\frac{\sqrt{\sin^2\theta_i - (n_2/n_1)^2}}{\cos\theta_i}\right) \tag{7.85}$$

Inserting this phase into (7.81), and rearranging, we have

$$\tan\left(\frac{a\beta_1\cos\theta_i}{2} - \frac{m\pi}{2}\right) = \frac{\sqrt{\sin^2\theta_i - (n_2/n_1)^2}}{\cos\theta_i} \tag{7.86}$$

This transcendental equation cannot be solved analytically. Instead, we can find a graphical solution. As an example, let's consider a 50-mm-thick slab of $\varepsilon_{r_1} = 4$ ($n_1 = 2$) dielectric sandwiched by air. For a 4.5-GHz operating frequency, we want to evaluate (7.86) for all possible values of m over all possible incident angles.

First, we can use (7.78) to calculate a critical angle of 30° for this waveguide, so we will plot over a range from 90° down to this critical angle. Then, inserting the appropriate values into (7.86), we can generate the plot shown in Figure 7.16a. The intersection of the right side of (7.86) with the left side gives the allowed incidence angle for the mth mode. We see that for this example only three TE modes are possible: TE$_0$ at $\theta_i = 74.4°$, TE$_1$ at $\theta_i = 57.9°$, and TE$_2$ at $\theta_i = 39.8°$.

It is instructive to show what happens as the frequency is varied. In Figure 7.16b, the left side of (7.86) for $m = 0$ is plotted at several frequencies. We notice that there is no upper

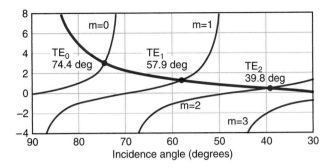

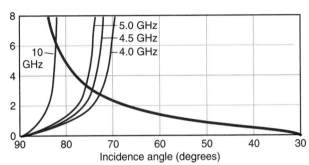

Figure 7.16 (a) The dielectric waveguide TE modes for a 50-mm-thick dielectric of $\varepsilon_r = 4$ operating at 4.5 GHz. The bold line plots the value of the right side of (7.86) on the vertical axis against the angle. The other lines plot the value of the left side of (7.86) on the vertical axis versus angle for different values of m. (b) TE mode plots at $m = 0$ for several different frequencies.

frequency limit. As frequency is increased, the angle of incidence becomes closer and closer to 90°. We can also intuit from this figure that decreasing the frequency will result in fewer supported propagating modes in the guide. This is confirmed by studying (7.86). Note, however, that there will always be a TE_0 mode.

TM Mode

Finding the TM modes proceeds in a similar manner. From Chapter 5,

$$\Gamma_{TM} = \frac{\eta_2 \cos\theta_t - \eta_1 \cos\theta_i}{\eta_2 \cos\theta_t + \eta_1 \cos\theta_i} \tag{7.87}$$

which can be written, for nonmagnetic media,

$$\Gamma_{TM} = \frac{\eta_1 \cos\theta_t - \eta_2 \cos\theta_i}{\eta_1 \cos\theta_t + \eta_2 \cos\theta_i} \tag{7.88}$$

This leads to the phase expression

$$\tan\left(\frac{a\beta_1 \cos\theta_i}{2} - \frac{m\pi}{2}\right) = \frac{\sqrt{\sin^2\theta_i - (n_2/n_1)^2}}{(n_2/n_1)^2 \cos\theta_i} \tag{7.89}$$

The TM modes for the 50-mm-thick dielectric at 4.5 GHz in our example are shown in Figure 7.17a. As with the TE case, decreasing a or f will result in fewer modes, but there will always be a TM_0 mode. In Figure 7.17b, we see what happens to the right side of (7.89)

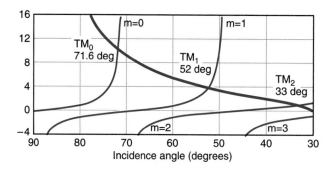

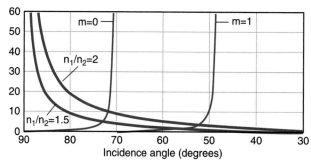

Figure 7.17 (a) The dielectric waveguide TM modes for a 50-mm-thick dielectric of $\varepsilon_r = 4$ operating at 4.5 GHz. The bold line plots the right side of (7.89) and the other lines are the left side for different values of m. (b) The right side of (7.89) plotted for two different n_1/n_2 ratios.

as the difference between n_1 and n_2 is varied. A larger difference results in a lower critical angle and therefore more propagating modes. This will also be the case for the TE modes. If n_1 and n_2 are close in value, the TE and TM modes will occur at about the same angles.

It can be shown that single-mode operation occurs for

$$\frac{a}{\lambda_o} < \frac{1}{2} \frac{1}{\sqrt{n_1^2 - n_2^2}} \tag{7.90}$$

Drill 7.6 Suppose a polyethylene dielectric slab of thickness 100. mm exists in air. What is the maximum frequency at which this slab will support only one mode? (*Answer:* 1.33 GHz)

Field Equations

Figure 7.18 shows the geometry of the dielectric waveguide for which the TE mode field equations will now be given. We want to find the field E_y as a function of x across the guide. The equations will depend on whether the mode is even ($m = 0, 2, 4,...$) or odd ($m = 1, 3, 5,...$). We have

$$E_y = E_o \cos(\beta_1 x \cos \theta_i)e^{-j\beta_1 z \sin \theta_i} \ (m = 0, 2, 4,...)$$

$$E_y = E_o \sin(\beta_1 x \cos \theta_i)e^{-j\beta_1 z \sin \theta_i} \ (m = 1, 3, 5,...) \tag{7.91}$$

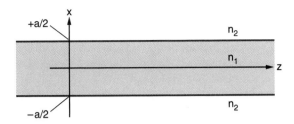

Figure 7.18 Cross-sectional view of dielectric waveguide.

Note that the fields are not zero at $x = \pm a/2$. Instead, they attenuate into the surrounding media. For the even modes we have

$$E_y = \begin{cases} E_o \cos (\beta_1(a/2) \cos \theta_i)e^{-\alpha_2(x-a/2)}e^{-j\beta_1 z \sin \theta_i} & \text{for } x > a/2 \\ E_o \cos (\beta_1(a/2) \cos \theta_i)e^{+\alpha_2(x+a/2)}e^{-j\beta_1 z \sin \theta_i} & \text{for } x < -a/2 \end{cases}$$
(7.92)

For the odd modes we have

$$E_y = \begin{cases} E_o \sin (\beta_1(a/2) \cos \theta_i)e^{-\alpha_2(x-a/2)}e^{-j\beta_1 z \sin \theta_i} & \text{for } x > a/2 \\ -E_o \sin (\beta_1(a/2) \cos \theta_i)e^{\alpha_2(x+a/2)}e^{-j\beta_1 z \sin \theta_i} & \text{for } x < -a/2 \end{cases}$$
(7.93)

The attenuation in medium 2 is

$$\alpha_2 = \beta_1 \sqrt{\sin^2 \theta_i - (n_2/n_1)^2}$$
(7.94)

With propagation in the $+z$ direction, from inspection of (7.91)–(7.93) we can define an effective guide phase constant

$$\beta_e = \beta_1 \sin \theta_i$$
(7.95)

so an effective wavelength in the guide is

$$\lambda_e = \frac{2\pi}{\beta_1 \sin \theta_i} = \frac{\lambda_o}{n_1 \sin \theta_i}$$
(7.96)

We can also determine the propagation velocity in the guide to be

$$u_p = \frac{\omega}{\beta_e} = \frac{c}{n_1 \sin \theta_i}$$
(7.97)

To use these equations, we must first find the θ_i corresponding to a particular mode m.

Drill 7.7 Find λ_e and u_p at 4.5 GHz for the TE_0 mode in a 50.-mm-thick $n_1 = 2.0$ dielectric in air. (*Answer:* 35 mm and 1.6×10^8 m/s)

▶ **MATLAB 7.3**

The program ML0703 is used to draw the E_y field pattern for the $m = 0$ mode shown in Figure 7.19. It can be easily modified to draw the $m = 2$ mode. A different routine is required to draw the $m = 1$ mode (see Problem 7.23).

```
%    M-File: ML0703
%
%    Plot even-mode field patterns for dielectric guide
%    at z = 0.
%
%    The theta angle must be entered for a particular
%    mode.  The "hold on" function allows the results of
%    multiple runs to be placed on one plot.
%
%    Wentworth, 11/26/02
%
%    Variables
%  m            even mode (0 or 2)
%  a            dielectric thickness (m)
%  b            phase constant
%  thdeg        angle theta in degrees
%  th           angle theta in radians
%  n1,n2        indices of refraction
%  n21          the ratio n2/n1
%  f            frequency (Hz)
%  c            speed of light (m/s)
%  w            radian frequency (rad/s)
%  Eo           initial amplitude (V/m)
%  alpha        atten. in medium 2 (Np/m)

clc              %clears the command window
clear            %clears variables

% initialize variables
%m=0;
a=50e-3;
thdeg=74.4;    %corresponds to m = 0
th=pi*thdeg/180;
n2=1;
n1=2;
n21=n2/n1;
f=4.5e9;
w=2*pi*f;
c=2.998e8;
Eo=1;
b=(w/c)*n1;
alpha=b*sqrt(sin(th)^2-n21^2);

x=-a/2:a/40:a/2;
Ey=Eo*cos(b*cos(th)*x);
hold on
plot(x,Ey,'k')
grid on
```

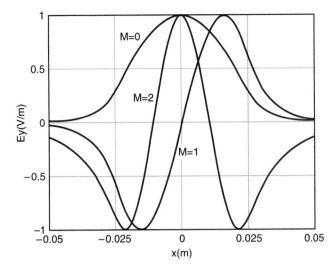

Figure 7.19 E_y field patterns for the first three TE modes of a 5-cm-thick dielectric guide ($n = 2$) in air. The dielectric extends from $x = -0.025$ m to $x = +0.025$ m.

```
xlow=-2*a/2:a/40:-a/2;
Eylow=Eo*cos(b*cos(th)*a/2)*exp(alpha*(xlow+a/2));
plot(xlow,Eylow,'k')

xhi=a/2:a/40:2*a/2;
Eyhi=Eo*cos(b*cos(th)*a/2)*exp(-alpha*(xhi-a/2));
plot(xhi,Eyhi,'k')
hold off
```

▷ 7.4 OPTICAL FIBER

Having looked at the propagating modes in dielectric waveguide, we are now ready to look at signal propagation in optical fibers. The first widespread application of optical fibers was for telephone links. They are now also used in cable television systems to transmit signals to central distribution locations and are expected to eventually replace the coaxial cable running to the home. Optical fiber also finds use in local area networks (LANs) interconnecting computers and their peripherals. The three primary transmission windows are centered around 850, 1300, and 1550 nm.

A typical optical fiber is represented by Figure 7.20. The fiber *core* is completely encased in a fiber *cladding* that has a slightly lesser value of refractive index. Signals propagate along the core by total internal reflection at the core–cladding boundary. Both core and cladding are typically made of silicon dioxide (*silica*), with appropriate additives to control the index of refraction. Plastic fibers may be used for shorter length transmission applications where the higher attenuation of plastic is not a problem. Outside the cladding is usually a plastic jacket, typically polyethylene or Kevlar, used to protect the fiber from scratches and moisture and to provide an opaque shield. Dimensions of the fiber are often specified by the diameter of the core and the diameter of the cladding separated with a slash. For instance, a 50/125 fiber has a 50-μm-diameter core sheathed in a 125-μm-diameter cladding. The core diameter of silica optical fibers ranges from 5 to 200 μm, and the diameter of the cladding ranges from 125 to 240 μm.

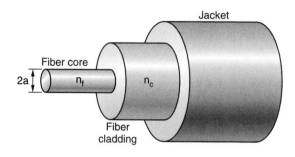

Figure 7.20 Typical optical fiber consists of a core surrounded by cladding and sheathed in a protective jacket.

Because optical fiber operates at optical frequencies, it has much greater information-carrying capability than does coaxial cable. It is also much smaller and lighter and more flexible than coax. In addition, optical fiber is fairly immune to electromagnetic interference. Finally, optical fiber has about an order of magnitude less attenuation than coaxial cable and its attenuation is relatively independent of frequency, whereas that in coax grows exponentially with frequency. On the downside, it can be harder to repair breaks in an optical fiber line than in coaxial cable, and optical connectors are expensive.

A basic understanding of signal propagation in optical fiber is afforded by a geometric optics approach similar to that done in Section 7.1 for rectangular waveguide. However, this approach is not as accurate as rigorous electromagnetic theory. For instance, there is some penetration of the field into the cladding, a result not shown by geometric optics. The cylindrical nature of the problem greatly complicates the electromagnetic field analysis and lies beyond the scope of this text.

Figure 7.21 shows a cross section of the fiber with rays traced for two different incident angles. If the phase-matching condition is met, these rays each represent propagating modes. The index of refraction profile is also shown. The abrupt change in n is characteristic of a *step-index* fiber. Optical fiber designed to support only one propagating mode is termed *single-mode fiber*. More than one mode propagates in *multimode fiber*.

As we saw for dielectric waveguide, the lowest order propagating mode is always present, having no associated cutoff wavelength. In step-index optical fiber, only one mode will propagate if the wavelength is big enough such that

$$\lambda > \frac{2\pi a\sqrt{n_f^2 - n_c^2}}{k_{01}} \tag{7.98}$$

where k_{01} is the first root of the zeroth-order Bessel function, equal to 2.405. We see that a single mode of propagation is supported by keeping the difference in index of refraction between the core and cladding small and by using a small-diameter fiber.

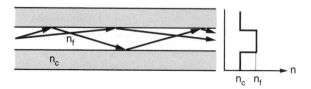

Figure 7.21 Cross section and index of refraction profile of a step-index fiber with rays for two propagating modes traced.

For step-index multimode fiber, the total number of propagating modes is approximately

$$N = 2\left(\frac{\pi a}{\lambda}\right)^2 \left(n_{\mathrm{f}}^2 - n_{\mathrm{c}}^2\right)$$

(7.99)

Table 7.2 compares typical characteristics for glass optical fibers. The terminology used in this table is explained in the remainder of this section.[7.5]

▶ **EXAMPLE 7.3**

Suppose we have an optical fiber core of index 1.465 sheathed in cladding of index 1.450. What is the maximum core radius allowed if only one mode is to be supported at a wavelength of 1550 nm? Approximately how many modes are supported at this maximum radius for a source wavelength of 850 nm?

First, we rearrange (7.98) in terms of fiber radius a:

$$a < \frac{k_{01}\lambda}{2\pi\sqrt{n_{\mathrm{f}}^2 - n_{\mathrm{c}}^2}}$$

Solving, we find

$$a < \frac{(2.405)\left(1550 \times 10^{-9}\ \mathrm{m}\right)}{2\pi\sqrt{(1.465)^2 - (1.450)^2}}$$

or

$$a < 2.84\ \mu\mathrm{m}$$

The number of propagating modes at 850 nm is estimated using (7.99):

TABLE 7.2 Typical Characteristics of Glass Optical Fiber

Type[a]	λ (nm)	Core diameter (μm)	NA	Attenuation (dB/km)	Chromatic dispersion [(ns/nm)/km]	Intermodal dispersion (ns/km)
SMF: SI	850	5	0.10	4	100	—
	1300	10	0.10	0.6	0.003	—
	1550	10	0.10	0.2	0.003	—
MMF: SI	850	50	0.24	4	0.10	15
MMF: GRIN	850	50	0.24	4	0.10	3
	1300	50	0.20	1	0.003	0.5

[a]SMF, single-mode fiber; SI, step-index fiber; GRIN, graded-index fiber.

Source: Adapted from J. Palais, *Fiber Optic Communications,* 4th Ed., Prentice-Hall, 1998, p. 140.

[7.5]A basic animated tutorial, provided by Corning, Inc., one of the leading manufacturers of optical fibers, can be found at *www.corning.com/opticalfiber/discovery_center/tutorials/fiber_101/.*

$$N = 2 \left(\frac{\pi \left(2.84 \times 10^{-6} \text{ m} \right)}{850 \times 10^{-9} \text{ m}} \right)^2 \left((1.465)^2 - (1.450)^2 \right) = 9.6$$

So, we conclude that about 9 modes are supported.

Drill 7.8 Suppose we want to support a single mode at 850. nm in a 4.000-μm-radius fiber of index 1.465. What cladding index is required? (*Answer:* 1.463)

Numerical Aperture

Light must be fed into the end of the fiber to initiate mode propagation. As Figure 7.22 shows, upon incidence from air (n_o) to the fiber core (n_f) the light is refracted by Snell's law:

$$n_o \sin \theta_a = n_f \sin \theta_b \tag{7.100}$$

This light goes on to make an angle θ_c with a normal to the core–cladding boundary. A necessary condition for propagation is that θ_c exceed the critical angle $(\theta_i)_{crit}$, where

$$\sin \left(\theta_i \right)_{crit} = \frac{n_c}{n_f} \tag{7.101}$$

Let's find the maximum acceptance angle θ_a that will define a cone of acceptance over which light will propagate along the fiber. Letting $\theta_c = (\theta_i)_{crit}$, we can observe from the geometry that

$$\sin \theta_c = \cos \theta_b \tag{7.102}$$

Then, relating $\sin \theta_b$ to $\cos \theta_b$ by

$$\sin^2 \theta_b + \cos^2 \theta_b = 1$$

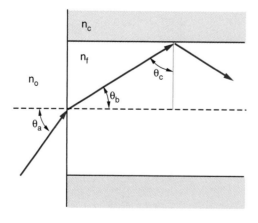

Figure 7.22 Expanded view of the cross section of an optical fiber at one end for determining the acceptance angle.

We can rewrite (7.100) as

$$n_o \sin \theta_a = n_f \sqrt{1 - \cos^2 \theta_b} = n_f \sqrt{1 - \sin^2 \theta_c} \qquad (7.103)$$

At $\theta_c = (\theta_i)_{crit}$, (7.103) can be manipulated using (7.101) to find

$$\boxed{\sin \theta_a = \frac{\sqrt{n_f^2 - n_c^2}}{n_o}} \qquad (7.104)$$

This term is commonly referred to as the *numerical aperture*, given by

$$NA = \sin \theta_a = \frac{\sqrt{n_f^2 - n_c^2}}{n_o} \qquad (7.105)$$

Numerical aperture is a parameter given by the fiber manufacturer and along with core diameter is important in determining coupling of light to the fiber.

▶ **EXAMPLE 7.4**

Let's find the critical angle within the fiber described in Example 7.3. Then we'll find the acceptance angle and the numerical aperture.

The critical angle is found from (7.101) as

$$(\theta i)_{crit} = \sin^{-1}\left(\frac{n_c}{n_f}\right) = \sin^{-1}\left(\frac{1.450}{1.465}\right) = 81.8°$$

The acceptance angle, from (7.104), is given by

$$\theta_a = \sin^{-1}\left(\frac{\sqrt{(1.465)^2 - (1.450)^2}}{1}\right) = 12.1°$$

Finally, the numerical aperture is

$$NA = \sin \theta_a = 0.209$$

Drill 7.9 Determine the acceptance angle and numerical aperture for the fiber of Drill 7.8. (*Answer:* $\theta_a = 4.4°$, $NA = 0.076$)

Signal Degradation

Consider a single-frequency source (called a *monochromatic* source) launching a pulse onto a multimode fiber. We will assume the power of the pulse is evenly divided among the N modes of the fiber. Each mode will travel at a different angle, and therefore each mode will travel at a different propagation velocity. When the pulse is collected at the receiving end, it will have spread out owing to the different mode velocities. We call this effect *intermodal dispersion*, and we express its value in terms of how much a pulse will spread in time (in nanoseconds) as it travels a kilometer. Single-mode fiber doesn't suffer from this kind of signal distortion, which explains why it is preferred over multimode fiber for long distance applications.

No light source is truly monochromatic; it will always have at least some bandwidth ($\Delta\lambda$) associated with it. For instance, the typical $\Delta\lambda$ for a 1300-nm laser source is 3 nm. The finite bandwidth gives us two more sources of signal degradation: *waveguide dispersion* and *material dispersion*.

We've seen that the propagation velocity of a particular mode for a wave traveling in waveguide is a function of frequency. Since the light source has a finite bandwidth, there will be a spread in the propagating signal known as waveguide dispersion. Also, the index of refraction for optical materials is generally a function of frequency. A band of frequencies representing a pulse will therefore spread out as it propagates along the line, a phenomenon called material dispersion.

Since both waveguide and material dispersion are proportional to the optical bandwidth, they are often lumped together as *chromatic dispersion*. This is expressed as the amount of pulse spread in nanoseconds per nanometer of $\Delta\lambda$ as the signal travels a kilometer. For well-engineered materials, dispersion can be kept very small. In fact, between 1300 and 1600 nm it is possible to construct the fiber such that the waveguide dispersion cancels the material dispersion, and the chromatic dispersion becomes almost zero.

Attenuation

As light propagates along an optical fiber, some of its power is lost by interaction with the fiber material. The primary mechanisms for this loss are electronic and vibrational absorption and scattering.

In electronic absorption, photonic energy at short wavelengths may have the right amount of energy to excite crystal electrons to higher energy states. Subsequent relaxation of these electrons is by *phonon* emission (i.e., heating the crystal lattice). In vibrational absorption, atoms vibrate depending on their arrangement in the crystal. If the photonic energy matches the vibrational energy (at longer wavelength), energy is lost to vibrational absorption. Light is also attenuated by scattering at imperfections in the crystal lattice. This tends to be more a problem at longer wavelengths and is caused by imperfections and local variations in the refractive index.

Figure 7.23 shows the attenuation in a typical silica fiber. The peak at 1400 nm is a result of vibrational absorption by OH^- (hydroxyl ion) contamination. Care is taken in fiber fabrication to minimize this impurity. It can be seen that the lowest attenuation occurs at about 1550 nm.

Graded-Index Fiber

Coupling a light source to the very small diameter required for a single-mode step-index fiber proves to be quite difficult, often requiring a highly directional and relatively expensive laser light source. Making the core diameter bigger results in a multimode fiber that suffers from intermodal dispersion since the modes all travel with different speeds along the line.

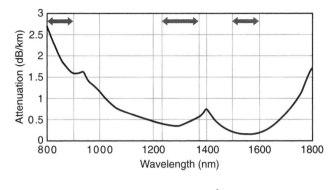

Figure 7.23 Typical attenuation in silica fiber with the three common usage bands indicated.

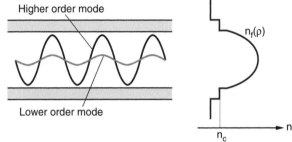

Figure 7.24 Graded-index fiber shown with a parabolic index profile.

One approach to minimize dispersion in a multimode fiber is to use a *graded-index fiber* (or GRIN, for short). The index of refraction in the core has an engineered profile like the one shown in Figure 7.24. Here, higher order modes have a longer path to travel, but they spend most of their time in lower index of refraction material, which has a faster propagation velocity. Lower order modes have a shorter path, but they travel mostly in the higher index material near the center of the fiber. The result is the different modes all propagate along the fiber at close to the same speed. The GRIN therefore exhibits less of a dispersion problem than a multimode step-index fiber.

Graded-index fibers of dimension 50/125 or 85/125 are common. Such fibers can use less costly LED light sources and are often used in data links and LAN applications.

▶ 7.5 FIBER OPTIC COMMUNICATION SYSTEMS

The basic components of a fiber optic communication system are shown in Figure 7.25. An electrical signal, either analog or digital, modulates the input current to a light source, which in turn modulates the intensity of the light emitted. This light is coupled to an optical fiber and propagates along the line until it is coupled to an optical detection device. Upon demodulation, the original electrical signal is recovered.

Optical fibers can operate out to about 50 km before the signal level drops too much to be recovered. If longer distance communication is needed, the signal must be amplified. In Figure 7.25, the signal is boosted by passing it through a *repeater*, made up of an optical detector, signal amplifier (that may include signal conditioning electronics), and an optical source.

We now describe in basic terms the components of the system.

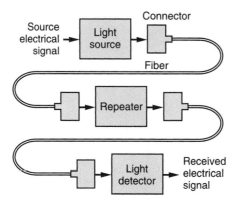

Figure 7.25 Typical optical fiber communication system.

Optical Sources

We saw in the previous section that optical fiber has low attenuation in the 0.8- to 1.8-μm wavelength range. Two basic types of light source are available for this range: light emitting diodes and laser diodes.

In a forward biased p–n junction (Figure 7.26a), electrons are excited to a higher energy state. When they relax, or fall back to the lower energy state, they may do so by a so-called direct path that releases a *photon* (light), or by an indirect path that releases a phonon (a lattice vibration, or heat). Silicon, used for constructing the vast majority of integrated circuits, is an indirect semiconductor and it therefore isn't easy to get light out of a silicon

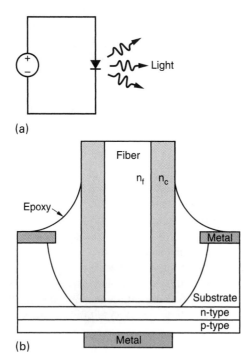

Figure 7.26 (a) Forward-biased photodiode emits photons. (b) Simplified cross section of a Burrus surface-emitting diode.

device. Gallium arsenide (GaAs), a binary compound, is a direct semiconductor that is very useful for producing photons.

The light emitted by a GaAs p–n junction has a wavelength of about 0.9 μm. By carefully adding aluminum, the ternary compound $Ga_xAl_{1-x}As$ can be created to give control over the emitted wavelength from 0.8 to 0.9 μm depending on the value of x. Other GaAs-based compounds extend the wavelength range from as low as 0.64 μm to as high as 1.7 μm.

LEDs may be constructed in either a surface-emitting configuration, as is shown for the Burrus LED of Figure 7.26b, or an edge-emitting configuration. For the Burrus LED, a well is etched into the substrate to bring the fiber close to the p–n junction where light is emitted. The fiber is held in place by epoxy of similar refractive index. The emitted light for this structure has a beamwidth of approximately 120°. This is too broad to couple efficiently with the fiber. Adding a focusing lens between the p–n junction and the fiber can improve the coupling.

An edge-emitting structure, similar to the laser diode configuration of Figure 7.27, can emit light at a beamwidth of around 30°. Compared to a Burrus LED, light can be much more efficiently coupled to the fiber from an edge-emitting structure. It should come as no surprise that LEDs with narrower beamwidths and brighter light outputs are more expensive.

An ideal light source from the performance standpoint is the semiconductor laser[7.6] diode. A simplified version of such a laser diode is shown in Figure 7.27. The n+ GaAs and p+ GaAs are heavily doped layers that provide a good conductive path from the metal contacts to the active portion of the structure. The p-AlGaAs and n-AlGaAs layers form the diode. Forward biasing the diode provides energy to pump energetic electrons into the GaAs *lasing* region. It is a characteristic of this lasing region that electrons are reluctant to relax to lower energy; they need a nudge. When an electron does drop back to the lower energy state, a photon is generated. This photon stimulates the relaxation of a second electron, thus providing a second photon in phase with the first. These two photons in turn stimulate the emission of two more photons, and emission grows geometrically as long as there is an abundance of excited electrons to draw from.

Another convenient feature of this structure is that the semiconductor layers adjacent to the lasing region are of a lower index of refraction. Thus, the in-phase photons find themselves in a dielectric waveguide and propagate out the side of the device.

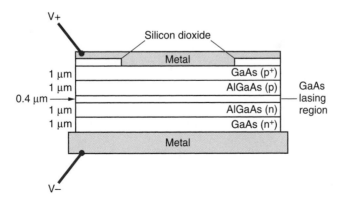

Figure 7.27 Simplified cross section of a GaAs laser diode.

[7.6]Laser is the acronym for light amplification by stimulated emission of radiation.

Lasers offer more intense light sources with narrower beamwidths than LEDs. They can be modulated at about an order of magnitude higher frequency than LEDs, leading to much higher data rates. But lasers require higher drive currents than LEDs and tend to wear out faster. They are also more expensive. Table 7.3 compares some typical values of the key operational characteristics of the LED and laser diode.

Optical Detectors

Detectors of optical radiation must be fast and capable of detecting very weak signals. The most used structure is the PIN photodiode, followed closely by the avalanche photodiode.

A PIN photodiode is shown in Figure 7.28. The PIN in the name comes from the structure layers: An intrinsic (very pure or undoped) layer of semiconductor is sandwiched by p-type and n-type regions. The large intrinsic region provides ample room for the capture of photons. When a photon is captured, it generates an electron–hole pair. Since the p–n junction is reverse-biased, electrons are quickly swept into the n side and holes are quickly swept to the p side, thereby producing a weak current proportional to the light intensity. An amplification stage generally follows.

An avalanche photodiode (APD) is a heavily doped structure with a large reverse-bias voltage. When a photon is captured in the junction region, an electron–hole pair is produced. The field in such a heavily biased junction rapidly accelerates the electrons and holes such that they can slam into atoms and release additional electron–hole pairs. A runaway "avalanche" effect results in a fairly strong signal. You could say the avalanche photodiode has built-in amplification. A downside of this device is that it is very noisy.

TABLE 7.3   Property Comparison for LEDs and Laser Diodes

Property	LED	Laser diode
Optical wavelength (nm)	850, 1300	1300, 1550
($\Delta\lambda$) Spectral width (nm)		
range	20–100	1–5
typical	50	3
Rise time (ns)		
range	2–20	0.1–1
typical	10	0.4
Power output (mW)		
range	0.1–10	10^{-6}–10
typical	1	1
Coupling efficiency	low	moderate
Lifetime (hr)	10^5–10^7	10^4–10^5
Cost	low	high
Primary use	short paths, moderate data rates	long paths, high data rates

Source: Adapted from J. Palais, *Fiber Optic Communications,* 4th Ed., Prentice-Hall, 1998.

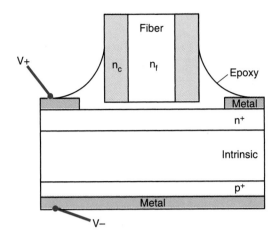

Figure 7.28 Simplified cross section of a PIN photodiode.

Table 7.4 compares a few of the characteristics of optical detectors. Device responsivity indicates how many amps of current are generated by the detector per watt of absorbed optical power. The avalanche photodiode has considerably higher response than the PIN diode, which requires an amplifier at its output. How fast the device can respond to an optical sensor is indicated by the *rise time*. The PIN diode and the avalanche photodiode have very similar speeds. Sensitivity refers to the minimum detectable signal power level. The PIN diode has a slight advantage, since the avalanche photodiode must overcome noise. Finally, it may be noted that the responsivity advantage of the avalanche photodiode is somewhat offset by its higher noise level.

Repeaters and Optical Amplifiers

It is very difficult for an optical system to operate much beyond 50 km without using repeaters or optical amplifiers to boost the signal. In a basic repeater, represented by Figure 7.29, the optical signal is converted to an electrical signal in an optical detector. It is then

TABLE 7.4   Comparison of Optical Detectors

	PIN photodiode	Avalanche photodiode
Responsivity (A/W)		
range	0.5–0.7	10–100
typical	0.6	20
Rise time (ns)		
range	0.1–0.5	0.25–1
typical	0.3	0.3
Sensitivity (dB$_m$)		
range	−40 to −30	−40 to −20
typical	−35	−30
Noise	low	high

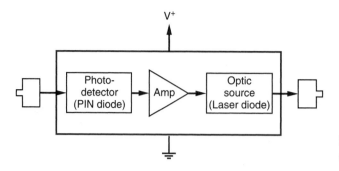

Figure 7.29   Simplified view of a repeater.

amplified and used to modulate the output of an optical source. Repeaters are most often used for digital signals and contain additional circuitry to remove noise and recover the digital signal.

If possible, repeaters are to be avoided as they add construction and maintenance cost, and they also require their own source of power. The source of power may be local or (as is the case for undersea repeaters) it may be transmitted via copper wires that are included in the cable containing the optical fibers.

A useful alternative to the repeater is the optical amplifier, in particular the *erbium-doped fiber amplifier* (EDFA), represented by Figure 7.30. Erbium-doped silica fiber contains electron energy states that enable direct amplification of an optical signal. This is in contrast to repeater operation, which requires conversion of the optical signal to electrical form prior to amplification, conditioning, and conversion back to optical form.

The input optical signal is combined with the output of a semiconductor laser pump. The pump excites electrons in the erbium-doped fiber to a higher energy state. Then, the optical signal stimulates relaxation of these excited electrons,[7.7] generating photons that augment the signal. The pump power is therefore converted to signal power. The amplified signal out of the erbium-doped fiber section is passed through another coupler to extract any unconverted laser pump signal.

The EDFA features high gain (as much as 40 dB) and high output power capability (up to 50 mW) without introducing excessive noise to the signal.

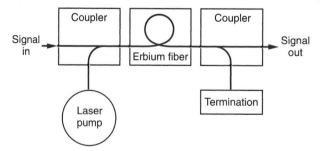

Figure 7.30   Erbium-doped fiber amplifier

[7.7]This stimulated emission is also the mechanism for laser operation.

Connections

Connections are made from the optical source to the fiber, from the fiber to the optical detector, and between lengths of fiber. In any of these connections, care must be taken to carefully align the optical paths and to reduce reflections at air gaps, for instance by using a matching refractive index epoxy as shown in Figure 7.26.

LEDs transmit their optical power in a broader beam than can be efficiently accepted by the narrow acceptance cone of the fiber. Consequently, about 12 dB loss is typical between an LED and the relatively large core of a multimode fiber. Attempting to launch into a single-mode fiber can result in a prohibitive extra 20 dB of loss. Focusing lenses can reduce this loss but will not eliminate it. Laser diodes, in contrast, have a much more focused beam, and coupling to a single-mode fiber can be accomplished with as little as 2 dB of loss.

At the detector end of the system, the fiber emits radiation in a narrow cone within the fiber's acceptance angle. As long as the detector area is as large as the fiber core, and they are closely connected with no significant air gap, coupling will be very efficient with less than 1.5 dB of loss.

Joining a pair of fibers can be accomplished using special connectors or by making splices. Connectors are a nonpermanent connection. Attenuation arises because of mis-alignment of the fiber axes (for instance, mating different size fibers or having one core off-set or at an angle to the other) and reflection at the fiber ends. A good connector will suffer no more than 1 dB of loss, with 0.7 dB being typical.

Splices are meant to be a permanent connection and tend to have less attenuation than a connector, generally no more than 0.1 dB, with 0.05 dB being typical. In a common splic-ing approach, the fiber ends are carefully aligned and then fused together with heat. Long runs of optical fiber are realized by splicing together 2-km lengths, so a 500-km fiber can contain as many as 250 splices. Typical losses associated with connectors are summarized in Table 7.5.

For both the source-to-fiber and fiber-to-detector connections, it is often the case that a short length of fiber is carefully attached and epoxied into place (as shown in Figures 7.26 and 7.28) to minimize the loss. These *pigtailed* devices are then connected to fiber using a low-loss splice.

▶ 7.6 OPTICAL LINK DESIGN

We now wish to apply what we've learned about optical fibers and basic optical compo-nents to the design of an optical fiber system. The design proceeds in two parts. First, a power budget is analyzed to ensure that the optical source provides enough power to result

TABLE 7.5 Typical Losses Associated with Connections

LED to MMF	12 dB
LED to SMF	>32 dB
Laser to SMF	2 dB
Fiber to detector	1.5 dB
Fiber-to-fiber connector	0.7 dB
Fiber-to-fiber splice	0.05 dB

in a detectable signal at the receiving end. Second, a rise-time budget is performed to verify that the received signal has not been fatally distorted.

Power Budget

The optical source must provide enough power to overcome source-to-fiber loss, connector and splice loss, and fiber-to-detector loss and still deliver at least the minimum detectable power to the optical detector.

▷ **EXAMPLE 7.5**

Suppose we require an optical link to transmit data over a 1.0-km distance. We choose an 850-nm LED source with 1.0-mW power (0 dB_m). We'll launch into a 850-nm step-index multimode fiber and assume a 12-dB source-to-fiber loss. The loss in the fiber itself from Table 7.2 for 1 km of fiber is 4 dB. At the detector end, we assume a 1.5-dB loss from fiber to detector. Finally, we'll assume optical connectors are used at each end (0.7 dB each) to connect the fiber to the pigtailed source and detector.

Our system must also have a margin of extra power to account for unexpected losses, such as extra splices, and to ensure the system will work even after the components begin to age. A system *margin* ranging from 3 to 10 dB is typical. For our design, let's include an 8 dB margin. Our power budget is then as follows:

Source	0 dB_m
Source-to-fiber	−12 dB
Fiber	−4 dB
Fiber-to-detector	−1.5 dB
Two connectors	−1.4 dB
Margin	−8 dB
Power available at detector:	−26.9 dB_m (2.04 μW)

If we select a PIN photodiode with typical sensitivity of −35 dB_m from Table 7.4, we see that the power budget can very easily be satisfied.

How far could we lengthen this link without changing the source, detector, or fiber type? From the previous calculations above, we see we can lose an additional 8.1 dB in the fiber. Neglecting splice loss we can therefore support propagation in a 3.0-km length of fiber.

Rise-Time budget

Data may be transmitted in analog or digital form. For short distances where noise is not an issue an analog transmission is acceptable. For longer connections, or high information rates, digital transmission has the ability to withstand much more noise.

Digital signals can be broadcast in a variety of forms, one of which is the *return-to-zero format* evidenced by Figure 7.31a. Each bit of information occupies the first half of the period T, with the signal level at zero for the other half. The pulse width t_{pw} is measured across the pulse where it is at half power (Figure 7.31b).

Figure 7.32 shows the spreading of a signal as it passes through an optical system. The rise time of the source and the detector as well as the effects of dispersion in the fiber cause

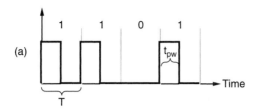

(a)

(b)

Figure 7.31 (a) In the return-to-zero data format, the first half of a period T is occupied by either a 1 or a 0. The pulse width t_{pw} is measured across the pulse at half power.

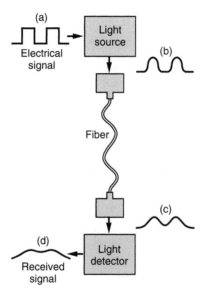

Figure 7.32 A pulsed electrical signal modulating the light source (a) is distorted by the rise time of the optical source (b) and is further distorted by dispersion in the fiber (c) and finally by the rise time of the optical detector (d).

the pulse to spread.[7.8] As evidenced in the figure, the signal may be barely recognizable by the time it exits the detector. Excessive distortion can result in errors in the received bit stream. A generally accepted *bit error rate* (BER) is 1 error in 10^9 bits, or a BER of 10^{-9}.

▶ **EXAMPLE 7.6**

Suppose the data of the previous example is in the return-to-zero format at a rate of 20×10^6 bits per second (bps), or 20 Mbps. This rate corresponds to a time period T of

$$T = \frac{1}{\text{bps}} = 50\,\text{ns}$$

We'll assume the -35-dB_m sensitivity of the receiver is sufficient for a 10^{-9} BER. In performing the rise-time budget, response times must be found for the transmitter (Δt_t), the receiver (Δt_r), and the fiber (Δt_f). The fiber response time may consist of both intermodal and chromatic dispersion effects.

[7.8]Our task is somewhat simplified in that connectors and splices do not significantly affect the rise time.

The total system response time Δt_s is the square root of the sum of the squares of each response time, that is,

$$\Delta t_s = \sqrt{\Delta t_t^2 + \Delta t_f^2 + \Delta t_r^2} \tag{7.106}$$

This system response is the spreading out of the signal pulse. For an input signal of pulse width $(t_{pw})_{in}$, the system response spreads the pulse such that the output signal $(t_{pw})_{out}$ is

$$\left(t_{pw}\right)_{out} = \sqrt{\left(t_{pw}\right)_{in}^2 + \Delta t_s^2}$$

For acceptable transmission, a common practice is to require the spread to be less than half of the signal period:

$$\Delta t_s < \frac{1}{2}T \tag{7.107}$$

Note that this criterion can also be used when evaluating analog signals where the highest frequency analog signal is related to the period by $T = 1/f$.

For our example, using the typical values from Tables 7.3 and 7.4, we have $\Delta t_t = 10$ ns and $\Delta t_r = 0.3$ ns. The fiber dispersion can be calculated from the data of Table 7.2. A typical value for chromatic dispersion from Table 7.2 is 5 ns for $\Delta \lambda = 50$ nm over a 1-km distance. Also from this table intermodal dispersion is seen to be 15 ns over this same distance. The total rise time associated with a 1-km length of this fiber is therefore

$$\Delta t_f = \sqrt{\Delta t_{intermodal}^2 + \Delta t_{chromatic}^2}$$
$$\sqrt{(5 \text{ ns})^2 + (15 \text{ ns})^2} = 15.8 \text{ ns}$$

Using (7.106), we get the system rise time of $\Delta t_s = 18.7$ ns. Since this is less than half of the period, the rise-time budget is satisfied.

From (7.106) and (7.107), we have that the system can handle a fiber dispersion of no more than

$$\Delta t_f = \sqrt{\left(\frac{1}{2}T\right)^2 - \Delta t_t^2 - \Delta t_r^2}$$

For our example, Δt_f must be less than 22.9 ns. Since the fiber rise time is 15.8 ns per kilometer of length, the rise-time budget tells us that no more than a 1.45-km length of fiber can be supported. This is considerably less than the 3.0 km that could be supported from power considerations.

Drill 7.10 What distance link would be supported if we replace the step-index multimode fiber of Examples 7.5 and 7.6 with a graded-index multimode fiber? (*Answer:* 2.9 km)

▶ **EXAMPLE 7.7**

Let us now apply our design technique to a more challenging problem. Suppose data must be transmitted at the rate of 500 Mbps between a pair of stations 50. km apart. Such a distance requires the low attenuation afforded by single-mode fiber along with a laser source that can efficiently couple power to the fiber. At one splice every 2 km, and assuming splices to a pigtailed laser diode and a pigtailed PIN photodiode, 26 splices will be needed.

We'll initially choose step-index single-mode fiber for operation at 1300 nm. Using our tabulated typical values, a power budget can be constructed:

Source	$0\ dB_m$ (1 mW)
Source-to-fiber	−2 dB
Fiber	−30 dB
Splices (26 at 0.05 dB each)	−1.3 dB
Fiber-to-detector	−1.5 dB
Power available at detector with no margin:	$-34.8\ dB_m$

This would be sufficient power for a typical PIN photodiode, but it leaves us with practically no margin.

The loss can be reduced by using a 1550-nm fiber with an attenuation of only 0.2 dB/km. Our new power budget gives $-14.8\ dB_m$ available at the detector, leaving us with ~20-dB margin.

Looking at the rise-time budget, we see that a data rate of 500 Mbps has a period $T = 2$ ns. The system rise time must be less than half this value. The chromatic dispersion in the fiber for a laser source with a 3-nm spectral width amounts to $\Delta t_f = 0.45$ ns. Using rise times of $\Delta t_t = 0.4$ ns and $\Delta t_r = 0.3$ ns for the laser diode source and the PIN diode receiver, respectively, we arrive at a system rise time $\Delta t_s = 0.67$ ns. Thus, the rise-time budget is satisfied.

Drill 7.11   Determine the maximum length fiber that could be supported at 1300 nm in Example 7.7 if an 8-dB margin is required. (*Answer:* 37.5 km)

▶ SUMMARY

- Rectangular waveguide supports propagation of TE_{mn} and TM_{mn} modes. The modes have a cutoff frequency given by

$$f_{c_{mn}} = \frac{1}{2\sqrt{\mu\varepsilon}}\sqrt{\left(\frac{m}{a}\right)^2 + \left(\frac{n}{b}\right)^2}$$

where a and b are the cross-sectional dimensions of the guide and m and n are integer values.

- Rectangular waveguide is most often operated in its lowest order mode. This TE_{10} mode is often referred to as the dominant mode or the fundamental mode.

- The phase velocity u_p and the group velocity u_G are functions of frequency related to the wave propagation in the unbounded media u_u by

$$u_p = \frac{u_u}{\sqrt{1-\left(\frac{f_c}{f}\right)^2}}$$

and

$$u_G = u_u\sqrt{1-\left(\frac{f_c}{f}\right)^2}$$

The wavelength in the guide is also frequency dependent, given by

$$\lambda = \frac{\lambda_u}{\sqrt{1-\left(\frac{f_c}{f}\right)^2}}$$

- Waveguide impedance is a function of frequency and mode type. For TE mode,

$$Z_{mn}^{TE} = \frac{\eta_u}{\sqrt{1-\left(\frac{f_c}{f}\right)^2}}$$

and for TM mode,

$$Z_{mn}^{TM} = \eta_u \sqrt{1 - \left(\frac{f_c}{f}\right)^2}$$

where η_u is the intrinsic impedance of the propagating media.

- It is customary to cite an optical material's index of refraction n, which for nonmagnetic materials is written

$$n = \sqrt{\varepsilon_r}$$

Snell's law of refraction for a wave incident from material n_1 to material n_2 at an angle to a boundary normal θ_i can then be written

$$\frac{n_1}{n_2} = \frac{\sin\theta_t}{\sin\theta_i}$$

where θ_t is the angle of the transmitted wave to a boundary normal.

- Waves propagate in dielectric waveguide with index of refraction n_1 when the angle the wave makes with a normal to a boundary (with index of refraction n_2) exceeds the critical angle, given by

$$(\theta_i)_{critical} = \sin^{-1}\left(\frac{n_2}{n_1}\right)$$

Because of an additional requirement for phase matching, only a finite number of modes may propagate in the dielectric guide. The number of modes depends on the indices of refraction n_1 and n_2 along with the frequency and guide dimensions.

- Optical fiber of circular cross section consists of a fiber core (n_f) and a fiber cladding (n_c), where $n_f > n_c$. Compared to coaxial cable, much higher frequencies with much lower attenuation are realized with optical fiber.

- The term *numerical aperture* (NA) describes the cone of acceptance for light incident at the end of a fiber. It is related to the acceptance angle θ_a and to the indices of refraction of the fiber (n_f), the cladding (n_c), and the medium it is incident from (n_o) by

$$NA = \sin\theta_a = \frac{\sqrt{n_f^2 + n_c^2}}{n_o}$$

Single-mode fiber has a considerably smaller numerical aperture ($\sim$0.10) than that of multimode fiber ($\sim$0.24).

- An optical fiber communications link basically consists of an LED or laser diode optical source, the fiber, and a PIN diode or avalanche photodiode optical detector. Special connectors, fiber splices, and repeaters can also be included. Design requires study of both a power budget and a rise-time budget.

▶ SUGGESTED REFERENCES

Hecht, J., *Understanding Fiber Optics,* 4th ed., Prentice–Hall, 2001.

Inan, U. S., and Inan, A. S., *Electromagnetic Waves,* Prentice–Hall, 2000.

Palais, J. C., *Fiber Optic Communications,* 4th ed., Prentice–Hall, 1998.

Rogers, A., *Understanding Optical Fiber Communications,* Artech House, 2001.

Yeh, C., *Handbook of Fiber Optics: Theory and Applications,* Academic Press, 1990.

▶ PROBLEMS

7.1 Rectangular Waveguide Fundamentals

7.1 Find the cutoff frequency for the first eight modes of WR430.

7.2 Calculate the cutoff frequency for the first eight modes of a waveguide that has $a = 0.900$ in and $b = 0.600$ in.

7.3 Calculate the cutoff frequency for the first eight modes of a waveguide that has $a = 0.900$ in and $b = 0.300$ in.

7.4 Calculate u_G, the wavelength in the guide, and the wave impedance at 10 GHz for WR90.

7.5 Consider WR975 filled with polyethylene. Find (a) u_u, (b) u_p, and (c) u_G at 600 MHz.

7.6 Plot u_p and wavelength in the guide as a function of frequency over the cited useful frequency range for WR90.

7.7 WR90 waveguide is to be operated at 16 GHz. Tabulate the values of the guide wavelength, phase velocity, group velocity, and impedance for each supported mode.

7.8 Modify MATLAB 7.1 by plotting u_G and u_p versus frequency for the same guide over the same frequency range.

7.9 Plot the TE_{10} wave impedance for WR430 waveguide versus frequency if the guide is filled with Teflon. Choose a suitable frequency range for your plot.

7.10 Suppose a length of WR137 waveguide operated at 7.0 GHz is terminated in a short circuit. At what distance from this short circuit does the input impedance appear infinite?

7.2 Waveguide Field Equations

7.11 Manipulate (7.41) to get (7.1).

7.12 Find expressions for the phasor field components of the TE_{01} mode.

7.13 Find an expression for the magnetic field of the TE_{11} mode.

7.14 Modify MATLAB 7.2 to look at the H_z field for the TE_{02} mode.

7.15 You are to create a movie showing how a surface plot of H_z over a cross section of rectangular waveguide changes with position. Use WR284 waveguide operating at 5 GHz and animate H_z for the TE_{11} mode.

7.3 Dielectric Waveguide

7.16 Start with (7.83) and derive (7.84) and (7.85).

7.17 Compose a program that will plot the left and right side of (7.86) versus all possible values of θ_i for $m = 0$. Test the program using the following values: $a = 1$ mm, $f = 100$ GHz, $n_1 = 3$, and $n_2 = 1$.

7.18 Compose a program that will plot the left and right side of (7.89) versus all possible values of θ_i for $m = 0$. Test the program using the following values: $a = 1$ mm, $f = 100$ GHz, $n_1 = 3$, and $n_2 = 1$.

7.19 Devise a Newton–Raphson iterative technique to solve for θ_i from (7.86). Test the program for $m = 0$, 1, and 2 using the following values of Figure 7.16: $a = 50$ mm, $f = 4.5$ GHz, and $\varepsilon_r = 4$. (*Hint: the tangent argument must be between $-\pi/2$ and $\pi/2$.*)

7.20 Find $(\theta_i)_{\text{critical}}$ for a wave incident from distilled water into air.

7.21 Suppose a Teflon slab of thickness 60 mm exists in air. What is the maximum frequency at which this slab will support only one mode?

7.22 Suppose a polystyrene dielectric slab is sandwiched between thick slabs of polyethylene. How thin must the polystyrene slab be such that only one propagating mode is supported at 1 GHz?

7.23 Modify MATLAB 7.3 to find the odd TE mode field patterns. Use the example information and duplicate the $m = 1$ plot for Figure 7.19.

7.24 Generate a figure similar to Figure 7.17(b) for the $m = 0$ mode of 10-mm-thick $\varepsilon_{r_1} = 9$ dielectric at 4.5 GHz for $n_1/n_2 = 1.5$, 2, and 3.

7.4 Optical Fiber

7.25 A 100/240 silica optical fiber has a core index of 1.460 and a cladding index of 1.450. Estimate the number of propagating modes for source wavelengths of (a) 850 nm, (b) 1300 nm, and (c) 1550 nm.

7.26 Given a fiber of core index 1.478 and cladding index 1.445, find the numerical aperture for a source of light incident from (a) air and from (b) distilled water.

7.27 Suppose $n_f = 1.475$ and $n_c = 1.470$. Determine the numerical aperture for a source of light incident from air. What is the maximum core diameter allowed to support only a single propagating mode if the source wavelength is 1300 nm?

7.28 Given a step-index fiber with $n_f = 1.480$ and a source wavelength of 1550 nm, determine the minimum value of n_c that will allow only one propagating mode for a core radius of 2 μm.

7.29 At a source wavelength of 1550 nm for a 5/125 silica fiber with $n_f = 1.470$, what is the minimum value n_c can be and only allow one propagating mode?

7.6 Optical Link Design

7.30 A 10.-km optical link is established between a typical LED and a typical PIN photodiode using a 1300-nm graded-index multimode fiber. Find the power margin and the maximum frequency analog signal that can be supported by this link. Assume two connectors and four splices.

7.31 Assume that the coupling efficiency from a typical LED to a graded-index multimode fiber can be approximated as $(NA)^2$. What is the power received by a typical PIN diode if 2 km of 850-nm fiber is used? Repeat for 2.0 km of 1300-nm fiber.

7.32 Calculate the maximum data rate, in bits per second, that could be supported for each case of the previous problem.

7.33 A 10.-km optical link is established between a typical laser diode and a typical PIN photodiode using a 1300-nm step-index single-mode fiber. Find the power margin and the maximum frequency analog signal that can be supported by this link. Assume two connectors and four splices. Compare your answer with that of Problem 7.30.

CHAPTER ▶ **8**

Antennas

Learning Objectives

▷ Introduce antenna terminology and describe antennas used for wireless communications

▷ Derive field relations for dipole and loop antennas

▷ Use image theory to describe operation of monopole antennas

▷ Develop field relations for antenna arrays

▷ Use the Friis transmission equation to study signal transfer between a pair of antennas

▷ Derive and utilize the radar equation

Wires passing an alternating current emit, or *radiate*, electromagnetic energy. The shape and size of the current-carrying structure determine how much energy is radiated as well as the direction of radiation. If the structure is designed to efficiently radiate in a preferred direction, it is called a *transmitting antenna*.

We also know that an electromagnetic field will induce current in a wire. The shape and size of the structure determine how efficiently the field is converted into current, or put another way, they determine how well the radiation is captured. The shape and size also determine from which direction the radiation is preferentially captured. In this case, the structure is considered a *receiving antenna*. In most cases, the efficiency and directional nature for an antenna are the same whether it is transmitting or receiving.[8.1]

Heinrich Hertz constructed the first antennas in 1886. He built a dipole antenna for the first radio transmitter and a loop antenna for the first radio receiver. Since then, a wide variety of antennas have been devised. Figure 8.1 shows some of the more common types.

In Figure 8.2, a source network (an AC voltage v_s in series with a source impedance Z_s) launches guided waves along a T-line terminated in a dipole antenna element. The antenna acts as a transition region, or transducer, between the guided waves and the waves radiated

---

[8.1]Most antennas are *reciprocal* devices, with the same patterns for reception and transmission. This is convenient as it is much easier to calculate a transmission pattern than a receiving pattern. Special solid-state and ferrite-based antennas can be *nonreciprocal*.

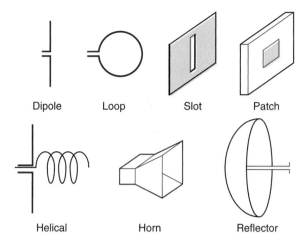

Dipole   Loop   Slot   Patch

Helical   Horn   Reflector

Figure 8.1 Common single-element antennas.

into space. Efficient operation requires that the complex antenna impedance Z_{ant} be matched to the system impedance.

Highly efficient antennas with appropriate radiation patterns are a critical component of wireless communication systems. In this chapter, we begin in Section 8.1 by introducing common antenna terminology to describe radiation patterns and antenna performance. In Section 8.2 we discuss the electromagnetic theory of small antennas, primarily the *Hertzian dipole*. The performance of larger dipole antennas can be accurately predicted by integrating a collection of these Hertzian dipoles, the topic of Section 8.3. Section 8.4 describes image theory and monopole antennas. In Section 8.5, we discuss simple arrays of antenna elements. By careful control of each element in an array, it is possible to steer the radiation or receiving pattern of an antenna, a process called beam steering. Design of a wireless communication link requires both a transmitting and a receiving antenna. Power transferred between these antennas is concisely described by the Friis transmission equation, the topic of Section 8.6. The Friis transmission equation is used to derive the radar equation in Section 8.7. Finally, Section 8.8 describes some of the other antennas useful for wireless communications.

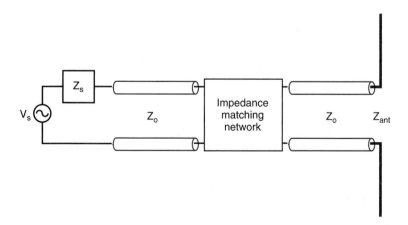

Figure 8.2 Generic antenna network. The antenna acts as a transducer between guided waves on the T-line and waves propagating in space.

▶ 8.1 GENERAL PROPERTIES

Before delving into the performance of actual antennas, we will first discuss some of the general terminology. The radiated power, beam pattern, directivity, antenna impedance, and efficiency are all important parameters in characterizing antennas.

Radiated Power

Suppose a transmitting antenna (transmitter) is located at the origin of a spherical coordinate system. As we will see in the next section, there are three components of the radiated field. The intensities of these three components vary with radial distance as $1/r$, $1/r^2$, and $1/r^3$. For almost all practical applications, a receiving antenna (receiver) will be located far enough away that the transmitter appears as a point source of radiation. At such distances, termed the *far-field* region, the intensity of the $1/r^2$ and $1/r^3$ field components are insignificant compared to that of the $1/r$ component. A distance r from the origin is generally accepted as being in the far-field region if

$$r \geq \frac{2L^2}{\lambda} \tag{8.1}$$

where L is the length of the largest dimension on the antenna element. Here it is assumed that $L > \lambda$. For smaller L, r should be at least as large as λ.

In the far field, the radiated waves resemble plane waves propagating in the $\mathbf{a}_r$ direction and time-harmonic fields can be related by the Chapter 5 equations (5.33)

$$\mathbf{E}_s = -\eta_o \mathbf{a}_r \times \mathbf{H}_s, \quad \mathbf{H}_s = \frac{1}{\eta_o} \mathbf{a}_r \times \mathbf{E}_s \tag{8.2}$$

where $\eta_o = 120\pi \ \Omega$ in free space. The time-averaged power density vector of the wave is found by the Poynting theorem,

$$\mathbf{P}(r,\theta,\phi) = \frac{1}{2} \text{Re}\left[\mathbf{E}_s \times \mathbf{H}_s^*\right] \tag{8.3}$$

Here it is indicated explicitly in parentheses that this power density is most generally[8.2] a function of r, θ, and ϕ. In the far field,

$$\mathbf{P}(r, \theta, \phi) = P(r, \theta, \phi)\mathbf{a}_r \tag{8.4}$$

The total power radiated by the antenna, P_{rad}, is found by integrating $\mathbf{P}(r, \theta, \phi)$ over a closed spherical surface,

$$P_{\text{rad}} = \oint \mathbf{P}(r,\theta,\phi) \cdot d\mathbf{S} = \iint P(r,\theta,\phi)r^2 \sin\theta \, d\theta \, d\phi \tag{8.5}$$

[8.2]In many cases, such as for dipole antennas, the power density will be invariant with ϕ and will therefore be reported as $\mathbf{P}(r, \theta)$.

Drill 8.1 In free space, suppose a wave propagating radially away from an antenna at the origin has

$$\mathbf{H}_s = \frac{I_s}{r} \sin\theta \, \mathbf{a}_\phi$$

where the driving current phasor $I_s = I_o e^{j\alpha}$. Find (a) $\mathbf{E}_s$, (b) $\mathbf{P}(r,\theta,\phi)$, and (c) P_{rad}.

(*Answer:* (a) $\mathbf{E}_s = \frac{\eta_o I_s}{r} \sin\theta \, \mathbf{a}_\theta$

(b) $\mathbf{P}(r,\theta,\phi) = \mathbf{P}(r,\theta) = \frac{1}{2}\eta_o \frac{I_o^2}{r^2} \sin^2\theta \, \mathbf{a}_r$

(c) $P_{rad} = \frac{4}{3}\pi\eta_o I_o^2$)

Radiation Patterns

Although the field intensity decreases with increasing r, the shape or pattern of the radiated field is independent of r in the far field. Radiation patterns usually indicate either electric field intensity or power intensity. Magnetic field intensity has the same radiation pattern as the electric field intensity, related by η_o. The polarization, or orientation, of the electric field vector is an important consideration in an electric field intensity plot. A transmit–receive pair of antennas must share the same polarization for the most efficient communication. In the discussion to follow, we will focus on the more useful power intensity radiation patterns.

Since the actual field intensity or power level depends not only on radial distance but also on how much power is delivered to the antenna, it is customary to divide the field or power component by its maximum value and to plot a normalized function. For our discussion of radiation patterns we will consider the *normalized* power function

$$P_n(\theta,\phi) = \frac{P(r,\theta,\phi)}{P_{max}} \tag{8.6}$$

where at particular values for θ and ϕ, $P(r, \theta, \phi)$ will reach its maximum value of P_{max}. The function $P_n(\theta, \phi)$ is also referred to as the *normalized radiation intensity*.

If the antenna radiates electromagnetic waves equally in all directions, it is termed an *isotropic antenna*. As shown in Figure 8.3, such a hypothetical antenna has a spherical radiation pattern independent of θ and ϕ. Here, then, the normalized power function $P_n(\theta, \phi)$ is equal to one, that is,

$$P_n(\theta, \phi)_{iso} = 1 \tag{8.7}$$

where the *iso* subscript indicates the function is for an isotropic antenna.

In contrast to an isotropic antenna, a *directional* antenna radiates and receives preferentially in some direction. Figure 8.4 shows the normalized radiation patterns for a generic antenna. A three-dimensional plot of the radiation pattern can be difficult to generate and work with, especially by hand. It is customary, then, to take slices of the pattern and generate two-dimensional plots. In Figure 8.4a, a polar plot is shown, where a slice has been

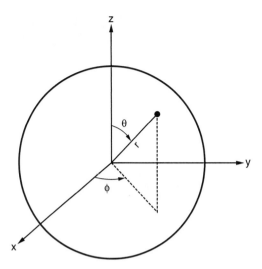

Figure 8.3 The spherical radiation pattern for an isotropic antenna.

taken and the pattern plotted over all θ for $\phi = \pi/2$ (right half of plot) and for $\phi = 3\pi/2$ (left half of plot). In Figure 8.4b, the same slice is shown in a rectangular plot[8.3] of the power level, in decibels, versus angle θ.

The polar plot can also be in terms of decibels. It is interesting to note that a normalized electric field pattern

$$E_n(\theta, \phi) = \frac{E(r, \theta, \phi)}{E_{max}} \tag{8.8}$$

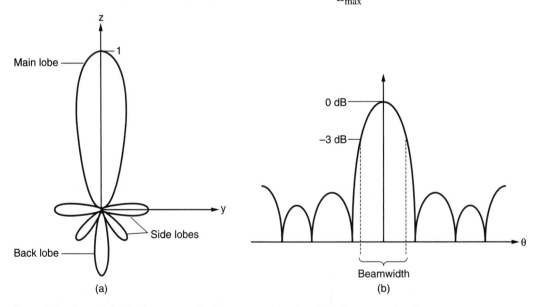

Figure 8.4 General far-field antenna radiation pattern: (a) polar plot; (b) rectangular plot.

[8.3]In such a plot, it is customary to plot negative θ angles on the left axis. Although this is technically incorrect since $0 \le \theta \le 180°$, it is useful for showing both sides of the two-dimensional slice.

in decibels will be identical to the power pattern in decibels. This is because power is proportional to the square of the electric field intensity, and for E we plot

$$E_n(\theta, \phi)(dB) = 20 \log [E_n(\theta, \phi)] \tag{8.9}$$

whereas

$$P_n(\theta, \phi)(dB) = 10 \log [P_n(\theta, \phi)] \tag{8.10}$$

It is clear in Figure 8.4 that in some very specific directions there are zeros, or *nulls*, in the pattern, indicating no radiation. The protuberances between the nulls are referred to as *lobes*, and the main, or major, lobe lies in the direction of maximum radiation. There are also side lobes and back lobes. Because these other lobes divert power away from the main beam, a good antenna design will seek to minimize the side and back lobes.

One measure of a beam's directional nature is the *beamwidth*, also called the half-power beamwidth or 3-dB beamwidth. As Figure 8.4b shows, this is the angular width of the beam measured at the half power, or −3 dB, points. If the beam cross section is elliptical, the half-power beamwidth is the average of the beamwidths measured on the major and the minor elliptical axes.

Directivity

It is often desirable to radiate most of the power fed to an antenna into the main lobe, rather than to the side or back lobes. A measure of how well an antenna does this is termed the *directivity D*. Before defining directivity, we will first describe the antenna's *pattern solid angle* (sometimes referred to as *beam solid angle*).

A radian is defined with the aid of Figure 8.5a. It is the angle subtended by an arc along the perimeter of the circle with length equal to the radius. In a like manner, a *steradian* may be defined using Figure 8.5b. Here, one steradian (sr) is subtended by an area r^2 at the surface of a sphere of radius r. A *differential solid angle* $d\Omega$, in sr, is defined as

$$d\Omega = \sin \theta \, d\theta \, d\phi \tag{8.11}$$

For a sphere, the solid angle is found by integrating $d\Omega$:

$$\Omega = \int_{\phi=0}^{2\pi} \int_{\theta=0}^{\pi} \sin \theta \, d\theta \, d\phi = 4\pi(\text{sr})$$

An antenna's pattern solid angle Ω_p is given by

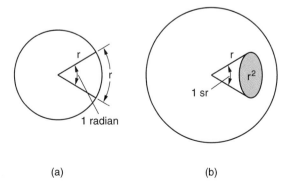

(a)

(b)

Figure 8.5 (a) An arc with length equal to a circle's radius defines a radian. (b) An area equal to the square of a sphere's radius defines a steradian.

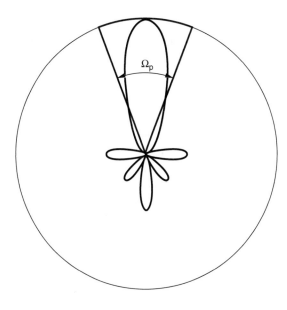

Figure 8.6 The pattern solid angle, in steradians, for a typical antenna radiation pattern.

$$\boxed{\Omega_p = \iint P_n(\theta,\phi)\,d\Omega} \tag{8.12}$$

as illustrated in Figure 8.6. Here, all of the radiation emitted by the antenna is concentrated in a cone of solid angle Ω_p over which the radiation is constant and equal to the antenna's maximum radiation value.

If we want to find the normalized power's average value taken over the entire spherical solid angle, we have

$$P_n(\theta,\phi)_{ave} = \frac{\iint P_n(\theta,\phi)\,d\Omega}{\iint d\Omega} = \frac{\Omega_p}{4\pi} \tag{8.13}$$

The *directive gain* $D(\theta,\phi)$ of an antenna is the ratio of the normalized power in a particular direction to the average normalized power,

$$D(\theta,\phi) = \frac{P_n(\theta,\phi)}{P_n(\theta,\phi)_{ave}} \tag{8.14}$$

The *directivity* D_{max} is the maximum directive gain,

$$D_{max} = D(\theta,\phi)_{max} = \frac{P_n(\theta,\phi)_{max}}{P_n(\theta,\phi)_{ave}} \tag{8.15}$$

It is apparent from (8.6) that $P_n(\theta,\phi)_{max} = 1$, and with (8.13) we conclude

$$\boxed{D_{max} = \frac{4\pi}{\Omega_p}} \tag{8.16}$$

Directivity is often expressed in decibels as

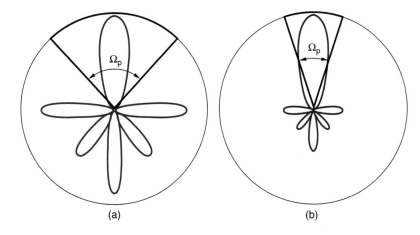

(a) (b)

Figure 8.7 Comparing Ω_p for two radiation patterns.

$$D_{max}(\text{dB}) = 10 \log (D_{max}) \tag{8.17}$$

A useful relation gleaned from (8.14) and (8.15) is

$$\boxed{D(\theta,\phi) = D_{max}P_n(\theta,\phi)} \tag{8.18}$$

Figure 8.7 compares two radiation patterns. In Figure 8.7a, considerable power gets radiated to the side and back lobes. As a result, the pattern solid angle is large and the directivity is small. In Figure 8.7b, almost all of the power gets radiated to the main beam, so Ω_p is small and the antenna has a high directivity.

From (8.5), (8.6), and (8.12) we can also write the total radiated power as

$$P_{rad} = r^2 P_{max} \iint P_n(\theta,\phi)\, d\Omega$$

or

$$\boxed{P_{rad} = r^2 P_{max} \Omega_p} \tag{8.19}$$

▶ **EXAMPLE 8.1**

To clarify some of these points, let's consider the normalized radiation intensity of a given antenna to be

$$P_n(\theta) = \begin{cases} \cos^2\theta & \text{for } 0 \le \theta \le \pi/2 \\ (-\cos\theta)/10 & \text{for } \pi/2 \le \theta \le \pi \end{cases}$$

We see here that $P_n(\theta)$ is independent of both r and ϕ. A polar plot for this beam pattern is given in Figure 8.8 (see MATLAB 8.1).

To find the beamwidth for the main beam, we need to calculate the values of θ that correspond to $P_n(\theta) = \frac{1}{2} P_n(\theta)_{max}$, or

$$\cos^2\theta = \frac{1}{2}$$

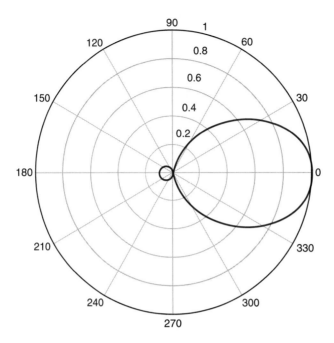

Figure 8.8 MATLAB-generated polar plot for the antenna of Example 8.1. Note that, in this representation, θ is shown from 0° to 360°.

A value of $\theta = \pm 45°$ results in a 90° beamwidth.

We can calculate the pattern solid angle using the following integral:

$$\Omega_p = \int_{\phi=0}^{2\pi}\int_{\theta=0}^{\pi/2} \cos^2\theta \, d\Omega + \int_{\phi=0}^{2\pi}\int_{\theta=\pi/2}^{\pi} \frac{-\cos\theta}{10} \, d\Omega$$

Solving, we find $\Omega_p = 23\pi/30$ sr. The average normalized power level is then

$$P_n(\theta,\phi)_{ave} = \frac{\Omega_p}{4\pi} = \frac{23}{120} = 0.19$$

The directivity is simply the inverse of the average normalized power, or

$$D_{max} = \frac{4\pi}{\Omega_p} = 5.2$$

Drill 8.2 Suppose $P_n(\theta,\phi) = 1$ for $0 < \theta < \pi/3$ and $P_n(\theta,\phi) = 0$ otherwise. Calculate the beamwidth, pattern solid angle, and directivity. (*Answer:* beamwidth = 120°, $\Omega_p = \pi$ sr, $D_{max} = 4$)

▶ **MATLAB 8.1**

Plot the beam pattern for Example 8.1.

```
%    M-File:ML0801
%    Polar plot for example 8.1
%
```

```
%   Wentworth, 12/6/02
%
%   Variables
%   theta        angle in radians
%   Pn           normalized power function

for i=1:100
    theta(i)=(-pi/2)+i*pi/100;
    Pn(i)=(cos(theta(i)))^2;
end
for j=101:200
    theta(j)=(-pi/2)+j*pi/100;
    Pn(j)=-(cos(theta(j)))/10;
end
polar(theta,Pn)

%   note that in a polar plot, the theta angle, in
%   radians, is automatically converted to degrees.
```

This is plotted in Figure 8.8.

Impedance and Efficiency

Power is fed to an antenna through a T-line (see Figure 8.9a), and the antenna appears as a complex impedance Z_{ant} (Figure 8.9b). The antenna impedance is a resistance R_{ant} in series with an antenna reactance jX_{ant}, so

$$Z_{ant} = R_{ant} + jX_{ant} \qquad (8.20)$$

This impedance can be modeled in Figure 8.9c, where the antenna resistance consists of radiation resistance R_{rad} and a dissipative resistance R_{diss} that arises from ohmic losses in the metal conductor.

For an antenna driven by phasor current $I_s = I_o e^{j\alpha}$, we can relate P_{rad} to R_{rad} by

$$P_{rad} = \frac{1}{2} I_o^2 R_{rad} \qquad (8.21)$$

So, for maximum radiated power we desire R_{rad} to be as large as possible without being too large to easily match with the feed line.[8.4]

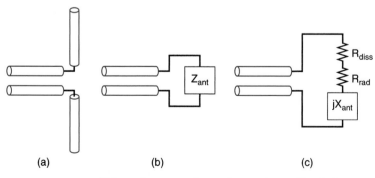

(a)　　　(b)　　　(c)

Figure 8.9 (a) A T-line terminated in a dipole antenna can be modeled with an antenna impedance (b) consisting of resistive and reactive components (c).

[8.4]The total antenna resistance R_{ant} must be considered when constructing an impedance matching network for the antenna.

We also have power dissipated by ohmic losses,

$$P_{\text{diss}} = \frac{1}{2} I_o^2 R_{\text{diss}} \tag{8.22}$$

An antenna efficiency e can be defined as the ratio of the radiated power to the total power fed to the antenna, that is,

$$e = \frac{P_{\text{rad}}}{P_{\text{rad}} + P_{\text{diss}}} = \frac{R_{\text{rad}}}{R_{\text{rad}} + R_{\text{diss}}} \tag{8.23}$$

The power gain $G(\theta,\phi)$ of an antenna is very much like its directive gain, but it also takes into account efficiency. It is given by

$$G(\theta,\phi) = eD(\theta,\phi) \tag{8.24}$$

and the maximum power gain $G_{\text{max}} = eD_{\text{max}}$. The maximum power is often expressed in dB$_i$, where the i subscript indicates dB with respect to an isotropic antenna.

Drill 8.3 Suppose the antenna of Drill 8.2 has $R_{\text{rad}} = 40\ \Omega$ and $R_{\text{diss}} = 10\ \Omega$. Find antenna efficiency and maximum power gain. (*Answer:* $e = 0.80$, $G_{\text{max}} = 3.2$).

A Commercial Antenna

Figure 8.10 shows the specifications and beam pattern for a commercially available antenna. This is a helical antenna (metal spring) embedded in a rubber or plastic protective enclosure. These compact and rugged antennas are commonly used in LAN systems and

PAWIN24-5RD Specifications

Frequency Range	2400–2485 MHz
Gain	5.5 dB$_i$
VSWR	1.5:1
Impedance	50 Ω
Input Power	10 W (max)
Operating Temperature	−10°C < T < 60°C
Weight	25 g
Length	197 mm, 203 mm

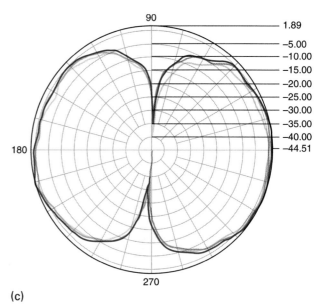

(a)   (b)   (c)

Figure 8.10 (a) A pair of rubber duck antennas, (b) the commercial specifications, and (c) beam pattern at 2450 MHz. Courtesy of Pacific Wireless.

cell phones and go by the generic name *rubber duck* antennas. With the antenna held vertical, rubber duck antennas are omnidirectional; that is, the beam pattern features a consistent
360° horizontal beam pattern. The beam pattern in Figure 8.10c shows the broadside direction performance corresponding to a 5.5-dB$_i$ gain.

▶ 8.2   ELECTRICALLY SHORT ANTENNAS

If the current distribution of a radiating element is known, it is possible to calculate the radiated fields by a direct integration. However, the integral can be very complex, so an intermediate step is usually inserted to make the job easier. For time-harmonic fields, integration
is performed to find the phasor $\mathbf{A}_{os}$, called the *retarded vector magnetic potential*. This is
followed by a relatively simple differentiation to find the magnetic field.

We begin this section with a derivation of the retarded vector magnetic potential $\mathbf{A}_{os}$.
Then we will find the radiated fields for a fictitious radiating element known as the *Hertzian
dipole*. This infinitesimally short element is assumed to have a uniform current along its
length. The Hertzian dipole serves as a differential radiating element for which the fields
from longer structures can be solved via integration, as we will see in the next section.
Finally, the radiated fields for a small loop antenna will be determined.

Vector Magnetic Potential

In working with electric fields, we have that a scalar electric potential V is related to $\mathbf{E}$ by
the gradient equation

$$\mathbf{E} = -\nabla V$$

An analogous term for magnetic fields is the vector magnetic potential $\mathbf{A}$, often used for
antenna calculations.

The point form of Gauss's law for magnetic fields is

$$\nabla \cdot \mathbf{B} = 0$$

A vector identity states that the divergence of the curl of any vector $\mathbf{A}$ is zero, that is,

$$\nabla \cdot (\nabla \times \mathbf{A}) = 0 \tag{8.25}$$

We can therefore define $\mathbf{B}$ in terms of $\mathbf{A}$ as

$$\mathbf{B} = \nabla \times \mathbf{A} \tag{8.26}$$

We now seek a relation between $\mathbf{A}$ and a current source. Referring to Figure 8.11, we
can write the Biot–Savart law from Chapter 3 as

$$\mathbf{B}_o = \frac{\mu_o}{4\pi} \int \mathbf{J}_d \times \frac{\mathbf{a}_{do}}{R_{do}^2} \, dv_d \tag{8.27}$$

where the vector $\mathbf{R}_{do} = R_{do}\,\mathbf{a}_{do}$ goes from the source point (or driving point), designated with
a d subscript, to the observation point, designated with an o subscript. These subscripts are
also used to indicate that the field is sought at the observation point ($\mathbf{B}_o$) from a source at
the driving point ($\mathbf{J}_d$). Finding the total field at the observation point requires integration
over the entire v_d volume.

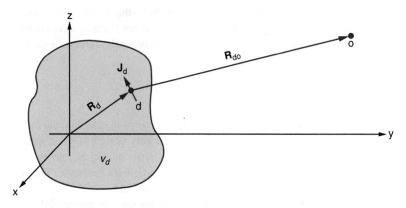

Figure 8.11 The vector magnetic potential at the observation point (o) results from a current density distributed about the volume v_d.

If we take the gradient at the observation point of $1/R_{do}$ we get $-\mathbf{a}_{do}/R_{do}^2$. We indicate the gradient is taken at the observation point by using the o subscript:

$$\nabla_o\left(\frac{1}{R_{do}}\right) = -\frac{\mathbf{a}_{do}}{R_{do}^2} \tag{8.28}$$

Replacing $\mathbf{a}_{do}/R_{do}^2$ in (8.27) we get

$$\mathbf{B}_o = \frac{-\mu_o}{4\pi}\int \mathbf{J}_d \times \nabla_o\left(\frac{1}{R_{do}}\right)dv_d \tag{8.29}$$

The next step is to apply a vector identity. For a general vector $\mathbf{M}$ and scalar field N,

$$\mathbf{M} \times \nabla N = N\nabla \times \mathbf{M} - \nabla \times N\mathbf{M} \tag{8.30}$$

If we let $\mathbf{M} = \mathbf{J}_d$ and $N = 1/R_{do}$, then we have

$$\mathbf{B}_o = \frac{-\mu_o}{4\pi}\int\left[\frac{1}{R_{do}}\nabla_o \times \mathbf{J}_d - \nabla_o \times \left(\frac{1}{R_{do}}\right)\mathbf{J}_d\right]dv_d \tag{8.31}$$

Notice that, in the first term inside the brackets, we are asking for the curl taken at the observation point of the current density at the source point. Since the source isn't changing with respect to the observation point, this term is zero. So (8.31) becomes

$$\mathbf{B}_o = \frac{\mu_o}{4\pi}\int \nabla_o \times \left(\frac{1}{R_{do}}\right)\mathbf{J}_d dv_d \tag{8.32}$$

Here, the integration is with respect to the source, whereas the curl operation is with respect to the observation point. It makes no difference which order we perform these operations, so (8.32) can be written

$$\mathbf{B}_o = \nabla_o \times \frac{\mu_o}{4\pi}\int \frac{\mathbf{J}_d}{R_{do}}dv_d \tag{8.33}$$

From this equation and (8.26) we see that

$$\mathbf{A}_o = \frac{\mu_o}{4\pi} \int \frac{\mathbf{J}_d}{R_{do}} dv_d \tag{8.34}$$

More properly, we can write (8.34) as

$$\mathbf{A}_o(x_o, y_o, z_o) = \frac{\mu_o}{4\pi} \int \frac{\mathbf{J}_d(x_d, y_d, z_d)}{R_{do}} dv_d \tag{8.35}$$

since so far all we have considered is that the vector potential at point o is a function of the position of the current element, (x_d, y_d, z_d). We have yet to consider time dependence. It would be tempting to write

$$\mathbf{A}_o(x_o, y_o, z_o, t) = \frac{\mu_o}{4\pi} \int \frac{\mathbf{J}_d(x_d, y_d, z_d, t)}{R_{do}} dv_d$$

but this would imply that the magnetic vector potential at an observation point well removed from the source changes the instant in time that the source changes, which is clearly impossible. Instead, there will be a finite time R_{do}/u_p for any change in $\mathbf{J}_d$ to be felt at the observation point. So a general time-dependent form of (8.35) can be written

$$\mathbf{A}_o(x_o, y_o, z_o, t) = \frac{\mu_o}{4\pi} \int \frac{\mathbf{J}_d(x_d, y_d, z_d, t - R_{do}/u_p)}{R_{do}} dv_d \tag{8.36}$$

In terms of phasors, the current density is

$$\mathbf{J}_d(x_d, y_d, z_d, t - R_{do}/u_p) = \mathbf{J}_d(x_d, y_d, z_d)e^{j\omega(t - R_{do}/u_p)}$$

$$= \mathbf{J}_d(x_d, y_d, z_d)e^{j\omega t}e^{-j\omega R_{do}/u_p} = \mathbf{J}_{ds}e^{j\omega t} \tag{8.37}$$

where $\mathbf{J}_{ds}$ is the retarded phasor quantity

$$\mathbf{J}_{ds} = \mathbf{J}_d(x_d, y_d, z_d)e^{-j\beta R_{do}}$$

since $\beta = \omega/u_p$. Using this current density, the phasor form of (8.36) is

$$\boxed{\mathbf{A}_{os} = \frac{\mu_o}{4\pi} \int \frac{\mathbf{J}_{ds}}{R_{do}} dv_d} \tag{8.38}$$

This is the time-harmonic equation for the retarded vector magnetic potential.

In phasor notation the vector magnetic potential is related to the magnetic flux density by

$$\mathbf{B}_{os} = \nabla \times \mathbf{A}_{os} \tag{8.39}$$

Following the integration of (8.38) to find $\mathbf{A}_{os}$, (8.39) can be solved for $\mathbf{B}_{os}$, and $\mathbf{H}_{os} = \mathbf{B}_{os}/\mu_o$ in free space. Because the radiation is propagating radially away from the source, it is then a simple matter to find $\mathbf{E}_{os}$ using (5.33),

$$\mathbf{E}_{os} = -\eta_o \mathbf{a}_r \times \mathbf{H}_{os} \tag{8.40}$$

Finally, the time-averaged power radiated is

$$\mathbf{P}(r, \theta, \phi) = \frac{1}{2} \text{Re}\left[\mathbf{E}_{os} \times \mathbf{H}_{os}^*\right] \tag{8.41}$$

The Hertzian Dipole

Suppose a short line of current, $i(t) = I_o\cos(\omega t + \alpha)$, is placed along the z-axis as shown in Figure 8.12. Here, the phasor current is $I_s = I_o e^{j\alpha}$. To maintain constant current over its entire length, it is helpful to imagine a pair of plates at the ends of the line that can store charge. The stored charge at the ends resembles an electric dipole, and the short line of oscillating current is then referred to as a *Hertzian dipole*.

For I_s conducted in the $+\mathbf{a}_z$ direction through a cross-sectional area S, the current density at the source seen by the observation point is

$$\mathbf{J}_{ds} = \frac{I_s}{S} e^{-j\beta R_{do}} \mathbf{a}_z$$

A differential volume of this current element is

$$dv_d = S dz$$

so

$$\mathbf{J}_{ds} dv_d = I_s e^{-j\beta R_{do}} dz \mathbf{a}_z$$

and the vector magnetic potential equation can be written

$$\mathbf{A}_{os} = \frac{\mu_o}{4\pi} \int_{-\ell/2}^{\ell/2} \frac{I_s dz \mathbf{a}_z e^{-j\beta R_{do}}}{R_{do}} \tag{8.42}$$

A key assumption for the Hertzian dipole is that it is very short, so that $\mathbf{R}_{do}$ is approximately equal to $\mathbf{r}$. Replacing R_{do} with r in (8.42) and integrating, we find

$$\mathbf{A}_{os} = \frac{\mu_o I_s \ell}{4\pi} \frac{e^{-j\beta r}}{r} \mathbf{a}_z \tag{8.43}$$

The unit vector $\mathbf{a}_z$ can be converted to its equivalent direction in spherical coordinates using the transformation equations in Appendix B, giving

$$\mathbf{a}_z = \cos\theta\, \mathbf{a}_r - \sin\theta\, \mathbf{a}_\theta,$$

so that (8.43) becomes

$$\mathbf{A}_{os} = \frac{\mu_o I_s \ell}{4\pi} \frac{e^{-j\beta r}}{r} (\cos\theta\, \mathbf{a}_r - \sin\theta\, \mathbf{a}_\theta) \tag{8.44}$$

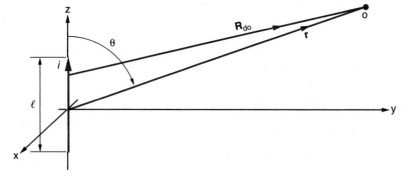

Figure 8.12 The vector magnetic potential is sought at point o from a z-directed Hertzian dipole at the origin.

This is the retarded vector magnetic potential at the observation point resulting from the Hertzian dipole element oriented in the $+\mathbf{a}_z$ direction at the origin. It is now a relatively straightforward matter to find $\mathbf{B}_{os}$ from (8.39), and then

$$\mathbf{H}_{os} = \frac{I_s\ell}{4\pi}\frac{e^{-j\beta r}}{r}\left(j\beta + \frac{1}{r}\right)\sin\theta\,\mathbf{a}_\phi \tag{8.45}$$

It is useful to group β and r together, and (8.45) becomes

$$\mathbf{H}_{os} = \frac{I_s\ell\beta^2 e^{-j\beta r}}{4\pi}\left[\frac{j}{\beta r} + \frac{1}{(\beta r)^2}\right]\sin\theta\,\mathbf{a}_\phi \tag{8.46}$$

The second term in brackets, with $(\beta r)^2$ in the denominator, drops off with increasing radius much faster than the first term. We specify

$$\frac{1}{\beta r} \gg \frac{1}{(\beta r)^2} \tag{8.47}$$

as being a far-field condition. Manipulation of (8.47) reveals the condition is met if

$$r \gg \frac{\lambda}{2\pi} \tag{8.48}$$

In the far field, we can neglect the second bracketed term in (8.46) and arrive at the following expression for $\mathbf{H}_{os}$:

$$\mathbf{H}_{os} = j\frac{I_s\ell\beta}{4\pi}\frac{e^{-j\beta r}}{r}\sin\theta\,\mathbf{a}_\phi \tag{8.49}$$

The far-field value of the electric field is found using (8.40) to be

$$\mathbf{E}_{os} = j\eta_o\frac{I_s\ell\beta}{4\pi}\frac{e^{-j\beta r}}{r}\sin\theta\,\mathbf{a}_\theta \tag{8.50}$$

Although we will spend most of our time with plots of the power density, (8.50) is important for telling us the polarization of the field. As we will see in Section 8.6, maximum power transfer between a transmitting and receiving antenna requires equivalent polarization.

Finally, the time-averaged power density at the observation point is found using the Poynting theorem:

$$\mathbf{P}(r,\theta) = \left(\frac{\eta_o\beta^2 I_o^2\ell^2}{32\pi^2 r^2}\right)\sin^2\theta\,\mathbf{a}_r \tag{8.51}$$

The term in brackets is the maximum power density:

$$P_{max} = \frac{\eta_o\beta^2 I_o^2\ell^2}{32\pi^2 r^2} \tag{8.52}$$

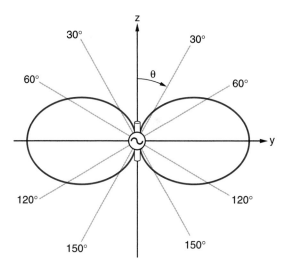

Figure 8.13 Polar plot of the Hertzian dipole's normalized radiation intensity. In three dimensions the pattern would appear toroidal.

The $\sin^2\theta$ term is the normalized radiation intensity $P_n(\theta)$ from (8.6) plotted in Figure 8.13.

The normalized radiation intensity can be inserted into (8.12) to find the pattern solid angle,

$$\Omega_p = \iint \sin^2\theta \, d\Omega = \iint \sin^2\theta \sin\theta \, d\theta \, d\phi \tag{8.53}$$

Upon integration we have $\Omega_p = 8\pi/3$. The directivity is then calculated as

$$D_{\text{max}} = \frac{4\pi}{\Omega_p} = 1.5 \tag{8.54}$$

The total power radiated by a Hertzian dipole can be calculated by inserting (8.52) into (8.19), giving

$$P_{\text{rad}} = r^2\left(\frac{\eta_o \beta^2 I_o^2 \ell^2}{32\pi^2 r^2}\right)\Omega_p = 40\pi^2\left(\frac{\ell}{\lambda}\right)^2 I_o^2 \tag{8.55}$$

Also, R_{rad} is found by equating (8.55) with (8.21), or

$$R_{\text{rad}} = 80\pi^2\left(\frac{\ell}{\lambda}\right)^2 \tag{8.56}$$

For Hertzian dipoles, where the dimension $\ell \ll \lambda$, R_{rad} will be fairly small and the antenna will not efficiently radiate power. Larger dipole antennas, described in the next section, have much higher R_{rad} and are therefore more efficient.

▶ **EXAMPLE 8.2**

A 3.00-mm-length dipole antenna is made of AWG#20 copper wire (0.406-mm radius). Suppose we drive the center of this antenna with a 1.00-GHz sinusoidal current. We want to estimate the efficiency and maximum power gain by treating this short antenna as a Hertzian dipole.

We first find the skin depth of copper at 1 GHz:

$$\delta_{Cu} = \frac{1}{\sqrt{\pi f \mu_o \sigma}} = \frac{1}{\sqrt{\pi \left(10^9 \frac{1}{s}\right)\left(4\pi \times 10^{-7} \frac{H}{M}\right)\left(5.8 \times 10^{-7} \frac{1}{\Omega\text{-}m}\right)\left(\frac{V\text{-}s}{H\text{-}A}\right)\left(\frac{\Omega\text{-}A}{V}\right)}}$$

$$= 2.09 \times 10^{-6} \text{ m}$$

This is much smaller than the wire radius ($a = .406 \times 10^{-3}$ m), so we can estimate the wire area over which current is conducted by

$$S = 2\pi a \delta_{Cu} = 5.33 \times 10^{-9} \text{m}^2$$

Now, the ohmic resistance of the small dipole can be calculated as

$$R_{diss} = \frac{1}{\sigma}\frac{l}{S} = 9.7 \text{ m}\Omega$$

To find the radiation resistance from (8.56), the wavelength at 1 GHz is 0.3 m and we have

$$R_{rad} = 80\pi^2 \left(\frac{3 \times 10^{-3} \text{ m}}{0.3 \text{ m}}\right)^2 = 79 \text{ m}\Omega$$

The efficiency from (8.23) is

$$e = \frac{79 \text{ m}\Omega}{79 \text{ m}\Omega + 9.7 \text{ m}\Omega} = 0.89$$

and with a directivity of 1.5, the power gain from (8.24) is

$$G_{max} = eD_{max} = (0.89)(1.5) = 1.34$$

Drill 8.4 A Hertzian dipole of length $\lambda/100$ is excited by a current $i = 1 \cos(2\pi \times 10^9 t)$ A. (a) Determine the maximum power density radiated by this antenna at a distance of 100 m. (b) What is the time-averaged power density at the point P(100, $\pi/4$, $\pi/2$)? (c) Find the radiation resistance. (*Answer:* 470 nW/m², (b) 236 nW/m², (c) 79 mΩ)

The Small Loop Antenna

Figure 8.14 shows a small loop of current located in the *x–y* plane centered at the origin. Such a small loop is known as a *small loop antenna* or sometimes a *magnetic dipole*. Recalling the general formula for the vector magnetic potential, we can make the substitution

$$\mathbf{J}_{ds}dv_d = I_s e^{-j\beta R_{do}} a d \phi\, \mathbf{a}_\phi \tag{8.57}$$

giving us

$$\mathbf{A}_{os} = \frac{\mu_o I_s a}{4\pi} \oint \frac{e^{-j\beta R_{do}}}{R_{do}} d\phi\, \mathbf{a}_\phi \tag{8.58}$$

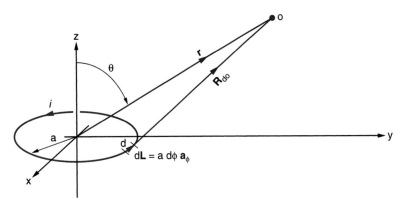

Figure 8.14 A small loop antenna (magnetic dipole).

We assume the loop is electrically small, or $a \ll \lambda$, and $\mathbf{A}_{os}$ is found in the far field. Even with these assumptions, solving the integral remains quite complicated, involving a pair of series expansions and some dexterous coordinate transformations. Eventually, it is discovered that

$$\mathbf{A}_{os} = \frac{\mu_o I_s S}{4\pi r^2}(1 + j\beta r)e^{-j\beta r}\sin\theta\,\mathbf{a}_\phi \tag{8.59}$$

where $S = \pi a^2$. Evaluating the curl of $\mathbf{A}_{os}$ and making the far-field assumption leads to

$$\mathbf{H}_{os} = \frac{-\omega\mu_o I_s S\beta}{4\pi\eta_o r}\sin\theta\,e^{-j\beta r}\mathbf{a}_\theta \tag{8.60}$$

and

$$\mathbf{E}_{os} = \frac{\omega\mu_o I_s S\beta}{4\pi r}\sin\theta\,e^{-j\beta r}\mathbf{a}_\phi \tag{8.61}$$

The power density vector is then

$$\mathbf{P}(r,\theta) = \left(\frac{\omega^2\mu_o^2 I_o^2 S^2\beta^2}{32\eta_o\pi^2 r^2}\right)\sin^2\theta\,\mathbf{a}_r \tag{8.62}$$

where

$$P_{max} = \frac{\omega^2\mu_o^2 I_o^2 S^2\beta^2}{32\eta_o\pi^2 r^2} \tag{8.63}$$

and since the normalized power function is the same as for the Hertzian dipole, $\Omega_p = 8\pi/3$ and $D_{max} = 1.5$.

Calculation of P_{rad} (in watts per square meter) and R_{rad} (Problem 8.15) yields

$$P_{rad} = \frac{4\eta_o\pi^3 I_o^2}{3}\left(\frac{S}{\lambda^2}\right)^2 \tag{8.64}$$

and

$$R_{rad} = 320\pi^4\left(\frac{S}{\lambda^2}\right)^2\,\Omega \tag{8.65}$$

The fields for the small loop antenna are very similar to that of a Hertzian dipole. Since it is the dual for the Hertzian (electric) dipole, the small loop antenna is often called a *magnetic dipole*.

The magnetic dipole equations are also valid for a multiturn loop, as long as the loops remain small compared to wavelength. For an *N*-loop coil, $S = N\pi a^2$ in the equations just derived. The loops are not required to be circular. To use the equations for a square coil of *N* loops, each of side length *b*, $S = Nb^2$.

Increasing the diameter of the loop antenna results in an increase in the radiation resistance and hence in the efficiency. Wrapping the loops around a ferrite core (i.e., a *ferrite-loop* antenna) concentrates magnetic flux in the loops and makes them appear larger. This is a common approach for constructing compact receiving antennas for AM radio.

Drill 8.5 A small loop antenna of radius $\lambda/100$ is excited by a current $i = 1 \cos(2\pi \times 10^9 t)$ A. (a) Determine the maximum power density radiated by this antenna at a distance of 100 m. (b) What is the time-averaged power density at the point $P(100, \pi/4, \pi/2)$? (c) Find the radiation resistance. (*Answer:* (a) 18 nW/m², (b) 9 nW/m², (c) 3.1 mΩ)

Drill 8.6 Repeat Drill 8.5 if 10 of the loops form a coil. (*Answer:* (a) 1.8 µW/m², (b) 0.92 µW/m², (c) 310 mΩ)

▷ 8.3 DIPOLE ANTENNAS

A drawback to the Hertzian dipole as a practical antenna is its very small radiation resistance. A longer dipole antenna, such as the one shown in Figure 8.15a, will have higher radiation resistance and will be a more efficient antenna. The T-line-fed dipole antenna in this

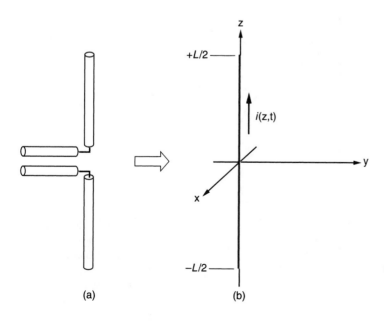

(a) (b)

Figure 8.15 (a) A T-line feeding a dipole antenna. (b) The dipole resembles a continuous line of current of length *L*.

figure is modeled in Figure 8.15b as an *L*-long conductor conveniently placed along the *z*-axis and supporting a current distribution $i(z, t)$.

Our analysis begins with division of the *L*-long dipole into a series of infinitesimal Hertzian dipoles. Over the differential length dz of any one Hertzian dipole, the current is assumed constant with value found from $i(z, t)$. The total magnetic field is then determined at a far-field point by integrating the field from the series of Hertzian dipoles. Finally, the electric field is found from $\mathbf{H}_{os}$ by

$$\mathbf{E}_{os} = -\eta_o \mathbf{a}_r \times \mathbf{H}_{os}$$

This section begins with a derivation of the far fields resulting from an arbitrary length dipole antenna. The fields and antenna properties are then discussed for several specific length dipoles, most notably the half-wave dipole antenna.

Derivation of Fields

Figure 8.16 shows the model for a general *L*-long dipole antenna. The first step in the analysis involves choosing a suitable current distribution. We know the current must go to zero at the bare ends of the conductor, but determining the exact distribution in the rest of the conductor is extremely complicated. Fortunately, a sinusoidal current distribution on each arm of the dipole, as shown in Figure 8.16, makes a good approximation. This distribution is best written as

$$i(z, t) = I_s(z)\cos\omega t \tag{8.66}$$

where

$$I_s(z) = \begin{cases} I_o e^{j\alpha} \sin\beta\left(\dfrac{L}{2} - z\right) & \text{for } 0 < z < L/2 \\[3mm] I_o e^{j\alpha} \sin\beta\left(\dfrac{L}{2} + z\right) & \text{for } -L/2 < z < 0 \end{cases} \tag{8.67}$$

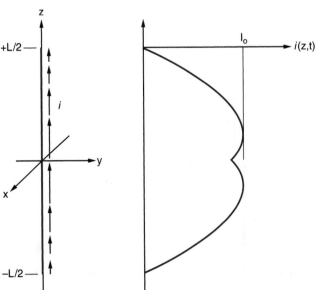

Figure 8.16 The current distribution on each arm of a dipole antenna is well approximated by a sinusoidal function.

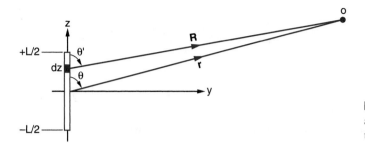

Figure 8.17 Dipole antenna parameters used to evaluate the far fields.

For simplicity in the following arguments, we will assume the phase term $\alpha = 0$.

To set up the integral, we make use of Figure 8.17, the current distribution of (8.67), and the magnetic field equation (8.49) for a Hertzian dipole. We find

$$\mathbf{H}_{os} = \frac{j\beta I_o \mathbf{a}_\phi}{4\pi} \left[\int_{-\frac{L}{2}}^{0} \frac{e^{-j\beta R}}{R} \sin\left(\beta \left(\frac{L}{2} + z \right) \right) \sin\theta' dz \right.$$

$$\left. + \int_{0}^{+\frac{L}{2}} \frac{e^{-j\beta R}}{R} \sin\left(\beta \left(\frac{L}{2} - z \right) \right) \sin\theta' dz \right]$$

(8.68)

Solving this integral will require some simplifying assumptions. In the far field, the vectors $\mathbf{r}$ and $\mathbf{R}$ appear to be parallel such that $\theta' \approx \theta$ and $R \approx r$. But we cannot substitute r for R in the phase term $e^{-j\beta R}$ where small differences are critical. In Figure 8.18, we see that the difference in length between R and r is $z\cos\theta$, or

$$R = r - z\cos\theta$$

We can then write

$$e^{-j\beta R} = e^{-j\beta(r - z\cos\theta)} = e^{-j\beta r} e^{j\beta z\cos\theta}$$

(8.69)

Now, after pulling the components that don't change with z (along the antenna's length) out of the integral, we have

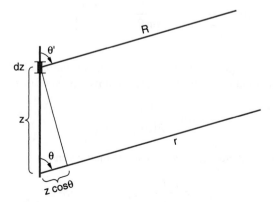

Figure 8.18 Expanded geometry near the dipole.

$$\mathbf{H}_{os} = \frac{j\beta I_o}{4\pi} \frac{e^{-j\beta r}}{r} \sin\theta\, \mathbf{a}_\phi \left[\int_{-\frac{L}{2}}^{0} e^{j\beta z \cos\theta} \sin\left(\beta\left(\frac{L}{2}+z\right)\right) dz \right.$$
$$\left. + \int_{0}^{+\frac{L}{2}} e^{j\beta z \cos\theta} \sin\left(\beta\left(\frac{L}{2}-z\right)\right) dz \right] \tag{8.70}$$

From the table of integrals in Appendix D we have

$$\int e^{ax}\cos(c+bx)\,dx = \frac{e^{ax}}{a^2+b^2}\left[a\sin(c+bx)-b\cos(c+bx)\right]$$

In our case, $x = z$, $a = j\beta\cos\theta$, $c = \beta L/2$, and $b = +\beta$ for the first integral and $b = -\beta$ for the second. After considerable algebra, including application of Euler's identity, we arrive at

$$\boxed{\mathbf{H}_{os} = H_{os}\mathbf{a}_\phi = \left(\frac{jI_o}{2\pi}\frac{e^{-j\beta r}}{r}\right)\left(\frac{\cos\left(\frac{\beta L}{2}\cos\theta\right)-\cos\left(\frac{\beta L}{2}\right)}{\sin\theta}\right)\mathbf{a}_\phi} \tag{8.71}$$

The vector $\mathbf{E}_{os}$ is then easily found from

$$\mathbf{E}_{os} = -\eta_o \mathbf{a}_r \times \mathbf{H}_{os}$$

to be

$$\boxed{\mathbf{E}_{os} = \eta_o H_{os}\mathbf{a}_\theta} \tag{8.72}$$

Finally, the time-averaged power radiated, using (8.41), is

$$\mathbf{P}(r,\theta) = \frac{15 I_o^2}{\pi r^2} F(\theta)\mathbf{a}_r \tag{8.73}$$

where the pattern function $F(\theta)$ is given by

$$F(\theta) = \left[\frac{\cos\left(\frac{\beta L}{2}\cos\theta\right)-\cos\left(\frac{\beta L}{2}\right)}{\sin\theta}\right]^2 \tag{8.74}$$

This pattern function is not generally equivalent to the normalized power function $P_n(\theta)$ since $F(\theta)$ can be greater than one. Instead, the normalized power function is

$$P_n(\theta) = \frac{F(\theta)}{F(\theta)_{max}} \tag{8.75}$$

where the maximum time-averaged power density is then given by

$$P_{max} = \frac{15 I_o^2}{\pi r^2} F(\theta)_{max} \tag{8.76}$$

Antenna Properties

The most straightforward way to determine characteristics of an arbitrary length dipole antenna is to use numerical (computer) routines, as detailed in the next three MATLAB examples.

As a first step, we can plot the normalized power function (8.75). In a numerical routine, $F(\theta)$ is calculated over the full range of θ for a length L given in terms of wavelength. Then a maximum is found ($F(\theta)_{max}$), and $P_n(\theta)$ is calculated. Examples of this procedure are seen in MATLAB 8.2 and MATLAB 8.3.

Next, the pattern solid angle Ω_p must be found. An analytical solution proves to be rather involved. We are much better served by using numerical integration to evaluate (8.12):

$$\Omega_p = \iint P_n(\theta,\phi)\,d\Omega$$

With a little manipulation, for a dipole antenna this becomes

$$\Omega_p = \frac{2\pi}{F(\theta)_{max}} \int \frac{\left[\cos\left(\frac{\beta L}{2}\cos\theta\right) - \cos\left(\frac{\beta L}{2}\right)\right]^2}{\sin\theta}\,d\theta \tag{8.77}$$

MATLAB 8.4 shows how numerical integration of (8.77) is carried out. Directivity follows since

$$D_{max} = \frac{4\pi}{\Omega_p}$$

We can find the radiation resistance by considering

$$P_{rad} = \frac{1}{2}I_o^2 R_{rad} = r^2 P(r,\theta)_{max}\,\Omega_p \tag{8.78}$$

which leads to

$$R_{rad} = \frac{30}{\pi}F(\theta)_{max}\,\Omega_p \tag{8.79}$$

> **MATLAB 8.2**

Let's devise a routine to plot the current distribution on a dipole antenna along with a polar plot of the normalized power radiated. In our program we select $L = 1.25\lambda$ for our test case.

```
%   M-File: ML0802
%
%   This plots the current distribution along an
%   arbitrary length dipole antenna as well as the
%   normalized radiation pattern.
%
%   Wentworth, 11/26/02
%
%   Variables
%   L             dipole length (in wavelengths)
%   bL2           phase constant * length/2
```

(continued)

```
% N          number of theta points
% th,thr     angle theta in deg.,radians
% z1,z2      <0 and >0 z position
% iz1,iz2    <0 and >0 z-dependent current
% F          un-normalized power function
% Fmax       maximum value for F
% Pn         normalized power function

clc          %clears the command window
clear        %clears variables
clf          %clear figure

%  Initialize variables
L=1.25;
bL2=pi*L;
N=360;

%  Calculate current distribution
z1=-(L/2):(L/200):0;
iz1=sin(2*pi*((L/2)+z1));
z2=0:(L/200):(L/2);
iz2=sin(2*pi*((L/2)-z2));

%  Calculate normalized power function
th=1:.1:N;
thr=th*pi./180;
F=((cos(bL2.*cos(thr))-cos(bL2))./sin(thr)).^2;
Fmax=max(F);
Pn=F./Fmax;

%  Generate Plots
subplot(211)
plot(z1,iz1,'-k',z2,iz2,'-k')
xlabel('z(in wavelengths)')
ylabel('current')
axis([-L/2 L/2 -1 1])
grid on
subplot(212),
polar(0,1)
hold on
polar(thr,Pn)
T=num2str(L);
S=strvcat('Length',T,'wavelengths');
text(1.2,.8,S)
```

The result is plotted in Figure 8.19.

▷ MATLAB 8.3

The previous MATLAB example can be enhanced to show a movie of the power function $F(\theta)$ as the length of the antenna grows from 0.1λ to 2.1λ. We choose to look at $F(\theta)$ rather than $P_n(\theta)$ so we can see the relative power level of the antenna as the length is changed.

Notice the use of the text command and a couple of string operators to place wavelength information on the plot.

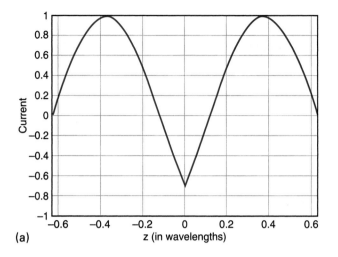

(a)

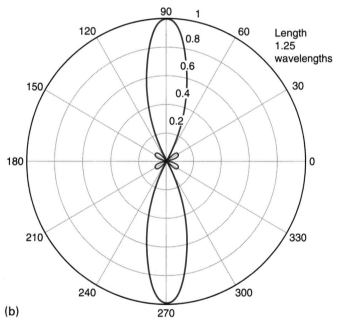

(b)

Figure 8.19 (a) The current distribution and (b) normalized radiation pattern as a function of θ for an $L = 1.25\lambda$ dipole antenna.

```
%   M-File: ML0803
%
%   Dipole antenna movie shows radiation pattern as
%   dipole length grows from 0.1 lambda to 2.1 lambda.
%
%   Wentworth, 11/26/02
%
%   Variables
%   L          dipole length (in wavelengths)
%   bL2        phase constant * length/2
```

(*continued*)

```
% N          number of theta points
% th,thr     angle theta in deg.,radians
% num,den    temporary variables
% F          un-normalized power function

clc          %clears the command window
clear        %clears variables

%   Initialize variables
N=360;
th=1:1:N;
thr=th*pi./180;

%   Generate Reference Frame
L=0.1;
polar(0,6);%sets scale for polar plot
T=num2str(L);
S=strvcat('Length',T,'wavelengths');
text(6,6,S)
axis manual
title('Linear Antenna Radiation Pattern')
hold on
pause

%   Make the Movie
L=0.1:0.02:2.1;
for n=1:100
    polar(0,6)
    axis manual
    title('Linear Antenna Radiation Pattern')
    T=num2str(L(n));
    S=strvcat('Length',T,'wavelengths');
    text(6,6,S)
    hold on
    num=cos(pi*L(n)*cos(thr))-cos(pi*L(n));
    den=sin(thr);
    F=(num./den).^2;
    polar(thr,F)
    hold off
    M(:,1)=getframe;
end
```

Figure 8.20 shows snapshots from the movie at $L = 0.5\lambda$, 1.0λ, and 1.5λ.

MATLAB 8.4

We want to numerically integrate (8.77). The integral can be written as a summation:

$$\Omega_p = \frac{2\pi}{F(\theta)_{max}} \sum_{i=1}^{N} \frac{\left[\cos\left(\frac{\beta L}{2}\cos\theta_i\right) - \cos\left(\frac{\beta L}{2}\right)\right]^2}{\sin\theta_i}\Delta\theta$$

where N is the number of increments and $\Delta\theta$ is the angular increment, given as π/N. The angle θ_i is calculated as

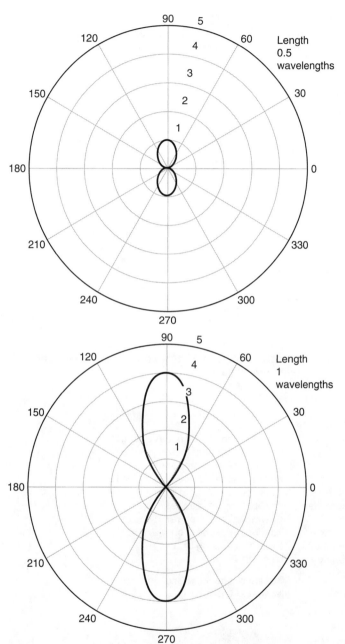

Figure 8.20 *x–y* plane snapshots of $F(\theta)$ taken at three dipole antenna lengths from ML0803.

$$\theta_i = i\,\pi/N$$

The routine is as follows:

```
%    M-File: ML0804
%
%    Perform numerical integration to find beam solid
```

(*continued*)

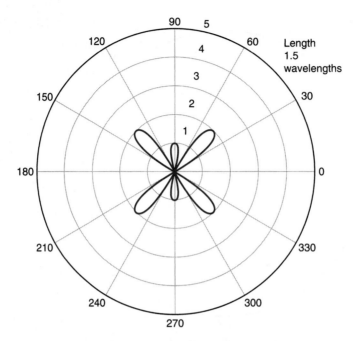

Length
1.5
wavelengths

Figure 8.20 (*continued*)

```
%    angle, directivity, and the maximum power function
%    for a given length dipole.
%
%    Wentworth, 11/26/02
%
%    Variables
%    L          dipole length (in wavelengths)
%    bL2        phase constant * length/2
%    N          number of theta points
%    th,thr     angle theta in degrees,radians
%    dth        differential theta
%    num,den    temporary variables
%    F          un-normalized power function
%    Fmax       max power function (W/m^2)
%    omegaP     beam solid angle (sr)
%    Dmax       Directivity

clc              %clears the command window
clear            %clears variables

%   Initialize variables
L=1.25;
bL2=pi*L;
N=90;

%   Perform calculations
i=1:1:N;
dth=pi/N;
th(i)=i*pi/N;
num(i)=cos(bL2.*cos(th(i)))-cos(bL2);
```

```
den(i)=sin(th(i));
F(i)=((num(i)).^2)./den(i);
Fmax=max(F);
Pn=F./Fmax;

omegaP=2*pi*dth*sum(Pn)
Dmax=4*pi/omegaP
Fmax
```

An accurate solution is found by repeating the routine for an increasing number of increments (N). With only 90 increments, for $L = 1.25\lambda$ we find $\Omega_p = 3.83$.

▷ **EXAMPLE 8.3**

Let's find the directivity and radiation resistance for a dipole of length 1λ.

From ML0802 we find the current distribution and normalized power pattern. To find D_{max} and R_{rad}, we must first use ML0804. Running this program at $L = 1\lambda$, we obtain

$$\Omega_p = 5.2121$$

$$D_{max} = 2.4110$$

$$F_{max} = 4$$

So we see that this antenna is more directive than the Hertzian dipole. We find the radiation resistance from (8.79), where

$$R_{rad} = \frac{30}{\pi} F_{max}\Omega_p = \frac{30}{\pi}(4.0)(5.21) = 200\ \Omega$$

Half-Wave Dipole

Because of its convenient radiation resistance, and because it is the smallest resonant dipole antenna, the half-wavelength (or simply half-wave) dipole antenna merits special attention. With $(\beta L/2) = \pi/2$, the field equations reduce to simpler expressions leading to the time-averaged power radiated,

$$\mathbf{P}(r,\theta) = \left(\frac{15I_o^2}{\pi r^2}\right)\left(\frac{\cos^2\left(\frac{\pi}{2}\cos\theta\right)}{\sin^2\theta}\right)\mathbf{a}_r \tag{8.80}$$

The maximum value of the $F(\theta)$ term in (8.80) is 1, so we see that the maximum power density is

$$P_{max} = \frac{15I_o^2}{\pi r^2} \tag{8.81}$$

and the normalized power density is

$$P_n(\theta) = \frac{\cos^2\left(\frac{\pi}{2}\cos\theta\right)}{\sin^2\theta} \tag{8.82}$$

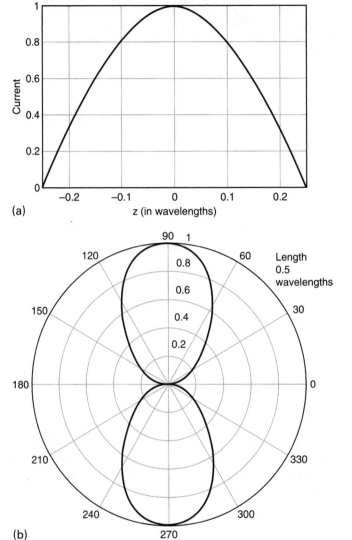

Figure 8.21 (a) The current distribution and (b) the normalized radiation pattern as a function of θ for a half-wave dipole antenna.

This pattern is the first one plotted in Figure 8.20. The current distribution and normalized power function from MATLAB 8.2 are shown in Figure 8.21. Another view of this pattern, with antenna included, is shown in Figure 8.22. Note the similarity to the pattern for a Hertzian dipole (Figure 8.13).

Numerical integration in MATLAB 8.4 gives us $\Omega_p = 7.658$ for the $L = \lambda/2$ case. We can then find directivity as

$$D_{max} = \frac{4\pi}{\Omega_p} = 1.640$$

which is only slightly higher than the directivity of a Hertzian dipole antenna.

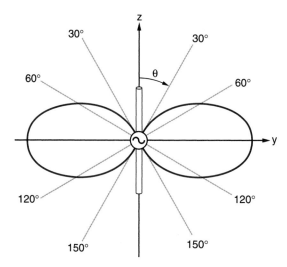

Figure 8.22 Radiation pattern for a half-wave dipole antenna.

Next, we can determine the radiation resistance by

$$P_{rad} = \frac{1}{2}I_o^2 R_{rad} = r^2 P_{max}\Omega_p \tag{8.83}$$

which leads to

$$R_{rad} = \frac{30}{\pi}\Omega_p = 73.2\,\Omega \tag{8.84}$$

With this R_{rad} much higher than that of a Hertzian dipole, the half-wave dipole radiates much more efficiently. In addition, it is much easier to construct an impedance matching network for this antenna impedance.

The antenna impedance also contains a reactive component, X_{ant}. This component depends on the behavior of the fields in the near-field region of the antenna, and its derivation for dipole antennas lies well beyond the scope of this text. However, for a $\lambda/2$ dipole antenna it is equal to $42.5\,\Omega$. Therefore, the impedance of a half-wave dipole antenna, neglecting R_{diss}, is

$$Z_{ant} = 73.2 + j42.5\,\Omega \tag{8.85}$$

For impedance matching, it is very convenient to operate the antenna where its reactive component is zero (i.e., in a *resonant* condition). This can be achieved by making the antenna slightly shorter. Although the exact length depends on the antenna's wire radius, a dipole antenna of about 0.485λ length will have an approximately zero reactive component and its real part will be close to $73\,\Omega$.

▶ **EXAMPLE 8.4**

Let's find the efficiency and maximum power gain of a $\lambda/2$ dipole antenna constructed with AWG#20 copper wire operating at 1.0 GHz.

At 1.0 GHz, the wavelength is 0.30 m and the $\lambda/2$ dipole is 0.15 m long. From Example 8.2, we found the area over which current is conducted in AWG#20 wire at 1 GHz to be 5.33×10^{-9} m^2. The ohmic resistance is then

$$R_{\text{diss}} = \frac{1}{\sigma}\frac{\ell}{S} = 0.485 \ \Omega$$

Since the radiation resistance is 73.2 Ω, we have

$$e = \frac{73.2 \ \Omega}{73.2 \ \Omega + 0.485 \ \Omega} = 0.99$$

and a gain of

$$G_{\text{max}} = eD_{\text{max}} = (0.993)(1.640) = 1.63$$

This antenna is clearly more efficient with a higher gain than the short dipole of Example 8.2.

▷ **EXAMPLE 8.5**

Suppose a 0.485λ dipole transmitting antenna's power source is a 12-V amplitude voltage in series with a 25-Ω source resistance as shown in Figure 8.23a. We want to determine the total power radiated from the antenna with and without insertion of a matching network.

Referring to Figure 8.23b, we have

$$I_S = \frac{V_S}{R_S + R_{\text{rad}}} = \frac{12 \ \text{V}}{98 \ \Omega} = 122 \ \text{mA}$$

Then,

$$P_{\text{rad}} = \frac{1}{2}|I_S|^2 R_{\text{rad}} = \frac{1}{2}(122 \ \text{mA})^2 (73 \ \Omega) = 547 \ \text{mW}$$

With matching in place as shown in Figure 8.23c, we have $Z_{\text{in}} = 25 \ \Omega$ and the current in the network is

$$I_S = \frac{12 \ \text{V}}{50 \ \Omega} = 240 \ \text{mA}$$

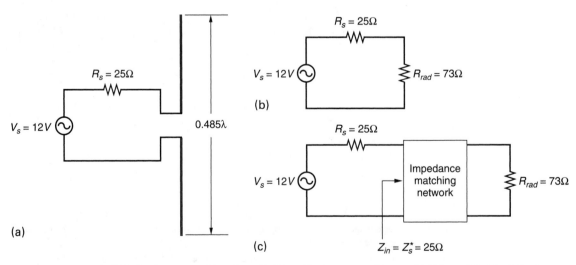

Figure 8.23 (a) A 0.485λ dipole antenna driven by a 12-V amplitude source with $R_s = 25 \ \Omega$ is modeled in (b) by its antenna impedance. In (c), a matching network is added.

The power entering the matching network is all radiated:

$$P_{rad} = \frac{1}{2}(240 \text{ mA})^2(73 \text{ }\Omega) = 2.1 \text{ W}$$

Thus, for this particular example, insertion of a matching network results in almost a quadrupling of the radiated power.

Drill 8.7   Suppose a λ/2 dipole antenna is used in the previous problem. Recalculate the radiated power with and without the matching network. (*Answer:* 459 mW without matching, 2.1 W with matching)

▷ 8.4   MONOPOLE ANTENNAS

Let us consider the construction of a half-wave dipole antenna for an AM radio station broadcasting at 1 MHz. At this frequency, a wavelength is 300 m long and the half-wave dipole antenna must be 150 m tall. We can cut this length in half, and save considerable expense, by employing *image theory* to build a quarter-wave *monopole antenna* that is only 75 m tall.

Following a brief description of image theory, we will look at the radiation pattern and antenna properties of a quarter-wave monopole antenna.[8.5]

Image Theory

Consider the pair of charges, +Q and –Q (termed an *electric dipole* as described in Example 2.20) in Figure 8.24a. The dashed line shows the location of a zero-potential surface. The field lines are normal to this surface. If we slide a conductive plane over the zero-potential surface, we see in Figure 8.24b that the field lines in the upper half-plane are unchanged. Conversely, the field for a charged object over a conductive plane appears exactly the same as if the plane were removed and a mirror image of opposite charge were inserted.

Figure 8.25a shows some charge +Q distributed in the shape of a tilted letter "A" over a conductive plane. The field above the conductive plane is exactly the same as the field above the zero-potential surface of Figure 8.25b, where the opposite charge image has been

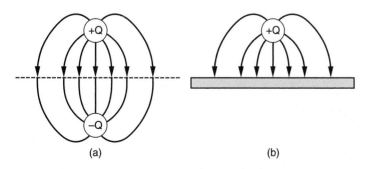

(a)   (b)

Figure 8.24   (a) The zero-potential surface between a pair of opposite but equal charges can be replaced by a conductive plane, as shown in (b).

[8.5]Such an antenna is sometimes referred to as a "Marconi antenna."

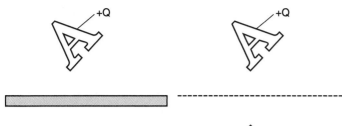

(a)   (b)

Figure 8.25 The charge over a conductive surface shown in (a) can be modeled in (b) by replacing the conductor with a mirrored image charge of opposite polarity.

inserted as shown. Note that the charge can be in any distribution (point charge, line charge, surface charge, or volume charge), and the image charge is a mirror image of opposite polarity.

Antenna Properties

A monopole antenna is excited by a current source at its base. By image theory, the current in the image monopole will be in the same direction as the current in the actual monopole. The pair of monopoles thus resembles a dipole antenna.

A monopole antenna placed over a conductive plane and half the length of a corresponding dipole antenna will have identical field patterns in the upper half-plane. The most common realization of this type of antenna is the quarter-wave monopole antenna, shown in Figure 8.26.

For the upper half-plane ($0 \leq \theta \leq 90°$), the time-averaged power for the quarter-wave monopole is exactly the same as for the half-wave dipole (8.79), with the same maximum power density and normalized power density. However, the pattern solid angle is different. Since the normalized power density $P_n(\theta) = 0$ for $90° \leq \theta \leq 180°$, the pattern solid angle found by

$$\Omega_p = \iint P_n(\theta)\, d\Omega$$

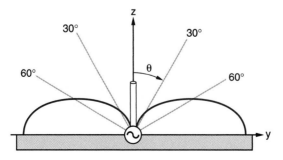

Figure 8.26 Quarter-wave monopole antenna.

integrated over all space will be half the value of Ω_p for the corresponding half-wave dipole. So for the quarter-wave monopole antenna we have $\Omega_p = 3.829$. The directivity is therefore doubled, or

$$D_{max} = \frac{4\pi}{\Omega_p} = 3.28$$

Finally, recalling the relationship between Ω_p and R_{rad} in (8.79), we see that the radiation resistance is halved:

$$R_{rad} = \frac{30}{\pi}\Omega_p = 36.6 \ \Omega$$

The reactive part of the antenna impedance is also halved, so for a quarter-wave monopole antenna we have

$$Z_{ant} = 36.6 + j21.25 \ \Omega$$

Like the half-wave dipole, a slight decrease in the monopole length will result in a zero reactive component with little effect on R_{rad}.

▷ **EXAMPLE 8.6**

Let's find the pattern solid angle, directivity, and radiation resistance for a $5\lambda/8$ monopole antenna.

This monopole has the same field and radiation pattern as a 1.25λ dipole antenna for $0 < \theta < 90°$. We can execute ML0804 to find

$$\Omega_p\big|_{1.25\lambda \text{ dipole}} = 3.828$$

The monopole will have half this value:

$$\Omega_p\big|_{0.625\lambda \text{ monopole}} = 1.914$$

The directivity is then calculated as

$$D_{max} = 6.67$$

or twice the value of the corresponding dipole antenna.

Calculating the radiation resistance requires that we know the maximum power, given by

$$P_{max} = \frac{15I_o^2}{\pi r^2}F_{max}$$

where F_{max} for the 0.625λ monopole is the same as for the 1.25λ dipole and is equal to 2.914 from ML0804. Then, from the relation

$$P_{rad} = r^2 P_{max}\Omega_p = \frac{1}{2}I_o^2 R_{rad}$$

we find

$$R_{rad} = \frac{30}{\pi}F_{max}\Omega_p = 53.3 \ \Omega$$

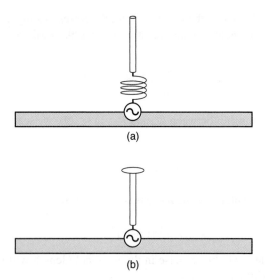

Figure 8.27 Shortening a monopole antenna by using (a) an inductive coil and (b) a top-hat capacitor.

Practical Considerations

The best operation of a monopole antenna is achieved when the ground beneath it is highly conductive, a condition well met by damp ground. A buried grid of wires can also be used to improve the conductivity. If the antenna is remote from actual ground (for instance, on top of a building), a conductive screen of diameter roughly twice the antenna height (called a *counterpoise*) may be placed under the antenna. For best operation this counterpoise must be isolated from true ground.

Monopole antennas have been widely used for hand-held devices. Making these antennas shorter is desirable for such applications. A problem that arises with antennas shorter than a quarter wavelength is that the input impedance becomes highly capacitive, which reduces efficiency. One approach to solving this problem is to place an inductive loading coil at the base of the monopole, as shown in Figure 8.27a. This series inductance offsets or tunes out the capacitance of the monopole. However, the coil may suffer from significant resistive losses. Also, the current amplitude will be highest at this end of the monopole. This doesn't necessarily help our monopole's radiated fields since the inductive coil doesn't radiate in the same direction.

Another approach is to place a capacitive plate atop the monopole, as shown in Figure 8.27b. This is sometimes referred to as a *top-hat* monopole. The plate appears as a series capacitance to the capacitive reactance of the monopole, thus lowering the total capacitive reactance. Also, the capacitor prevents current amplitude from going to zero at the end of the wire and that improves radiation.

Table 8.1 provides a summary of the key parameters for four types of simple antennas.

▶ 8.5 ANTENNA ARRAYS

The antennas we have studied so far have all been omnidirectional—there has been no variation with ϕ. A properly spaced collection of antennas, such as the Ka-band array shown in Figure 8.28, can have significant variation with ϕ, leading to dramatic improvements in

TABLE 8.1 Summary of Key Antenna Parameters

Parameter	Symbol	Hertzian dipole	Magnetic dipole (small loop)	Half-wavelength dipole	Quarter-wavelength monopole ($0 < \theta < 90°$)
Magnetic field intensity (A/m) (far field)	$\mathbf{H}_{os}$	$\dfrac{jI_s\beta\ell}{4\pi}\dfrac{e^{-j\beta r}}{r}\sin\theta\,\mathbf{a}_\phi$	$\dfrac{\omega\mu_o I_s\beta S}{-4\pi\eta_o}\dfrac{e^{-j\beta r}}{r}\sin\theta\,\mathbf{a}_\theta$	$\dfrac{jI_o}{2\pi}\dfrac{e^{-j\beta r}}{r}\dfrac{\cos\left(\frac{\pi}{2}\cos\theta\right)}{\sin\theta}\mathbf{a}_\phi$	$\dfrac{jI_o}{2\pi}\dfrac{e^{-j\beta r}}{r}\dfrac{\cos\left(\frac{\pi}{2}\cos\theta\right)}{\sin\theta}\mathbf{a}_\phi$
Time-averaged power density vector (W/m²)	$\mathbf{P}(r,\theta)$	$P_{max}P_n(\theta)\mathbf{a}_r$	$P_{max}P_n(\theta)\mathbf{a}_r$	$P_{max}P_n(\theta)\mathbf{a}_r$	$P_{max}P_n(\theta)\mathbf{a}_r$
Maximum power density (W/m²)	P_{max}	$\dfrac{\eta_o I_o^2\beta^2\ell^2}{32\pi^2 r^2}$	$\dfrac{1}{32\eta_o}\left(\dfrac{\omega\mu_o I_o\beta S}{\pi r}\right)^2$	$\dfrac{15 I_o^2}{\pi r^2}$	$\dfrac{15 I_o^2}{\pi r^2}$
Normalized power function	$P_n(\theta)$	$\sin^2(\theta)$	$\sin^2(\theta)$	$\left(\dfrac{\cos\left(\frac{\pi}{2}\cos\theta\right)}{\sin\theta}\right)^2$	$\left(\dfrac{\cos\left(\frac{\pi}{2}\cos\theta\right)}{\sin\theta}\right)^2$
Pattern solid angle (sr)	Ω_p	$\dfrac{8\pi}{3}$	$\dfrac{8\pi}{3}$	7.658	3.829
Directivity	D_{max}	1.5	1.5	1.64	3.28
Total radiated power (W)	P_{rad}	$40\pi^2\left(\dfrac{\ell}{\lambda}\right)^2 I_o^2$	$\dfrac{4\eta_o\pi^3 I_o^2}{3}\left(\dfrac{S}{\lambda^2}\right)^2$	$r^2 P_{max}\Omega_p$	$r^2 P_{max}\Omega_p$
Radiation resistance (Ω)	R_{rad}	$80\pi^2\left(\dfrac{\ell}{\lambda}\right)^2$	$320\pi^4\left(\dfrac{S}{\lambda^2}\right)^2$	73.2	36.6

Figure 8.28 A Ka-band array. Courtesy of Harris Corporation.

directivity. An antenna array can also be designed to give a particular shape to the radiating pattern. Moreover, control of the phase and amplitude of the current driving each array element, along with spacing of the elements of the array, can provide beam steering capability.

It should be understood that advanced arrays can be and are constructed using different antenna elements driven by current of varying phase and amplitude and that the main beam can be pointed in practically any direction. However, to simplify our treatment of arrays we will consider the following:

1. All antenna elements in the array are identical.
2. The current amplitude is the same feeding each element.
3. The radiation pattern lies in the x–y plane only ($\theta = \pi/2$).

We will then control the radiation pattern by

1. controlling the spacing between elements and/or
2. controlling the phase of the current driving each element.

Typically, a common current source is equally divided and fed into each element. Also, each feed line may contain a microwave component called a *phase shifter* for control of the current phase.

As a simple example, consider a pair of dipole antennas driven by in-phase current sources and separated by $\lambda/2$ on the x-axis as represented by Figure 8.29a. We assume each antenna radiates independently, and in fact the radiation pattern for a single dipole element is shown by the dashed circle around the origin.

At a far-field point along the x-axis (point P), the fields from the two antennas will be 180° out of phase owing to the extra $\lambda/2$ distance traveled by the wave from the farthest antenna. The fields cancel in this direction. But along the y-axis (point Q), the fields are in phase and thus add. The electric field intensity is then twice that of a single dipole, and a doubling of the field results in a four-fold increase in the power. This is a simple example

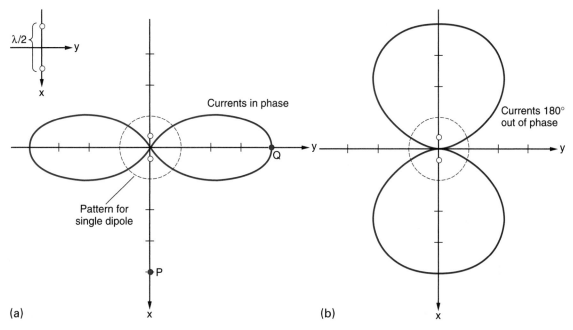

Figure 8.29 A pair of dipole antennas are oriented in the z direction and separated by $\lambda/2$ in the x direction. The radiated power pattern in the x–y plane ($\theta = \pi/2$) is shown for current sources that are (a) in-phase and (b) out of phase by 180°.

of a *broadside* array, so called because the maximum radiation is directed broadside to the axis containing the elements of the array.

We can modify this example by now driving the pair of dipoles with current sources that are 180° out of phase. Then, along the x-axis the fields will now be in phase and along the y-axis they will be out of phase. The resulting beam pattern will appear as in Figure 8.29b. This is a simple example of what is termed an *endfire* array, since the maximum radiation is directed at the ends of the axis containing the array elements.

We will begin by deriving the pattern functions for a pair of Hertzian dipoles. This will be accomplished simply by adding the electric field intensity for each dipole, and then finding the power. The result is then applicable for a pair of any arbitrary length dipole antennas. Then we will look at n-element linear arrays.

Pair of Hertzian Dipoles

We recall from (8.50) that the far-field value of the electric field resulting from a Hertzian dipole at the origin is

$$\mathbf{E}_{os} = j\eta_o \frac{I_s \ell \beta}{4\pi} \frac{e^{-j\beta r}}{r} \sin\theta \, \mathbf{a}_\theta$$

Since we are confining our discussion to the x–y plane where $\theta = \pi/2$, we have

$$\mathbf{E}_{os} = j\eta_o \frac{I_s \ell \beta}{4\pi} \frac{e^{-j\beta r}}{r} \, \mathbf{a}_\theta \tag{8.86}$$

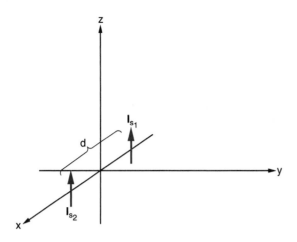

Figure 8.30 A pair of z-oriented Hertzian dipole antennas separated by a distance d on the x-axis.

Now consider a pair of z-oriented dipoles separated by a distance d on the x-axis as shown in Figure 8.30. The total field is the vector sum of the fields for both dipoles:

$$\mathbf{E}_{os(tot)} = \mathbf{E}_{os_1} + \mathbf{E}_{os_2} = j\eta_o \frac{I_{s_1}\ell\beta}{4\pi}\frac{e^{-j\beta r_1}}{r_1}\mathbf{a_\theta} + j\eta_o \frac{I_{s_2}\ell\beta}{4\pi}\frac{e^{-j\beta r_2}}{r_2}\mathbf{a_\theta} \qquad (8.87)$$

where r_1 and r_2 are as shown in Figure 8.31a. In accordance with our simplifying assumptions, we will keep the magnitudes of the currents the same but will insert a phase shift α between them. We have

$$I_{s_1} = I_o$$
$$I_{s_2} = I_o e^{j\alpha} \qquad (8.88)$$

To further develop our total field equation, we assume that since the far-field observation point is far away, the distance vectors can be accurately modeled as parallel lines as shown in Figure 8.31b. We can therefore assume

$$\phi_1 \approx \phi \approx \phi_2 \qquad (8.89)$$

We can further assume

$$r_1 \approx r \approx r_2 \qquad (8.90)$$

for the distances in the denominator of (8.87), since a small difference will have very little effect on the resultant magnitude. But the exponential term represents a phase, and small differences in distance can result in sizable differences in phase. By inspection of the geometry we see that

$$r_1 = r + \frac{d}{2}\cos\phi, \qquad r_2 = r - \frac{d}{2}\cos\phi \qquad (8.91)$$

Inserting (8.88), (8.89), and (8.91) into (8.87) we now have

$$\mathbf{E}_{os(tot)} = j\eta_o \frac{I_o\ell\beta}{4\pi}\frac{e^{-j\beta r}}{r} e^{j\alpha/2}\left[e^{-j\left(\beta\frac{d}{2}\cos\phi+\frac{\alpha}{2}\right)} + e^{j\left(\beta\frac{d}{2}\cos\phi+\frac{\alpha}{2}\right)} \right]\mathbf{a_\theta} \qquad (8.92)$$

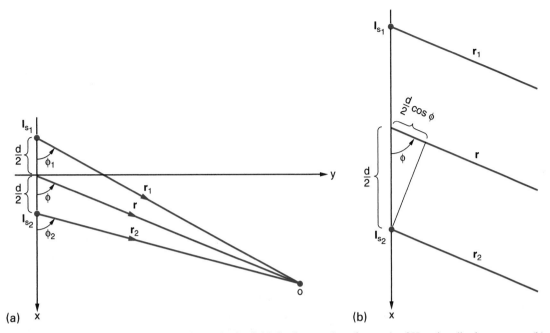

Figure 8.31 (a) Parameters used to evaluate the far fields in the *x–y* plane for a pair of Hertzian dipole antennas. (b) Expanded geometry near the dipoles.

The exponentials inside the brackets are converted using Euler's identity to give us

$$\mathbf{E}_{os(tot)} = E_{os(tot)}\mathbf{a}_\theta = j\eta_o \frac{I_o\ell\beta}{4\pi}\frac{e^{-j\beta r}}{r}\left[2e^{j\frac{\alpha}{2}}\cos\left(\beta\frac{d}{2}\cos\phi + \frac{\alpha}{2}\right)\right]\mathbf{a}_\theta \tag{8.93}$$

This is the expression for the total electric field at a far-field observation point resulting from our two-element Hertzian dipole array.

It is then a simple matter to find the radiated power using

$$\mathbf{P}\left(r,\frac{\pi}{2},\phi\right) = \frac{1}{2}\mathrm{Re}\left[\mathbf{E}_s\times\mathbf{H}_s^*\right] = \frac{1}{2}\eta_o E_{os(tot)}^2\mathbf{a}_r$$

We then have

$$\mathbf{P}\left(r,\frac{\pi}{2},\phi\right) = \left(\frac{\eta_o\beta^2 I_o^2\ell^2}{32\pi^2 r^2}\right)\left[4\cos^2\left(\beta\frac{d}{2}\cos\phi + \frac{\alpha}{2}\right)\right]\mathbf{a}_r \tag{8.94}$$

This expression for radiated power can be written

$$\boxed{\mathbf{P}\left(r,\frac{\pi}{2},\phi\right) = F_{unit}F_{array}\mathbf{a}_r} \tag{8.95}$$

where the radiated power is divided into a *unit factor* F_{unit} and an *array factor* F_{array}. The unit factor F_{unit} is the maximum time-averaged power density for an individual antenna element, or unit, at $\theta = \pi/2$:

$$F_{\text{unit}} = \left(\frac{\eta_o \beta^2 I_o^2 \ell^2}{32\pi^2 r^2} \right) \tag{8.96}$$

The array factor F_{array} is

$$F_{\text{array}} = 4\cos^2\left(\frac{\psi}{2}\right) \tag{8.97}$$

where

$$\psi = \beta d \cos\phi + \alpha \tag{8.98}$$

This is the pattern function resulting from an array of two isotropic radiators. It only depends on the separation between the antenna units and the relative current phases driving them. In other words, we can use this particular array factor for any pair of identical arbitrary length dipoles.

As evidenced by (8.95), the resulting radiation pattern for an array of identical elements is the product of the unit factor and the array factor. This is called the principle of *pattern multiplication*.

▷ **EXAMPLE 8.7**

Let's use the principle of pattern multiplication to find the far-field radiation pattern for a pair of half-wave dipole antennas as shown in Figure 8.32a. The $\lambda/2$-long antennas are driven in phase and are $\lambda/2$ apart. Also we'll find the maximum power density 1 km away from the array if each antenna is driven by a 1-mA amplitude current source at 100 MHz.

At 100 MHz, $\lambda = 3$ m, so 1 km away is definitely in the far field. For a half-wave dipole, we have from (8.80)

$$P(r,\theta) = P(r,\theta)a_r = \left(\frac{15I_o^2}{\pi r^2}\right)\left(\frac{\cos^2\left(\frac{\pi}{2}\cos\theta\right)}{\sin^2\theta}\right)a_r$$

Here, the unit factor is found by evaluating $P(r,\theta)$ at $\theta = \pi/2$, or

$$F_{\text{unit}} = \frac{15I_o^2}{\pi r^2}$$

Now, for a pair of dipoles we evaluate F_{array} from (8.97), where $d = \lambda/2$ and $\alpha = 0$ (since the antennas are driven at the same phase). This gives us

$$F_{\text{array}} = 4\cos^2\left(\frac{\pi}{2}\cos\phi\right)$$

For the array, we now have

$$P\left(r,\theta=\frac{\pi}{2},\phi\right) = F_{\text{unit}} F_{\text{array}} = \frac{60I_o^2}{\pi r^2}\cos^2\left(\frac{\pi}{2}\cos\phi\right)$$

The normalized power function is

$$P_n\left(\frac{\pi}{2},\phi\right) = \frac{P\left(r,\frac{\pi}{2},\phi\right)}{P\left(r,\frac{\pi}{2},\phi\right)_{\max}} = \cos^2\left(\frac{\pi}{2}\cos\phi\right)$$

which is plotted in Figure 8.32b.

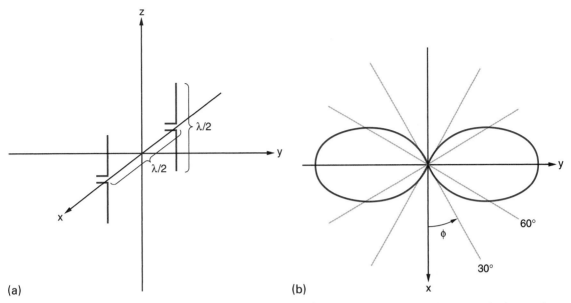

Figure 8.32 (a) A pair of half-wave dipole antennas separated by $\lambda/2$, (b) The resulting radiation pattern in the x–y plane.

The maximum radiated power density at 1000 m is

$$P_{\max} = \frac{60I_o^2}{\pi r^2} = \frac{60\left(10^{-3}\right)^2}{\pi(1000)^2} = 19 \ \frac{\text{pW}}{\text{m}^2}$$

The ϕ location of nulls in the radiated power pattern can be analytically determined by finding the values of ϕ that make $\cos(\psi/2) = 0$. Likewise, the ϕ locations of maximum radiation are found where the derivative with respect to ϕ is zero, that is,

$$\frac{d\left(\cos\left(\dfrac{\psi}{2}\right)\right)}{d\phi} = 0$$

N-Element Linear Arrays

The procedure for the two-element array can be extended for an arbitrary number of array elements. We make the following additional simplifying assumptions:

1. The array is *linear*, meaning the antenna elements are evenly spaced along a line.
2. The array is *uniform*, meaning each antenna element is driven by the same magnitude current source, with constant phase difference between adjacent elements.

Figure 8.33 shows a general uniform linear array of N elements along the x-axis with spacing d between each element. The phase increases by α for each element, so

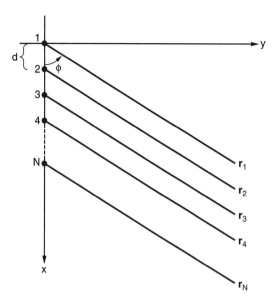

Figure 8.33 A uniform linear array.

$$I_{S_1} = I_o, I_{S_2} = I_o e^{j\alpha}, I_{S_3} = I_o e^{j2\alpha}, \ldots, I_{S_N} = I_o e^{j(N-1)\alpha} \tag{8.99}$$

Summing the contributions to the far-field electric field intensity, we have

$$\mathbf{E}_{os(tot)} = \left(j\eta_o \frac{I_o l \beta}{4\pi} \frac{e^{-j\beta r}}{r} \right) \left[1 + e^{j\psi} + e^{j2\psi} + \cdots + e^{j(N-1)\psi} \right] \mathbf{a}_r \tag{8.100}$$

where again

$$\psi = \beta d \cos \phi + \alpha$$

Manipulation of this series eventually gives

$$F_{array} = \frac{\sin^2 \left(\dfrac{N\psi}{2} \right)}{\sin^2 \left(\dfrac{\psi}{2} \right)} \tag{8.101}$$

Determining F_{array} at the limit as ψ approaches zero yields a maximum value

$$(F_{array})_{max} = N^2 \tag{8.102}$$

The normalized power pattern for these elements is therefore

$$P_n(\phi) = \frac{F_{array}}{(F_{array})_{max}} = \frac{1}{N^2} \frac{\sin^2 \left(\dfrac{N\psi}{2} \right)}{\sin^2 \left(\dfrac{\psi}{2} \right)} \tag{8.103}$$

▷ **EXAMPLE 8.8**

Let's verify that (8.101) for a pair of Hertzian dipoles reduces to (8.97).

For $N = 2$, (8.101) becomes

$$F_{array} = \frac{\sin^2 \psi}{\sin^2 \left(\dfrac{\psi}{2} \right)}$$

The numerator can be written as

$$\sin^2 \psi = 1 - \cos^2 \psi = (1 + \cos\psi)(1 - \cos\psi)$$

and the denominator can be written, via the half-angle formula for the squared sine (see Appendix D), as

$$\sin^2 \left(\frac{\psi}{2} \right) = \frac{1}{2}(1 - \cos\psi)$$

Dividing, we have

$$F_{array} = 2(1 + \cos\psi)$$

Now, using the half-angle formula for a squared cosine, this becomes

$$F_{array} = 4\cos^2 \left(\frac{\psi}{2} \right) = 4\cos^2 \left(\frac{\beta d}{2} \cos\phi + \frac{\alpha}{2} \right)$$

▷ **EXAMPLE 8.9**

Let's plot the normalized radiation pattern for five antenna elements spaced $\lambda/4$ apart with progressive phase steps of 30°. The antennas are assumed to be a linear array of z-oriented dipoles on the x-axis, and we want the normalized radiation pattern in the x–y plane.

To find the array factor we first evaluate ψ as

$$\psi = \beta d \cos\phi + \alpha$$

$$= \frac{2\pi}{\lambda} \frac{\lambda}{4} \cos\phi + 30° \left(\frac{\pi \text{ radians}}{180°} \right) = \frac{\pi}{2} \cos\phi + \frac{\pi}{6}$$

Inserting this ratio into (8.101) we have

$$F_{array} = \frac{\sin^2 \left(\dfrac{5\pi}{4} \cos\phi + \dfrac{5\pi}{12} \right)}{\sin^2 \left(\dfrac{\pi}{4} \cos\phi + \dfrac{\pi}{12} \right)}$$

and

$$(F_{array})_{max} = N^2 = 25$$

The normalized radiation pattern is then

$$P_n = \frac{1}{25} \frac{\sin^2 \left(\dfrac{5\pi}{4} \cos\phi + \dfrac{5\pi}{12} \right)}{\sin^2 \left(\dfrac{\pi}{4} \cos\phi + \dfrac{\pi}{12} \right)}$$

This is plotted via MATLAB in Figure 8.34.

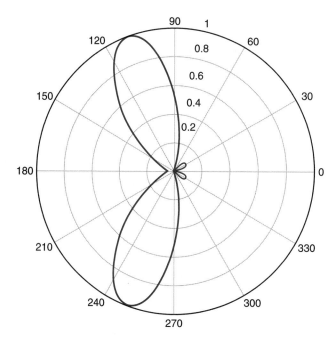

Figure 8.34 *x–y* plane radiation pattern versus ϕ for the five-element dipole array of Example 8.9.

Parasitic Arrays

Not all the elements in an array need be directly driven by a current source. A *parasitic array* features typically one driven element and several parasitic elements. The parasitic elements are indirectly driven by currents induced in them from the driven element.

The best-known parasitic array is a linear one called the *Yagi–Uda* antenna, shown in Figure 8.35. On one side of the driven element is a *reflector*, whose length and spacing are chosen to cancel most of the radiation in that direction and to enhance the radiation in the forward, or main-beam, direction. Several *directors* (typically four to six) may be added that help focus the main beam in the forward direction. The Yagi–Uda antenna features high gain and is fairly easy to construct. Many rooftop television antennas are of this type.

Parasitic elements tend to pull down the radiation resistance of the driven element. For instance, the radiation resistance of a half-wave dipole antenna would drop from 73 Ω to

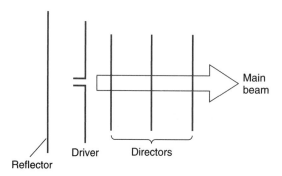

Figure 8.35 The Yagi–Uda antenna.

about 20 Ω when used as the driven element in a Yagi–Uda antenna. Since a higher radiation resistance is more efficient, often a half-wavelength *folded dipole* antenna is used as the driven element as it has roughly four times the radiation resistance of the half-wave dipole. The folded dipole antenna will be described in Section 8.8.

► 8.6 THE FRIIS TRANSMISSION EQUATION

Antennas are used in communications systems to both transmit and receive electromagnetic energy. How well the energy is exchanged is quantified by the Friis transmission equation.

Consider a pair of horn antennas as shown in Figure 8.36. A horn antenna is a transition element that launches electromagnetic energy from a waveguide into space. The particular antennas in Figure 8.36 are shown aligned with each other and with the same polarization so that power transmission for a particular separation distance R is maximized. The radiated power density from horn 1 at the location of horn 2 is

$$P_1(R,\theta,\phi) = \frac{P_{\text{rad}_1}}{4\pi R^2} D_{\text{max}_1} \tag{8.104}$$

The power received by horn 2 is a product of this power density and the cross-sectional area, or capture area, A_2, of the second horn. This received power is written

$$P_{\text{rec}_2} = P_1(R,\theta,\phi)A_2 = P_{\text{rad}_1} \frac{D_{\text{max}_1} A_2}{4\pi R^2} \tag{8.105}$$

We could also see how much power is received at horn 1 resulting from power emitted by horn 2 and by similar arguments arrive at

$$P_{\text{rec}_1} = P_{\text{rad}_2} \frac{D_{\text{max}_2} A_1}{4\pi R^2} \tag{8.106}$$

When we were first introduced to antennas in this chapter, it was pointed out that the transmission pattern of a reciprocal antenna is the same as its receive pattern. Another consequence of this reciprocity property is that the ratio of received power to radiated power will be the same regardless of which of the pair is transmitting and which is receiving. In other words we have

$$\frac{P_{\text{rec}_2}}{P_{\text{rad}_1}} = \frac{P_{\text{rec}_1}}{P_{\text{rad}_2}} \tag{8.107}$$

Figure 8.36 Pair of horn antennas aligned for maximum power transfer.

which leads to the relation

$$D_{\text{max}_1} A_2 = D_{\text{max}_2} A_1 \tag{8.108}$$

or

$$\frac{D_{\text{max}_1}}{A_1} = \frac{D_{\text{max}_2}}{A_2} \tag{8.109}$$

Now, since the directivity and area of one antenna are independent of any second antenna, the ratio in (8.109) must be equal to a constant. This constant is found to be

$$\frac{D_{\text{max}}}{A} = \frac{4\pi}{\lambda^2} \tag{8.110}$$

The capture area of a horn antenna can be considered equivalent to its physical cross-sectional area. But what can we say about the capture area for a dipole antenna? If in this case the capture area is the dipole length multiplied by the very small wire diameter, the capture area will be very small and these types of antennas will be useless. However, these antennas have an *effective area* A_e that is much larger than the physical cross section.[8.6] The definition for effective area is then the received power divided by the power density at the receiver, or in terms of the effective area of the second antenna, from (8.105) we have

$$A_{e_2} = \frac{P_{\text{rec}_2}(\text{W})}{P_1(R,\theta,\phi)\left(\text{W}/\text{m}^2\right)} \tag{8.111}$$

► **EXAMPLE 8.10**

Let's find the effective area of a $\lambda/2$ dipole antenna.

Recalling that the directivity of a half-wave dipole antenna is 1.64, we can manipulate (8.110) to find

$$A = \frac{D_{\text{max}}\lambda^2}{4\pi} = 0.13\lambda^2$$

For the half-wave dipole antenna, we can model the effective area as a rectangle of height 0.5λ and width 0.26λ (from $0.13\lambda^2/0.5\lambda$) as shown in Figure 8.37.

Drill 8.8   What is the effective area of a $\lambda/2$ dipole antenna operating in air at 2.4 GHz? (*Answer:* 20 cm^2)

For the more general communication situation of Figure 8.38, we find

$$P_{\text{rec}} = \frac{P_{\text{rad}}}{4\pi R^2} D_t(\theta,\phi) A_r(\theta,\phi) \tag{8.112}$$

[8.6]The effective area is also called the *effective aperture* or the *receiving cross-section*.

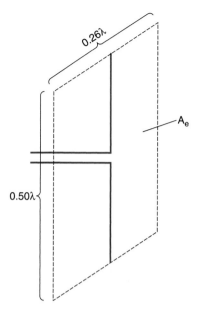

Figure 8.37 Conceptual view of the effective area of a half-wavelength dipole antenna.

where the t and r subscripts indicate values for the transmitter and the receiver, respectively. The ratio of (8.110) is also valid even if the antennas are not in line, so that you have

$$\frac{D(\theta,\phi)}{A_e(\theta,\phi)} = \frac{4\pi}{\lambda^2} \tag{8.113}$$

where both the directive gain and the effective area depend on the direction of interest. Using (8.113), we can replace $A_r(\theta,\phi)$ to get

$$\frac{P_{rec}}{P_{rad}} = D_t(\theta,\phi) D_r(\theta,\phi)\left(\frac{\lambda}{4\pi R}\right)^2 \tag{8.114}$$

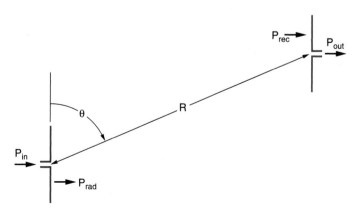

Figure 8.38 The general case of a communication link not oriented for maximum power transfer.

Finally, considering that $P_{rad} = e_t P_{in}$ and $P_{out} = e_r P_{rec}$, and that $G_t = e_t D_t$ and $G_r = e_r D_r$, we have

$$\boxed{\frac{P_{out}}{P_{in}} = G_t(\theta,\phi)\,G_r(\theta,\phi)\left(\frac{\lambda}{4\pi R}\right)^2} \tag{8.115}$$

This is the general form of the *Friis transmission equation*. It relates the *power transfer ratio*, P_{out}/P_{in}, to the antenna gains, wavelength, and separation distance. It assumes a matched impedance condition between the transmitter circuitry and the transmitting antenna (i.e., none of the power P_{in} is reflected) and between the receiver circuitry and receiving antenna. It also assumes the antenna polarizations are aligned. We will see how the Friis equation can be modified to account for polarization differences, but first let's work a couple of examples.

▶ **EXAMPLE 8.11**

Consider a pair of half-wavelength dipole antennas, separated by 1.0 km and aligned for maximum power transfer as shown in Figure 8.39. We drive the transmission antenna with 1.0 kW of power at 1.0 GHz. Assuming the antennas are 100% efficient, we want to find the receiving antenna's output power P_{out}.

For 100% efficiency and antennas optimally aligned, the Friis equation becomes

$$\frac{P_{out}}{P_{in}} = D_{max_t}\,D_{max_r}\left(\frac{\lambda}{4\pi R}\right)^2$$

For the $\lambda/2$ dipole antennas we have $D_{max_t} = D_{max_r} = 1.64$, and at 1.0 GHz $\lambda = 0.30$ m. Therefore,

$$\frac{P_{out}}{P_{in}} = (1.64)^2\left(\frac{0.30 \text{ m}}{4\pi 1\times10^3 \text{ m}}\right)^2 = 1.5\times10^{-9} \ \frac{W}{W}$$

In terms of decibels, we have

$$\frac{P_{out}}{P_{in}}(\text{dB}) = 10\log\left(1.5\times10^{-9}\right) = -88 \text{ dB}$$

Finally,

$$P_{out} = (1.5\times10^{-9})\,(1 \text{ kW}) = 1.5\mu\text{W}$$

Drill 8.9 What is the output power for Example 8.11 if we double the separation distance? (*Answer:* 0.38 μW)

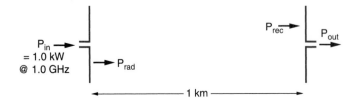

Figure 8.39 A pair of $\lambda/2$ dipole antennas for Example 8.11.

▷ **EXAMPLE 8.12**

Consider a pair of half-wavelength dipole antennas, separated by 1.0 km and with identical polarizations. The transmitter is fixed at the origin and aligned with z; the second antenna remains 1.0 km distant but makes an angle $\theta = 60°$ with respect to the transmitter. The situation is well represented by Figure 8.38. If we drive the transmission antenna with 1.0 kW of power at 1.0 GHz and assume the antennas are 100% efficient, what is the power received?

In this case we have

$$\frac{P_{out}}{P_{in}} = D_t(\theta) D_r(\theta) \left(\frac{\lambda}{4\pi R} \right)^2$$

and we need to find $D_t(\theta = 60°) = D_r(\theta = 120°)$. From (8.13) and (8.14) we have

$$D(\theta) = D_{max} P_n(\theta)$$

where $D_{max} = 1.64$ for a $\lambda/2$ dipole antenna.

The normalized power from (8.82) is

$$P_n\left(\theta = 60°\right) = \frac{\cos^2\left(\frac{\pi}{2}\cos\theta\right)}{\sin^2\theta} = 0.667$$

so

$$D(\theta = 60°) = (1.64)(0.667) = 1.11$$

Now we can calculate the power ratio as

$$\frac{P_{out}}{P_{in}} = 700 \times 10^{-12}$$

or

$$\frac{P_{out}}{P_{in}}(dB) = -92 \text{ dB}$$

The output power is then $P_{out} = 680$ nW.

Polarization Effects

For linear polarized waves, maximum power transfer occurs when the electric fields for the transmitting and receiving antennas have the same polarization. No power will be transferred if the electric field polarizations are 90° misaligned. We can express a *polarization efficiency* term e_p as

$$e_p = \left| \mathbf{a}_{E_t} \cdot \mathbf{a}_{E_r} \right|^2 \tag{8.116}$$

Here the maximum transmission occurs when the polarization vectors are in the same direction, that is, $\mathbf{a}_{E_t} = \mathbf{a}_{E_r}$.

For the general case of elliptic polarization, (8.116) must be expressed with a complex conjugate on one of the unit vectors:

$$e_p = \left| \mathbf{a}_{E_t} \cdot \mathbf{a}_{E_r}^* \right|^2 \tag{8.117}$$

As an example, consider the case of a transmitting antenna that radiates a right-hand circular polarized wave with polarization vector $\mathbf{a}_{E_t} = \left(\mathbf{a}_x - j\mathbf{a}_y\right)/\sqrt{2}$. A receiving antenna must have this same polarization in order to receive maximum power. If the receiving antenna has a polarization vector $\mathbf{a}_{Et} = \left(\mathbf{a}_x - j\mathbf{a}_y\right)/\sqrt{2}$ it will receive maximum power.

The efficiency term can be included in the Friis transmission equation, giving

$$\frac{P_{\text{out}}}{P_{\text{in}}} = e_p G_t(\theta, \phi) G_r(\theta, \phi)\left(\frac{\lambda}{4\pi R}\right)^2 \tag{8.118}$$

▷ **EXAMPLE 8.13**

Let us try to improve upon the power transfer ratio of Example 8.12 by orienting the receiving antenna as shown in Figure 8.40. Here we again have a pair of half-wave dipole antennas 1.0 km apart with angle $\theta = 60°$.

With this new orientation, we see that the directive gain of the receiving antenna will be maximized (equal to the antenna's directivity):

$$D_r(\theta) = D_{\text{max}_r} = 1.64$$

Neglecting polarization effects we would then have

$$\frac{P_{\text{out}}}{P_{\text{in}}} = (1.11)(1.64)\left(\frac{0.30}{4\pi\left(1 \times 10^3\right)}\right)^2 = 1.0 \times 10^{-9}$$

or –90 dB. So we would see a 2-dB improvement over the arrangement of Example 8.12.

However, we really cannot neglect the polarization effects in this problem. A close examination of the geometry of Figure 8.40 reveals that

$$\mathbf{a}_{E_1} \cdot \mathbf{a}_{E_2} = \cos(90° - \theta) = 0.866$$

for $\theta = 60°$. So we have

$$e_p = (0.866)^2 = 0.75$$

Factoring this into the power transfer ratio, we find

$$\frac{P_{\text{out}}}{P_{\text{in}}} = 0.77 \times 10^{-9}$$

or –91 dB. So we see that reorienting the receiver antenna results in only a 1-dB improvement for this case.

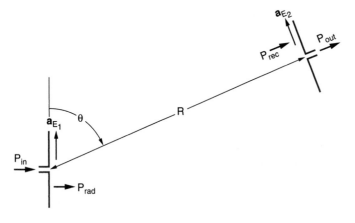

Figure 8.40 Antenna arrangement to demonstrate the effect of polarization on the power transfer ratio.

Drill 8.10 Using $\theta = 45°$, find the power transfer ratio in decibels for the arrangements of Example 8.12 and 8.13. (*Answer*: –96 dB, –95 dB)

Receiver Matching

In Example 8.5 we saw how the addition of an impedance matching network offered improved radiation for a transmitting antenna. Because of the reciprocal nature of most antennas, the same matching network will also improve receiver performance.

Figure 8.41a shows an impedance matching network between the receiving antenna and the load. A consequence of the matching network is that $Z_{in} = Z_{ant}{}^*$ (see Problem 8.49). The power received by the antenna appears as an open-circuit voltage amplitude V_{oc} in series with the antenna impedance, as shown in Figure 8.41b. V_{oc} is related to the received power calculated from the Friis equation by

$$P_{rec} = \frac{V_{oc}^2}{2(Z_{ant} + Z_{in})} = \frac{V_{oc}^2}{4R_{rad}} \tag{8.119}$$

Since the receiver is matched, half the received power is dissipated in the load:

$$\frac{1}{2}P_{rec} = \frac{V_L^2}{2|Z_L|} \tag{8.120}$$

The value of V_{oc} is related to the P_{rec} calculated from the Friis equation assuming a matched load, but it is itself independent of the matching condition. With the impedance matching network removed, as shown in Figure 8.41c, we have

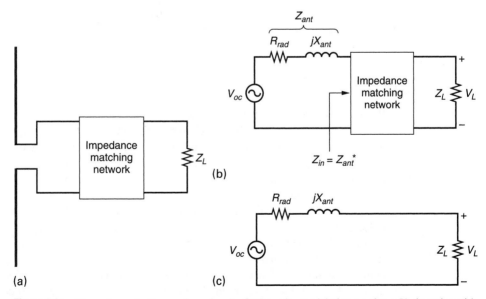

Figure 8.41 The antenna in the receiver circuit of (a) can be modeled as a voltage V_{oc} in series with the antenna impedance in (b). In (c), the matching network has been removed.

$$V_L = \left| \frac{Z_L}{Z_L + Z_{ant}} \right| V_{oc} \qquad (8.121)$$

▷ **EXAMPLE 8.14**

In Example 8.5, a 0.485λ dipole antenna was used as a transmitter powered by a 12-V source with a 25-Ω impedance. Now suppose we receive our signal 2.0 km away using an identical 0.485λ dipole antenna with a 25-Ω load. We want to find the voltage amplitude across this load when the transmitter is impedance matched and both antennas are aligned for maximum power transfer ratio. We'll assume a 300-MHz frequency.

We can apply the Friis transmission equation to find

$$\frac{P_{rec}}{P_{trans}} = (D_{max})^2 \left(\frac{\lambda}{4\pi R} \right)^2$$

Each antenna's directivity is approximately that of a half-wavelength dipole, or $D_{max} = 1.64$. With the impedance-matched transmitter of Example 8.5 emitting 2.1 W of power, we have $P_{rec} = 8.9$ nW. From (8.119) we solve for V_{oc} as

$$V_{oc} = \sqrt{4R_{rad}P_{rec}} = \sqrt{4(73\ \Omega)(8.9 \times 10^{-9}\ W)} = 1.61\ mV$$

Without impedance matching, we find from (8.121)

$$V_L = \frac{25\ \Omega}{25\ \Omega + 73\ \Omega}(1.61\ mV) = 410\ \mu V$$

Drill 8.11 Calculate the voltage across the 25-Ω load of Example 8.14 if a matching network is in place. (*Answer:* 470 μV)

▷ 8.7 RADAR

Radar (the term derived from *ra*dio *d*etection *a*nd *r*anging) was invented primarily by the British in the Second World War and led to their victory in the Battle of Britain. In addition to numerous military applications, today radar has uses in a variety of commercial areas including meteorology (weather radar), law enforcement (speed guns), air traffic control, automobile collision avoidance, and astronomy (space-imaging radar).

Figure 8.42 depicts the operation of a *monostatic* radar system, where the transmitter and the receiver are at the same location. In Figure 8.42a, the radar antenna transmits a pulse of electromagnetic energy toward a target. Some of this energy is reflected, or *scattered*, by the target. As Figure 8.42b indicates, the scattering is assumed to be *isotropic*, radiating equally in all directions. Some of the scattered energy, the *echo signal*, is received at the radar antenna. The direction of the antenna's main beam determines the location of the target (azimuth and elevation). The distance (or range) to the target corresponds to the time between transmitting and receiving the electromagnetic pulse. The speed of the target,

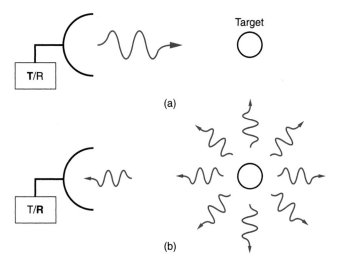

(a)

Target

(b)

Figure 8.42   A monostatic radar system. (a) A radar antenna transmits a signal to the target. (b) The target scatters this signal, some of which is received by the radar antenna.

relative to that of the radar antenna, can be determined by observing any frequency shift in the electromagnetic energy (i.e., the *Doppler effect*).

This description of radar operation can now be put in mathematical terms by first finding how much power is incident from the radar antenna to the target. If P_{rad_1} is transmitted by the radar antenna, then the radiated power density $P_1(R,\theta,\phi)$ at the target, a distance R away, is

$$P_1(R,\theta,\phi) = \frac{P_{rad_1}}{4\pi R^2}\, D(\theta,\phi) \tag{8.122}$$

The amount of power reflected by the target (i.e., scattered isotropically by the target) is determined by its *radar cross section*, or $\sigma_s(m^2)$. Note that an object's radar cross section may not necessarily correspond to the object's size. A stealth aircraft, for instance, has a large surface area but tends to absorb the radar energy rather than reflect it, so its radar cross section is extremely small. The power scattered by the target is then

$$P_{rad_2} = \sigma_s P_1(R,\theta,\phi) \tag{8.123}$$

This scattered power results in a radiated power density at the radar antenna of

$$P_2(R,\theta,\phi) = \frac{P_{rad_2}}{4\pi R^2} = \sigma_s \frac{P_{rad_1}}{\left(4\pi R^2\right)^2}\, D(\theta,\phi) \tag{8.124}$$

The amount of this power received at the radar antenna is then

$$P_{rec_1} = P_2(R,\theta,\phi)A_e \tag{8.125}$$

where A_e is the antenna's effective area. Manipulating this equation and using (8.113) we find the *radar equation*,

$$\frac{P_{rec_1}}{P_{rad_1}} = \frac{\sigma_s \lambda^2}{(4\pi)^3 R^4}\, D(\theta,\phi)^2 \tag{8.126}$$

A more popular expression (again using (8.113)) in terms of an effective area of the radar antenna is

$$\frac{P_{rec_1}}{P_{rad_1}} = \frac{\sigma_s}{4\pi R^4 \lambda^2} A_e^2$$

(8.127)

The strongest received power will of course occur when the antenna's main beam is pointing at the target, that is, when $D(\theta, \phi) = D_{max}$. The received power must also be detectable over the noise in the system. A *minimum detectable power* is therefore specified for a radar system.

▶ **EXAMPLE 8.15**

A radar system with minimum detectable power specified as 1 pW is 1 km distant from a target with a 1 m² radar cross section. Operated at 1 GHz the antenna has a directivity of 100. We want to determine how long it will take a pulse to travel to the target and back and to find how much power must be radiated to enable detection of the target.

First, the round-trip travel time for an electromagnetic pulse in air is

$$t = 2\frac{R}{c} = \frac{2(1000 \text{ m})}{3\times10^8 \text{ m/s}} = 6.7 \text{ μs}$$

Next, we solve the radar equation in terms of P_{rad_1}:

$$P_{rad_1} = P_{rec_1} \frac{(4\pi)^3 R^4}{\sigma_s \lambda^2} \frac{1}{D_{max}^2}$$

At 10 GHz we have $\lambda = 0.3$ m. Solving for P_{rad_1}, we get

$$P_{rad_1} = \left(10^{-12} \text{ W}\right)\frac{(4\pi)^3 (1000 \text{ m})^4}{(1 \text{ m}^2)(0.3 \text{ m})^2} \frac{1}{(100)^2} = 2.2 \text{ W}$$

Drill 8.12 Repeat Example 8.15 if the target is 2 km distant. (*Answer:* 13 μs, 35 W).

Drill 8.13 What is the effective area for the antenna of Example 8.15? Recalculate the required P_{rad_1} using (8.126). (*Answer:* 0.72 m², 2.2 W)

▶ **8.8 ANTENNAS FOR WIRELESS COMMUNICATIONS**

In this section we describe several types of antennas that find use in wireless communications. Analysis is generally beyond the scope of this text. Students interested in seeking more information about these and other antennas are directed to the references at the end of this chapter.

Parabolic Reflectors

One of the most recognizable types of antenna is the parabolic reflector, like the one in Figure 8.43, widely used for satellite communications owing to its very high directivity. Figure 8.44 shows one version of this type of antenna. Operation is based on the geometric optics principle that a point source of radiation placed at the focal point of a parabolic reflector will radiate the energy incident on the *dish* in a narrow, collimated beam. The high directivity is exhibited by gains in excess of 30 dB and beamwidths less than 2°.

Figure 8.43 A parabolic dish antenna. Courtesy of Harris Corporation.

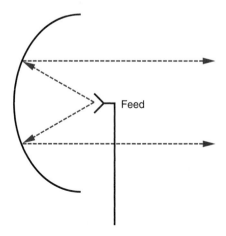

Figure 8.44 Parabolic reflector antenna.

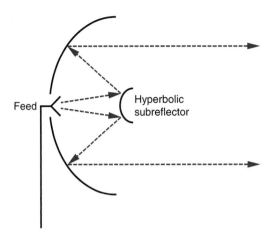

Figure 8.45   Cassegrain reflector antenna.

For the most efficient operation, the dish must be significantly larger than the radiation wavelength, and the feed element itself must be directive. Ideally, all of the radiation from the feed element would be incident on the dish, but there is a compromise between illuminating the entire dish (some radiation spills over) and reflecting all of the radiation. Horn antennas are routinely used as feed elements for parabolic reflectors.

A slightly different version of this antenna is the Cassegrain reflector, shown in Figure 8.45. Here the feed element is located in a hole in the middle of the parabolic reflector and its radiation is reflected off of a hyperbolic subreflector.

The reflective dish need not be a continuous conductor. Excellent performance can be achieved with a meshed conductor.

Patch Antennas

The microstrip patch antenna as shown in Figure 8.46 is very simple to construct using conventional microstrip fabrication techniques. It is commonly a very thin square patch of metal approximately $\lambda/2$ on a side supported by a thin, low-loss, low-permittivity substrate. Here λ is the wavelength calculated in the dielectric. Although square and rectangular patches are the most prevalent, other shapes such as circles, triangles, and annular rings have also been used.

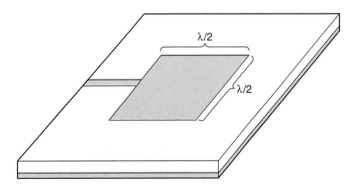

Figure 8.46   An edge-fed patch antenna.

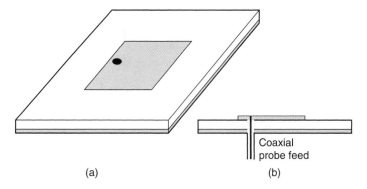

(a)

Coaxial
probe feed

(b)

Figure 8.47 (a) A probe-fed patch antenna and (b) its cross section.

The patch antennas can be excited using an edge feed (Figure 8.46) or a probe feed (Figure 8.47). The exact location of the feedpoint for a probe-fed patch antenna is chosen to provide an impedance match between the cable and the antenna. Substrate parameters, size of ground plane, and location of nearby structures are all considered in choosing this feedpoint.

Patch antennas are relatively narrow band devices, typically with bandwidth 10% of the resonant frequency; they also have relatively poor radiation efficiency. Accurate prediction of performance requires significant computer analysis. But patch antennas are still popular because they are so easy to fabricate and are compatible with integrated-circuit fabrication techniques such that electronics can be easily integrated with the antenna. Large arrays of patch antennas may be cheaply fabricated with support electronics (such as phase shifters) placed on the other side of the ground plane where they don't interfere with the radiation.

Slot Antennas

A *complementary* structure to the half-wavelength dipole antenna is the half-wavelength *slot*, cut in a conductive sheet as shown in Figure 8.48. By complementary, we mean that the pattern and impedance of a dipole antenna can be used to predict the pattern and impedance of a slot antenna. The slot antenna oriented as shown in Figure 8.48 generates $\mathbf{a}_x$ polarized radiation in the $\mathbf{a}_z$ direction.

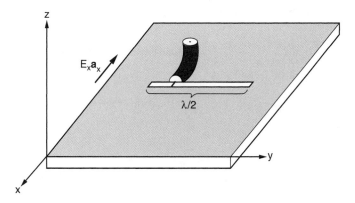

Figure 8.48 A slot antenna.

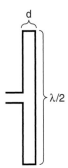

d

λ/2

Figure 8.49   The two-wire folded dipole antenna.

There are a variety of ways to feed a slot antenna, including the coaxial feed arrangement of Figure 8.48. The feed may be off-center to provide an impedance match between the cable and the slot. An array of slots may also be cut into the walls of rectangular waveguide to achieve a particular radiation pattern.

Folded Dipole Antennas

Twin-lead T-line such as the kind commonly used for FM and VHF receivers has an impedance of close to 300 Ω. Although the conventional half-wave dipole with its input impedance of around 73 Ω can be impedance matched to the twin-lead T-line, perhaps a better way is to use a variation of the half-wavelength dipole antenna called the *folded dipole*, as shown in Figure 8.49. Here, a pair of half-wavelength dipole elements are joined at the ends and fed from the center of one of the pair. If the two λ/2 sections are close together (d on the order of λ/64), analysis reveals an impedance four times greater than that of a regular λ/2 dipole antenna, or ~288 Ω. The directivity is the same as the λ/2 dipole antenna, but the bandwidth of the folded dipole antenna is significantly broader.

These antennas are very easy to construct by stripping about a wavelength of insulation off of the end of a twin-lead cable, soldering the ends together, and stretching the wire out to get the shape shown in Figure 8.49.

Folded dipoles may also be made with other lengths and with more than two wires.

▶ SUMMARY

- Table 8.2 lists the common terms and their units for parameters relating to antennas.

- The radiation pattern, or shape of the radiated field, is viewed in the far field. The far-field radius r is related to the length L of the radiating element and the radiation wavelength λ by

$$r \ge \frac{2L^2}{\lambda}$$

where $L > \lambda$.

- The time-averaged power density of an electromagnetic wave is found via the Poynting theorem as

$$P(r,\theta,\phi) = \frac{1}{2}\mathrm{Re}\left[E_s \times H_s^*\right]$$

and the total power radiated by the antenna is

$$P_{\mathrm{rad}} = \oint P(r,\theta,\phi) \cdot dS$$

- The normalized power function is

$$P_n(\theta,\phi) = \frac{P(r,\theta,\phi)}{P_{\max}}$$

For an antenna that radiates equally in all directions (an isotropic antenna)

Table 8.2 Antenna Parameters

Parameter	Units	Definition
$P(r,\theta,\phi)$	W/m^2	Time-averaged power density vector
P_{rad}	W	Total power radiated
$P_n(\theta,\phi)$	—	Normalized power function, or normalized radiation intensity
$P_n(\theta,\phi)_{ave}$	—	Average normalized power function
P_{max}	W/m^2	Maximum time-averaged power density
Ω_p	sr	Pattern solid angle
$D(\theta,\phi)$[a]	dB	Directive gain
D_{max}[a]	dB	Directivity
Z_{ant}	Ω	Antenna impedance
R_{ant}	Ω	Real part of Z_{ant}
X_{ant}	Ω	Imaginary part of Z_{ant}
R_{rad}	Ω	Radiation resistance
R_{diss}	Ω	Ohmic resistance
P_{diss}	W	Power dissipated by resistive losses
e	—	Efficiency
$G(\theta,\phi)$[a]	dB	Power gain
G_{max}[a]	dB	Maximum power gain
σ_s	m^2	Radar cross section

[a]$D(\theta,\phi)$, D_{max}, $G(\theta,\phi)$, and G_{max} are unitless fractions that may be represented by dB or dB$_i$.

$$P_n(\theta,\phi)_{iso} = 1$$

- A measure of an antenna's ability to focus its main beam in a particular direction is the directivity D_{max}, which is related to the pattern solid angle Ω_p by

$$D_{max} = \frac{4\pi}{\Omega_p}$$

Here, the pattern solid angle is given by

$$\Omega_p = \iint P_n(\theta,\phi)\,d\Omega$$

where $d\Omega = \sin\theta\,d\theta\,d\phi$.

- The directive gain is related to the time-averaged power function by

$$D(\theta,\phi) = D_{max}P_n(\theta,\phi)$$

The total power radiated from an antenna is given by

$$P_{rad} = r^2 P_{max}\Omega_p = \frac{1}{2}I_o^2 R_{rad}$$

- The efficiency of an antenna is the ratio of the radiated power to the total power fed to the antenna, or

$$e = \frac{P_{rad}}{P_{rad} + P_{diss}} = \frac{R_{rad}}{R_{rad} + R_{diss}}$$

and the antenna impedance is

$$Z_{ant} = R_{ant} + jX_{ant} = (R_{rad} + R_{diss}) + jX_{ant}$$

- Based on the current distribution of a radiating element, we can determine the retarded, or time-delayed, vector magnetic potential $\mathbf{A}_{os}$. Then, magnetic field intensity is found by

$$\mathbf{H}_{os} = \frac{1}{\mu_o}\nabla \times \mathbf{A}_{os}$$

and the electric field intensity is

$$\mathbf{E}_{os} = -\eta_o\mathbf{a}_r \times \mathbf{H}_{os}$$

- If a differential length line of current known as a Hertzian dipole is placed on the z-axis, the far-field value of the electric field intensity is

$$\mathbf{E}_{os} = j\eta_o\frac{I_s\ell\beta}{4\pi}\frac{e^{-j\beta r}}{r}\sin\theta\,\mathbf{a}_\theta$$

and the time-averaged power density vector is

$$P(r,\theta) = \left(\frac{\eta_o \beta^2 I_o^2 \ell^2}{32\pi^2 r^2}\right)\sin^2\theta\, a_r$$

The directivity of the Hertzian dipole is $D_{max} = 1.5$, and the radiation resistance R_{rad} is

$$R_{rad} = 80\pi^2\left(\frac{\ell}{\lambda}\right)^2$$

- The small loop antenna of area $S \ll \lambda^2$ located at $z = 0$ in the x–y plane has

$$E_{os} = \frac{\omega\mu_o I_s S\beta}{4\pi r}\sin\theta\, e^{-j\beta r} a_\phi$$

and

$$P(r,\theta) = \left(\frac{\omega^2\mu_o^2 I_o^2 S^2\beta^2}{32\eta_o\pi^2 r^2}\right)\sin^2\theta\, a_r$$

It has the same directivity as a Hertzian dipole antenna, and it has a radiation resistance

$$R_{rad} = 320\pi^4\left(\frac{S}{\lambda^2}\right)^2 \Omega$$

- An L-long dipole antenna with phasor current distribution

$$I_s(z) = \begin{cases} I_o e^{j\alpha}\sin\beta\left(\dfrac{L}{2}-z\right) & \text{for } 0 < z < L/2 \\ I_o e^{j\alpha}\sin\beta\left(\dfrac{L}{2}+z\right) & \text{for } -L/2 < z < 0 \end{cases}$$

has in the far field

$$\mathbf{H}_{os} = H_{os}a_\phi = \left(\frac{jI_o}{2\pi}\frac{e^{-j\beta r}}{r}\right)\left[\frac{\cos\left(\dfrac{\beta L}{2}\cos\theta\right) - \cos\left(\dfrac{\beta L}{2}\right)}{\sin\theta}\right]a_\phi$$

and

$$\mathbf{E}_{os} = \eta_o H_{os}a_\theta$$

The time-averaged power density vector is

$$P(r,\theta) = \frac{15 I_o^2}{\pi r^2}F(\theta)a_r$$

where the pattern function $F(\theta)$ is given by

$$F(\theta) = \left[\frac{\cos\left(\dfrac{\beta L}{2}\cos\theta\right) - \cos\left(\dfrac{\beta L}{2}\right)}{\sin\theta}\right]^2$$

For an antenna of length L (in λ), numerical analysis is employed to find Ω_p and D_{max}.

- The time-averaged power density for a half-wave dipole antenna is

$$P(r,\theta) = \left(\frac{15 I_o^2}{\pi r^2}\right)\left(\frac{\cos^2\left(\dfrac{\pi}{2}\cos\theta\right)}{\sin^2\theta}\right)a_r$$

with a directivity $D_{max} = 1.64$ and antenna impedance

$$Z_{ant} = 73.2\Omega + j42.5\ \Omega$$

A dipole slightly shorter than a half wavelength ($L = 0.485\lambda$) has a resonant condition where the reactive portion of Z_{ant} is zero and R_{rad} is approximately 73 Ω.

- From image theory we find that a monopole antenna, where a linear antenna is placed over a ground plane and is excited at its base, has the same radiation characteristics in the region above the ground plane as a dipole antenna twice its length. Its impedance is halved, so for a quarter-wavelength monopole antenna

$$Z_{ant} = 36.6 + j21.25\ \Omega$$

- Beam shaping and focusing is possible by arraying antenna elements. For radiation only in the x–y plane ($\theta = \pi/2$), the time-averaged power density vector is

$$P\left(r,\frac{\pi}{2},\phi\right) = F_{unit}F_{array}a_r$$

where the unit factor F_{unit} is the radiated power for an antenna element and F_{array} is the array factor. For an N-element linear array

$$F_{array} = \frac{\sin^2\left(\dfrac{N\psi}{2}\right)}{\sin^2\left(\dfrac{\psi}{2}\right)}$$

where $\psi = \beta d\cos\phi + \alpha$.

- A parasitic array, such as the Yagi–Uda antenna, can have a single driven element and several undriven parasitic elements.

• The Friis transmission equation quantifies how well energy is transmitted between antennas and is expressed as

$$\frac{P_{out}}{P_{in}} = G_t(\theta,\phi)G_r(\theta,\phi)\left(\frac{\lambda}{4\pi R}\right)^2$$

A factor can be included in the Friis equation to account for differences in polarization. Given polarization unit vectors $\mathbf{a}_{E_t}$ and $\mathbf{a}_{E_r}$ for the transmitted and the received waves, respectively, the factor is

$$e_p = \left|\mathbf{a}_{E_t}\cdot\mathbf{a}_{E_r}^*\right|^2$$

• The Friis transmission equation can be adapted to yield the radar equation

$$\frac{P_{rec_1}}{P_{rad_1}} = \frac{\sigma_s}{4\pi R^4\lambda^2}A_e^2$$

where σ_s is the radar cross section of the target and A_e is the radar antenna's effective area. The round-trip travel time of an electromagnetic pulse determines the range of the target, and the location is discerned using a highly directive beam.

• In addition to dipole and loop antennas, other popular antennas include parabolic reflectors, patch antennas, slot antennas, and folded dipole antennas. The patch, slot, and folded dipole antennas may be fabricated cheaply using printed-circuit techniques.

► SUGGESTED REFERENCES

Balanis, C., *Antenna Theory,* 2nd ed., Wiley, 1996.

Godara, L. C., ed., *Handbook of Antennas in Wireless Communications,* CRC Press, 2002.

Johnson, R. C., and Jasik, H., eds., *Antenna Engineering Handbook,* 3rd ed., McGraw–Hill, 1992.

Kraus, J., and Marhefka, R. J., *Antennas,* 3rd ed., McGraw–Hill, 2001.

Skolnik, M. I., *Introduction to Radar Systems,* 3rd ed., McGraw–Hill, 2001.

Stutzman, W. L., and Thiele, G. A., *Antenna Theory and Design,* 2nd ed., Wiley, 1998.

► PROBLEMS

8.1 General Properties

8.1 In free space, a wave propagating radially away from an antenna at the origin has

$$\mathbf{H}_s = \frac{-I_s}{r}\cos^2\theta\,\mathbf{a}_\theta$$

where the driving current phasor $I_s = I_o e^{j\alpha}$. Determine (a) $\mathbf{E}_s$, (b) $\mathbf{P}(r,\theta,\phi)$, and (c) R_{rad}.

8.2 What is the pattern solid angle and the directivity for (a) an isotropic antenna and for (b) a *semiisotropic* antenna, which radiates equally in all directions above $\theta = \pi/2$ but is zero otherwise?

8.3 Sketch an appropriate cross section of the radiation pattern and determine the beamwidth BW, pattern solid angle, and directivity for the following normalized radiation intensities:

(a) $P_n(\theta,\phi) = \cos\theta$ for $0\le\theta\le\pi/2$, 0 otherwise;

(b) $P_n(\theta,\phi) = \cos^2\theta$ for $0\le\theta\le\pi/2$, 0 otherwise;

(c) $P_n(\theta,\phi) = \cos^3\theta$ for $0\le\theta\le\pi/2$, 0 otherwise.

8.4 Sketch an appropriate cross section of the radiation pattern and determine the beamwidth, pattern solid angle, and directivity for the following normalized radiation intensities:

(a) $P_n(\theta,\phi) = \sin\theta$;

(b) $P_n(\theta,\phi) = \sin^2\theta$;

(c) $P_n(\theta,\phi) = \sin^3\theta$.

8.5 You are given the following normalized radiation intensity:

$$P_n(\theta,\phi) = \begin{cases} \sin^2\theta\,\sin^3\phi & \text{for } 0\le\phi\le\pi \\ 0 & \text{otherwise} \end{cases}$$

Find the beamwidth BW, pattern solid angle, and directivity.

8.6 You are given the following normalized radiation intensity:

$$P_n(\theta,\phi) = \sin^2\theta\,\sin\frac{\phi}{2}$$

Determine the beamwidth, direction of maximum radiation, pattern solid angle, and directivity.

8.2 Electrically Short Antennas

8.7 Use the phasor form of Ampère's circuital law,

$$\mathbf{E}_{os} = \frac{1}{j\omega\varepsilon_o}\nabla\times\mathbf{H}_{os}$$

to find $\mathbf{E}_{os}$ from (8.46) without assuming the far-field condition. Then, show that this value of $\mathbf{E}_{os}$ reduces to (8.50) in the far field.

8.8 Suppose, for a particular antenna in free space,

$$\mathbf{A}_{os} = \mu_o I_o e^{-j\beta y}\mathbf{a}_z$$

Find $\mathbf{H}_{os}$, $\mathbf{E}_{os}$, and the time-averaged power density vector $\mathbf{P}$.

8.9 Suppose a Hertzian dipole antenna is 1.0 cm long and is excited by a 10.-mA amplitude current source at 100. MHz. (a) What is the maximum power density radiated by this antenna at a 1.0-km distance? (b) What is the antenna's radiation resistance?

8.10 A 1.0-cm-long, 1.0-mm-diameter copper wire is used as a Hertzian dipole radiator at 1.0 GHz. (a) Find R_{rad}. (b) Estimate R_{diss} by considering the skin effect resistance of the wire. (c) Find the efficiency e. (d) Find the maximum power gain G_{max}.

8.11 Evaluate the curl of $\mathbf{A}_{os}$ for the small loop antenna (8.59) to find $\mathbf{H}_{os}$. Now apply a far-field approximation to verify (8.60).

8.12 Neglecting resistive losses in the wire, how much current must drive a loop antenna of radius 2.0 cm at 60 MHz to radiate 1.0 W of power? Repeat for a 20-turn loop.

8.13 Suppose in the far field for an antenna at the origin,

$$H_{os} = \frac{\beta I_s}{4\pi}\frac{e^{-j\beta r}}{r}\sin\theta\cos\phi \mathbf{a}_\phi$$

where $I_s = I_o e^{j\alpha}$. What is the radiation resistance of this antenna at 100 MHz?

8.14 Suppose in the far field for a particular antenna at the origin, the electric field is

$$\mathbf{E}_{os} = \eta_o I_o \frac{e^{-j\beta r}}{\pi r}\sin\theta\, \mathbf{a}_\theta$$

What is the radiation resistance of this antenna?

8.15 Derive the expressions for radiated power (8.64) and radiation resistance (8.65) for a small loop antenna.

8.3 Dipole Antennas

8.16 Develop a routine to calculate the beamwidth for a dipole antenna of arbitrary length between 0.1λ and 1λ.

8.17 How long is a 1.5λ-long dipole antenna at 1.0 GHz? Suppose this antenna is constructed using AWG#20 (0.406-mm radius) copper wire. Determine R_{diss}, e, and G_{max}.

8.18 Find the half-power beamwidth of a $\lambda/2$ dipole antenna.

8.19 A 2.45-GHz $\lambda/2$ dipole antenna is driven by a 2.0-A amplitude current source. Find the maximum power density at a distance of 1.0 km.

8.20 Given a z-polarized half-wave dipole antenna at the origin, and a driving current $i(t) = 10\cos(2\pi \times 10^9 t)$ A, find the instantaneous electric and magnetic fields at a point 2.0 km distant and angle $\theta = 60°$.

8.21 Modify MATLAB 8.4 to calculate directivity and radiation resistance for an arbitrary length dipole antenna. Evaluate these properties for a 0.75λ dipole antenna.

8.22 Find a 3.0-m-long dipole antenna's directivity and radiation resistance if it is operated at (a) 250 MHz, (b) 500 MHz, and (c) 750 MHz.

8.23 A 50-Ω impedance line is terminated in a 3.0-m-long dipole antenna at 50 MHz. What is the VSWR looking into this antenna? Design a shorted shunt-stub network to impedance match the antenna to the 50-Ω line.

8.24 Use MATLAB 8.2 to generate plots like those of Figure 8.19 for a dipole antenna of length 3λ.

8.25 A 0.485λ dipole antenna is constructed for operation at 4.0 GHz. (a) How long is the antenna? (b) What impedance is required of a quarter-wave transformer to match this antenna to a 50-Ω impedance line?

8.26 Modify MATLAB 8.3 to run the movie from 0.1λ up to 4λ.

8.27 Using MATLAB 8.4, generate data of the pattern solid angle versus number of increments N to see the function convergence. Consider a 1.25λ dipole. Try $N = 2, 4, 8, 16, 32, 64,$ and 128.

8.4 Monopole Antennas

8.28 Consider a 1.0-nC charge at (0.0, 0.0, 5.0 m) above a conductive sheet occupying the x–y plane at $z = 0$. Use image theory to find the electric field intensity at the point (0.0, 5.0 m, 5.0 m).

8.29 Find the half-power beamwidth for a quarter-wave monopole antenna.

8.30 Devise a routine to give a polar plot of the normalized power radiated for an arbitrary length monopole antenna. Use your program to generate the polar plot for a half-wave monopole.

8.31 Determine the pattern solid angle, directivity, and radiation resistance for a half-wave monopole antenna.

8.32 How long is a 0.75λ monopole antenna at 1.0 GHz? Suppose this antenna is constructed using AWG#20 (0.406-mm radius) copper wire. Determine R_{diss}, e, and G_{max}. Compare your results with the 1.5λ dipole antenna of Problem 8.17.

8.33 What is the *VSWR* looking into a quarter-wave monopole antenna if the feed line has a 50-Ω impedance? Design an open-ended shunt-stub matching network to match this antenna to the line.

8.34 Given a 1-GHz quarter-wave monopole antenna at the origin, excited by a 1.0-A amplitude current, find the amplitudes for the electric and magnetic field intensities at a point 1.0 km distant at an angle $\theta = 80°$.

8.5 Antenna Arrays

8.35 Find and plot the far-field radiation pattern at $\theta = \pi/2$ for a two-element dipole antenna array given the following:

1. The dipoles are driven in phase.

2. Each dipole is 1λ in length oriented in the z-direction.

3. The pair of dipoles are 1λ apart on the x-axis.

Also find the maximum time-averaged power density, in watts per square meter, 1.0 km away from the array if each antenna is driven by a 1.0-A amplitude current source at 1.0 GHz.

8.36 Repeat Problem 8.35 if the dipoles are 180° out of phase.

8.37 Repeat Problem 8.35 for the case where the dipoles are 90° out of phase, 1.5λ in length, and separated by λ/2.

8.38 Two z-polarized λ/2 dipole antennas are spaced λ/4 apart, centered at the origin on the x-axis. (a) If the dipole located at $x = -\lambda/8$ is driven by $I_{s_1} = I_o e^{j0°}$, what phase shift α would you employ on the other dipole ($I_{s_2} = I_o e^{j\alpha}$) to get maximum power at a far-field point on the $+x$ axis? (b) If the dipole antennas are each driven by 1.0-A amplitude currents at 500 MHz, with the phase shift from part (a), find the time-averaged power density vector at 2.0 km on the x-axis.

8.39 Two small loop antennas, each oriented in the x–y plane, are centered at $\pm\lambda/2$ on the x-axis. They each have a 1.0-cm radius and are driven in phase by a 10.-mA current source at 500. MHz. Find and plot the radiation pattern at $\theta = \pi/2$ and determine the maximum time-averaged power density at a distance 100. m from the array.

8.40 Given a pair of dipole antennas separated by λ/4 and driven in phase, determine, for $\theta = \pi/2$, (a) the values for ϕ at the nulls in the radiation pattern and (b) the values of ϕ where the radiated power is maximum.

8.41 Create a movie to plot the radiated power pattern in the x–y plane for the pair of dipoles in Example 8.7 as the separation distance varies from λ/10 to 4λ.

8.42 Plot the normalized radiation pattern at $\theta = \pi/2$ for three dipole antenna elements spaced λ/2 apart with progressive phase steps of 90°.

8.43 A particular broadside antenna array consists of ten λ/2 dipole antenna elements spaced λ/2 apart with all currents driven at the same phase. Plot the radiation pattern, and find the maximum broadside power density (i.e., at $\theta = \phi = \pi/2$) at a distance of 10. km if the antenna is driven by 10.-A current sources at 2.45 GHz.

8.44 A particular endfire antenna array consists of ten λ/2 dipole antenna elements spaced λ/2 apart with a progressive phase shift of 90° to each antenna. Plot the radiation pattern, and find the maximum endfire power density (i.e., at $\theta = \pi/2$ and $\phi = 0°$) at a distance of 10. km if the antenna is driven by 10.-A current sources at 2.45 GHz.

8.6 The Friis Transmission Equation

8.45 Consider a pair of half-wave dipole antennas operating at 2.45 GHz, separated by 50. m and aligned for maximum power transfer. If the output power must be at least -35 dB$_m$ to be detectable, calculate how much power is required to drive the transmitting antenna. Assume the antennas are 100% efficient.

8.46 A half-wave dipole transmitting antenna is centered on the z-axis oriented in the z direction. Show in a sketch where you would place a small loop antenna, 100 m distant, to receive the maximum power. (*Hint: Consider both radiation pattern and polarization to achieve maximum power transfer.*) Calculate the power transfer ratio for the maximum power transfer case at 800 MHz if the small loop antenna has a 2.0-cm diameter.

8.47 A pair of z-polarized dipole antennas with lengths indicated is shown in Figure 8.50. If the 3.0-m dipole is driven by a 50.-MHz source, calculate the power transfer ratio.

8.48 Consider a pair of half-wave dipole antennas operating at 1.0 GHz and separated by 100. m on the y-axis. Initially, both antennas are aligned in the z direction for maximum power transfer. To test the effect of polarization, the antenna at the origin is allowed to rotate an angle θ in the x–z plane as shown in Figure 8.51. Plot the power transfer ratio versus θ from $\theta = 0°$ (maximum transfer case) to $\theta = 90°$.

8.49 Design an open-ended shunt-stub matching network to match a half-wave dipole transmitting antenna to a source with 50-Ω impedance. Now suppose this antenna network is to be used as a receiver. Use a Smith Chart to determine the impedance looking into the matching network from the antenna.

8.50 Referring to Figure 8.52, suppose a source voltage with amplitude 12. V and source resistance 50 Ω drives a half-wave dipole transmitting antenna at 500 MHz. An identical receiving antenna, 100. m away and aligned for

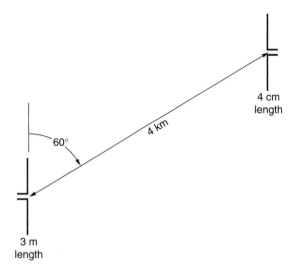

4 cm
length

4 km

60°

3 m
length

Figure 8.50 Pair of different length dipole antennas for Problem 8.47.

maximum power transfer, is coupled to a 50-Ω load resistance. Clearly neither antenna is impedance matched to the transmit and receive circuitry. Calculate the voltage amplitude across the load resistor.

8.51 Design open-ended shunt-stub matching networks for both the transmitter and receiver of Problem 8.50. Now recalculate the voltage amplitude across the load resistor.

8.7 Radar

8.52 Manipulate (8.126) using (8.113) to arrive at (8.127).

8.53 Suppose a 2-GHz radar antenna of effective area 6.0 ▶ EMAG SOLUTI m² transmits 100 kW. If a target with a 12 m² radar cross section is 100 km away, (a) what is the round-trip travel time for the radar pulse? (b) What is the received power? (c) What is the maximum detectable range if the radar system has a minimum detectable power of 2.0 pW?

8.54 A half-wave dipole antenna is used in a radar system to determine range to a target that has a 1.0 m² radar cross section. Consider that 1.0 kW is available to drive the antenna at 300 MHz. What power is received if the target is (a) 100 m distant and (b) 1.0 km distant?

8.55 Suppose a 10-GHz radar antenna of effective area 100 m² is to be used to determine the distance to the moon. The moon, with radius 1.74×10^6 m, has a measured radar cross section of 6.64×10^{11} m². A 27-pW echo signal is received 2.56 s after transmission. (a) What is the distance to the moon, and (b) approximately how much power was radiated?

Figure 8.51 Friis transmission with mismatched polarities for Problem 8.48.

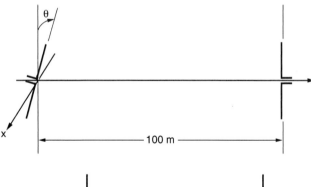

θ

x

—— 100 m ——

Figure 8.52 Transmitting and receiving antennas for Problems 8.50 and 8.51.

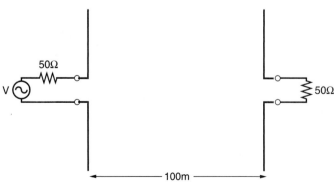

50Ω

V

50Ω

—— 100m ——

Electromagnetic Interference

Learning Objectives

▷ Define the sources of noise that lead to electromagnetic interference

▷ Develop circuit models to account for nonideal behavior of conductors, resistors, capacitors, and inductors

▷ Use Fourier series to determine the frequency components present in a digital signal

▷ Define electrical grounds and provide guidelines for their design

▷ Use shields and filters to decrease electromagnetic interference

We saw in the previous chapter how to design efficient radiators of electromagnetic energy. But any circuit component containing a time-varying current will radiate, and conversely it will also receive unintentional radiation. Unintentional signals received by a circuit are termed *noise*. Noise can arise from other circuit components or systems or from natural phenomena such as lightning or sunspots. If the intercepted noise degrades operation of the circuit or system, it is termed *interference*.

We can define *electromagnetic interference*, or *EMI*, as the degradation in the performance of an electrical circuit or system resulting from electromagnetic noise. An electrical circuit or system has achieved *electromagnetic compatibility*, or *EMC*, if it can function in an electromagnetic environment without being susceptible to noise and without causing interference in other circuits.

For EMI to exist requires a source that generates the noise, a coupling path to transmit the noise, and a receptor that is susceptible to the noise. This simple concept is represented by the block diagram of Figure 9.1. In designing for EMC, we can avoid interference by reducing noise emissions at the source, by blocking the transmission path, and by rendering the receptor less susceptible to the noise. Reducing the source noise emission in a circuit or system has the reciprocal benefit of rendering the circuit or system less susceptible to noise.

From a cost standpoint, it is far less expensive to take action at the design and prototyping stage of product development to prevent and reduce EMI than it is to correct problems in postproduction.

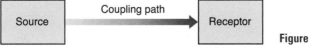

Figure 9.1   EMI block diagram.

EMI in digital circuits is of particular concern for two reasons. First, high-speed digital signals with short rise times generate a broad spectrum of noise that can lead to interference. Compounding the problem is the lower threshold voltages in modern digital logic that renders digital circuits more susceptible to noise.

We begin this chapter by looking at common sources of noise responsible for EMI. Then, in Section 9.2, we'll look at high-frequency models for commonly used passive circuit elements (resistors, inductors, and capacitors). These circuit elements are routinely used in high-speed circuits, but as frequency increases their performance can degrade, leading to interference problems. In particular, high-speed digital signals comprise frequency components much higher than one might expect given a particular clock speed, so we'll examine the spectral content of digital signals in Section 9.3. Then in Section 9.4 we'll take a look at how circuits are tied to ground. Improper grounding can lead to serious EMI. However the interference is generated, it may be controlled somewhat through the use of shields and filters, the topics of Sections 9.5 and 9.6.

▶ 9.1 INTERFERENCE SOURCES

There are a host of potential sources for EMI, as evidenced from Table 9.1. The sources range from natural, transient phenomena such as lightning strikes to manufactured, continuous emissions such as those from fluorescent lights. The interference may result from intentional emissions, such as those from cellular phones and radio transmitters. It may originate within the system being interfered with, or it may have an origin external to the system.

Many of these sources feature a sudden change in current, for instance from lightning strikes, electrostatic discharge, or the sharp rise and fall times of digital signals. These sudden changes, or *transients*, produce a spectrum of frequency components that may cause interference. The transients may be analyzed via Fourier analysis or using Laplace transforms to ascertain the amplitude of the frequency components in the generated spectra. In the case of digital signals, frequency components can be measured with a spectrum analyzer.

This section briefly describes some of the sources of EMI, namely lightning, electrostatic discharge, power source disturbances, and radio transmitters.

Lightning

A bolt of lightning can involve the rapid transfer of over 20 C of charge with stroke current in excess of 50 kA. A direct strike can often prove fatal to a piece of electrical equipment (or to an unfortunate but dedicated golfer). The rapid current pulse rise time (typically 1 μs) is responsible for spectral components ranging from several hertz to over 100 MHz that can induce interference in nearby electrical equipment.

Electrostatic Discharge

We are perhaps most familiar with electrostatic discharge from shuffling our feet on carpet on a cold, dry day and touching a metal doorknob. The zap we see and feel is a dramatic instance of electrostatic discharge, or ESD. When two materials are rubbed together, positive charge accumulates on one material and negative charge on the other. The degree to which this occurs is a function of the material and can be gleaned from the *triboelectric series* shown in Table 9.2.

For instance, suppose you are wearing nylon-soled shoes as you shuffle across an acrylic carpet. Considerable charge can be transferred to the shoes and hence to your body (a fairly decent conductor). Then, when you come in contact with a circuit board or piece of electrical equipment, the sudden release of charge can cause serious damage. Wearing shoes with a sole made from a material closer to acrylic in the triboelectric series, rubber for instance, can help reduce the severity of ESD.

Since ESD from humans is a primary culprit in damage to sensitive electronic components, manufacturers of such components have stringent guidelines in place to reduce ESD. One requirement is the use of ground straps worn by technicians as they handle the electronics. Other precautions include use of ionizing fans, shoe-grounding straps, and anti-ESD mats.

Power Disturbance Sources

Have you ever seen the lights dim briefly when the air conditioner or other large appliance switches on? A sudden change in the power load can cause the voltage to drop, or *sag*, briefly. Likewise, sudden removal of a large load can result in a temporary *swell* in the

TABLE 9.1 Common Sources of EMI

Lightning strikes
Fluorescent lights
Radar transmitters
Radio and TV transmitters
Cellular telephones
Computers
Computer peripherals
Power supplies
Power lines
Motors
Electrical relays

TABLE 9.2 The Triboelectric Series

| **More electropositive** |
| Air |
| Skin |
| Glass |
| Mica |
| Human hair |
| Nylon |
| Wool |
| Silk |
| Aluminum |
| Paper |
| Cotton |
| Wood |
| Steel |
| Rubber |
| Copper |
| Silver |
| Gold |
| Acrylic |
| Polyester |
| Polyethylene |
| PVC |
| Silicon |
| Teflon |
| **More electronegative** |

Source: Adapted from W. K. Kimmel and D. D. Gerke, *Electromagnetic Compatibility in Medical Equipment,* IEEE Press, 1995.

power level. The sags and swells of the power supply can cause interference in an electrical component.

Sometimes power substations will switch large voltage levels. The resulting transient may contain frequency components in excess of 200 MHz. These components can cause interference in equipment tied to the power line.

The power network is increasingly used for certain kinds of communications, where the signal rides on the same lines as the lower frequency power. An example of this is a home signaling system that communicates within different locations in a home via the power lines. These *mains signaling systems* are another source of potential interference from the power source.

Radio Transmitters

There is obviously considerable intentional emission of electromagnetic waves from a host of radio transmitters including radio stations, cellular phones, and wireless local area networks (WLANs). Bands of frequencies have been allocated for specific use (see Table 9.3), but emission will also include harmonics of the fundamental frequency that can lie well outside the intended band. In this way, for instance, radio transmission may interfere with cell phone reception.

Radio transmitters are generally designed to curtail the emission of harmonic frequencies. A 70-dB drop in the harmonic power level compared to the fundamental frequency is not uncommon.

TABLE 9.3 Some Frequency Bands of Interest

Frequency band	Application[a]
148.5–283.5 kHz	broadcasting
526.5–1606.5 kHz	AM broadcasting
13.533–13.567 MHz	ISM
26.957–27.283 MHz	ISM
27.6–28 MHz	Citizen's band
54–88 MHz	TV: lower VHF
88–108 MHz	FM broadcasting
174–216 MHz	TV: upper VHF
470–806 MHz	TV: UHF
886–906 MHz	ISM
824–894 MHz	Advanced mobile phone service
902–928 MHz	WLAN
931–932 MHz	US paging
934–935 MHz	Citizen's band
1.71–1.88 GHz	Personal communication services
2.400–2.484 GHz	ISM and WLAN
5.725–5.850 GHz	ISM and WLAN

[a]Acronyms: ISM, industrial, scientific and medical bands; UHF, ultra high frequency; VHF, very high frequency; WLAN, wireless local area network.

Source: Adapted from D. M. Pozar, *Microwave and RF Design of Wireless Systems,* Wiley, 2001.

A local oscillator (LO) is routinely employed at the receiving end of radio transmission. The LO signal is mixed with the received RF signal to down-convert the received signal to a more manageable, controlled frequency. Adjustment of the LO frequency also allows for tuning of the radio. However, the LO will also generate harmonics of the fundamental frequency and can interfere with its own nearby electronics.

9.2 PASSIVE CIRCUIT ELEMENTS

Resistors, inductors, capacitors, and connecting wire are considered *passive* circuit elements to distinguish them from active elements such as transistors. Passive elements are used for a wide variety of reasons. For EMI control, for instance, inductors and capacitors are used in filters to remove undesirable frequency components from a signal. Passive elements are also used in lumped-element matching networks (to be discussed in Chapter 10), where inductors and capacitors replace the tuning stubs described in Chapter 6 for impedance matching.

At low frequencies, passive elements are considered *ideal* since they behave as expected, but as frequency increases, nonideal behavior is encountered. For instance, an inductive coil has ohmic resistance along its length as well as capacitance between its windings. At high frequency, such unintentional (or *parasitic*) effects can render the element useless.

In this section we will account for the element parasitics in simple circuit models. From these models the useful frequency range for a given element can be determined.

Conductors

Despite what we remember from our circuits classes, conductive wire and conductive traces connecting circuit elements are really not perfect conductors of negligible length. The wires have some finite amount of resistance. Moreover, at high frequencies, since the current is mostly carried near the surface of the conductor, resistance becomes even higher.

For a length ℓ of wire with radius a and conductivity σ, the DC resistance is

$$R_{\text{dc}} = \frac{\ell}{\sigma \pi a^2} \qquad (9.1)$$

At high frequencies we can consider the current to be carried near the periphery of the wire, through an area $2\pi a\delta$, where the skin depth δ is

$$\delta = \frac{1}{\sqrt{\pi f \mu \sigma}}$$

If $\delta \ll a$, the AC resistance is then given by

$$R_{\text{ac}} = \frac{\ell}{\sigma 2\pi a\delta} \qquad (9.2)$$

Next, we want to find the inductance for a wire conductor. We saw in Chapter 3 that the internal inductance for a length ℓ of wire is given by

$$L_i = \frac{\mu\ell}{8\pi}$$

The inductance is based on current equally distributed over the wire's cross-sectional area, clearly not the case at high frequency. In fact, as frequency increases the current gets confined more and more to the periphery of the conductor and the internal inductance decreases. This is approximated by

$$L_i = \frac{\mu \ell}{4\pi a}\delta \tag{9.3}$$

Equation 9.3 is accurate for $\delta \ll a$.

Whereas the internal inductance may be negligible at high frequency, the external inductance certainly is not. Calculating this inductance accurately requires being able to identify the complete loop traveled by the current. Not knowing this loop, we can still estimate the external inductance for a segment of wire using the equation

$$L = 2\times10^{-7}\left(\frac{H}{m}\right)\ell\left[\ln\left(\frac{2\ell}{a}\right)-1\right] \tag{9.4}$$

This formula is based on a "partial inductances" concept, a subject beyond the scope of this text.

The wire length is of special concern, since as frequency increases even a short length of wire can become a significant fraction of a wavelength and must then be treated as a transmission line. Also, as an approximation, when wire lengths reach $\lambda/20$, antenna behavior can be assumed.

▶ **EXAMPLE 9.1**

Let us find the impedance for a 1.00-cm length of AWG24 copper wire operating at 1.00 GHz.

If we consider the AC wire resistance in series with the wire's external inductance, the impedance will be

$$Z = R_{ac} + j\omega L$$

From Appendix E, we see that AWG24 wire has a 20.10-mil diameter, corresponding to a radius $a = 0.255$ mm. The skin depth is calculated as

$$\delta = \left(\sqrt{\pi\left(10^9\frac{1}{s}\right)\left(4\pi\times10^{-7}\frac{H}{m}\right)\left(5.8\times10^7\frac{1}{\Omega\text{-m}}\right)}\right)^{-1} = 2.09\times10^{-6}\text{ m}$$

where we use the unit conversion factors V-s = H-A and Ω-A = V. Then, from (9.2) we calculate $R_{ac} = 51.5$ mΩ. This is considerably larger than the DC resistance for the wire segment, calculated from (9.1) as $R_{dc} = 0.85$ mΩ.

Now we can use (9.4) to calculate the external inductance for this wire segment. We find $L = 6.7$ nH, and therefore our impedance is

$$Z = 0.0515\,\Omega + j2\pi\left(10^9\frac{1}{s}\right)\left(6.7\times10^{-9}\text{ H}\right)\left(\frac{H\text{-}A}{V\text{-}s}\right)\left(\frac{A\text{-}\Omega}{V}\right) = 0.05 + j42\,\Omega$$

The inductive behavior of the wire at this frequency clearly overshadows the resistive behavior.

It may be noted that at 1 GHz, the wavelength is 30 cm. The wire is short enough so that antenna behavior can be safely ignored.

Figure 9.2 A DIP component is shown along with a number of SMT (surface mount technology) components.

Drill 9.1 Find R_{ac}, L, and Z for a 2.0-cm length of AWG20 copper wire at 500. MHz. (*Answer:* $R_{ac} = 46$ mΩ, $L = 14$ nH, $Z = 0.046 + j45$ Ω)

Most of the wire used in electronics is copper with an insulative sheath. The insulation is nonmagnetic ($\mu_r = 1$); hence it has no effect on the magnetic field properties of the wire.

Conductive traces on circuit board are often realized in copper, though other metals are also used. Sometimes copper traces are electroplated with a very thin layer of gold to prevent oxidation. Another popular circuit-board metallization approach is to use screen-printed conductive inks.[9.1] Trace resistance for these inks tends to be at least an order of magnitude greater than that of solid conductors, however.

At high frequency or high clock speeds, it is very important to minimize connector inductance by keeping the connections as short as possible. For instance, a surface mount component offers shorter, flatter leads with much less inductance than the wires of a larger DIP (dual-in-line package) component (see Figure 9.2).

Resistors

The three general types of resistors are wire wound, carbon composite, and film.

Wire-wound resistors are made by taking a specific length of fine wire, often nickel, and winding it to make it compact. The DC resistance is very accurate, but at higher frequencies such a resistor has severe parasitics. After all, its construction resembles that of an inductive coil.

Carbon-composite resistors, as shown in Figure 9.3, are routinely used and are typically accurate to within 10%. They are typically made from densely packed carbon granules. These resistors are visually the same as wire-wound resistors.

For high-frequency applications, films of resistive material (such as nichrome or tantalum nitride) are available as chip resistors. Typical chip resistors are shown in Figure 9.3. Figure 9.4 shows a cross-sectional view of a chip resistor made with resistive film. Chip resistors are usually specified by their length and width dimensions in mils. For instance, an "0804" chip resistor is 8 mils long by 4 mils wide.

[9.1] Recent developments allow use of conductive inks to be dispensed from standard ink-jet printers.

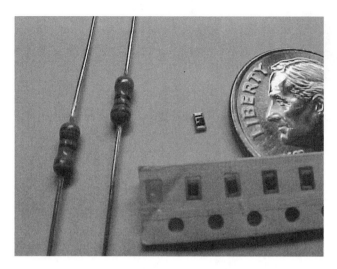

Figure 9.3 A surface mount chip resistor shown between a dime and a pair of $\frac{1}{4}$-W carbon composite resistors. A section of surface mount resistor reel is shown.

A disadvantage of thin-film resistors is their susceptibility to damage from high voltage levels, such as those encountered from ESD. Carbon-composite resistors are much more robust.

Film resistors may also be fabricated directly on or within the circuit board. Fairly long traces of resistive film can be meandered to fit in a small space, and these films can achieve high, precise values of resistance. These meandering (or serpentine) resistors, as represented by Figure 9.5, have considerable inductance along with capacitance between the traces; they are therefore impractical except at low frequencies.

The equivalent circuit model for a resistor is shown in Figure 9.6. Here, the parasitic element inductance L_x arises because current is present through some length of resistor. This is often considered negligible except when dealing with wire-wound resistors. The parasitic element capacitance C_x arises from charge separation (a voltage drop) across the resistor. This is typically 1–2 pF for carbon-composite resistors and may be as low as 50 fF for chip resistors. Additional parasitic lead inductance L_L is also present and is represented by a total lumped value at one end of the resistor. This is typically 10–20 nH for carbon-composite resistors. The much smaller leads of chip resistors reduce L_L to about 0.4 nH.

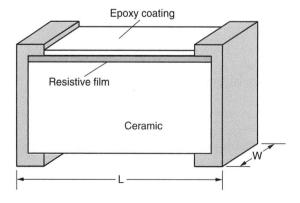

Epoxy coating

Resistive film

Ceramic

L

W

Figure 9.4 Detail of a typical resistive film SMT chip resistor.

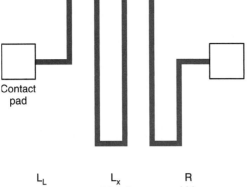

Figure 9.5 Top-down depiction of a serpentine resistor made with a meandering line of resistive film between two contact pads.

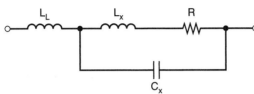

Figure 9.6 Resistor equivalent circuit model.

► MATLAB 9.1

Suppose a 200-Ω carbon-composite resistor operated at 1.00 GHz can be characterized as having negligible element inductance and 2.00 pF of element capacitance. It has 1.00-cm leads at each end, giving it a total lead inductance of 13.4 nH at 1.00 GHz (from Example 9.1). Assuming C_x and L_L remain fixed from 10 MHz up to 10 GHz, use MATLAB to produce a plot of the magnitude of the impedance versus frequency.

First, because we will be calculating parallel impedances in a number of circuits in this section, let's create a function to do this:

```
function y=parallel(A,B)
%   Calculates the parallel combination of a pair of
%   impedances, A and B.
y=(A.*B)./(A+B);
```

Notice that we are using the ".*" and "./" operations so we can use this function in array calculations. This function will also handle complex numbers used in the following routine for finding impedance for the resistance model.

```
%   M-File: ML0901
%
%   Z-f plot for the equivalent circuit of a resistor.
%
%   Wentworth, 12/2/02
%
%   Variables
%   R           resistance (ohms)
%   LX          element inductance (nH)
%   LL          lead inductance (nH)
%   CX          element capacitance (pF)
```

(continued)

```
%   f,w          freq and ang freq (1/s)
%   XLX,XLL,XCX  element impedances
%   Z            overall impedance
%   Zmag         impedance magnitude

clc              %clears the command window
clear            %clears variables

%   Initialize variables
R=200;
LX=0;
LL=13.4;
CX=2;

%   Perform calculations
f=10e6:10e6:10e9;
w=2*pi*f;
XLX=complex(0,w*LX*1e-9);
XLL=complex(0,w*LL*1e-9);
XCX=complex(0,-1./(w*CX*1e-12));
Z=XLL+parallel(R+XLX,XCX);
Zmag=abs(Z);

%   Generate plot
loglog(f,Zmag)
grid on
xlabel('frequency (Hz)')
ylabel('Z magnitude (ohms)')
```

The plot for this example is given in Figure 9.7. In this particular case we see that the 200-Ω resistor has an impedance magnitude of only about 30 Ω at the stated frequency of operation, 1 GHz.

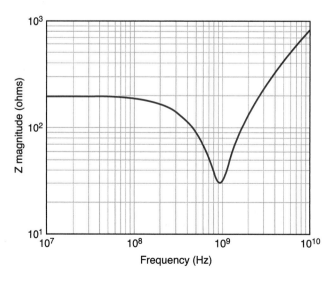

Figure 9.7 Plot for the resistor circuit of MATLAB 9.1.

Inductors

As we have seen, a simple length of conductor can function as an inductor. Increasing inductance is accomplished by lengthening the conductor and creating loops or coils. Figure 9.8 shows some of these approaches. Planar loops can form inductors by wrapping them in a spiral pattern. But the highest inductance is achieved by wrapping loops around a ferromagnetic or ferrite core, as shown in Figure 9.9.

Inductors find use as tuning elements and high-frequency *chokes*.[9.2] They may be used in EMC as series elements in wires where they can choke off noise currents.

Figure 9.10 shows an equivalent circuit model for a typical inductor. Parasitic resistance, R_x, arises from ohmic losses in the wire. If a magnetic core is used, R_x also accounts for hysteresis and eddy-current losses. The parasitic capacitance, C_x, arises from electric fields established between the windings. Notice the absence of lead inductance in this model. Since the component is intended as an inductor, the element L will most likely be much larger than any lead inductance.

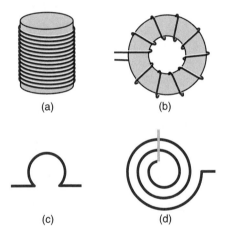

(a) (b)

(c) (d)

Figure 9.8 Several types of inductors: (a) ferrite-rod inductor, (b) ferrite toroidal inductor, (c) single-loop inductor, and (d) spiral inductor.

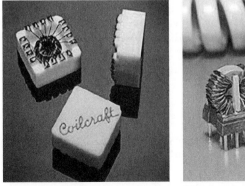

(a)

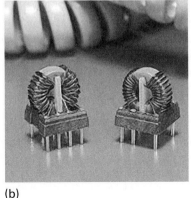

(b)

Figure 9.9 Inductive coils used as filter elements. Courtesy of Coilcraft, Inc.

[9.2]An inductor appears as an open circuit at high frequency and is said to choke or restrict high-frequency signals.

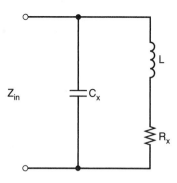

Figure 9.10   Inductor equivalent circuit model.

Analysis of the circuit in Figure 9.10 shows that the impedance of an actual inductor is

$$Z_{in} = \frac{R_x + j\omega L}{1 - \omega^2 L C_x + j\omega R_x C_x} = |Z_{in}| e^{j\phi} \tag{9.5}$$

The *self-resonant frequency* (*SRF*, or f_{SRF}) occurs when $\omega^2 L C_x = 1$, or

$$f_{SRF} = \frac{1}{2\pi \sqrt{L C_x}} \tag{9.6}$$

As a suggested practice, an inductor or capacitor should be operated at no more than half of its *SRF*.

The values of R_x and C_x are largely determined by the details of construction. Increasing L by increased windings also increases C_x, thus lowering the *SRF*. Typically, a 1-µH air-core inductor has a *SRF* on the order of 100 MHz. A way to increase the inductance without requiring numerous windings is to employ magnetic materials. It is usual practice to use wire-wound inductors for frequencies below 50 MHz and change to ferrite-core inductors for higher frequencies. One potential drawback of the ferrite inductors is the tendency of the magnetic material to saturate, or reach a maximum field value, at high current levels (as described in Chapter 3).

Open magnetic circuits (using an air or ferrite-rod core for instance) generate magnetic fields and are also good receivers of such fields. In fact, AM radio antennas often consist of open magnetic circuits (ferrite-core antennas). In contrast, closed magnetic circuits (ring cores) generate little and receive very little magnetic field. The closed magnetic circuits are more prone to saturation, however.

One ferrite material often chosen for EMI use is NiZn, because it is actually very lossy between 100 and 400 MHz.

▷ **EXAMPLE 9.2**

An inductor is formed by evenly wrapping 20 turns of AWG30 insulated copper wire around a 300-mil-long Teflon rod of diameter 0.50 cm, as shown in Figure 9.11a. Neglecting the wire resistance, estimate the coil's inductance and its self-resonance frequency.

The inductance equation from Chapter 3 for an N-turn coil is

$$L = \frac{\mu N^2 \pi a^2}{h}$$

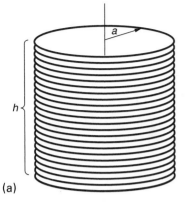

(a)

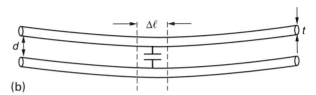

(b)

Figure 9.11 (a) Coil inductor. (b) Close-up of a pair of windings to find C_x.

which is valid when the coil core radius a is much less than the core length h. In our case, a 0.5-cm core has about a 100-mil radius. With a core length of only 300 mils, we do not really have the case of $a \ll h$. Nevertheless, we will use the given inductance equation as an approximation.

Since the Teflon rod is nonmagnetic, we calculate

$$L = \left(4\pi \times 10^{-7} \, \frac{H}{m} \right)(20)^2(\pi)\left(\frac{(100 \text{ mil})^2}{300 \text{ mil}} \right)\left(\frac{25.4 \text{ μm}}{\text{mil}} \right)\left(\frac{m}{10^6 \text{ μm}} \right) = 1.3 \text{ μH}$$

Calculating the *SRF* requires an estimate of the interwinding capacitance C_x.

For a short section of a pair of the coil windings, as shown in Figure 9.11b, we can estimate the capacitance as that of a parallel-plate capacitor:

$$C = \frac{\varepsilon S}{d}$$

We can estimate the gap d between the windings by considering the relationship between the height h of the rod, the number N of windings, and the wire diameter t:

$$h = Nt + (N-1)d$$

or

$$d = \frac{h - Nt}{N - 1}$$

For AWG30 wire, $t = 10$ mils, so $d = 5.3$ mils (0.130 mm). The surface to consider for the short section is estimated as $\Delta S = t\Delta\ell$. For the total surface we estimate

$$S = t(2\pi a)(N-1)$$

or approximately 120×10^3 mil^2 (77×10^{-6} m^2) for our case.

The dielectric between the windings is mostly air, although a significant portion of the field will cut through the $\varepsilon_r = 2.2$ Teflon material. Still, we will assume $\varepsilon_r = 1$ for this estimate and find

$$C_x = \left(\frac{10^{-9} \text{ F}}{36\pi \text{ m}} \right)\left(\frac{77 \times 10^{-6} \text{ m}^2}{130 \times 10^{-6} \text{ m}} \right) = 5.2 \text{ pF}$$

The *SRF* is then calculated as

$$f_{SRF} = \frac{1}{2\pi\sqrt{\left(1.3\times10^{-6}\ H\right)\left(5.2\times10^{-12}\ F\right)}} = 60\ \text{MHz}$$

Drill 9.2 Recalculate L, C_x, and f_{SRF} if the AWG30 wire for the coil of Example 9.2 is replaced with AWG28 wire. (*Answer:* L is unchanged, $C_x = 14$ pF, $f_{SRF} = 38$ MHz)

Capacitors

Capacitors are charge storage devices realized in a variety of configurations. Electrolytic capacitors can achieve very high capacitance (1–1000 μF) but only at low frequencies. Multilayer ceramic capacitors as shown in Figure 9.12a are capable of much higher frequency operation, but they have more limited capacitance (typically 5 pF–1 μF). Still higher frequency operation, with even further limited capacitance, is afforded by interdigitated capacitors as shown in Figure 9.12b. These are often used as tuning elements in high-frequency circuits.

In typical amplifier circuits, coupling capacitors (for DC blocking), bypass capacitors, and feedback capacitors are routinely used. For EMC, capacitors are widely used for filtering and for power decoupling. They are also used as local charge reservoirs in digital circuits to help overcome the problem of ΔI *noise*. This problem occurs because, during

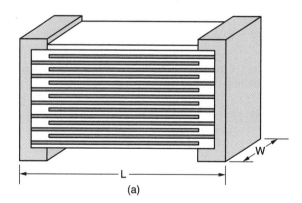

(a)

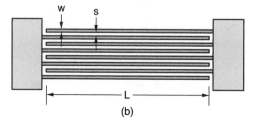

(b)

Figure 9.12 (a) A ceramic multilayer capacitor consists of multiple layers of metal electrodes and ceramic dielectric. (b) Top-down view of an interdigitated capacitor.

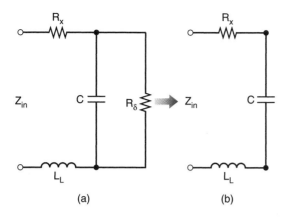

Figure 9.13 (a) Equivalent electrical circuit for a capacitor. (b) Circuit neglecting the typically very small R_δ.

switching, relatively large currents must pass through the input/output (I/O) ports of the digital integrated circuit (IC). The connecting wire inductance causes a temporary voltage drop for this current feeding the IC. Sufficient voltage drop can result in an erroneous reading by the IC. Local capacitors placed very close to the I/O pads act as charge reservoirs to minimize this source of noise.

An equivalent circuit model for the typical capacitor is shown in Figure 9.13a. Here R_x is an equivalent series resistance that can be several ohms for an electrolytic capacitor but is practically negligible for ceramic capacitors; L_L is primarily inductance from the leads, which should be kept as short as possible; and the resistance R_δ is related to the dielectric loss tangent as

$$R_\delta = \frac{1}{\omega C \tan \delta} \tag{9.7}$$

For good dielectrics and typical capacitance values, R_δ is very large and can be ignored, as shown in the circuit model of Figure 9.13b. Input impedance for this model is

$$Z_{\text{in}} = R_x + j\left(\frac{\omega^2 L_L C - 1}{\omega C}\right) \tag{9.8}$$

and the *SRF* is given by

$$f_{SRF} = \frac{1}{2\pi\sqrt{L_L C}} \tag{9.9}$$

In some EMI situations, it is desired to shunt a noise frequency to ground using a capacitor. In such a case selecting capacitance such that the *SRF* matches the noise frequency attains the lowest impedance.

▷ **EXAMPLE 9.3**

Suppose we have a 2.2-nF mica capacitor with a pair of 0.50-cm-long AWG26 copper leads. We want to compare the estimated performance of this capacitor with and without inclusion of the element resistance R_δ. For simplicity, we will assume a constant $R_x = 0.010 \, \Omega$.

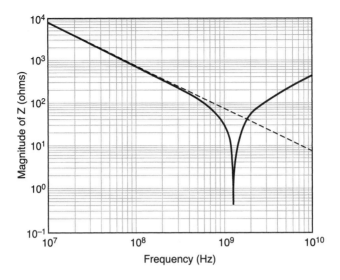

Figure 9.14 Plot for the capacitor circuit of Example 9.3.

For a total 1.0-cm length of AWG26 copper lead, we can use (9.4) to determine $L_L = 7.2$ nH. The impedance is

$$Z = R_x + j\omega L_L - \frac{j}{\omega C}$$

A plot of the magnitude of this Z versus frequency is given as the bold line in Figure 9.14. The dashed line represents an ideal capacitor with no parasitics.

For inclusion of R_δ, we find from Appendix E that $tan\ \delta = 0.0003$ for mica, and use (9.7). For the low loss tangent of mica there is no discernible difference in the plots with and without inclusion of R_δ.

▷ **MATLAB 9.2**

The following routine is used to generate Figure 9.14.

```
%    M-File: ML0902
%
%    plot for the capacitor equivalent circuit of
%    Example 9.3.
%
%    Wentworth, 12/2/02
%
%    Variables
%    C          capacitance (F)
%    LL         lead inductance (H)
%    Rx         series resistance (ohms)
%    Rd         diel loss resistance (ohms)
%    tand       dielectric loss tangent
%    f,w        freq and ang freq (1/s)
%    XC,CRC,XL  element impedances
%    Z1         mag of ideal impedance (ohms)
```

```
%   Z2           mag of impedance with RL (ohms)
%   Z3           mag of impedance without RL

clc              %clears the command window
clear            %clears variables

%   Initialize variables
C=2.2e-12;
LL=7.2e-9;
Rx=0.01;
tand=0.0003;

%   Perform calculations
f=10e6:10e6:10e9;
w=2*pi*f;
Rd=1./(w.*C*tand);
XC=complex(0,-1./(w.*C));
Z1=abs(XC);
XRC=parallel(Rd,XC);
XL=complex(0,w.*LL);
Z=Rx+XL+XRC;
Z2=abs(Z);
Z=Rx+XC+XL;
Z3=abs(Z);

%   Generate plot
loglog(f,Z1,'-k',f,Z2,'k',f,Z3)
grid on
xlabel('frequency (Hz)')
ylabel('magnitude of Z (ohms)')
```

Drill 9.3   Calculate the self-resonance frequency for a 4.7-nF mica capacitor with a pair of 0.50-cm-long AWG26 copper leads. (*Answer:* 27 MHz)

► 9.3   DIGITAL SIGNALS

In Section 9.1 we mentioned that transients or short rise and fall times in digital signals can generate a broad spectra of frequency components that may cause interference. Higher frequency signals in particular tend to couple or radiate to nearby components and cause interference. Wire connections may appear as transmission lines at high frequency, generating reflections and establishing standing waves that make the line more radiative. Moreover, circuit components, such as capacitors, only perform as expected well below their self-resonant frequency.

In this section, the frequency components of a digital signal are determined using Fourier series. It will be seen that the frequency component amplitudes are functions of rise and fall times. The shorter the rise and fall times, the greater the amplitude of higher frequency components.

A digital pulse can be represented by a linear superposition of its Fourier components. Consider a digital signal of the form shown in Figure 9.15. For simplicity, we let the rise time t_r equal the fall time t_f, where $t_f = t_2 - t_1$. Also, the rising and falling voltages are a

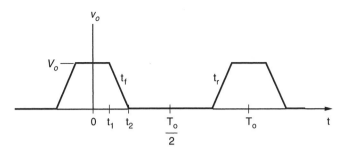

Figure 9.15 A typical digital pulse modeled as a trapezoidal function.

linear function of time. We align the signal with the time axis to give us even-mode symmetry. The equations for the Fourier coefficients (see Chapter 6 Section 6.8) derived for this case are

$$a_o = \frac{V_o}{T_o}(t_1 + t_2) \tag{9.10}$$

and

$$a_n = \frac{2V_o}{n^2 \pi \omega_o t_f}\left(\cos(n\omega_o t_1) - \cos(n\omega_o t_2)\right) \tag{9.11}$$

The voltage function is

$$v(t) = a_o + \sum_{n=1}^{\infty}\left(a_n \cos(n\omega_o t)\right) \tag{9.12}$$

This function is evaluated in MATLAB 9.3.

MATLAB 9.3

We want to model a digital signal like the one in Figure 9.15 using (9.10–9.12) in a MATLAB routine. Here, we will adjust the rise and fall times by modifying the statements for t_1 and t_2 within the program. We will plot the digital signal over one period and add a second plot that shows the amplitudes for each harmonic.

```
%   M-File: ML0903
%
%   Decomposes a realistic digital signal into its
%   Fourier components.
%
%   Wentworth, 12/2/02
%
%   Variables
%   N           upper limit on Fourier components
%   fo          fundamental freq. (Hz)
%   wo          fund angular freq. (rad/s)
%   To          period (s)
%   Vo          max voltage level (V)
%   t1,t2       times to determine tf (s)
```

```
%   tf        fall time(s), equal to rise time
%   a0        first coeff.
%   a(i)      ith coeff.

clc                %clears the command window
clear              %clears variables

%   Initialize variables
N=80;
fo=10e6;
wo=2*pi*fo;
To=1/fo;
Vo=1;
t1=0.20*To;
t2=0.30*To;
tf=t2-t1;

%   Determine the coefficients
a0=(Vo/To)*(t1+t2);
for i=1:N
    n(i)=i;
    a(i)=((2*Vo)/(pi*wo*tf*i^2))*(cos(i*wo*t1)-cos(i*wo*t2));
end

%   Determine the function components
for j=1:1000
    t(j)=j*To/1000;
    for i=1:N
        Fn(i)=a(i)*cos(i*wo*t(j));
    end
    F(j)=a0+sum(Fn);
end

%   Generate the plot
subplot(2,1,1)
plot(t,F)
xlabel('time(sec)')
ylabel('voltage(V)')
grid on
subplot(2,1,2)
m=1:40;
b=a(m);
bar(m,b,'-k')
xlabel('n')
ylabel('an')
```

The values shown in this program were used to generate Figure 9.16a. Note that N, the number of Fourier components, can be increased if needed to more accurately portray a digital signal. This was in fact done in Figure 9.16b, and the program was slightly modified to only show the frequency components out to $n = 40$ for comparison purposes.

Using MATLAB 9.3 we generate the results shown in Figure 9.16. Here, we choose a fundamental frequency of 10 MHz and compare the frequency components for a 10-ns rise time (Figure 9.16a) with the components for a 1-ns rise time (Figure 9.16b). These values were chosen since they are fairly typical for digital clock rates. In the first case, significant

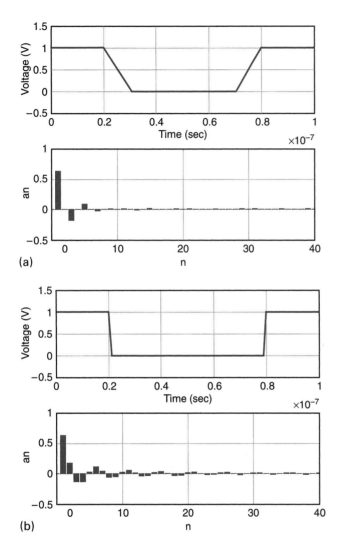

(a)

(b)

Figure 9.16 One period of a digital signal and its corresponding frequency components are plotted using MATLAB 9.3 for two cases where the fundamental frequency is set at 10 MHz. In (a) $t_f = t_r = 10$ ns, and in (b) $t_f = t_r = 1$ ns.

frequency components are observed out to about the 7th harmonic. However, in the second case, significant components are seen to exist beyond the 20th harmonic!

A general approximation relating rise time to the spectral bandwidth is

$$BW \approx \frac{1}{t_r} \tag{9.13}$$

Using this rule, we estimate a bandwidth of 100 MHz for the 10-ns rise time and a 1-GHz bandwidth for the 1-ns rise time. Although this rule significantly overestimates the spectral content shown in Figure 9.16, it does give the engineer a quick estimator of the bandwidth.

Several general rules may be used to minimize potential EMI problems from the source. First and foremost, the lowest possible operating frequency with the longest possible rise time should be chosen to achieve a given task. This is, however, counter to the direction of technology. Clock rates at present exceed 3 GHz and will continue to increase. Second, minimizing the changes in voltage and current during switching will reduce the amplitude of the frequency components. Here, the semiconductor industry is aiming for 1.5-V supply voltages. And third, in multiple-frequency systems, care should be taken in choosing frequencies that have dissimilar harmonics.

One approach used to minimize EMI problems has been to employ spread-spectrum techniques such as frequency hopping. In this approach, transmission at a particular frequency only occurs for a very short time before a different frequency is used, resulting in far fewer problems with interference.

▷ 9.4 GROUNDS

When we speak of electrical ground we are actually talking about a path for current to return from some load back to the source. A ground to earth (a ground at *earth potential*) is required of most electrical equipment to protect against shock hazard. Without this *safety ground*, the equipment frame could unintentionally be at a high voltage, and an unfortunate person could touch the frame to provide the conductive path for current to ground.

But proper grounding is also extremely important for minimizing EMI problems. If done well, grounding along with judicious use of shields (Section 9.5) can eliminate most noise and EMI problems without our having to resort to relatively expensive filters (Section 9.6).

A signal ground is often needed on a circuit board to provide a reference for the signal voltage. What is desired here is an *equipotential reference* for the signals. However, ground wires, planes, or grids are not perfect conductors, and as they must conduct current, potential differences will be present. The goal then of a good signal ground is to maintain a low impedance over the entire frequency range of interest to minimize differences in the reference potential.

Several types of grounds are indicated by the circuit symbols shown in Figure 9.17. The first one, in Figure 9.17a, is a commonly used symbol for signal or earth ground. Figure 9.17b shows the symbol for a connection made to the chassis of a piece of electrical equipment. The chassis ground itself may be tied to earth ground. Figure 9.17c shows a special ground symbol that may be used for DC power ground or perhaps a common connection that is isolated from earth or chassis ground. Too often, circuit schematics contain such symbols without indicating their physical location. For EMI reduction, however, it is crucial that we know the current paths.

Bond Wires

The circuit or system is connected to ground using bond wires or straps. Here it is important to keep the leads short (preferably less than $\lambda/20$ at the highest frequency of interest) and to keep the impedance low. We saw in Section 9.2 that conductors of circular cross section can have considerable impedance at higher frequency owing to inductance. Using (9.4)

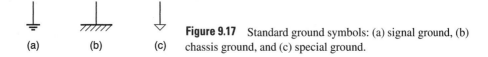

Figure 9.17 Standard ground symbols: (a) signal ground, (b) chassis ground, and (c) special ground.

(a)　　　(b)　　　(c)

in Example 9.1, we saw that a 1-cm length of AWG24 copper wire at 1 GHz would have an impedance $Z = 0.05 + j42\ \Omega$. Even at lower frequency, impedance can be considerable for lengths of wire typically used for connecting to ground.

Lower inductance is attained using flat, wide-cross-section conductive straps. The impedance of a strap with a length to width ratio of five or less can have less than half the impedance of a similar length of wire.

Signal Grounds

In creating a signal ground (or *reference ground*) we want to create a low-impedance return path between loads and sources. The variation in potential over a good reference ground will be insignificant compared to the signal potential.

Figure 9.18 shows signal grounding arrangements for three parallel circuits. The ground is connected in a single-point arrangement in Figures 9.18a and 9.18b. The reason single-point ground should be used can best be described with the aid of Figure 9.19. The facility ground may not be equipotential. A large ground current I_G, initiated perhaps by a lightning strike, can cause significant difference in the ground potentials for the two signal lines of Figure 9.19a. The presence of uneven ground potentials can establish a *ground loop*. Ground loops can especially be a problem in audio or video equipment if multiple outlets are used to plug in interconnected electronic equipment. The difference in potential at the various grounds can result in some of the 60-Hz current returning through the audio system, resulting in a loud hum. The problem can be avoided by tying the grounds together at a single point so that all the signal grounds will be at the same potential. This approach is represented by Figure 9.19b.

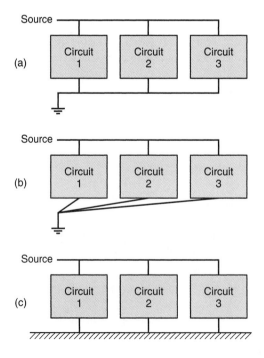

Figure 9.18 Signal ground topologies: (a) series-connected single-point, (b) parallel-connected single-point, and (c) multipoint arrangements.

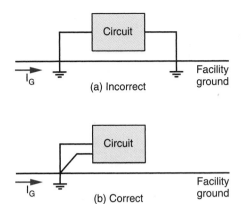

Figure 9.19 The motivation for single-point connection.

(a) Incorrect

(b) Correct

The series-connected single-point approach of Figure 9.18a is very commonly used. A problem with this method is that the bond wire between circuits 1 and 2, and that between circuits 2 and 3, can see a potential difference at the ground level that can affect operation and cause EMI problems. For this reason, the bond leads must be kept as short as possible. Despite this shortcoming, this daisy-chain approach is widely used for convenience.

The parallel-connected single-point configuration of Figure 9.18b offers much better EMI control. Still, the circuit operates best at low frequency where the wire inductance (from the circuits to the single-point ground) is low. At high frequency the ground wires will have high impedance and may also act as antennas. Finally, the parallel-connected single-point approach can require an excessive amount of wire for a large system.

▷ **EXAMPLE 9.4**

In Figure 9.20a, AWG24 copper wire is used to interconnect three circuits in a series-connected single-point arrangement. If each circuit is feeding 1.0 mA of 100-MHz current to ground, we want to determine the voltage at points A, B, and C.

Using (9.4) we find for each 4.0-cm section of bond wire a 38-nH inductance. At 100 MHz this corresponds to an impedance $j24\ \Omega$. Calculating the skin depth (6.6 μm) and AC resistance (7.6 mΩ) shows that the impedance can be considered entirely reactive.

The equivalent circuit is shown in Figure 9.20b. Here it is easy to see that the voltage amplitudes at A, B, and C are, respectively, 72, 120, and 144 mV.

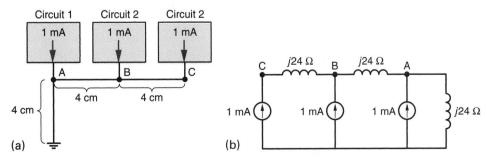

Figure 9.20 (a) Series-connected single-point ground arrangement for Example 9.4. (b) The equivalent circuit.

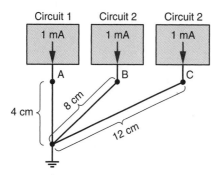

Figure 9.21 Parallel-connected single-point ground arrangement for Example 9.5.

Drill 9.4 Find the voltage amplitudes at points A, B, and C in Example 9.4 if the frequency is 500 MHz. (*Answer: V*$_A$ = 360 mV, *V*$_B$ = 600 mV, and *V*$_C$ = 720 mV)

▷ **EXAMPLE 9.5**

In Figure 9.21, AWG24 copper wire is used to interconnect three circuits in a parallel-connected single-point arrangement. If each circuit is feeding 1 mA of 100 MHz current to ground, we want to determine the voltage at points A, B, and C.

Again using (9.4), we find the wires to A, B, and C have inductances of 38, 87, and 140 nH, respectively. At 100 MHz, and continuing to neglect the small AC resistance, we then have impedances $j24$, $j55$, and $j88$ Ω, respectively, for the three bond wires. This leads to voltage amplitudes V_A = 24 mV, V_B = 55 mV, and V_C = 88 mV.

Drill 9.5 Find the voltage amplitudes at points A, B, and C in Example 9.5 if the frequency is 500 MHz. (*Answer: V*$_A$ = 120 mV, *V*$_B$ = 270 mV, and *V*$_C$ = 440 mV)

A better approach at high frequency utilizes the multipoint connection shown in Figure 9.18c. Here it is critical that the ground leads from circuit to the common ground plane be as short as possible, preferably less than $\lambda/20$. For circuit boards, it is typical to use a ground plane on the component side of the board with the signal traces on the other. Holes, or *vias*, are placed in the board to connect the component side to the trace side. The ground plane used in the multipoint topology is itself tied to the board's single-point ground.

The multipoint approach works best for frequencies above 10 MHz. For lower frequency, higher power situations the reference plane may have too much potential variation and should be avoided in favor of a single-point connection.

Loop Area

A circuit containing a source, connecting wire, load, and ground creates a current loop that resembles a loop antenna. It can serve as an unintentional source of radiated emissions and is also susceptible to reception of incident fields, for instance from radio transmitters.

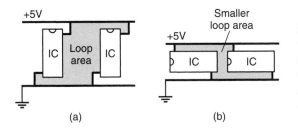

Figure 9.22 Pair of identical circuits with the topology of (b) having a smaller loop area than the topology of (a).

Adapted from C. Paul, *Introduction to Electromagnetic Compatibility*, Wiley, 1992, p. 750, Figure 13.41.

Recall from Chapter 8 that radiated power from a loop antenna is proportional to the loop area squared. Care should therefore be taken to minimize the area enclosed by these current loops to reduce EMI.

As an example, consider the pair of circuit topologies shown in Figure 9.22. The circuit functions are identical, but the loop area is significantly smaller in Figure 9.22b. Consequently, this second topology is preferred for reducing EMI.

► 9.5 SHIELDS

Some circuits, such as AM and FM transmitters, are bound to radiate and cause interference with other devices. Other circuits, such as receivers, may be very sensitive to interference. In such cases it may be necessary to block radiation using *shields*. A shield can be a metal or metal screen box placed around a circuit, or it can be a metal or metal braid wrapping around a cable or connector. It is common practice to tie the zero reference of the shielded circuit to a single point on the shield.

As depicted in Figure 9.23, a shield can attenuate an electromagnetic wave and can reflect waves at both interfaces. What is sought is a negligible amount of transmitted radiation. The performance of a shield is given by its *shielding effectiveness*, defined as the decibel reduction in radiated power or field strength resulting from the shield:

$$SE = 10 \log\left(\frac{P_{\text{ns}}}{P_{\text{s}}}\right) \tag{9.14}$$

where P_{ns} is the transmitted power with no shield in place and P_{s} is the transmitted power with the shield. Equation (9.14) can also be written in terms of electric and magnetic fields as

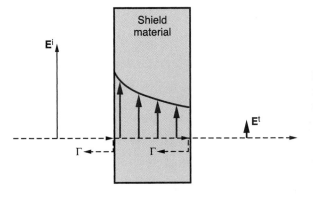

Figure 9.23 Some of a field $\mathbf{E}^{\text{i}}$ incident on a shield will be reflected at the air–shield boundary, some will be transmitted and attenuated within the shield, some will be reflected at the shield–air boundary, and finally some will be transmitted as $\mathbf{E}^{\text{t}}$.

$$SE = 20 \log\left(\frac{|E_{ns}|}{|E_s|}\right) = 20 \log\left(\frac{|H_{ns}|}{|H_s|}\right) \tag{9.15}$$

Typically, 40–60 dB of shielding is sufficient for most EMI problems.

▶ EXAMPLE 9.6

The electric field amplitude on the source side of a shielded enclosure is measured as 6.0 V/m. On the outside, near the shield, the electric field amplitude is measured as 1.0 mV/m. We want to look at the power levels on both sides of the shield and determine the shielding effectiveness.

The power density is related to the field amplitude and media impedance by

$$P = \frac{1}{2}\frac{E^2}{\eta}$$

Since air is assumed on either side of the shield, we have

$$P_i = \frac{1}{2}\frac{\left(6\,{}^V\!/_m\right)^2}{120\pi\ \Omega} = 48 \text{ mW}$$

and

$$P_o = \frac{1}{2}\frac{\left(1\,{}^{mV}\!/_m\right)^2}{120\pi\ \Omega} = 1.33 \text{ nW}$$

The shielding effectiveness is found by comparing power levels just inside to just outside of the shield, or

$$SE = 10 \log\left(\frac{P_i}{P_o}\right) = 10 \log\left(\frac{48 \text{ mW}}{1.33 \text{ nW}}\right) = 76 \text{ dB}$$

This is equivalent to taking the field levels on each side of the shield as

$$SE = 20 \log\left(\frac{E_i}{E_o}\right) = 20 \log\left(\frac{\left(6\,{}^V\!/_m\right)}{\left(1\,{}^{mV}\!/_m\right)}\right) = 76 \text{ dB}$$

Drill 9.6 The field amplitude inside a shield with a shielding effectiveness of 40 dB is 8 mV/m. What is the electromagnetic power density outside the shield? (*Answer:* 8.5 pW/m^2)

Attenuation, or absorption, converts the electromagnetic energy to heat in the shield material. A material with high permeability, such as steel or iron, makes an especially good absorption material at high-frequency since

$$\alpha = \sqrt{\pi f \mu \sigma} \tag{9.16}$$

Although higher μ_r materials are available (e.g., mumetal) they are expensive and can be difficult to handle. Fortunately, adequate absorption is also accomplished using cheap, versatile conductors such as copper and aluminum.

For reflective shielding, any good metal will do. Conductive spray paints may even be used over plastic. This paint is typically 80% metal (copper, nickel, or silver flakes) with 20% binding material. Even a thin coating (50 μm or so) can provide adequate shielding for most purposes.

▷ **EXAMPLE 9.7**

Let us evaluate the shielding effectiveness by using a T-line analogy for a typical aluminum shield of 10.-μm thickness. For simplicity, we assume the electromagnetic wave is a 1.0-GHz steady-state sinusoidal signal that is normally incident on the shield.

The situation is shown in Figure 9.24a. In our T-line analogy we can represent the fields E^i, E^r, and E^t by voltages v^i, v^r, and v^t, respectively. The intrinsic impedances η_1 and η_o are represented by characteristic impedances Z_1 and Z_o. The result is the T-line model for our problem shown in Figure 9.24b. To find the shielding effectiveness, we need to find the ratio v^i/v^t.

There are two approaches to solving this problem. The exact method uses the complete circuit approach described in Chapter 6 Section 6.3. An approximate method ignores rereflections within the shield.

Exact Method   Using the complete circuit approach, we must find the voltage wave v_o^+ at the shield side of the shield–air boundary. From Section 6.3 we find

$$v_d = \frac{Z_{in}}{Z_{in} + Z_o} v_s = v_o^+ \left(e^{\gamma d} + \Gamma_o e^{-\gamma d} \right)$$

Solution of v_o^+ in terms of v_s requires that we calculate Z_{in}, γ, and Γ_o. From (6.55), we have

$$Z_{in} = Z_1 \frac{Z_o + Z_1 \tanh(\gamma d)}{Z_1 + Z_o \tanh(\gamma d)}$$

The propagation constant is calculated from the constitutive parameters of the shield and the frequency as

$$\gamma = \sqrt{j\omega\mu(\sigma + j\omega\varepsilon)} = 3.87 \times 10^5 + j3.87 \times 10^5 \ \frac{1}{m}$$

Likewise, the characteristic impedance is calculated as

$$Z_1 = \sqrt{\frac{j\omega\mu}{\sigma + j\omega\varepsilon}} = 14.4 e^{j45°} \ m\Omega$$

Given these values, and with $Z_o = 120\pi \ \Omega$ and $d = 10$ μm, we find

$$Z_{in} = 14.4 e^{j45°} m\Omega$$

The reflection coefficient seen by the wave going from the shield to air is

$$\Gamma_o = \frac{Z_o - Z_1}{Z_o + Z_1} = 0.9999 - j0.0001$$

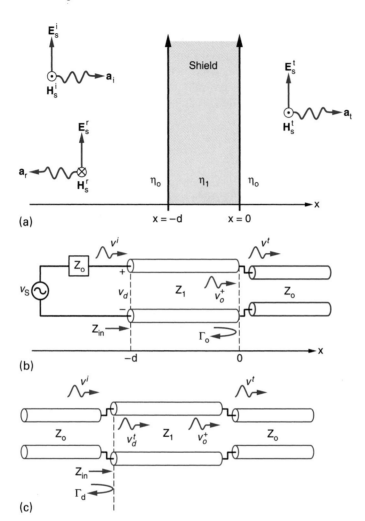

(a)

(b)

(c)

Figure 9.24 The shield problem of (a) is modeled by the T-line of (b). (c) Model used in the approximate method.

Now we can solve for v_o^+, finding

$$v_o^+ = \frac{Z_{in}}{Z_{in} + Z_o} \frac{1}{e^{\gamma d} + \Gamma_o e^{-\gamma d}} = -(796 + j45) \times 10^{-9} v_s$$

The voltage transmitted into the air is then

$$v^t = \tau_o v_o^+ = (1 + \Gamma_o)v_o^+ \approx 2v_o^+ = -(1592 + j90) \times 10^{-9} v_s$$

The incident voltage wave v^i calculated with no shield in place is simply

$$v^i = \frac{1}{2}v_s$$

The shielding effectiveness is then calculated as

$$SE = 20\log\left(\left|\frac{v_{ns}}{v_s}\right|\right) = 20\log\left(\left|\frac{v^i}{v^t}\right|\right) = 110 \text{ dB}$$

Approximate Method First we can calculate how much of v^i is transmitted into the shield at $x = -d$. Referring to Figure 9.24c, we'll call this value v_d^t; we see that it is related to v^i by the transmission coefficient at $x = -d$, so

$$v_d^t = \tau_d v^i = (1 + \Gamma_d)v^i = \left(\frac{2Z_{in}}{Z_{in} + Z_o}\right)v^i$$

With the value of Z_{in} calculated before, we find

$$\tau_d = 76.48 \times 10^{-6} e^{j44.95°}$$

As the transmitted voltage propagates through the shield, it attenuates such that at $x = 0$ we have

$$v_0^+ = v_d^t e^{-\gamma d}$$

The amount transmitted into free space is the v^t we are looking for:

$$v^t = \tau_o v_o^+ = \left(\frac{2Z_o}{Z_o + Z_1}\right)v_o^+ \approx 2v_o^+$$

Putting all of these equations together, we arrive at the following formula for the ratio v^i/v^t:

$$\frac{v^i}{v^t} = \frac{1}{\tau_d \tau_o e^{-\gamma d}}$$

The shielding effectiveness is then

$$SE = -20 \log\left(\left|\tau_d \tau_o e^{-\gamma d}\right|\right) = 110 \text{ dB}$$

Even though we have ignored rereflections, this is the same result found using the exact method. The approximate method shows that most of the attenuation (82 dB) comes about from reflection on the interior surface of the shield, and another 34 dB arises from attenuation in the shield. A voltage *gain* is actually seen going from the shield to the outside air.

▶ **MATLAB 9.4**

This program carries out the calculations described by Example 9.7. The losses due to reflection and attenuation are broken out.

```
%   M-File: ML0904
%
%   Shield analysis using T-line analogy.
%
%   Wentworth, 12/02/02
%
%   Variables
%   d          shield thickness (m)
%   s          shield conductivity (S/m)
%   ur         rel permeability
%   uo         free space permeability
%   er         rel permittivity
%   eo         free space permittivity
%   f,w        freq. and ang. freq. (1/s)
```

(continued)

```
%   c          speed of light (m/s)
%   Zo         free space impedance (ohms)
%   A,B,C      calculation variables
%   prop       propagation constant (1/m)
%   Z1         impedance (ohms)
%   taud       transmission coeff at z = -d
%   tao0       trans coeff at z = 0
%   ratio      power ratio
%   Serefd     SE from front face reflection
%   Seabs      SE from atten in shield
%   SEref0     SE from back face reflection
%   Setot      total shielding effectiveness

clc              %clears the command window
clear            %clears variables

%   Initialize variables
d=10e-6;
s=3.8e7;
ur=1;
er=1;
f=1e9;
eo=8.854e-12;
uo=pi*4e-7;
c=2.998e8;
w=2*pi*f;
Zo=120*pi;

%   Perform calculations
A=i*w*ur*uo;
B=s+i*w*er*eo;
prop=sqrt(A*B);
Z1=sqrt(A/B);
C=tanh(prop*d);
Zin=(Z1*(Zo+Z1*C))/(Z1+Zo*C);
taud=2*Zin/(Zin+Zo);
tau0=2*Zo/(Zo+Z1);
ratio=abs(taud*tau0*exp(-prop*d));

SErefd=-20*log10(abs(taud))
SEabs=-20*log10(abs(exp(-prop*d)))
SEref0=-20*log10(abs(tau0))
SEtot=-20*log10(ratio)
```

In the command-line window we have, upon program execution,

```
Erefd =
 82.3294

SEabs =
 33.6423

SErefo =
 -6.0204

SEtot =
 109.9513
```

> **Drill 9.7** Calculate shielding effectiveness in Example 9.7 if the aluminum is replaced by nickel of the same thickness. (*Answer: SE* = 562 dB, mostly coming from absorption loss)

Obviously, shielding effectiveness depends on the composition and thickness of the shield. It also depends on the radiation frequency, the shape of the shield, and any discontinuities that may exist in the shield. Openings in the shield are required to run signal, power, and ground connections. Small holes are often placed in the shield for ventilation. Also, portions of the shield are joined together at seams, which may not be perfect seals. These breaches in the shield often compromise the shield integrity.

Ventilation holes must be small compared to the wavelength of the highest frequency being shielded. Feed-through holes should be minimized, and in no case should slotted openings be made in the shield.

Slots can be very efficient radiators (see Chapter 8 Section 8.8 for a brief description of slot antennas) and may be unintentionally formed at the edges if the shield box is not held together well. Shielding tape (made, for instance, out of embossed copper with a conductive adhesive backing) can be used to seal the seams in shield boxes.

▶ 9.6 FILTERS

Even after care is taken to reduce emitted noise at the source, it still may be necessary to block or reduce the remaining noise using filters. A filter, consisting most generally of inductors and capacitors, is designed to block or reflect some frequencies and pass others. If the filter is to block noise at high frequencies, the passive element parasitics mentioned in Section 9.2 should be considered.

Reducing passed noise is the job of EMI filters. These are much less demanding than the more exotic filter varieties, such as Chebyshev filters, encountered in the next chapter on microwave engineering. Filters may be designed to reflect the noise, attenuate the noise, or both. We will first look at reflective filters that present a sharp impedance discontinuity to the targeted noise frequency. Then we'll look at ferrite chokes.

Reflective Filters

A reflective filter presents a large impedance discontinuity to the targeted noise frequency. The most common type of EMI filter is the low-pass filter used to decrease high-frequency noise. The simplest version of this is the shunt capacitor shown in Figure 9.25a.

A common way to characterize a filter is by its *insertion loss IL*. The insertion loss is the ratio, in decibels, of the power to the load without the filter element in place, P_L, to the power to the load with the filter element in place, P_{Lf}, that is,

$$IL = 10 \log \left(\frac{P_L}{P_{Lf}} \right) \tag{9.17}$$

We can also write the insertion loss in terms of voltage as

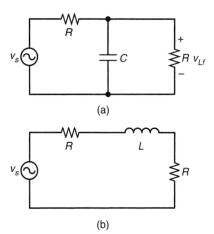

(a)

(b)

Figure 9.25 Low-pass filters using (a) a shunt capacitor and (b) a series inductor.

$$IL = 20 \log \left(\frac{|v_L|}{|v_{Lf}|} \right) \tag{9.18}$$

where v_{Lf} is the voltage across the load resistor with the filter element in place.

In Figure 9.25a, where the load and source resistances are of equal value R, we find

$$IL = 10 \log \left(1 + (\pi fRC)^2 \right) \tag{9.19}$$

The simple shunt capacitor is seen to work best for large source and load resistances.

▶ **EXAMPLE 9.8**

Let's derive the IL expression of (9.19).

With the capacitor removed from the circuit it is easy to see that

$$\frac{v_L}{v_S} = \frac{R}{R+R} = \frac{1}{2}$$

We now reinsert the capacitor to find v_{Lf}/v_S. The parallel impedance of the load resistance and shunt capacitance is

$$Z = \frac{R}{1 + j\omega RC}$$

and

$$\frac{v_{Lf}}{v_S} = \frac{Z}{Z+R}$$

This can be manipulated to find

$$\frac{v_{Lf}}{v_S} = \frac{1}{2 + j\omega RC} = \frac{1}{2} \left(\frac{1}{1 + j\pi fRC} \right)$$

Dividing v_L/v_S by v_{Lf}/v_S gives us

$$\frac{v_L}{v_{Lf}} = 1 + j\pi fRC$$

or

$$\frac{|v_L|}{|v_{Lf}|} = \sqrt{1 + (\pi fRC)^2}$$

Then we have

$$IL = 20\log\left(\sqrt{1 + (\pi fRC)^2}\right) = 10\log\left(1 + (\pi fRC)^2\right)$$

Whereas it is apparent that a good low-pass filter can be created using a shunt capacitor when the source and load resistances are large, if they are small then a series inductor may be used as shown in Figure 9.25b. The insertion loss for this case is calculated (see Problem 9.21) as

$$IL = 10\log\left(1 + \left(\frac{\pi fL}{R}\right)^2\right) \tag{9.20}$$

> **Drill 9.8** Suppose component values in Figure 9.25 are given as $R = 100.$ Ω, $C = 0.10$ nF, and $L = 10.$ nH. Determine the insertion loss at 1.0 GHz for (a) the shunt capacitor filter and (b) the series inductor filter. (*Answer:* (a) 30 dB, (b) 4 dB)

Reflective filters work by presenting a large impedance mismatch to the high-frequency noise. A high-impedance source or load should therefore see a shunt capacitor whereas a low-impedance source or load should see a series inductor. Sometimes the impedance level is not well known, in which case it is common to use a T-filter (Figure 9.26a) or π-filter (Figure 9.26b). If the source is of low impedance and the load is of high impedance, use can be made of an L-filter (Figure 9.26c), which presents a shunt capacitor to the high impedance side and a series inductor to the low-impedance side.

Recalling our discussion in Section 9.2, we see how critical it is for any shunt capacitor used in one of these EMI filters to have the shortest possible leads. Nonetheless, the filters will be characterized by a resonance frequency at which the transmitted noise can be greater than if no filter were present at all!

Reflected noise can still cause problems if, for instance, it is radiated out of some component or section of conductor. Adding a lossy component or some resistance to the reflective filters can attenuate this noise. A good dissipative filter is a short section of coaxial cable made with lossy ferrite material. Lossy ferrite beads can often be placed on the signal line to attenuate high-frequency noise.

▷ **EXAMPLE 9.9**

An L-filter as shown in Figure 9.27 is inserted between a source with 50. Ω impedance and a load with 10. kΩ impedance. Our task is to plot the insertion loss versus frequency from 10 MHz to 10 GHz.

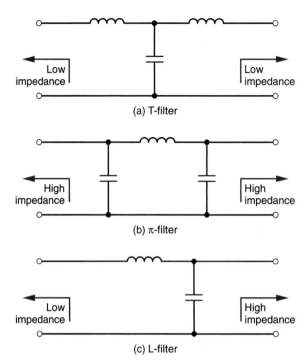

Figure 9.26   Other types of reflective filters: (a) T-filter, (b) π-filter, and (c) L-filter.

The parallel combination of R_L and C gives us an impedance

$$Z_C = \frac{R_L}{1 + j\omega R_L C}$$

The series combination of R_S and L gives us an impedance

$$Z_L = R_S + j\omega L$$

Figure 9.27b shows the simplified circuit with Z_L and Z_C in place. Here

$$\frac{v_{Lf}}{v_s} = \frac{Z_C}{Z_C + Z_L}$$

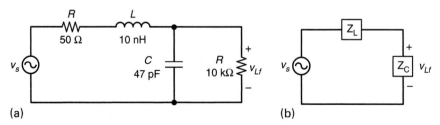

Figure 9.27   (a) The L-filter circuit for Example 9.9. (b) The simplified equivalent circuit.

With the filter removed from the circuit we would have

$$\frac{v_L}{v_s} = \frac{R_L}{R_L + R_S}$$

Dividing v_L/v_s by v_{Lf}/v_s gives us v_L/v_{Lf}. Taking the magnitudes we find insertion loss from (9.18). Inserting the given values, we are able to generate the *IL* versus *f* plot shown in Figure 9.28. MATLAB 9.5 provides the program details.

▶ **MATLAB 9.5**

Example 9.9 and Figure 9.28 are realized with the following routine.

```
%   M-File: ML0905
%
%   This routine plots the insertion loss versus
%   frequency for the simple L-filter of Example 9.9.
%
%   Wentworth, 12/2/02
%
%   Variables
%   RS          source resistance (ohms)
%   RL          load resistance (ohms)
%   L           inductance (H)
%   C           capacitance (F)
%   f,w         freq. and ang. freq. (1/s)
%   Zc,ZL       cap & ind impedance
%   vLf,vL      load voltage with, without filter
%   IL          insertion loss (dB)

clc             %clears the command window
clear           %clears variables

%   Initialize variables
RS=50;
RL=10e3;
L=10e-9;
C=47e-12;
f=.01e9:.01e9:10e9;
w=2*pi*f;

%   Perform calculations
Zc=RL./(complex(1,w*RL*C));
ZL=complex(RS,w*L);
vLf=abs(Zc./(Zc+ZL));
vL=RL/(RL+RS);
IL=20*log10(vL./vLf);

%   Generate plot
loglog(f,IL)
xlabel('frequency (Hz)')
ylabel('IL (dB)')
grid on
```

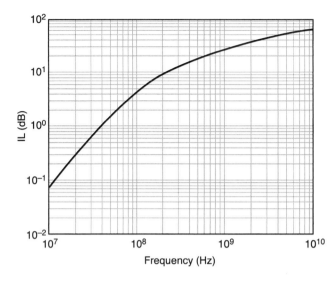

Figure 9.28 Plot of insertion loss as a function of frequency for the low-pass L-filter of Example 9.9.

Ferrite Chokes

Ferrite chokes are often placed on power lines entering EMI-sensitive equipment. One type in particular, the *common-mode choke*, is used on cable assemblies to kill common-mode currents. Some background is in order.

Consider the currents i_1 and i_2 on a pair of parallel conductors as shown in Figure 9.29a. The currents will consist of a common-mode current i_C and a differential-mode current i_D such that

$$i_1 = i_C + i_D, \qquad i_2 = i_C - i_D \qquad (9.21)$$

In ideal operation only the difference-mode currents would be present. The radiated fields for this mode mostly cancel since the parallel lines are in close proximity.

Common-mode currents are undesirable because their radiated fields add. A way to eliminate or at least greatly reduce these currents is to use common-mode chokes as shown in Figure 9.29b. The lines on these chokes are oppositely wound on toroidal ferrite cores as shown in the figure. If we assume that the flux is entirely confined to the ferrite core, then the inductance L of one of the windings will equal the mutual inductance M between the pair of windings. As Figure 9.29c shows, the common-mode choke has little effect on the differential-mode signal since L and M cancel. But as shown in Figure 9.29d, the choke can present considerable inductance to the common-mode currents.

The windings on the core are kept separate from each other to reduce interwinding capacitance that can lead to a resonance-frequency problem. It may be noted that a differential-mode choke could be made by making the windings go the same direction.

The toroidal ferrite cores are usually made with mixtures consisting of either manganese zinc (MnZn) or nickel zinc (NiZn). Manganese zinc has a higher relative permeability at low frequencies, reaching a maximum at about 100 kHz and then dropping

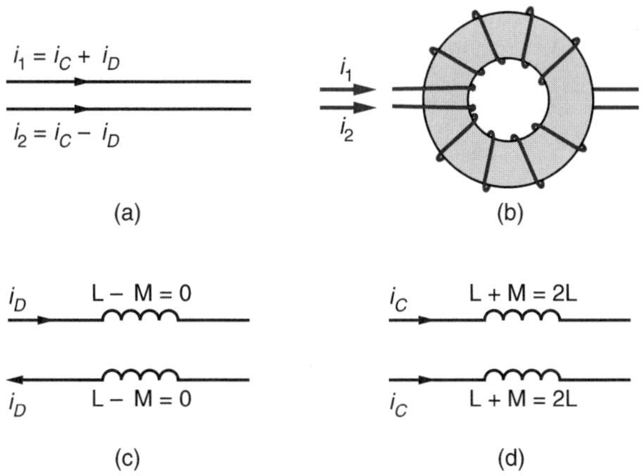

Figure 9.29 Operation of a common-mode filter.

rapidly. Although NiZn has a lower relative permeability, this value continues to increase and even surpasses that of MnZn at around a couple of megahertz.

As compared to the common-mode choke, ferrite beads placed on lines will attenuate both common-mode and differential-mode currents equally.

▶ SUMMARY

- Electromagnetic interference, or EMI, occurs in a circuit or electrical system when unintentional signals, called noise, degrade the operation of the circuit or system. Sources of interference include lightning, electrostatic discharge, power source swells and sags, and radio transmitters.

- The nonideal behavior of passive circuit elements (connecting wire, resistors, capacitors, and inductors) is modeled using parasitic circuit elements at high frequency. These parasitic elements are unintentional resistance, capacitance, and inductance that can adversely affect component performance.

- Connecting wire at high frequency may have appreciable skin-effect resistance, but this is generally overshadowed by the wire's self-inductance, estimated by

$$L = 2 \times 10^{-7} \left(\frac{H}{m}\right) \ell \left[\ln\left(\frac{2\ell}{a}\right) - 1\right]$$

where ℓ and a are the wire length and radius, respectively.

- Resistors suffer from both parasitic inductance and capacitance, as well as inductance in the connecting leads. For high frequencies, small, surface-mount chip resistors made with films of resistive material provide the best performance.

- A number of inductor types are available, with varying degrees of parasitic resistance and capacitance. Using multiple loops in a coil or spiral arrangement will increase the inductance at the expense of increased capacitance. The increased capacitance results in a drop in the self-resonance frequency (SRF). Using a limited number of loops around a toroidal ferrite core can achieve high inductance while maintaining a high SRF.

- Electrolytic capacitors offer the highest capacitance but at limited frequencies. Ceramic capacitors consisting of metallic layers separated by high-permittivity, low-loss dielectric can achieve high capacitance and can operate at fairly high frequencies. The SRF is mostly limited by connecting lead inductance. At very high frequencies, interdigitated metallic fingers are used for modest values of capacitance.

- Passive components should be operated at frequencies below half their SRF, unless their performance at resonance is required.

- A transient or periodic signal may be represented by a Fourier series of frequency components. Analysis reveals that very short rise and fall times in a digital signal result in a broad spectra of frequency components. As a rough approximation, the spectral bandwidth is related to rise time by

$$BW \approx \frac{1}{t_r}$$

- EMI from the source can be minimized by using the lowest possible operating frequency with the longest possible rise and fall times, minimizing voltage and current changes, and choosing frequencies with dissimilar harmonics in multifrequency systems.

- An electrical ground is a path for current to return from some load to a source. A safety ground ties electrical equipment to earth to prevent shock hazards. A signal ground on a circuit board is ideally an equipotential reference for signals.

- Bond wire or bond straps connecting an electric circuit or system to ground must be kept short ($< \lambda/20$) to keep impedance low. Flat, wide-cross-section conductive straps of minimal length present the least inductance.

- Bond wires tying multiple circuits to ground from an electrical system should be as short as possible and tied together at a single point. In this way the multiple circuits see the same ground potential and ground loops are avoided.

- At frequencies above 10 MHz, the leads connecting multiple circuits to a single-point ground can be excessively long. In this case it is common to connect multiple circuits to a common ground plane using short, multipoint connections.

- A loop of current on a circuit board consists of the source, load, ground, and connecting wire. The degree to which these loops radiate and receive radiation is determined by the area subtended by the loop. Therefore, current loops must be minimized to avoid EMI.

- Radiation may be reflected or absorbed by shields, typically consisting of a thin sheet of metal or ferrite material. The shielding effectiveness is

$$SE = 10 \log\left(\frac{P_{ns}}{P_s}\right)$$

where P_{ns} is the power transmitted when no shield is used and P_s is the power transmitted with the shield in place.

- Shields are most often compromised by cable feed-throughs, ventilation holes, and seams.

- Filters are used to block noise at undesirable frequencies while allowing signals to pass through. The most common types are low-pass, high-pass, and bandpass filters. Reflective filters present a large impedance discontinuity to the noise, whereas absorptive filters and ferrite chokes act to attenuate the noise.

- Filters are characterized by a plot of their insertion loss versus frequency. The insertion loss is given by

$$IL = 10 \log\left(\frac{P_L}{P_{Lf}}\right)$$

where P_L and P_{Lf} are without and with the filter in place, respectively.

▶ SUGGESTED REFERENCES

Carr, J., *The Technician's EMI Handbook*, Newnes, 2000.

Christopoulos, C., *Principles and Techniques of Electromagnetic Compatibility*, CRC Press, 1995.

Kimmel, W. D., and Gerke, D. D., *Electromagnetic Compatibility in Medical Equipment*, IEEE Press, 1995.

Ott, H. W., *Noise Reduction Techniques in Electronic Systems*, 2nd ed., Wiley Interscience, 1988.

Paul, C. R., *Introduction to Electromagnetic Compatibility*, Wiley Interscience, 1992.

Weston, D. A., *Electromagnetic Compatibility: Principles and Applications*, 2nd ed., Dekker, 2001.

▶ PROBLEMS

9.2 Passive Circuit Elements

9.1 Given a 2.0-cm length of AWG20 copper wire, (a) calculate R_{dc}, (b) calculate R_{ac} at 800 MHz, (c) estimate L.

9.2 Repeat MATLAB 9.1 for a typical 200-Ω chip resistor ($L_L = 0.40$ nH, $C_x = 50$ fF). Compare the resulting plot with that of Figure 9.7.

9.3 Recalculate L, C_x, and f_{SRF} if the AWG30 wire for the coil of Example 9.2 is replaced with AWG40 wire.

9.4 Estimate L and the *SRF* if a 99.8% iron core is inserted inside the coil of Example 9.2.

9.5 Consider a 99.8% iron toroidal core of inner diameter 0.50 cm and outer diameter 1.0 cm wrapped with 20 turns of evenly spaced AWG26 copper wire. Estimate inductance and the self-resonance frequency of this toroidal inductor.

9.6 Calculate the self-resonance frequency for a 47-nF mica capacitor with a pair of 1.0-cm-long AWG24 copper leads.

9.7 A thin film capacitor is made by sandwiching a 0.10-μm-thick layer of Teflon between copper conductive layers. Determine the capacitance per unit area and the maximum voltage that can be applied across such a capacitor.

9.8 If the 2.2-nF capacitor of Example 9.3 has an area of 20. mm^2, what thickness mica is used? What is the maximum voltage that can be applied across this capacitor?

9.9 Suppose a standard 300.-Ω twin-lead T-line is constructed with AWG24 wire separated by a center-to-center spacing of 0.800 cm. If this line is terminated in a short circuit realized using the shortest possible length of AWG24 wire, calculate the reflection coefficient looking into this "short" at 100 MHz, 1 GHz, and 10 GHz.

9.3 Digital Signals

9.10 What is the spectral bandwidth for a 4.0-ns rise time signal, using (9.13)? What rise time is required to achieve a 1-GHz bandwidth?

9.11 Suppose a 1.0-GHz clock rate is assumed. What is the minimum spectral bandwidth calculated using (9.13)?

9.12 Modify MATLAB 9.3 to look at a 1.0-GHz clock rate signal. Minimize the rise and fall times by letting the signal be a sawtooth wave.

9.4 Grounds

9.13 Repeat Example 9.4 using AWG22 wire and 200-MHz current.

9.14 Repeat Example 9.5 using AWG22 wire and 200-MHz current.

9.5 Shields

9.15 The field within a shielded enclosure is 12 kV/m. What shielding effectiveness is required such that the field outside the shield is no more than 1.0 nV/m?

9.16 Compare the attenuation in decibels at 1.0 GHz for 20 μm-thick layers of (a) copper, (b) aluminum, and (c) nickel.

9.17 A particular silver-filled paint is to be used as an absorptive layer. It has σ = 1.0×10^6 S/m with $μ_r$ and $ε_r$ assumed equal to one. Calculate the attenuation of a 100-MHz wave propagating through a 50.-μm-thick layer and compare with the attenuation through a pure silver layer of the same thickness.

9.18 Shielding low-frequency magnetic fields often requires a magnetic shield. What thickness of 99.8% iron is required to give 20-dB attenuation of a 1.0-kHz magnetic field?

9.19 Find the shielding effectiveness for the silver-filled paint shield of Problem 9.17 and compare the result with a pure silver shield of the same thickness.

9.20 Consider a 10.0-μm-thick copper shield. Plot the contributions to shielding effectiveness (and the total shielding effectiveness) from each of the reflective terms and from the absorption term from 1 MHz up to 1 GHz. Repeat for the same thickness nickel shield.

9.6 Filters

9.21 Derive the insertion loss expression (9.20) for the series inductor circuit of Figure 9.25b.

9.22 Suppose an L = 100. nH inductor is used in the series inductance filter of Figure 9.25b. Determine the insertion loss at 200 MHz if (a) R = 10. Ω and (b) R = 10. kΩ.

9.23 Suppose a C = 47. pF capacitor is used in the shunt capacitance filter of Figure 9.25a. Determine the insertion loss at 200. MHz if (a) R = 10. Ω and (b) R = 10. kΩ.

9.24 Determine the insertion loss at 1.0 GHz for a T-filter inserted between a 10.-Ω source impedance and a 10.-Ω load impedance. Consider L = 10. nH and C = 47. pF.

9.25 Determine the insertion loss at 40 MHz for a π-filter inserted between a 10.-kΩ source impedance and a 10.-kΩ load impedance. Consider L = 10. nH and C = 47. pF.

10

Microwave Engineering

Learning Objectives

▷ Describe the microwave components of a transceiver circuit

▷ Design impedance matching networks using lumped elements

▷ Introduce scattering parameters for describing multiport networks

▷ Discuss common microwave circuit components such as circulators, combiners, couplers, filters, and amplifiers

Microwaves are considered that part of the electromagnetic spectrum between 300 MHz and 300 GHz, although most microwave engineering takes place from 1 to 40 GHz. Table 10.1 shows the microwave bands of most interest. Microwaves are employed in a host of applications including microwave ovens, wireless communications, and radar. Microwave ovens are powered by a 500- to 1500-W magnetron tube that provides the 2.45-GHz radiation. In wireless communications, a wide band of frequencies can be transmitted without the severe attenuation encountered at optical frequencies, for instance from clouds. For radar, microwave beams may be focused for target tracking.

As mentioned in Chapter 1, electromagnetics is fundamental to the burgeoning field of wireless communications. Electromagnetic waves propagate on transmission lines or in space, where they are transmitted and received by antennas. Another aspect of high-frequency communications systems is the special circuit components required. The variety

Table 10.1   IEEE Frequency Band Designations

Band	Range (GHz)	Common applications
L	1–2	Global positioning satellites, personal communications
S	2–4	Microwave ovens (2.45 GHz), personal communications
C	4–8	Satellite communications
X	8–12.5	Ground radar for aircraft navigation
Ku	12.5–18	Radar, point-to-point radio
K	18–26.5	
Ka	26.5–40	

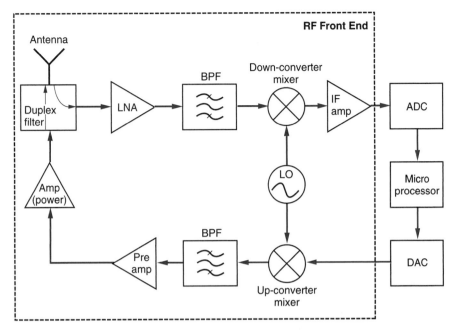

Figure 10.1   A typical transceiver configuration featuring the RF front end.

of special components is perhaps best viewed by looking at an *RF front end* (or *RF trans-ceiver*) shown in Figure 10.1.

Acting as a receiver, the antenna captures a weak electromagnetic wave and sends the signal to a low-noise amplifier (LNA) that will boost the signal without introducing much noise. Signals and noise outside the vicinity of the desired signal frequency are then removed using a bandpass filter (BPF). The next step is to mix the received signal with a similar frequency signal produced by a local oscillator. The output of the mixer contains a number of frequency components, but the primary one of interest is the *difference frequency* (or *intermediate frequency, IF*). The mixing operation to convert an RF signal to a lower IF signal is known as *down-conversion*. Down-conversion produces a relatively low frequency signal for which a very good IF amplifier can be constructed. The IF amplifier may be accompanied by a low-pass filter. The amplified signal passes to the analog-to-digital (A/D) converter, and then to the microprocessor where the information is acted upon in some way (passed to the speaker of a cell phone, for instance).

Transmission occurs in the other direction. The microprocessor sends a signal to a digital-to-analog (D/A) converter, and an IF signal is mixed with the LO signal to produce, among other things, a summed signal. This is known as *up-conversion*. The up-converted signal is passed through a bandpass filter and then on to some amplifier stages required to boost the signal level. The signal is then transmitted out of the same antenna that does the receiving. To avoid having the strong transmitted signal be picked up by the more delicate receiving circuit, a duplex filter is often employed; such a filter can typically provide 100-dB isolation between transmitter and receiver. [10.1]

[10.1] Isolation *I*, in decibels, is a power ratio. If P_1 is the transmitter power into the duplex filter and P_2 is the unintentional power leaked to the receiver, then $I = 10 \log (P_1/P_2)$.

In this chapter we will study some of the various RF front-end components to gain a good overview of the field known as microwave engineering. Impedance matching, discussed in Chapter 6 using T-line stubs, will be revisited in Section 10.1 with a lumped-element matching technique. In Section 10.2 we discuss the very important topic of scattering parameters (S-parameters). Design, testing, and understanding of microwave circuits are all made easier through the use of S-parameters. Couplers and dividers are discussed in Section 10.3, followed by filters in Section 10.4. These are all commonly used microwave components. Amplifiers, designed with the aid of S-parameters, are the topic of Section 10.5. The chapter concludes with a brief description of receiver design.

▷ 10.1 LUMPED-ELEMENT MATCHING NETWORKS

In Chapter 6 Section 6.5 we employed the Smith Chart to impedance match a complex load to a lossless T-line.[10.2] We saw that, in addition to the reflection coefficient, the Smith Chart also allows plotting of either the normalized impedance or the normalized admittance. The object of the matching network is to move to the center of the Smith Chart where $|\Gamma| = 0$.

Lossless T-line stubs provide pure reactance for matching networks. A disadvantage of these stubs is that they require a considerable amount of chip or board space. An option is to use lumped-element inductors and capacitors in the matching network. These elements can be much smaller than stubs, but care must be taken to operate them well below the component's self-resonant frequency.[10.3]

In this section we study a Smith Chart approach for the design of lumped-element (or *L-section*) matching networks. Adding a series element will be done in the normalized impedance Smith Chart, whereas adding a shunt element will be done in the normalized admittance Smith Chart.

To begin, Figure 10.2 shows a Smith Chart that has a rotated $1 \pm jx$ circle added to it. A point on the original $1 \pm jx$ circle (for example, point z in the figure) in a normalized impedance Smith Chart has a corresponding point on the rotated circle (point y in the figure) for the normalized admittance chart. Likewise, an admittance point on the $1 \pm jb$ circle of a normalized admittance Smith Chart would be transformed to an impedance point on a rotated $1 \pm jb$ circle.

There are three basic matching situations. First, suppose the load impedance lies within the $1 \pm jx$ circle. Adding a series reactive element will not bring you to the $1 \pm jx$ circle, so a shunt element must be used. Figure 10.3a shows the basic steps. Consider the normalized load impedance at point 1. Its corresponding admittance is found at point 1′. A shunt element of value jb moves the admittance to the rotated circle (point 2′). Then, the corresponding impedance point is found (point 2). Finally, a series reactive element of value jx is used to reach the center of the Smith Chart.

Now suppose the load impedance lies within the rotated $1 \pm jx$ circle. Figure 10.3b shows the basic steps. A series reactive element of value jx moves the impedance from the normalized load impedance at point 1 to the rotated circle at point 2. Then, the correspon-

---

[10.2] A review of Sections 6.4 and 6.5 is highly recommended.

[10.3] A general practice stated in Chapter 9 Section 9.2 was to operate the device at frequencies no higher than half the component's *SRF.*

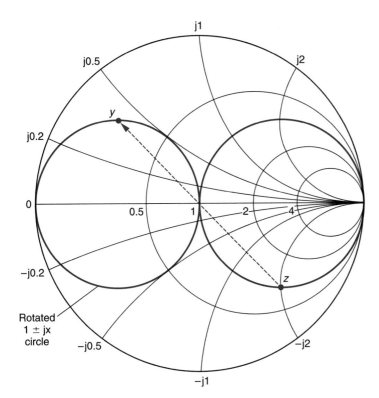

Figure 10.2 Smith Chart with rotated $1 \pm jx$ circle.

ding admittance point is found (point 2′). Finally, a shunt reactive element of value jb is used to reach the center of the Smith Chart.

If the normalized load impedance lies outside both the $1 \pm jx$ and the rotated $1 \pm jx$ circles, then either matching network configuration shown in Figure 10.3 can be used.

Design of an L-section matching network is best illustrated by the following examples.

▷ **EXAMPLE 10.1**

We want to design an L-section network to match a $250 - j250\ \Omega$ load to a 50-Ω line at 800 MHz.

Our first step is to locate the normalized load, $5.0 - j5.0$, at point 1 on the Smith Chart shown in Figure 10.4a. Since this is inside the $1 \pm jx$ circle, we use the matching network indicated in Figure 10.3a.

We will be adding a shunt element first, so we must convert the load impedance to a load admittance, which we find is $0.1 + j0.1$, or point 1′ in the figure. Now we want to move along the $0.1 \pm jx$ circle until we intersect the rotated $1 \pm jx$ circle. The shortest path is up to the point 2′, or $0.1 + j0.3$, requiring a shunt element value $jb = j0.2$.

Next we'll add a series element, so we need to convert the admittance $0.1 + j0.3$ to an impedance, located at point 2, or $1.0 - j3.0$. Adding a series element $jx = j3.0$ brings us to the matched condition at the center of the chart.

To find the element values, we make use of Table 10.2. For the normalized admittance element $jb = j0.2$, a capacitor is needed. We have

$$jb = j\omega C Z_o$$

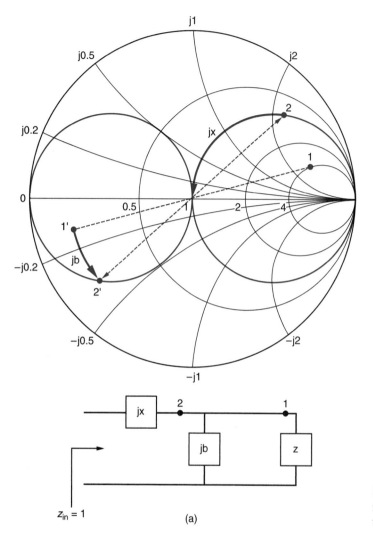

Figure 10.3 The two L-section matching networks: (a) network for z located within the $1 \pm jx$ circle.

and solving for C we find

$$C = \frac{b}{\omega Z_o} = \frac{(0.2)}{(2\pi)(800 \times 10^6 \, \frac{1}{s})(50\Omega)}\left(\frac{F\text{-}\Omega}{s}\right) = 0.80 \text{ pF}$$

For the normalized impedance element $jx = j3.0$, Table 10.2 shows that we need an inductor. We have

$$jx = \frac{j\omega L}{Z_o}$$

or solving for L, we get

$$L = \frac{xZ_o}{\omega} = \frac{(3.0)(50\Omega)}{(2\pi)(800 \times 10^6 \, \frac{1}{s})}\left(\frac{H}{\Omega\text{-}s}\right) = 30. \text{ nH}$$

The final circuit is shown in Figure 10.4b.

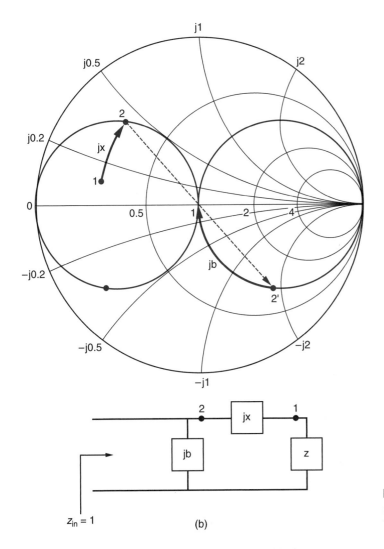

Figure 10.3 The two L-section matching networks: (b) network for *z* located within the rotated $1 \pm jx$ circle.

Drill 10.1 Rather than the shunt capacitor and series inductor in Example 10.1, we could have chosen another solution requiring a shunt inductor ($jb = -j0.5$) and a series capacitor ($jx = -j3.0$). Determine the values of these lumped elements at 800 MHz. (*Answer:* 25 nH, 1.3 pF).

▶ **EXAMPLE 10.2**

Let's design an L-section matching network for a 10.-Ω load to be matched to a 50.-Ω line at 1.0 GHz.

Showing our work on Figure 10.5, we first see that the normalized load impedance is found at point 1: $0.2 + j0$. We therefore follow the procedure suggested by Figure 10.3b and can go to either $0.2 + j0.4$ or to $0.2 - j0.4$ to reach the rotated $1 \pm jx$ circle. In this example we'll go to $0.2 - j0.4$ (point 2) and save the other solution for Drill 10.2. To make this move will require a series element of value $jx = -j0.4$, corresponding to a capacitor from Table 10.2.

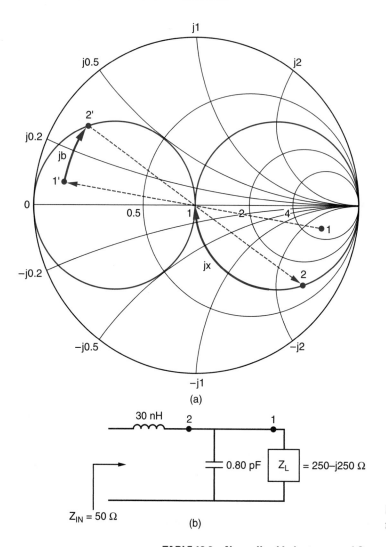

(a)

Z_IN = 50 Ω

(b)

Figure 10.4 (a) Smith Chart and (b) final matching network for Example 10.1.

TABLE 10.2 Normalized Inductance and Capacitance Values

	impedance chart	admittance chart
L	$\dfrac{j\omega L}{Z_o}$	$\dfrac{-jZ_o}{\omega L}$
C	$\dfrac{-j}{\omega C Z_o}$	$j\omega C Z_o$

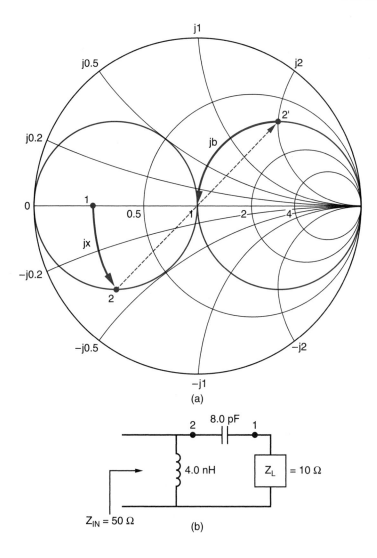

Figure 10.5 (a) Smith Chart and (b) final matching network for Example 10.2.

Next we add a shunt element, so we find the admittance point (2′) at $1 + j2.0$. We will need $jb = -j2.0$ to reach the matched condition, and this corresponds to an inductor.

Calculating the series capacitance and the shunt inductance values we find $C = 8.0$ pF and $L = 4.0$ nH. This is shown in the L-section matching network of Figure 10.5b.

Drill 10.2 Find the alternate solution for Example 10.2. (*Answer:* Replace the 8.0-pF capacitor with a 3.2-nH inductor, and replace the 4.0-nH inductor with a 6.4-pF capacitor.)

> **MATLAB 10.1**

Let's modify ML0603 to include the rotated $1 \pm jx$ circle as shown in Figure 10.2.

First we'll create the new function, "rotcirc(r)," which is identical to "realcirc(r)" except for a strategically placed sign change.

```
function [h]=rotcirc(r)
%ROTCIRC(r) draws the rotated 1+jx circle;
phi=1:1:360;
theta=phi*pi/180;
a=1/(1+r);
m=r/(r+1);
n=0;
Re=a*cos(theta)-m;
Im=a*sin(theta)+n;
z=Re+i*Im;
h=plot(z,'k');
axis('equal')
axis('off')
```

Then, in the Smith Chart routine of ML0603, we'll insert the lines

```
%add the rotated 1+jx circle
rotcirc(1)
```

just before the line

```
%now add +/- x circles
```

You may wish to save this new version of the Smith Chart routine as ML1001 to distinguish it from ML0603.

Running the program yields a Smith Chart with a rotated circle like the one shown in Figure 10.2.

It is apparent that we often have more than one choice of L-section matching networks. The particular choice may depend on availability of the lumped-element components. Also, in some cases it may be desirable to bias the load through the matching network, in which case an L-section with a series inductor and shunt capacitor would be required. Finally, the different solutions may have significantly different bandwidths.

▶ 10.2 SCATTERING PARAMETERS

Consider a circuit or device inserted into a T-line as shown in Figure 10.6a. We can refer to this circuit or device as a *two-port network*. The behavior of the network can be completely characterized by its *scattering parameters* (S-parameters), or its *scattering matrix*, [S], as shown in Figure 10.6b. Scattering matrices are frequently used to characterize multiport networks, especially at high frequencies. They are used to represent microwave devices, such as amplifiers and circulators, and are easily related to concepts of gain, loss, and reflection.

In the present treatment, we will deal with simple S-parameters that have the same characteristic impedance (Z_o) at all ports in the network. In such a treatment, the scattering

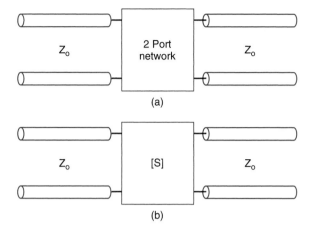

Figure 10.6 A two-port network inserted into a T-line (a) may be represented by a scattering matrix (b).

parameters represent ratios of voltage waves entering and leaving the ports. Referring to Figure 10.7, we represent a voltage wave entering a port with a + superscript and that exiting a port with a − superscript. A subscript indicates the port number. Using this terminology V_1^+ is the voltage wave entering port 1.

Some of V_1^+ is reflected at the port, contributing to the voltage wave exiting port 1, V_1^-, and some is transmitted out of port 2, contributing to V_2^-. Of the portion transmitted, some of this is reflected at the load and reenters port 2 as V_2^+. Of this V_2^+, some is reflected, contributing to V_2^-, and some is transmitted, contributing to V_1^-. A graphical representation of the situation is given by Figure 10.8.[10.4] We see that V_1^- consists of that portion of V_1^+ that is reflected along with that portion of V_2^+ that is transmitted. This can be written in terms of the scattering parameters as

$$V_1^- = S_{11}V_1^+ + S_{12}V_2^+ \tag{10.1}$$

Likewise, we can write V_2^- as

$$V_2^- = S_{21}V_1^+ + S_{22}V_2^+ \tag{10.2}$$

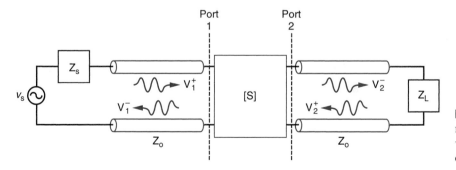

Figure 10.7 Two-port network to illustrate the voltage waves used to calculate the S-parameters.

[10.4]It must be emphasized that the voltage waves entering and leaving the ports are doing so across the T-lines. In other words, it is incorrect to assume V_1^+ entering from the top portion of the T-line and V_1^- leaving from the bottom portion.

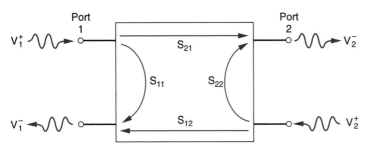

Port 1 ... Port 2

S_{21}, S_{11}, S_{22}, S_{12}

Figure 10.8 Scattering matrix formulation.

In matrix form this is written

$$\begin{bmatrix} V_1^- \\ V_2^- \end{bmatrix} = \begin{bmatrix} S_{11} & S_{12} \\ S_{21} & S_{22} \end{bmatrix} \begin{bmatrix} V_1^+ \\ V_2^+ \end{bmatrix} \tag{10.3}$$

or in abbreviated form

$$[V]^- = [S][V]^+ \tag{10.4}$$

Here, $[S]$ is a two-port scattering matrix.

A scattering parameter S_{ij} is defined as the fraction of the voltage wave entering port j that exits port i. From (10.1) we can solve for S_{11} if we set $V_2^+ = 0$, giving

$$S_{11} = \frac{V_1^-}{V_1^+}\bigg|_{V_2^+ = 0} \tag{10.5}$$

Using our definition of scattering parameter, we have that S_{11} represents how much of a voltage wave entering port 1 exits port 1. Setting $V_2^+ = 0$ in Figure 10.7 is a simple matter of terminating the line in a matched load ($Z_L = Z_o$). In such a case, S_{11} is the reflection coefficient looking into port 1.

In general for a multiport network,

$$S_{ij} = \frac{V_i^-}{V_j^+}\bigg|_{V_k^+ = 0, k \neq j} \tag{10.6}$$

We can think of the jth port as the source of the voltage and the ith port as the destination of the voltage. We can also consider that a general scattering parameter has both a magnitude and a phase:

$$S_{ij} = |S_{ij}| e^{j\phi_{ij}} \tag{10.7}$$

▶ **EXAMPLE 10.3**

We see that if we terminate port 2 in a matched load then the reflection coefficient looking into port 1 is simply equal to S_{11}. Now let's terminate port 2 in a short circuit as shown in Figure 10.9 and evaluate the reflection coefficient looking into port 1.

We know that the voltage across a short circuit is zero. So at the shorted port 2 we have the following relation for V_2^+ and V_2^-:

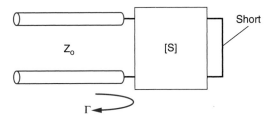

Figure 10.9 Finding the reflection coefficient in terms of the scattering parameters when port 2 is shorted for Example 10.3.

$$V_2^+ + V_2^- = 0$$

or

$$V_2^- = -V_2^+$$

We can insert this relation into (10.2), and solving for V_2^+, we get

$$V_2^- = S_{21}V_1^+ + S_{22}V_2^+ = -V_2^+$$

or

$$V_2^+ = \frac{-S_{21}V_1^+}{1 + S_{22}}$$

Inserting this into (10.1) gives a reflection coefficient of

$$\Gamma = \frac{V_1^-}{V_1^+} = S_{11} - \frac{S_{12}S_{21}}{1 + S_{22}}$$

Drill 10.3 Repeat Example 10.3 if port 2 is terminated in an open circuit. (*Answer:* $\Gamma = S_{11} + S_{12}S_{21}/(1 - S_{22})$)

▶ **EXAMPLE 10.4**

Let's calculate the S-parameters for the two-port network, consisting of a 100-Ω resistor inserted into a 50-Ω T-line, shown in Figure 10.10a.

We can set $V_2^+ = 0$ by terminating port 2 in a matched load. So that we won't lose sight of the impedance on each port, we'll leave a tiny section of 50-Ω line at port 2 before the termination, as shown in Figure 10.10b. Now the S_{11} scattering parameter is equal to the reflection coefficient Γ, calculated as

$$\Gamma = \frac{V_1^-}{V_1^+} = \frac{Z_L - Z_o}{Z_L + Z_o} = \frac{(100\ \Omega + 50\ \Omega) - 50\ \Omega}{(100\ \Omega + 50\ \Omega) + 50\ \Omega} = \frac{1}{2}$$

So we have $S_{11} = 1/2$.

With port 2 terminated in a matched load such that $V_2^+ = 0$, we can also calculate S_{21}. We find the transmission coefficient

$$\tau = 1 + \Gamma = \frac{3}{2}$$

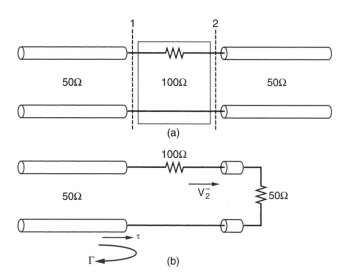

Figure 10.10 For Example 10.4, we want to find [S] for the two-port network shown in (a). In (b), we force $V_2^+ = 0$ by terminating port 2 in a matched load.

and realize that this is the amount of voltage dropped across the series combination of the 100-Ω resistor and the 50-Ω termination. To find S_{21}, we need to find how much of this voltage is dropped across the 50-Ω termination. Thus we have

$$S_{21} = \frac{V_2^-}{V_1^+}\bigg|_{V_2^+ = 0} = \tau\left(\frac{50\ \Omega}{100\ \Omega + 50\ \Omega}\right) = \frac{3}{2}\frac{1}{3} = \frac{1}{2}$$

From the symmetry of the problem we see that $S_{22} = S_{11}$ and $S_{12} = S_{21}$. So we have

$$[S] = \frac{1}{2}\begin{bmatrix} 1 & 1 \\ 1 & 1 \end{bmatrix}$$

Reciprocal Networks

Notice in Example 10.4 that the transmission characteristics are the same in both directions (i.e., $S_{21} = S_{12}$). We say that this two-port network is *reciprocal*. It is a property of passive circuits (circuits with no active devices or ferrites) that they form reciprocal networks.

A network is reciprocal if it is equal to its *transpose*. Stated mathematically, for a reciprocal network

$$[S] = [S]^t \tag{10.8}$$

where the transpose of a two-port scattering matrix is

$$\begin{bmatrix} S_{11} & S_{12} \\ S_{21} & S_{22} \end{bmatrix}^t = \begin{bmatrix} S_{11} & S_{21} \\ S_{12} & S_{22} \end{bmatrix} \tag{10.9}$$

▷ **EXAMPLE 10.5**

Let's evaluate the passive two-port network shown in Figure 10.11a and verify that it is reciprocal.

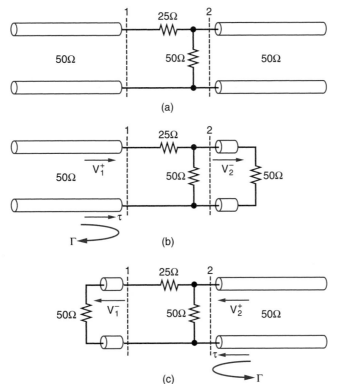

Figure 10.11 For Example 10.5, we want to find [S] for the network shown in (a). In (b) we set $V_2^+ = 0$ by terminating port 2 in 50 Ω, and in (c) we set $V_1^+ = 0$ by terminating port 1 in 50 Ω.

If we terminate port 2 in a matched load, $V_2^+ = 0$ and we can find S_{11} and S_{21}. Figure 10.11b shows this situation, where it is apparent that $\Gamma = 0$, and therefore $S_{11} = 0$. Also, $\tau = 1 + \Gamma = 1$. To find how much of the transmitted voltage gets dropped across the 50-Ω termination, we have a parallel combination of 50-Ω resistors (50 Ω//50 Ω) in series with a 25-Ω resistor. Evaluating the resistive divider circuit, we calculate

$$V_2^- = \left(\tau V_1^+\right)\left(\frac{50\ \Omega\|50\ \Omega}{25\ \Omega + 50\ \Omega\|50\ \Omega}\right) = \frac{1}{2}V_1^+$$

So we have

$$S_{21} = \left.\frac{V_2^-}{V_1^+}\right|_{V_2^+ = 0} = \frac{1}{2}$$

Finding S_{22} and S_{12} requires terminating port 1 in a matched load as shown in Figure 10.11c. Now we calculate $S_{22} = \Gamma = -1/4$, and $\tau = 3/4$. We then have

$$V_1^- = \left(\tau V_2^+\right)\left(\frac{50\ \Omega}{25\ \Omega + 50\ \Omega}\right) = \frac{1}{2}V_2^+$$

and therefore

$$S_{12} = \left.\frac{V_1^-}{V_2^+}\right|_{V_1^+ = 0} = \frac{1}{2}$$

Our scattering matrix for this two-port network is then

$$[S] = \begin{bmatrix} 0 & \frac{1}{2} \\ \frac{1}{2} & -\frac{1}{4} \end{bmatrix}$$

Clearly this network is reciprocal.

Lossless Networks

A lossless network does not contain any resistive elements and does not attenuate the signal. No real power is delivered to the network. Consequently, for any passive lossless network, what goes in must come out!

In terms of scattering parameters, a network is lossless if

$$[S]^t[S]^* = [U] \tag{10.10}$$

where $[U]$ is the *unitary matrix*, given by

$$[U] = \begin{bmatrix} 1 & 0 \\ 0 & 1 \end{bmatrix} \tag{10.11}$$

and $[S]^*$ is the complex conjugate of the $[S]$ matrix. Each element of $[S]^*$ is the complex conjugate of each element of $[S]$. For instance, if $S_{11} = \text{Re} + j\text{Im}$, then $S_{11}^* = \text{Re} - j\text{Im}$ and $S_{11}S_{11}^* = |S_{11}|^2$.

For a two-port network, the product of the transpose matrix and the complex conjugate matrix yields

$$[S]^t[S]^* = \begin{bmatrix} \left(|S_{11}|^2 + |S_{21}|^2\right) & \left(S_{11}S_{12}^* + S_{21}S_{22}^*\right) \\ \left(S_{12}S_{11}^* + S_{22}S_{21}^*\right) & \left(|S_{12}|^2 + |S_{22}|^2\right) \end{bmatrix} \tag{10.12}$$

▶ **EXAMPLE 10.6**

Let's look at the scattering matrix from Example 10.5 and determine if the network is lossless. We find

$$[S]^t[S]^* = \begin{bmatrix} \frac{1}{4} & \frac{1}{8} \\ \frac{1}{8} & \frac{5}{16} \end{bmatrix} \neq [U]$$

This network is clearly not lossless, nor should we have expected it to be so since it consists of resistive elements.

As another way of looking at lossy networks, consider that because a scattering parameter represents a voltage ratio, then a squared scattering parameter represents a power ratio. Suppose 1 W of power enters port 1 and we want to account for where this power goes. We see that S_{11}^2 of it is reflected, but since $S_{11} = 0$ in this example, 0 W is reflected. We also see that S_{21}^2 of it is transmitted out of port 2, or 1/4 W. That means that, of the 1 W entering, only 1/4 W leaves the network and therefore 3/4 W is dissipated.

Return Loss and Insertion Loss

Two-port networks are commonly described by their *return loss* and *insertion loss*. The return loss RL at the ith port of a network is defined as

$$RL_i = -20\log\left|\frac{V_i^-}{V_i^+}\right| = -20\log|\Gamma_i| \qquad (10.13)$$

For a perfectly matched system where the reflection coefficient into a particular port is zero, the return loss will be infinite. If a port is completely mismatched, for instance presenting a short or an open circuit, then all of the signal will be returned (none will be lost) and the return loss will be 0. Consequently, here is a situation where we usually want the parameter to be a large value.

The insertion loss IL defines how much of a signal is lost as it goes from a jth port to an ith port. In other words, it is a measure of the attenuation resulting from insertion of a network between a source and a load. Mathematically we have

$$IL_{ij} = -20\log\left|\frac{V_i^-}{V_j^+}\right| \qquad (10.14)$$

With insertion loss, a small value is generally desired. The exception would be for filtering or attenuating applications.

▷ **EXAMPLE 10.7**

Let's find the return and insertion losses characterizing the scattering matrix of Example 10.5, assuming each port is connected to a 50-Ω impedance.

Looking into port 1, we have

$$RL_1 = -20\log(0) = \infty$$

indicating no signal is returned. Looking into port 2, however, we find

$$RL_2 = -20\log\left(\frac{1}{4}\right) = 12 \text{ dB}$$

The insertion loss from port 1 to port 2 will be

$$IL_{21} = -20\log\left(\frac{1}{2}\right) = 6 \text{ dB}$$

Because the network is reciprocal, $IL_{12} = IL_{21}$.

▷ **EXAMPLE 10.8**

In the previous example we carried out calculations assuming the ports were impedance matched. A mismatched port requires a bit more effort. Let's calculate the return loss looking into port 1 of the network in Example 10.5 (shown in Figure 10.11) if port 2 is terminated in a short circuit.

From Example 10.3, for a network with port 2 terminated in a short, we found

$$\frac{V_1^-}{V_1^+} = S_{11} - \frac{S_{12}S_{21}}{1+S_{22}}$$

So, for the scattering matrix of Example 10.5 we calculate

$$\frac{V_1^-}{V_1^+} = 0 - \frac{(\frac{1}{2})(\frac{1}{2})}{1 + -\frac{1}{4}} = -\frac{1}{3}$$

Then, finding return loss we have

$$RL = -20\log\left|-\frac{1}{3}\right| = 9.5 \text{ dB}$$

Drill 10.4 Repeat Example 10.8 if port 2 is terminated in an open circuit. (*Answer:* 14 dB)

Shift in Reference Plane

The scattering parameters are, in general, complex quantities that depend on the specific location of the ports (or port reference planes). To begin this discussion, let's find the scattering matrix for a lossless length ℓ of T-line characterized by Z_o and β, as shown in Figure 10.12.

Since the line is impedance matched, we would correctly assume that the parameters S_{11} and S_{22} are zero. Also, since the line is lossless we'd expect the amplitude of the wave entering a port to be equal to its amplitude leaving the other port. This tells us the magnitudes of S_{21} and S_{12} must be unity. However, we also see that a phase shift must be involved. At any point along the T-line, the voltage $V(z)$ is related to its value at $z = 0$ by the expression

$$V(z) = V_1^+ e^{-j\beta(z-0)} \tag{10.15}$$

At port 2, where $z = \ell$, we have

$$V_2^- = V_1^+ e^{-j\beta\ell} \tag{10.16}$$

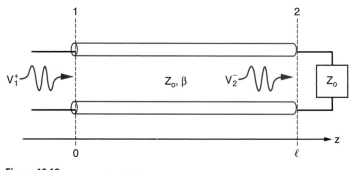

Figure 10.12 A matched T-line network.

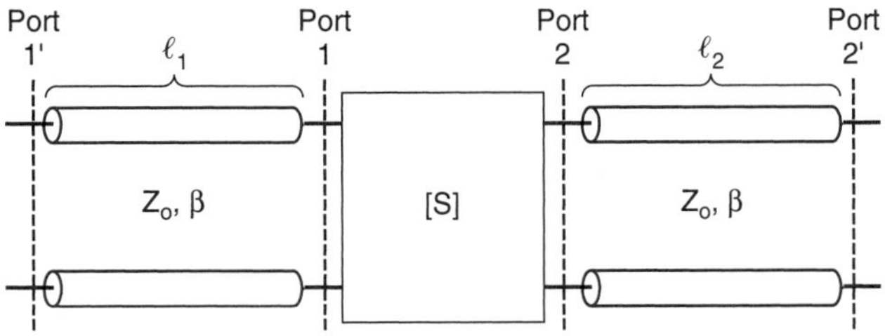

Figure 10.13 The scattering matrix for shifted reference planes is related to the original matrix by phase shifts.

and therefore

$$S_{21} = \frac{V_2^-}{V_1^+}\bigg|_{V_2^+ = 0} = 1e^{-j\beta\ell} \tag{10.17}$$

By the symmetry of the problem we see that the scattering matrix for this section of line is

$$[S] = \begin{bmatrix} 0 & e^{-j\beta\ell} \\ e^{-j\beta\ell} & 0 \end{bmatrix} \tag{10.18}$$

Now suppose we have a scattering matrix [S] referenced to ports 1 and 2 as shown in Figure 10.13. If we add ℓ_1 and ℓ_2 length sections of Z_o T-line with phase constant β, then the scattering parameters at the shifted reference planes (ports 1′ and 2′) are related to the original scattering parameters by

$$[S]' = \begin{bmatrix} S_{11}e^{-j2\beta\ell_1} & S_{12}e^{-j\beta(\ell_1+\ell_2)} \\ S_{21}e^{-j\beta(\ell_1+\ell_2)} & S_{22}e^{-j2\beta\ell_2} \end{bmatrix} \tag{10.19}$$

In these expressions, the phase term $\beta\ell$ is referred to as an *electrical length*. This is often expressed as a phase delay in degrees.

Drill 10.5 To the two-port network of Example 10.5, append a 30° length of Z_o line to port 1 and a 60° length of Z_o line to port 2. Calculate [S] accounting for this shift in the reference plane.

$$\left(Answer: [S]' = \begin{bmatrix} 0 & -j/2 \\ -j/2 & \frac{-1}{4}e^{-j120°} \end{bmatrix} \right)$$

The Vector Network Analyzer

The most powerful laboratory instrument for performing microwave measurements is the vector network analyzer, or VNA for short. VNAs like the one in Figure 10.14 had their start in 1967 with Hewlett–Packard's HP8410 Vector Network Analyzer. Present-day VNAs are capable of extremely accurate measurements of a device's scattering parameters over a frequency range from 10 MHz up to as high as 110 GHz. The analyzer presents the data in a

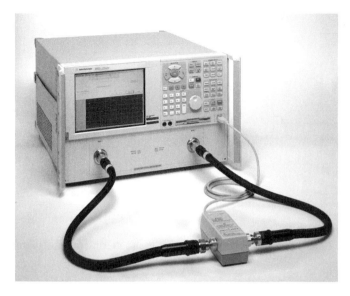

Figure 10.14 A vector network analyzer. Courtesy of Agilent Inc.

variety of ways, including a Smith Chart. The VNA can even determine the time-domain response (recall the discussion of *time-domain reflectometry* in Chapter 6), by calculating the inverse Fourier transform of the frequency-domain data.

The VNA measures a device over a swept frequency range. At a given frequency, a signal is sent out of port 1 of the VNA to the device under test (DUT). Some of this incident signal may be reflected back to port 1, some may be dissipated in the DUT (or radiated), and some may be transmitted to port 2. The VNA accurately measures the complex ratio of the reflected signal to the incident signal (S_{11}) as well as the ratio of the transmitted signal to the incident signal (S_{21}). Likewise, by sending a signal out of port 2 it is able to measure S_{12} and S_{22}.

The VNA's accuracy is due in part to a calibration procedure coupled with error-correction software. Calibration kits contain carefully measured matched terminations, shorts, opens, and through connections. By using the calibration kits, the ports are virtually extended through the various adapters and connecting cables right up to the DUT.

▶ 10.3 COUPLERS AND DIVIDERS

Microwave couplers and resistive power dividers or combiners are commonly used microwave elements. In some cases the power from a single microwave power amplifier may be insufficient to power, say, a radar transmitter. In such a case, several power amplifiers may be used with their power added using combiners. Also, using microwave couplers allows construction of balanced amplifiers that feature constant gain over a relatively wide bandwidth and that can have very low VSWR.

In this section we will describe three-port networks (circulators and resistive power dividers) and four-port networks (directional couplers and branch-line couplers). Our goal will be to introduce the terminology and common device types rather than to derive functional equations. Our focus will be on board level couplers and dividers rather than on the lesser used waveguide-type devices.

Circulators

A typical circulator is a layered construct consisting of disks of ferrite and permanent magnets. These materials impart a directional nature to the electromagnetic fields, resulting in decidedly nonreciprocal performance.

The schematic view of a typical circulator is shown in Figure 10.15. As indicated in the figure, microwave power entering port 1 will exit port 2, and power entering port 2 will exit port 3, and so on. For an ideal circulator, the scattering matrix is

$$[S] = \begin{bmatrix} 0 & 0 & 1 \\ 1 & 0 & 0 \\ 0 & 1 & 0 \end{bmatrix} \tag{10.20}$$

This circulator is ideal in that there is no power reflected at the ports ($S_{11} = S_{22} = S_{33} = 0$), with all the power coupled to the appropriate port (i.e., $S_{21} = 1$) and none to the isolated port (i.e., $S_{31} = 0$).

An actual, nonideal, circulator is characterized by its insertion loss, isolation, and VSWR. Consider the three-port scattering matrix

$$[S] = \begin{bmatrix} S_{11} & S_{12} & S_{13} \\ S_{21} & S_{22} & S_{23} \\ S_{31} & S_{32} & S_{33} \end{bmatrix} \tag{10.21}$$

The insertion loss is measured between the input port and the desired through port. For instance, from port 1 to port 2 we have

$$IL_{21} = -20 \log \left(|S_{21}| \right) \tag{10.22}$$

The *isolation I* is a measure of the amount of signal that makes it to the wrong port, or

$$I_{31} = -20 \log \left(|S_{31}| \right) \tag{10.23}$$

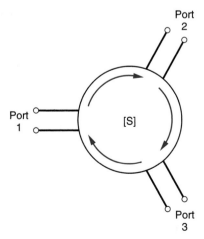

Figure 10.15   A circulator.

The amount of signal reflected at port 1 is measured by the voltage standing wave ratio

$$\text{VSWR} = \frac{1+|\Gamma_1|}{1-|\Gamma_1|} \qquad (10.24)$$

The standard assumption in calculation of insertion loss, isolation, and VSWR is that all ports are matched, in which case

$$\text{VSWR} = \frac{1+|S_{11}|}{1-|S_{11}|} \qquad (10.25)$$

For a symmetrical circulator, the scattering matrix is

$$[S] = \begin{bmatrix} S_{11} & S_{31} & S_{21} \\ S_{21} & S_{11} & S_{31} \\ S_{31} & S_{21} & S_{11} \end{bmatrix} \qquad (10.26)$$

▶ **EXAMPLE 10.9**

The specifications for a symmetrical three-port circulator are given as

$$IL = 1.0 \text{ dB}$$

$$I = 22. \text{ dB}$$

$$\text{VSWR} = 1.5$$

We want to determine the magnitude of the scattering parameters.
   Let's consider port 1 as our input and port 2 as our desired through port. We have

$$IL = IL_{21} = -20 \log (|S_{21}|) = 1 \text{ dB}$$

Solving, we find $|S_{21}| = 0.89$.
   Port 3 is isolated, and we have

$$I = I_{31} = -20 \log (|S_{31}|) = 22 \text{ dB}$$

Solving, we find $|S_{31}| = 0.079$.
   To find $|S_{11}|$ using the given VSWR, we assume that ports 2 and 3 are terminated in matched loads and have

$$|S_{11}| = |\Gamma_1| = \frac{\text{VSWR} - 1}{\text{VSWR} + 1} = 0.20$$

Drill 10.6   Calculate the return loss looking into port 1 of Example 10.9. (*Answer:* 14 dB).

Three-Port Dividers

A study of the scattering parameters reveals that it is not possible to construct a three-port network, matched at all ports, that is both reciprocal and lossless (see Problem 10.20). We

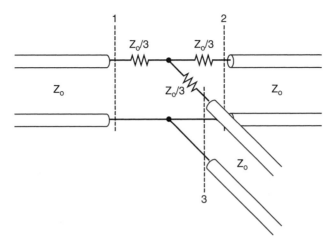

Figure 10.16 Resistive power divider.

can, however, relax the lossless requirement and build resistive networks that are useful for splitting or combining power.

Figure 10.16 shows a resistive divider. Analysis of this figure reveals the following scattering matrix:

$$[S] = \frac{1}{2} \begin{bmatrix} 0 & 1 & 1 \\ 1 & 0 & 1 \\ 1 & 1 & 0 \end{bmatrix} \qquad (10.27)$$

The resistive power divider is a symmetrical device. In some cases isolation may be desired between the output ports. A Wilkinson power divider, shown in Figure 10.17, evenly splits the transmitted power between the two isolated output ports. The scattering matrix for an ideal Wilkinson divider, although somewhat complicated to derive, turns out to be

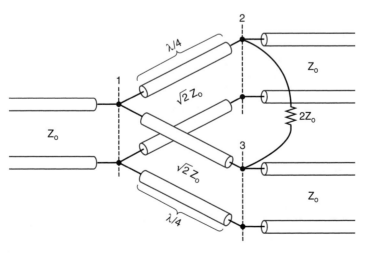

Figure 10.17 Wilkinson divider.

$$[S] = \frac{-j}{\sqrt{2}} \begin{bmatrix} 0 & 1 & 1 \\ 1 & 0 & 0 \\ 1 & 0 & 0 \end{bmatrix} \qquad (10.28)$$

We notice that the Wilkinson divider also has only a 3-dB insertion loss compared to 6 dB for the resistive divider.

Couplers

The common circuit symbols for a four-port coupler are shown in Figure 10.18. A coupler will transmit half or more of its power from its input (port 1) to its through port (port 2). A portion of the power will be drawn off to the coupled port (port 3), and ideally none will go to the isolated port (port 4). If the isolated port is internally terminated in a matched load, the coupler is most often referred to as a *directional coupler*.

Consider a four-port scattering matrix:

$$[S] = \begin{bmatrix} S_{11} & S_{12} & S_{13} & S_{14} \\ S_{21} & S_{22} & S_{23} & S_{24} \\ S_{31} & S_{32} & S_{33} & S_{34} \\ S_{41} & S_{42} & S_{43} & S_{44} \end{bmatrix} \qquad (10.29)$$

For an ideal coupler, this network will be reciprocal, matched at each port, and will have no power delivered to the isolated port. Applying these conditions, the scattering matrix becomes

$$[S] = \begin{bmatrix} 0 & S_{21} & S_{31} & 0 \\ S_{21} & 0 & 0 & S_{42} \\ S_{31} & 0 & 0 & S_{43} \\ 0 & S_{42} & S_{43} & 0 \end{bmatrix} \qquad (10.30)$$

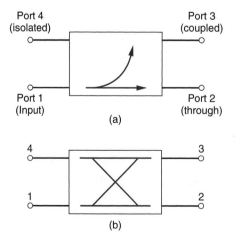

Figure 10.18 Common circuits symbols for four-port couplers.

There are two solutions for the ideal directional coupler. First, we have the *symmetrical* or *90° coupler* characterized by

$$[S] = \begin{bmatrix} 0 & \alpha & j\beta & 0 \\ \alpha & 0 & 0 & j\beta \\ j\beta & 0 & 0 & \alpha \\ 0 & j\beta & \alpha & 0 \end{bmatrix} \tag{10.31}$$

where α and β are the transmission and coupling coefficients, respectively, for the coupler. For a lossless network, we have

$$\alpha^2 + \beta^2 = 1 \tag{10.32}$$

Inspecting (10.31) we see that power to the coupled port is 90° out of phase with power to the through port.

The second solution is the *antisymmetrical* or *180° coupler* characterized by

$$[S] = \begin{bmatrix} 0 & \alpha & \beta & 0 \\ \alpha & 0 & 0 & -\beta \\ \beta & 0 & 0 & \alpha \\ 0 & -\beta & \alpha & 0 \end{bmatrix} \tag{10.33}$$

There are several terms used to characterize a coupler. First is the *coupling C*, also termed the *coupling coefficient*. The coupling relates the coupled power (P_3) to the input power (P_1), that is,

$$C = -10 \log \left(\frac{P_3}{P_1} \right) \tag{10.34}$$

If all ports are terminated in matched loads, the coupling coefficient becomes simply

$$C = -20 \log \left(|S_{31}| \right) \tag{10.35}$$

The insertion loss between the input and through ports is sometimes referred to as *main line loss*. For matched ports we have

$$IL = -20 \log \left(|S_{21}| \right) \tag{10.36}$$

The coupler's isolation is a measure of how much input power exits the isolated port. It is given by

$$I = -20 \log \left(|S_{41}| \right) \tag{10.37}$$

and is ideally ∞.

A term used to characterize a coupler's ability to direct energy only to the desired port is the *directivity D*. When all ports are matched, it is given by

$$D = 20 \log \left(\frac{|S_{31}|}{|S_{41}|} \right) \tag{10.38}$$

or

$$D = I - C \tag{10.39}$$

in decibels.

▷ **EXAMPLE 10.10**

Suppose an antisymmetrical coupler has the following characteristics:

$$C = 10.0 \text{ dB}$$

$$D = 15.0 \text{ dB}$$

$$IL = 2.00 \text{ dB}$$

$$\text{VSWR} = 1.30$$

We want to determine the scattering matrix. We'll assume all the ports are terminated in matched loads.

From (10.35) we can find S_{31} as a function of the coupling:

$$|S_{31}| = 10^{-10/20} = 0.316$$

Then, from (10.36) we can find S_{21}:

$$|S_{21}| = 10^{-2/20} = 0.794$$

The scattering parameter S_{11} is related to VSWR by

$$|S_{11}| = \frac{\text{VSWR} - 1}{\text{VSWR} + 1} = 0.130$$

Finally, to calculate S_{41} we need to know the isolation. Since we know D and C, we find I from (10.39) as

$$I = D + C = 25 \text{ dB}$$

and then

$$|S_{41}| = 10^{-25/20} = 0.056$$

Neglecting phases, we can write the scattering matrix as

$$[S] = \begin{bmatrix} 0.130 & 0.794 & 0.316 & 0.056 \\ 0.794 & 0.130 & 0.056 & -0.316 \\ 0.316 & 0.056 & 0.130 & 0.794 \\ 0.056 & -0.316 & 0.794 & 0.130 \end{bmatrix}$$

Drill 10.7   Suppose 1.000 W of power is incident on port 1 of the four-port coupler of Example 10.10. Determine the power out of each port and the power dissipated in the coupler. (*Answer:* $P_1 = 17$ mW, $P_2 = 630$ mW, $P_3 = 100$ mW, $P_4 = 3$ mW, $P_{\text{diss}} = 250$ mW)

Some of the more common four-port coupler types used in microwave circuits are the ring hybrid, the quadrature hybrid, and the Lange coupler. Top-down views of their microstrip patterns are shown in Figures 10.19–10.21.

A ring hybrid coupler (also known as the *rat-race coupler*) is shown in Figure 10.19. As may be gleaned from a study of the figure, a microwave signal into port 1 will split evenly in both directions, giving identical signals out of ports 2 and 3. But the split signals

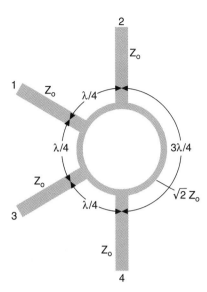

Figure 10.19 Ring hybrid (or rat-race) coupler.

are 180° out of phase at port 4, the isolated port, so they cancel and no power exits port 4. The scattering matrix for an ideal ring hybrid is

$$[S] = \frac{-j}{\sqrt{2}}\begin{bmatrix} 0 & 1 & 1 & 0 \\ 1 & 0 & 0 & -1 \\ 1 & 0 & 0 & 1 \\ 0 & -1 & 1 & 0 \end{bmatrix} \tag{10.40}$$

Here we see that the ring hybrid is an antisymmetrical coupler. The insertion loss and coupling are both equal to 3 dB. Not only can the ring hybrid split power to two ports, but it can add and subtract a pair of signals. If a pair of signals is fed to ports 2 and 3, their sum (multiplied by $-j/\sqrt{2}$) will exit port 1 and their difference will exit port 4 (again multiplied by $-j/\sqrt{2}$).

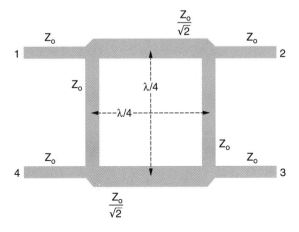

Figure 10.20 Quadrature hybrid (or branch-line) coupler.

Another 3-dB coupler is the quadrature hybrid (or branch-line hybrid) coupler, shown in Figure 10.20. The quadrature term comes from the 90° phase difference between the outputs at ports 2 and 3. Like the ring hybrid, the coupling and insertion losses are both equal to 3 dB. The scattering matrix is

$$[S] = \frac{-1}{\sqrt{2}} \begin{bmatrix} 0 & j & 1 & 0 \\ j & 0 & 0 & 1 \\ 1 & 0 & 0 & j \\ 0 & 1 & j & 0 \end{bmatrix}$$

(10.41)

so the quadrature hybrid is a symmetrical coupler.

Finally, a 3-dB coupler frequently employed in monolithic microwave integrated circuits and microwave circuits is the Lange coupler, shown in Figure 10.21. This coupler consists of interdigitated narrow lines, with alternate lines tied together as shown, typically by wire bonds. Although construction is obviously more challenging than the ring or quadrature hybrid couplers, the Lange coupler can operate over a much broader bandwidth (an octave[10.5]) than the others. It also tends to take up less space, since the interdigitated fingers can be very narrow. The length $\ell = \lambda/4$ is taken at the low-frequency limit of operation. Proper design must also consider the narrow finger width w and close spacing d, but these considerations are well beyond the scope of this text. The scattering matrix for the Lange coupler is the same as for the quadrature hybrid.

▷ **EXAMPLE 10.11**

Suppose to port 1 of an ideal ring hybrid coupler we apply the appropriate frequency voltage

$$V_1^+ = 1e^{j30°} \text{ V}$$

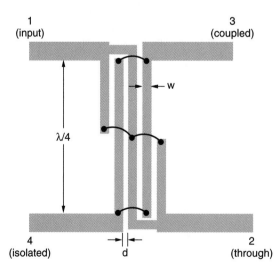

Figure 10.21 A Lange coupler features octave bandwidth operation.

[10.5]An octave represents the range of a doubling of frequency. For example, 1–2 GHz and 2.4–4.8 GHz are octave bandwidths.

We want to find the voltage exiting the other three ports.

Since the coupler is ideal, there will be no reflected voltage and no voltage exiting port 4, so

$$V_1^- = V_4^- = 0$$

From port 2, we have

$$V_2^- = S_{21}V_1^+ = \left(\frac{1}{\sqrt{2}}e^{-j90°}\right)1e^{j30°} V = \frac{1}{\sqrt{2}}e^{-j60°} \text{ V}$$

V_3^- will have the same value.

Drill 10.8   Repeat Example 10.11 for the quadrature hybrid coupler.

$$\left(Answer: V_1^- = V_4^- = 0, \; V_2^- = \left(1/\sqrt{2}\right)e^{-j60°}\text{ V}, \; V_3^- = \left(1/\sqrt{2}\right)e^{-j50°}\text{ V} \right)$$

▷ 10.4   FILTERS

Filters are two-port networks used to attenuate undesirable frequencies. As seen in Figure 10.1, microwave filters are commonly used in transceiver circuits. The four basic filter types are *low-pass*, *high-pass*, *bandpass*, and *bandstop*. Ideal performance and circuit symbols for each filter type are shown in Figure 10.22.

Of course, an actual filter doesn't have ideal characteristics. A low-pass filter is characterized by the insertion loss versus frequency plot in Figure 10.23. Notice that there may be *ripple* in the *passband* (the frequency range desired to pass through the filter) and a *roll off* in transmission above the cutoff or corner frequency f_c. Simple filters (like series inductors or shunt capacitors) feature 20 dB/decade roll off. Sharper roll off is available using active filters or multisection filters. Active filters employ operational amplifiers that are limited by performance to the lower RF frequencies. Multisection filters use passive components (inductors and capacitors) to achieve filtering. The two primary types are the Butterworth and the Chebyshev. A Butterworth filter has no ripple in the passband, whereas the Chebyshev filter features sharper roll off.

The insertion loss for a bandpass filter is shown in Figure 10.24. Here a small *passband ripple* is desired. The sharpness of the filter response is given by the *shape factor SF*, which is related to the filter bandwidth at 3 dB and 60 dB by

$$SF = \frac{BW_{60\,\text{dB}}}{BW_{3\,\text{dB}}} \tag{10.42}$$

As we saw in Chapter 9, a filter's insertion loss relates the power delivered to the load without the filter in place (P_L) to the power delivered with the filter in place (P_{Lf}):

$$IL = 10\log\left(\frac{P_L}{P_{Lf}}\right) \tag{10.43}$$

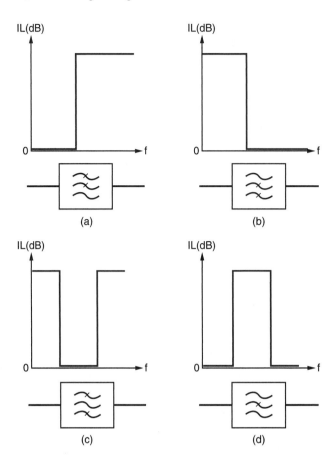

Figure 10.22 Ideal performance of the four filter types along with their circuit symbols: (a) low-pass, (b) high-pass, (c) bandpass, and (d) bandstop.

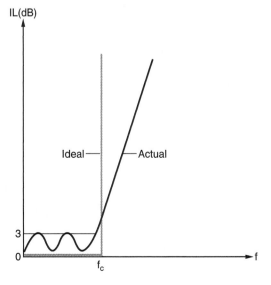

Figure 10.23 Typical performance for a low-pass filter compared with an ideal one.

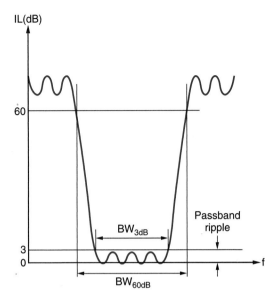

Figure 10.24 Typical filter performance for a bandpass filter.

Another way to represent the insertion loss is by using the *power loss ratio*, defined as

$$P_{LR} = \frac{P_A}{P_L} \tag{10.44}$$

where P_A is the maximum power available from the source and P_L is the power delivered to the load. The insertion loss becomes

$$IL = 10 \log \left(|P_{LR}| \right) \tag{10.45}$$

The maximum power P_A occurs when the source is terminated in a load impedance that is the complex conjugate of the source impedance, that is,

$$Z_L = Z_S^* \tag{10.46}$$

For the case where source and load are real and both equal to R_o, we have

$$P_A = \frac{v_1^2}{R_o} = \frac{\left(v_s/2 \right)^2}{R_o} = \frac{v_s^2}{4R_o} \tag{10.47}$$

The power delivered to the load is calculated based on the particulars of the filter network.

When the filter network is inserted in a system that has both a real source impedance and a real load impedance, as shown in Figure 10.25, we can also relate the insertion loss to the network's transmission S-parameter as

$$IL = -20 \log \left(|S_{21}| \right) \tag{10.48}$$

Simple Filters

Some simple lumped-element filter circuits are shown in Figure 10.26. Although these are most likely covered in sophomore-level circuits classes, it will be instructive to examine a couple of them.

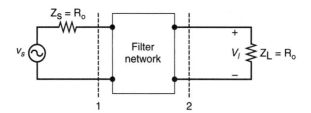

Figure 10.25 Filter network inserted in an R_o impedance system.

In Figure 10.26a, the power delivered to the load is

$$P_L = \frac{v_1^2}{R_o} \qquad (10.49)$$

where

$$v_1 = \frac{R_o}{2R_o + j\omega L} v_s \qquad (10.50)$$

Manipulating these equations with (10.44), (10.45), and (10.47) we arrive at

$$IL = 20\log\left(\left|1 + \frac{j\omega L}{2R_o}\right|\right) \qquad (10.51)$$

A plot of this relation for a simple low-pass filter is shown in Figure 10.27. The 3-dB cutoff frequency, also termed the corner frequency, occurs where insertion loss reaches 3 dB. Inspecting (10.51), we see that this occurs for the simple low-pass filter of Figure 10.25a when

$$\frac{\omega L}{2R_o} = 1$$

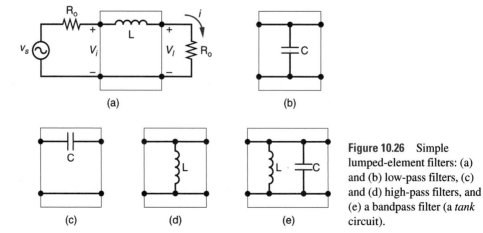

(a)

(b)

(c)

(d)

(e)

Figure 10.26 Simple lumped-element filters: (a) and (b) low-pass filters, (c) and (d) high-pass filters, and (e) a bandpass filter (a *tank* circuit).

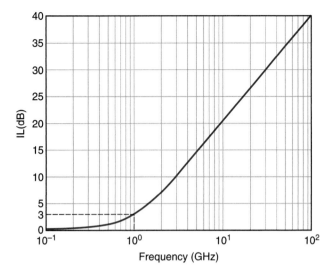

Figure 10.27 Characteristics for the simple low-pass filter of Example 10.12, generated via MATLAB 10.2.

so

$$f_c = \frac{R_o}{\pi L} \tag{10.52}$$

► **EXAMPLE 10.12**

Let us design a low-pass filter for a 50.0-Ω system using a series inductor. The 3-dB cutoff frequency is specified as 1.00 GHz.

Rearranging (10.52) we find

$$L = \frac{R_o}{\pi f} = \frac{50\Omega}{\pi\left(1\times10^9 \ 1/s\right)}\left(\frac{H}{\Omega\text{-}s}\right) = 15.9 \text{ nH}$$

The insertion loss, plotted in Figure 10.27, is from MATLAB 10.2.

MATLAB 10.2

We want to plot the insertion loss versus frequency for the low-pass filter of Example 10.12.

```
%   M-File: ML1002
%
%   This routine plots the insertion loss versus
%   frequency for the low-pass filter of Example 10.12.
%
%   Wentworth, 12/2/02
%
%   Variables
%   Ro          real line impedance (ohms)
%   L           inductance (H)
```

(continued)

```
%  f          freq. (GHz)
%  w          ang. freq. (rad/s)
%  PLR        power loss ratio
%  IL         insertion loss (dB)

clc           %clears the command window
clear         %clears variables

%  Initialize variables
Ro=50;
L=15.9e-9;
f=0.1:0.01:100;
w=2*pi*f*1e9;

%  Perform calculations
PLR=abs(complex(1,w*L/(2*Ro)));
IL=20*log10(PLR);

%  Generate plot
semilogx(f,IL)
grid on
xlabel('frequency (GHz)')
ylabel('IL(dB)')
```

Note again the use of the function "LOG10" to indicate base 10 log.

The plot is shown in Figure 10.27. Notice the 20 dB/decade roll off in the low-pass filter response.

Drill 10.9 Derive an expression for IL for a low-pass filter realized using a shunt capacitor. What capacitance value is needed for a 1.0-GHz cutoff frequency if $R_o = 50\ \Omega$?

$$\left(Answer\!: IL = 20 \log\left|1 + \frac{j\omega R_o C}{2}\right|, \quad 6.4 \text{ pF} \right)$$

Multisection Filters

Multisection filters are capable of broad bandwidth and sharp roll off in the transmission characteristics. Two popular approaches are the Butterworth filter and the Chebyshev filter. We will focus on how to design a Chebyshev filter, sacrificing the flat passband of the Butterworth for the sharper roll off of the Chebyshev.

Regardless of the type of filter desired (low-pass, high-pass, bandpass, or bandstop), Chebyshev filter design always begins with a low-pass filter, normalized with respect to frequency and impedance. This prototype filter is then frequency and impedance transformed to achieve the desired properties.

A low-pass prototype filter circuit is shown in Figure 10.28a. The *dual* of this circuit is shown in Figure 10.28b. Both circuits feature identical performance. The *order* of the filter is the number n of elements used. The element values g_o to g_{n+1} are values normalized to the

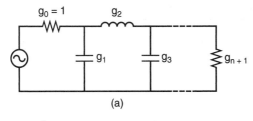

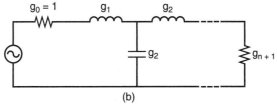

Figure 10.28 Two low-pass filter prototypes with elements normalized with respect to resistance and frequency.

resistance or conductance of the source. They are also normalized to the cutoff frequency. The parameters are defined as follows:

$$g_0 = \begin{cases} \text{source resistance (Figure 10.28a)} \\ \text{source conductance (Figure 10.28b)} \end{cases}$$

$$g_1 \text{ to } g_n = \begin{cases} \text{inductance for series inductors} \\ \text{capacitance for shunt capacitors} \end{cases}$$

$$g_{n+1} = \begin{cases} \text{load resistance if } g_n \text{ is a shunt capacitor} \\ \text{load conductance if } g_n \text{ is a series inductor} \end{cases}$$

Values of the normalized elements are listed in Table 10.3 for three values of allowed ripple: (a) 0.1 dB, (b) 1 dB, and (c) 3 dB.

The more elements that are chosen (higher n), the sharper the frequency response roll off will be. An odd number of elements is desired, as the terminating impedance will then be the same as the source impedance.

The design procedure for a low-pass filter to be inserted in a system with known impedance is as follows:

1. Determine the desired corner frequency and number of elements n. Also select the circuit type (Figure 10.28a or 10.28b) and the amount of ripple allowed in the passband (0.1, 1, or 3 dB).

2. Select values of the normalized elements from Table 10.3 for the appropriate amount of ripple.

3. Transform element values with regard to impedance. Here g_o is replaced with the system impedance $Z_o = R_o$. Likewise, if we've chosen an odd number of filter elements, then g_{n+1} is replaced with R_o as well.

   For inductors,

$$L' = LR_o \tag{10.53}$$

TABLE 10.3   Chebyshev Filter Coefficients for the Low-Pass Filter Prototype ($g_0 = 1$, $\omega_c = 1$, $n = 1$ to 10)

(a) 0.1 dB ripple

n	$g1$	$g2$	$g3$	$g4$	$g5$	$g6$	$g7$	$g8$	$g9$	$g10$	$g11$
1	.3052	1.0000									
2	.8430	.6220	1.3554								
3	1.0315	1.1474	1.0315	1.0000							
4	1.1088	1.3061	1.7703	0.8180	1.3554						
5	1.1468	1.3712	1.9750	1.3712	1.1468	1.0000					
6	1.1681	1.4039	2.0562	1.5170	1.9029	0.8618	1.3554				
7	1.1811	1.4228	2.0966	1.5733	2.0966	1.4228	1.1811	1.0000			
8	1.1897	1.4346	2.1199	1.6010	2.1699	1.5640	1.9444	0.8778	1.3554		
9	1.1956	1.4425	2.1345	1.6167	2.2053	1.6167	2.1345	1.4425	1.1956	1.0000	
10	1.1999	1.4481	2.1444	1.6265	2.2253	1.6418	2.2046	1.5821	1.9628	0.8853	1.3554

(b) 1-dB ripple

n	$g1$	$g2$	$g3$	$g4$	$g5$	$g6$	$g7$	$g8$	$g9$	$g10$	$g11$
1	1.0177	1.0000									
2	1.8219	0.6850	2.6599								
3	2.0236	0.9941	2.0236	1.0000							
4	2.0991	1.0644	2.8311	0.7892	2.6599						
5	2.1349	1.0911	3.0009	1.0911	2.1349	1.0000					
6	2.1546	1.1041	3.0634	1.1518	2.9367	0.8101	2.6599				
7	2.1664	1.1116	3.0934	1.1736	3.0934	1.1116	2.1664	1.0000			
8	2.1744	1.1161	3.1107	1.1839	3.1488	1.1696	2.9685	0.8175	2.6599		
9	2.1797	1.1192	3.1215	1.1897	3.1747	1.1897	3.1215	1.1192	2.1797	1.0000	
10	2.1836	1.1213	3.1286	1.1933	3.1890	1.1990	3.1738	1.1763	2.9824	0.8210	2.6599

(c) 3-dB ripple

n	$g1$	$g2$	$g3$	$g4$	$g5$	$g6$	$g7$	$g8$	$g9$	$g10$	$g11$
1	1.9953	1.0000									
2	3.1013	0.5339	5.8095								
3	3.3487	0.7117	3.3487	1.0000							
4	3.4389	0.7483	4.3471	0.5920	5.8095						
5	3.4817	0.7618	4.5381	0.7618	3.4817	1.0000					
6	3.5045	0.7685	4.6061	0.7929	4.4641	0.6033	5.8095				
7	3.5182	0.7723	4.6386	0.8039	4.6386	0.7723	3.5182	1.0000			
8	3.5277	0.7745	4.6575	0.8089	4.6990	0.8018	4.4990	0.6073	5.8095		
9	3.5340	0.7760	4.6692	0.8118	4.7272	0.8118	4.6692	0.7760	3.5340	1.0000	
10	3.5384	0.7771	4.6768	0.8136	4.7425	0.8164	4.7260	0.8051	4.5142	0.6091	5.8095

Source: from G.L. Matthaei et al., *Microwave Filters, Impedance-Matching Networks, and Coupling Structures*, Artech House, Inc., 1980).

where L is the g value of the inductor and L' is the inductor value transformed for impedance.

For capacitors,

$$C' = \frac{C}{R_o} \tag{10.54}$$

where C is the g value of the capacitor and C' is the capacitor value transformed for impedance.

4. Transform element values with respect to frequency.

For inductors,

$$L'' = \frac{L'}{\omega_c} \tag{10.55}$$

where L'' is the final inductance transformed with respect to frequency. For a cut-off frequency f_c, $\omega_c = 2\pi f_c$.

For capacitors,

$$C'' = \frac{C'}{\omega_c} \tag{10.56}$$

where C'' is the final capacitance transformed with respect to frequency.

▶ **EXAMPLE 10.13**

To get an idea of how to use the Chebyshev design procedure, let's design a simple first-order low-pass filter with 3-dB ripple and a corner frequency of 1.0 GHz in a 50.-Ω system. The situation is shown in Figure 10.29a.

From the problem statement, we have the desired corner frequency and the number of elements, so the next step is to select a circuit type and we arbitrarily choose Figure 10.28a.

For 3-dB ripple, we find from Table 10.3(c) that $g_1 = 1.9953$. Including the normalized element values for source and load, our filter circuit appears as Figure 10.29b.

Now the shunt capacitor is transformed with respect to impedance,

$$C' = \frac{C}{Z_o} = \frac{1.9953 \text{ F}}{50} = 39.9 \text{ mF}$$

Transforming with respect to frequency, we finally have

$$C'' = \frac{C'}{\omega_c} = \frac{39.9 \text{ mF}}{2\pi(1\times10^9)} = 6.4 \text{ pF}$$

The final circuit is shown in Figure 10.29c. The filter response is identical to that of Figure 10.27.

Drill 10.10 Rework Example 10.13 starting with Figure 10.28b and determine the value of the series inductance. (*Answer:* 16 nH)

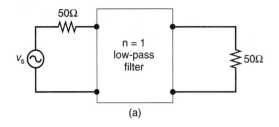

(a)

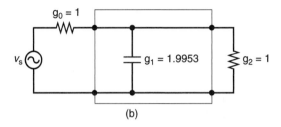

(b)

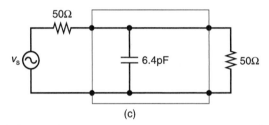

(c)

Figure 10.29 A first-order Chebyshev low-pass filter with 3-dB ripple designed for Example 10.13.

▶ **EXAMPLE 10.14**

Now let's design a third-order low-pass filter with $f_c = 2.0$ GHz for a 50.-Ω system where we will allow only 0.1 dB of ripple.

We choose to use the configuration shown in Figure 10.28b. Consulting Table 10.3(a), we find the normalized element values as indicated in Figure 10.30a.

The impedance transformation gives us

$$L' = LR_o = (1.0315)(50) = 51.6$$

and

$$C' = \frac{C}{R_o} = \frac{1.1474}{50} = 0.023$$

The circuit at this stage of design is indicated in Figure 10.30b. Now we perform the frequency transformation to get

$$L'' = \frac{L'}{\omega_c} = \frac{51.6}{2\pi\left(2\times10^9\right)} = 4.1 \text{ nH}$$

and

$$C'' = \frac{C'}{\omega_c} = \frac{0.023}{2\pi\left(2\times10^9\right)} = 1.8 \text{ pF}$$

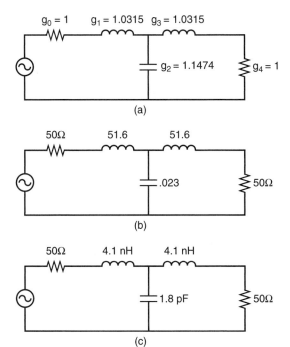

(a)

(b)

(c)

Figure 10.30 A third-order Chebyshev low-pass filter with 0.1-dB ripple designed for Example 10.14.

The resulting circuit is shown in Figure 10.30c.

   A plot of the insertion loss for this 0.1-dB ripple filter is compared with 1-dB and 3-dB ripple filter responses in Figure 10.31. Notice that the flatter passband is sacrificed for sharper roll off.

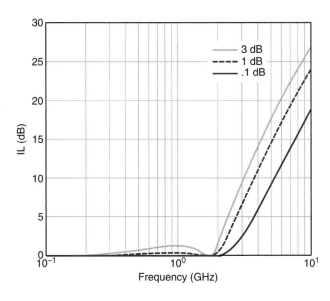

Figure 10.31 Insertion loss for third-order Chebyshev low-pass filters designed with different ripple tolerances.

> **Drill 10.11** Repeat Example 10.14 using the circuit configuration of Figure 10.28a. Determine the component values. (*Answer:* $C_1 = C_3 = 1.64$ pF, $L_2 = 4.56$ nH)

High-Pass Filters

We know that a low-pass filter using a series inductor becomes a high-pass filter if the inductor is replaced with a capacitor. We can transform the Chebyshev low-pass filter prototype into other filter types by making use of Table 10.4 in the frequency transformation step. We begin by designing a high-pass filter and then we will turn our attention to the bandpass filter. The bandstop filter will be left for the problems at the end of the chapter.

As before, we start with the impedance transformation, using (10.53) and (10.54). In the frequency transformation to a high-pass filter, we see from Table 10.4 that

$$C'' = \frac{1}{\omega_c L'} \tag{10.57}$$

TABLE 10.4   Filter Transformations

Impedance transformed low-pass prototype	$L' = g_i R_o$	$C' = \dfrac{g_i}{R_o}$
Low-pass filter	$\dfrac{L'}{\omega_o}$	$\dfrac{C'}{\omega_o}$
High-pass filter	$\dfrac{1}{\omega_o L'}$	$\dfrac{1}{\omega_o C'}$
Bandpass filter[a]	$\dfrac{L'}{BW_\omega}$; $\dfrac{BW_\omega}{\omega_o^2 L'}$	$\dfrac{BW_\omega}{\omega_o^2 C'}$; $\dfrac{C'}{BW_\omega}$
Bandstop filter[a]	$\dfrac{BW_\omega L'}{\omega_o^2}$; $\dfrac{1}{BW_\omega L'}$	$\dfrac{1}{BW_\omega C'}$; $\dfrac{BW_\omega C'}{\omega_o^2}$

[a] $BW_\omega = \omega_u - \omega_L$, $\omega_o = \sqrt{\omega_u \omega_L}$

Source: Adapted from R. Ludwig and P. Bretchko, *RF Circuit Design*, Prentice-Hall, 2000, p. 238.

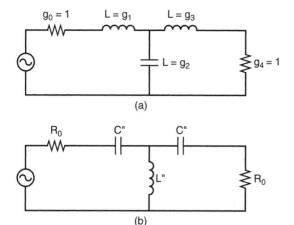

Figure 10.32 The prototype low-pass filter of (a) is frequency transformed to the high-pass filter of (b).

and

$$L'' = \frac{1}{\omega_c C'} \qquad (10.58)$$

Thus, the low-pass filter of Figure 10.32a is transformed into the high-pass filter of Figure 10.32b.

▷ **EXAMPLE 10.15**

Let us modify Example 10.14 to design a third-order high-pass filter with 0.1-dB ripple and 2.0-GHz corner frequency.

We proceed as before to Figure 10.30b. Now the frequency transformation gives us

$$C'' = \frac{1}{\omega_c L'} = \frac{1}{2\pi(2\times10^9)(51.6)} = 1.5 \text{ pF}$$

and

$$L'' = \frac{1}{\omega_c C'} = \frac{1}{2\pi(2\times10^9)(0.023)} = 3.5 \text{ nH}$$

The final high-pass circuit is shown in Figure 10.33a along with the filter response in Figure 10.33b.

Bandpass Filters

Design of bandpass and bandstop filters requires specification of the filter bandwidth. The difference between an upper and a lower angular frequency (ω_u and ω_l, respectively) constitutes the bandwidth,

$$BW_\omega = \omega_u - \omega_L \qquad (10.59)$$

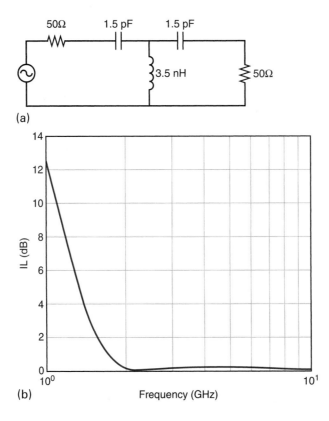

Figure 10.33 (a) Third-order Chebyshev high-pass filter circuit for Example 10.15. (b) The filter's insertion loss.

where the ω subscript indicates the bandwidth is with respect to angular frequency. The transforms also use a mean average angular frequency ω_o given by

$$\omega_o = \sqrt{\omega_u \omega_L} \qquad (10.60)$$

As seen from Table 10.4, an inductor from the low-pass prototype is transformed into a series combination of an inductor and a capacitor. A capacitor from the prototype transforms into a parallel combination.

▷ **EXAMPLE 10.16**

We want to design a third-order Chebyshev bandpass filter for a 50.-Ω system. The passband is to be from 900 to 1100 MHz with only 1 dB of ripple allowed.

To begin, we'll arbitrarily choose the low-pass filter prototype of Figure 10.28a. From Table 10.3(b) we find the element values as shown in Figure 10.34a.

Next, impedance transformation yields

$$C' = \frac{g_1}{R_o} = \frac{(2.0236)}{50} = 0.0405$$

and

$$L' = g_2 R_o = (0.9941)(50) = 49.7$$

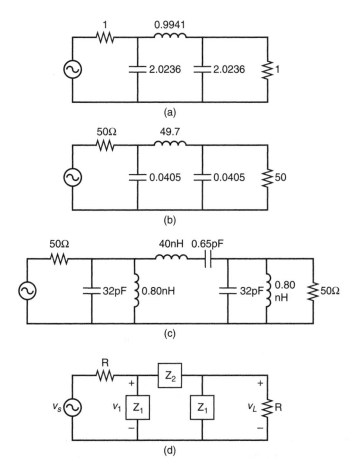

Figure 10.34 (a) The low-pass prototype for Example 10.16. (b) The following impedance transformation. (c) The following bandpass frequency transformation. (d) The simplified circuit for finding IL.

The prototype circuit after impedance transformation is shown in Figure 10.34b.

Performing the frequency transform requires that we know BW_ω, which is equal to $(2\pi)200 \times 10^6 = 1.257 \times 10^9$ radians/s, and the mean average angular frequency

$$\omega_o = 2\pi\sqrt{\omega_u\omega_L} = 2\pi\sqrt{(900 \text{ MHz})(1100 \text{ MHz})} = 6.25 \times 10^9 \text{ radians/s}$$

We first convert each 0.0405 capacitor into a parallel inductance–capacitance combination, where

$$C_C'' = \frac{C'}{BW_\omega} = \frac{0.0405}{1.257 \times 10^9} = 32 \text{ pF}$$

and

$$L_C'' = \frac{BW_\omega}{\omega_o^2 C'} = \frac{1.257 \times 10^9}{\left(6.25 \times 10^9\right)^2 (0.0405)} = 0.80 \text{ nH}$$

The "C" subscript is used to indicate that the components are transformed from a capacitor. Next we convert the 49.7 inductor to a series combination, where

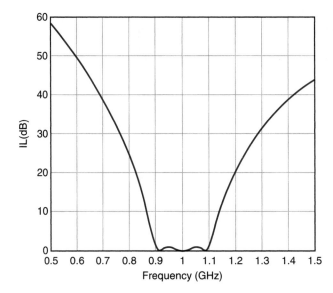

Figure 10.35 Insertion loss for bandpass filter of Example 10.16.

$$L_L'' = \frac{L'}{BW_\omega} = \frac{49.7}{1.257 \times 10^9} = 40 \text{ nH}$$

and

$$C_L'' = \frac{BW_\omega}{\omega_o^2 L'} = \frac{1.257 \times 10^9}{\left(6.25 \times 10^9\right)^2 (49.7)} = 0.65 \text{ pF}$$

These transformed values are shown in the bandpass filter circuit of Figure 10.34c.

To plot the bandpass filter insertion loss, we can simplify the circuit as shown in Figure 10.34d by replacing the parallel LC combination with an impedance Z_1 and replacing the series LC combination with an impedance Z_2. We then define $Z_3 = Z_1 \| R$, $Z_4 = Z_3 + Z_2$, $Z_5 = Z_1 \| Z_4$, and $Z_6 = Z_5 + R$. After some manipulation we arrive at

$$P_{LR} = \left(\frac{Z_4 Z_6}{2 Z_3 Z_5}\right)^2$$

and then

$$IL = 10 \log\left(\left|P_{LR}\right|\right)$$

This is plotted in Figure 10.35 using ML1003.

▶ **MATLAB 10.3**

```
%    M-File: ML1003
%
%    This routine plots the insertion loss versus
%    frequency for the bandpass filter of Example 10.16.
%
```

```
%    Wentworth, 12/2/02
%
%    Variables
%    R           real line impedance (ohms)
%    Lc          inductance (from cap) (H)
%    Cc          capacitance (from cap) (F)
%    LL          inductance (from ind) (H)
%    CL          capacitance (from ind) (F)
%    f           freq. (GHz)
%    w           ang. freq. (rad/s)
%    ZLc,ZCc,ZLL,ZCL
%                element impedances
%    Z1-Z6       impedance variables
%    PLR         power loss ratio
%    IL          insertion loss (dB)

clc             %clears the command window
clear           %clears variables

%    Initialize variables
R=50;
Lc=0.80e-9;
Cc=32e-12;
LL=40e-9;
CL=0.65e-12;
f=0.5:0.005:1.5;
w=2*pi*f*1e9;

%    Perform calculations
ZLc=i*w*Lc;
ZCc=-i./(w*Cc);
ZLL=i*w*LL;
ZCL=-i./(w*CL);
Z1=parallel(ZLc,ZCc);
Z2=ZLL+ZCL;
Z3=parallel(Z1,R);
Z4=Z3+Z2;
Z5=parallel(Z1,Z4);
Z6=Z5+R;
PLR=(Z4.*Z6./(2.*Z3.*Z5)).^2;
IL=10*LOG10(abs(PLR));

%    Generate plot
plot(f,IL)
grid on
xlabel('frequency(GHz)')
ylabel('IL(dB)')
```

The Chebyshev design procedure is a powerful approach for designing filters realized using discrete circuit components. In microwave circuits, these components are often transformed into T-line sections. To see how this is done, the interested student is directed to the texts by Pozar and by Ludwig and Bretchko listed at the end of this chapter.

▷ 10.5 AMPLIFIERS

Microwave amplifiers are a common and crucial component of wireless transceivers. They are constructed around a microwave transistor from the field effect transistor (FET) or bipolar junction transistor (BJT) families. For high-frequency, high-speed operation, several special transistor types have been developed including silicon germanium heterojunction bipolar transistors (SiGe HBTs), gallium arsenide field effect transistors (GaAs FETs), and GaAs high electron mobility transistors (GaAs HEMTs).

The task of an amplifier is, of course, to amplify a signal. The degree of amplification is given by the *gain G*, which relates the output power P_{out} to the input power P_{in} as

$$G = 10\log\left(\frac{P_{out}}{P_{in}}\right) \tag{10.61}$$

In addition to gain, amplifiers are also characterized by their *dynamic range, noise figure,* and VSWR at each port. The dynamic range is represented by Figure 10.36. It is basically the range over which the amplifier's gain is constant (i.e., the output is a linear function of the input). The range extends from the minimum signal discernible from the noise floor up to the 1-dB compression point, where the output has dropped 1 dB below its ideal value. Amplifier manufacturers most often cite the output power level as the 1-dB compression point (typically in dB$_m$).[10.6] The axes in Figure 10.36 are in terms of absolute power. Recall that dB$_m$ is referenced to 1 mW, so a 10-dB$_m$ output power level would correspond to

$$10\text{ dB}_m = 10\log\left(\frac{P_{out}}{1\text{ mW}}\right) \tag{10.62}$$

or P_{out} = 10 mW.

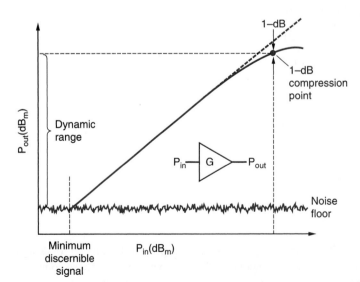

Figure 10.36 Amplifier dynamic range.

[10.6]To avoid confusion the term "1-dB output compression point" should be used in this case.

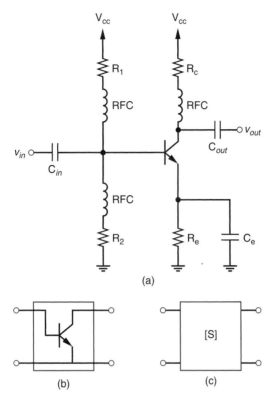

Figure 10.37 (a) A typical common-emitter amplifier circuit. The AC circuit resembles a two-port network (b) that can be represented by its S-parameters (c).

The noise figure is an indicator of how much noise is introduced to the signal by the amplifier. It is the ratio of the signal-to-noise ratio (*SNR*) at the input to the *SNR* at the output. For very weak input signals, it is critical that the amplifier have a very low noise figure.

The VSWR indicates how well the amplifier is impedance matched at the input and output.

For descriptive purposes, an amplifier circuit based on a BJT is shown in Figure 10.37a. The student is most likely familiar with the basic construction from a sophomore- or junior-level electronics course. Resistor values are chosen based on the desired DC operating point, which determines the AC performance of the transistor. The emitter resistance R_e, placed in the DC circuit to provide a stable operating point, is shorted to ground in the AC circuit by the *bypass capacitor* C_e to provide greatly improved gain. The input and the amplified output signals are coupled to the amplifier by C_{in} and C_{out} (*coupling capacitors*).

There is another element included in this circuit that is characteristic of microwave amplifiers: the RF choke (or RFC for short). The choke hides the DC biasing network from the high-frequency signals, making the AC circuit resemble the two-port network shown in Figure 10.37b. This two-port network can in turn be represented by a scattering matrix (Figure 10.37c), where the parameters are a function of the DC operating point and frequency.

A general microwave amplifier can be represented by the two-port S-matrix network between a pair of impedance matching networks as shown in Figure 10.38. The matching networks are necessary to minimize reflections seen by the source and to maximize power to the output.

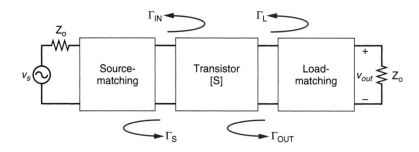

Figure 10.38 A microwave amplifier consists of the transistor sandwiched between source and load matching networks.

Some key parameters used in designing the matching networks and in calculating the amplifier gain are the reflection coefficients Γ_L, Γ_S, Γ_{IN}, and Γ_{OUT} shown in Figure 10.38. The reflection coefficient looking into the transistor from the source matching network, Γ_{IN}, is related to the transistor scattering parameters and Γ_L as

$$\Gamma_{IN} = S_{11} + \frac{S_{12}\Gamma_L S_{21}}{1 - S_{22}\Gamma_L} \tag{10.63}$$

Likewise,

$$\Gamma_{OUT} = S_{22} + \frac{S_{12}\Gamma_S S_{21}}{1 - S_{11}\Gamma_S} \tag{10.64}$$

The amplifier gain is termed the *transducer power gain*, given by

$$G_T = \frac{P_L}{P_A} \tag{10.65}$$

where P_L is the power delivered to the load and P_A is the maximum power available from the source. In terms of decibels,

$$G_T(dB) = 10 \log(G_T) \tag{10.66}$$

In a more advanced treatment on the subject matter, it can be shown that G_T is related to the two-port scattering parameters and the various reflection coefficients by the equation

$$G_T = \frac{1-|\Gamma_S|^2}{|1-\Gamma_{IN}\Gamma_S|^2}|S_{21}|^2\frac{1-|\Gamma_L|^2}{|1-S_{22}\Gamma_L|^2} \tag{10.67}$$

One of the first steps in a rigorous microwave amplifier design is to perform a stability analysis. This is accomplished by plotting stability circles on a Smith Chart. These circles are functions of the S-parameters and can be used as a guide to modify the circuit to achieve stability. This subject lies beyond the scope of this text, so we will make the simplifying assumption that our transistors are *unilateral*. The output signal at port 2 of a unilateral transistor has no effect on the input signal at port 1. Put another way, $S_{12} = 0$. This leads to

$$\Gamma_{IN} = S_{11} \tag{10.68}$$

and

$$\Gamma_{OUT} = S_{22} \tag{10.69}$$

The transistor is now unconditionally stable as long as $|S_{11}|$ and $|S_{22}|$ are less than 1, which is generally the case.

It is worth mentioning that S_{12} is close to zero for many transistors, and a unilateral assumption is often made to simplify amplifier design.

Maximum power is delivered from a source when the load is equal to the complex conjugate of the source impedance. A consequence of this relation is that maximum gain is achieved when $\Gamma_S = \Gamma_{IN}^*$ and $\Gamma_L = \Gamma_{OUT}^*$. For a unilateral transistor, we therefore have maximum gain when $\Gamma_S = S_{11}^*$ and $\Gamma_L = S_{22}^*$. For this case the maximum transducer power gain becomes

$$G_{Tu_{max}} = \frac{1}{1 - |S_{11}|^2} |S_{21}|^2 \frac{1}{1 - |S_{22}|^2} \tag{10.70}$$

▶ **EXAMPLE 10.17**

The scattering parameters for a unilateral transistor at a particular DC bias and frequency are given as

$$S_{11} = 0.560e^{j30°}$$

$$S_{12} = 0$$

$$S_{21} = 2.20e^{j66°}$$

$$S_{22} = 0.660e^{+j120°}$$

Inserting this transistor into a 50-Ω system, we want to determine the gain with and without optimized matching networks.

Without the matching networks, the transducer gain is simply $|S_{21}|^2$, or 4.84. In terms of decibels, $G_T(dB) = 20 \log (|S_{21}|) = 6.80$ dB.

With the matching networks, we have

$$G_{Tu_{max}} = \frac{1}{1 - |0.56|^2} |2.2|^2 \frac{1}{1 - |0.66|^2} = 12.$$

or $G_T(dB) = 11.0$ dB.

So in this example we see that the matching networks provide a 4.2-dB improvement in the amplifier power gain.

Maximizing gain is not the only criterion used in amplifier design. Some of the gain could be sacrificed for broader bandwidth or for improved noise performance. Although this is beyond the level of our coverage, it should be pointed out that the Smith Chart plays an important role in all of these design options. For example, circles of constant gain and of constant noise can be plotted on the same Smith Chart to aid in selecting an optimized solution.

Designing Matching Networks

In previous sections the task in designing an impedance matching network was to move from a mismatched load to the center of the Smith Chart. This is not the case for amplifier matching networks.

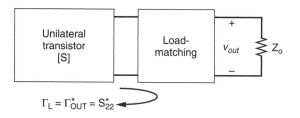

Figure 10.39 Load matching network at the output of a unilateral transistor.

Let's consider the load-matching network for an amplifier built with a unilateral transistor, depicted in Figure 10.39. For this case maximum gain is achieved when $\Gamma_L = S_{22}^*$. As can be gleaned from the figure, we will be moving from a matched load Z_o at the center of the Smith Chart to a location where $\Gamma_L = S_{22}^*$.

The same situation exists for the source-matching network. The examples to follow illustrate how the matching networks are designed.

▶ **EXAMPLE 10.18**

Let's design open-ended T-line stub matching networks for an amplifier constructed using a transistor with the S-parameters of Example 10.17.

We will start with the load-matching network, where we see that

$$\Gamma_L = S_{22}^* = 0.66e^{-j120°}$$

This point, located as point a on the Smith Chart of Figure 10.40a, represents the normalized impedance looking into the load-matching network. We must move from the center of the Smith Chart to this Γ_L point.

It is helpful to begin by drawing the constant-$|\Gamma_L|$ circle. We notice that this circle intercepts the $1 \pm jx$ circle at the points $1 \pm j1.8$. Our first step will be to move, in the normalized admittance circle (since we are adding a *shunt* stub), from the open end of the stub (point b' at WTG = 0 λ) to the point $0 + j1.8$ (point c' at WTG = 0.169λ). Adding the admittance looking into this stub to the admittance of our Z_o load we obtain $1 + j1.8$ (point d'). Now we need to add a through section of T-line to move to the point a on the normalized impedance chart. But point d' is an admittance point, so we first find the corresponding normalized impedance point d (at WTG = 0.433 λ) and move away from the load, clockwise toward the generator, to point a at WTG = 0.417λ. This corresponds to a through-line length of 0.484 λ. The resulting load-matching network is shown in Figure 10.40b.

We follow the same procedure to solve the source-matching network. First we locate

$$\Gamma_S = S_{11}^* = 0.560e^{-j30°}$$

This is point a on the chart shown in Figure 10.41a. Drawing the constant-$|\Gamma_S|$ circle we find the intersection points $1 \pm j1.35$. Then, we move from the open end of a shunt stub (point b') to the point c' where the admittance is $0 + j1.35$ located at WTG = 0.148 λ. The total admittance looking into the parallel combination of the shunt stub and the Z_o load is $1 + j1.35$ (point d'). Finally, we move from this point, at WTG = 0.172 λ, clockwise to point a' at WTG = 0.042 λ, a distance of 0.370 λ. Note that an equivalent approach for this last step would be to find the point d, the normalized impedance point corresponding to the normalized admittance at d', and move from this d point to the normalized impedance point a. The resulting source-matching network is given in Figure 10.41b.

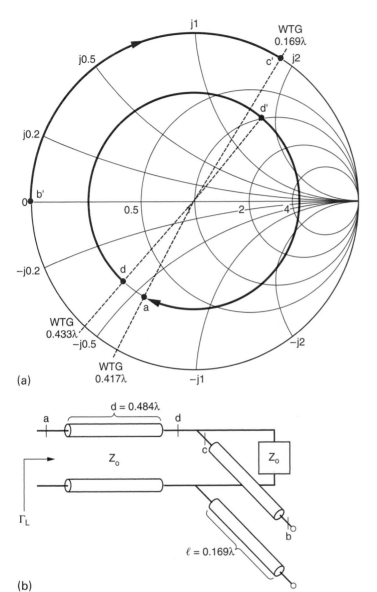

(a)

(b)

Figure 10.40 (a) Γ_L Smith Chart and (b) T-line stub load matching network for Example 10.18.

▶ **EXAMPLE 10.19**

Let us now realize the matching network of Example 10.18 in microstrip to be constructed on 20.0-mil-thick low-loss board with $\varepsilon_r = 10.2$. Assume 2.40-GHz operation in a 50.0-Ω system.

We can run the program ML0605 from Chapter 6 to determine the width of line needed to achieve a 50-Ω impedance. We get $w = 18.8$ mils. The program also tells us $\varepsilon_{eff} = 6.84$. This is used to determine guide wavelength:

$$\lambda_G = \frac{\lambda_o}{\sqrt{\varepsilon_{eff}}} = \frac{c/f}{\sqrt{\varepsilon_{eff}}} = \frac{3 \times 10^8 \text{ m}/\text{s}}{\left(2.4 \times 10^9 \text{ }1/\text{s}\right)\sqrt{6.84}} \left(\frac{1 \text{ in}}{0.0254 \text{ m}}\right) = 1.88 \text{ in}$$

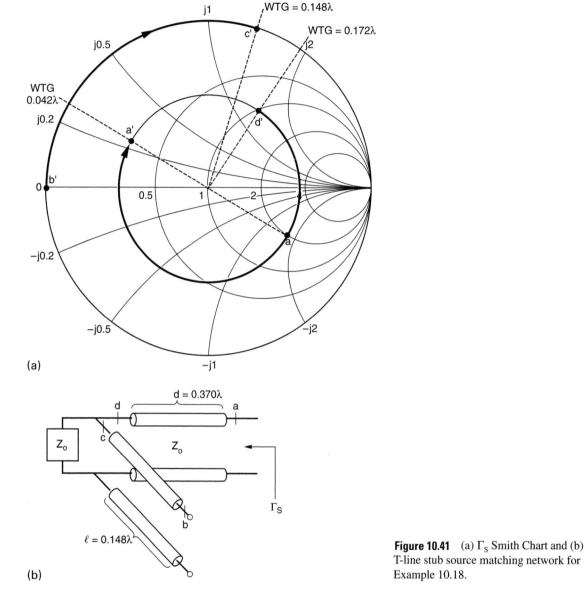

(a)

(b)

Figure 10.41 (a) Γ_S Smith Chart and (b) T-line stub source matching network for Example 10.18.

The line lengths are then calculated. For instance, the length of the open-ended microstrip stub in the load-matching network is

$$\ell = 0.169\lambda \left(\frac{1.88 \text{ in}}{\lambda} \right) = 0.318 \text{ in}$$

Figure 10.42 shows the final circuit.

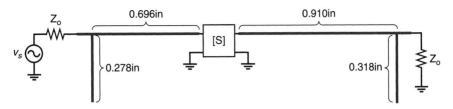

Figure 10.42 The top-down view of microstrip sections used for the matching networks of Example 10.19.

▷ **EXAMPLE 10.20**

Now let's design lumped-element matching networks for an amplifier constructed using a transistor with the S-parameters of Example 10.17. We'll again assume 2.4-GHz operation in a 50-Ω system.
As in Example 10.18, we have

$$\Gamma_L = S_{22}^* = 0.66e^{-j120°}$$

We locate this as point *a* on the Smith Chart shown in Figure 10.43a. Also on this Smith Chart we add the rotated $1 \pm jx$ circle. Our task will be to move from the matched load at the center of the Smith Chart to the Γ_L point *a*.

Referring to Figure 10.43a, we first work this problem backward to locate the key points on the Smith Chart. From *a* (at $z = 0.26 - j0.55$) we move along the constant $r = 0.26$ circle to point *b* ($z = 0.26 - j0.44$). Then we jump to the admittance chart point *b'* (at $y = 1 + j1.65$).

Now we can solve the load-matching network. We'll move in the admittance chart from the load (at $y = 1 + j0$) to point *b'*, requiring a normalized admittance value of $+j1.65$. From Table 10.2 we see that this corresponds to a shunt capacitance:

$$j1.65 = j\omega C Z_o$$

or

$$C = \frac{1.65}{2\pi(2.4 \times 10^9)(50)} = 2.2 \text{ pF}$$

Next we must move from point *b* to point *a*, requiring a normalized impedance value of $-j0.11$ corresponding to a series capacitance:

$$-j0.11 = \frac{-j}{\omega C Z_o}$$

or

$$C = \frac{1}{2\pi(2.4 \times 10^9)(50)(0.11)} = 12 \text{ pF}$$

The source-matching network is solved in the same way, moving from a matched load to the point on the Smith Chart where

$$\Gamma_S = 0.56e^{-j30°}$$

Again we work backward to find the key points indicated in Figure 10.43b. From the Γ_S point *a* (at $z = 2 - j1.7$), we convert to admittance point *a'* (at $y = 0.3 + j0.24$). We move from here to point *b'* (at $y = 0.3 + j0.46$), and then jump back to the impedance chart at point *b* (at $z = 1 - j1.5$).

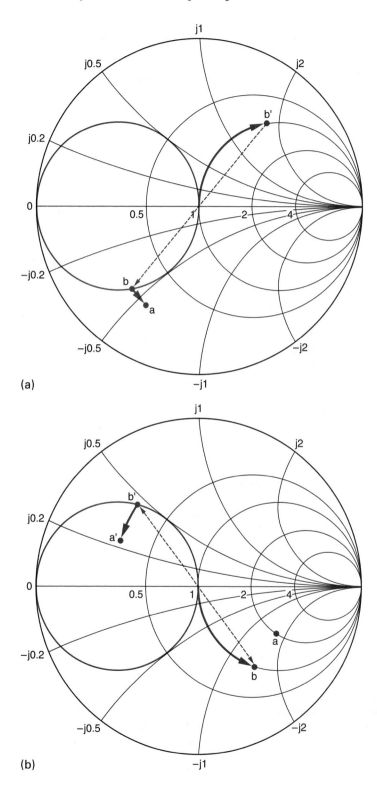

(a)

(b)

Figure 10.43 (a) The Γ_L Smith Chart for solution of the load-matching network. (b) The Γ_S Smith Chart for solution of the source-matching network.

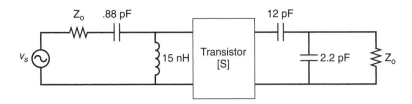

Figure 10.43 (c) The final solution for Example 10.20.

Now we solve the source-matching network by first moving in the impedance chart from the matched load to point *b*, requiring a normalized impedance of $-j1.5$, corresponding to a series capacitor of value 0.88 pF. Then we move from admittance point *b'* to point *a'*, requiring a normalized admittance of $-j0.22$. This corresponds to a shunt inductance of value 15 nH.

Our lumped-element matching network is shown in Figure 10.43c.

Balanced Amplifiers

One way to achieve a broader bandwidth amplifier is to sacrifice gain. However, this approach leads to increased VSWR at the input and output. The balanced amplifier, shown in Figure 10.44, eliminates reflection at the input and the output. It consists of 3-dB couplers (typically a Lange coupler if broad bandwidth is desired) at both the input and the output of a pair of identical amplifier stages. Each amplifier stage has its source and load matching networks identically optimized for gain, bandwidth, and noise performance.

The input signal passes through the coupler, where it is evenly split with a 90° phase shift between signals. Any reflection at the amplifier passes back through the coupler into ports 1 and 4. The reflected signals to port 1 are shifted an additional 90° from each other and therefore cancel. The reflected signals add at port 4, where they are dissipated in a matched termination.

An additional advantage of the balanced amplifier is the doubling of the output power compared to that of a single amplifier.

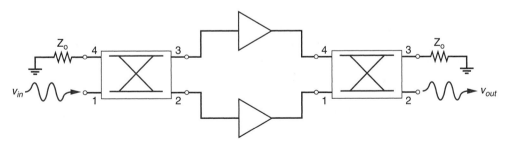

Figure 10.44 Balanced amplifier.

▷ 10.6 RECEIVER DESIGN

A block diagram of a microwave receiver is shown in Figure 10.45. We've discussed antennas, amplifiers, and filters, and this section will briefly describe oscillators and mixers.

The overall receiver must deliver high gain and have good selectivity. A typical power level of -100 dB$_m$ may be received at the antenna and this must be amplified by as much as

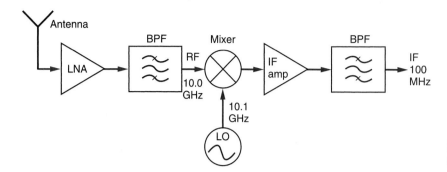

Figure 10.45 Microwave receiver block diagram.

100 dB by the receiver. Several amplifiers must be employed to do so, including one or more amplifiers for the IF signal.

The receiver must also be able to select a specific frequency while rejecting nearby frequencies. Selectivity is most often achieved by tuning the frequency of the oscillator for channel selection and using a narrow-band filter after the mixing stage.

Oscillators

Microwave oscillators convert DC power to RF power. They are characterized by their output power, output frequency, and *phase noise*. The phase noise is related to the bandwidth of the output; a wider bandwidth corresponds to more phase noise.

All wireless transceivers receive their microwave energy from solid state oscillators. The primary types are negative resistance oscillators and transistor oscillators.

Negative resistance oscillators use nonlinear devices that feature a negative resistance. That is, at a particular bias condition, an increase in voltage will result in a decrease in current. The first solid-state negative resistance devices used for oscillators (in the late 1960s) were the Gunn diode and the IMPATT diode. Gunn diode oscillators have been constructed to operate as high as several hundred gigahertz.

Transistor oscillators employ feedback to create instability, leading to oscillations. The typical output of a transistor oscillator is 30 dB$_m$ at 1 GHz, dropping to 20 dB$_m$ at 10 GHz, though some have been constructed to operate as high as 100 GHz.

S-parameters are employed in oscillator design, and operation depends on the careful design of the matching networks. However, because of the nonlinear nature of operation, exact design can be difficult.

Mixers

A microwave mixer uses a nonlinear device such as a diode or transistor to multiply a pair of RF signals. The output consists of signals at the input frequencies, at the difference and sum frequencies, and at the various harmonics. Figure 10.46 shows some of the frequencies exiting a mixer that is fed by an RF signal (at f_{RF}) and an LO signal (at f_{LO}). The difference or intermediate frequency is at a much lower frequency than the other components. These higher components are removed by passing them through either a low-pass or a bandpass filter.

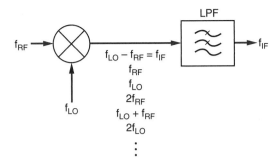

Figure 10.46 Frequency components related to mixing.

The important properties of a mixer are its conversion loss, isolation, and dynamic range. Conversion loss (*CL*) is a measure of how well the mixer converts an input RF signal to an output IF signal. It is given by

$$CL = 10 \log\left(\frac{P_{RF}}{P_{IF}}\right)$$ (10.71)

where P_{RF} is the available RF signal power and P_{IF} is the available IF power. For a diode mixer, 3 dB is the theoretical minimum conversion loss (with 3.5–8.5 dB being typical). The loss comes from mismatch and intrinsic loss in the diode.

Isolation (*I*) refers to how much RF or LO power makes it to the output. It is sometimes referred to as interport isolation or port-to-port isolation and is given by the maximum of

$$I = -10 \log\left(\frac{P_{RF_{out}}}{P_{RF_{in}}}\right) \quad \text{or} \quad I = -10 \log\left(\frac{P_{LO_{out}}}{P_{LO_{in}}}\right)$$ (10.72)

Isolation between 10 and 25 dB is typical.

Dynamic range for a mixer is similar to dynamic range for an amplifier. Here it relates to the range over which the output IF power is a linear function of the input RF power. The 1-dB compression point for a mixer refers to the RF input power level at which point the conversion loss increases 1 dB.

A mixer's performance is conveniently measured using a spectrum analyzer (Figure 10.47). This instrument provides a frequency-domain view of a mixer's output.

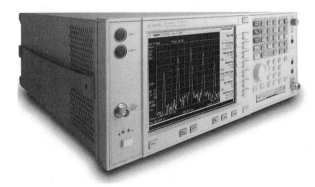

Figure 10.47 Spectrum analyzer. Courtesy of Agilent Inc.

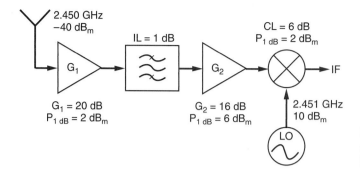

Figure 10.48 Portion of receiver circuit for Example 10.21.

▷ **EXAMPLE 10.21**

Let's analyze the portion of the receiver circuit shown in Figure 10.48.

The antenna receives 2.45 GHz at a -40-dB_m power level corresponding to 100 nW. This power passes through amplifier G_1 with a gain of 20 dB, so we have

$$P_{out} = -40 \text{ dB}_m + 20 \text{ dB} = -20 \text{ dB}_m$$

exiting G_1. Next, we lose 1 dB through the bandpass filter, so -21 dB_m is available entering amplifier G_2. We then have

$$P_{out} = -21 \text{ dB}_m + 16 \text{ dB} = -5 \text{ dB}_m$$

exiting G_2 and entering the mixer. This is below the 1-dB compression point, so we have

$$P_{IF} = -5 \text{ dB}_m - 6 \text{ dB} = -11 \text{ dB}_m$$

of IF power exiting the mixer. This corresponds to about 80 μW. With the local oscillator tuned to 2.451 GHz, the IF output of the mixer will be 1 MHz.

The analysis becomes more interesting if the power received by the antenna is boosted to -30 dB_m. Following the various stages we see that $+5$ dB_m of RF power now enters the mixer. But this exceeds the mixer's 1-dB compression point. With a 2-dB_m compression point and a 6-dB conversion loss, the most IF power we can expect out of the mixer is -4 dB_m (400 μW).

We should generally avoid operating a device beyond its compression point as the output will no longer be a linear function of the input.

Microwave CAD

We've used MATLAB to perform a number of straightforward simulations and calculations, but microwave and RF circuit design is most often accomplished with the assistance of specialized computer-aided design (CAD) software. Accurate simulation tools greatly reduce the need for expensive testing and tweaking of microwave circuits and components. The numerous CAD packages available offer a broad range of features.

One extremely powerful package is Agilent Eesof EDA's Advanced Design Suite. This system integrates multiple design tools to study both the DC network and the AC network with S-parameters. It can also perform both time-domain and frequency-domain simulations. Another feature, shared by a number of CAD packages, is a board layout tool; using this tool, designs can be quickly translated to a working structure.

Another powerful CAD suite is available from Sonnett Software. Of particular interest to the student is the free CAD package Sonnet Lite, available for download from Sonnet Software at http://www.sonetusa/products/lite/download.asp.

CAD software is also available for antenna design. One of the most economical packages is NEC WinPro, available from Nittany Scientific, Inc. (www.nittany-scientific.com). This package can handle a variety of wire antenna designs.

Practical Application: Radio Frequency Identification

Radio frequency identification (RFID) has found favor over other identification and inventory techniques chiefly due to its non-line-of-sight operation and capability of functioning in adverse environments. The system consists of a tag placed on the item to be identified and a reader used to interrogate the tag. The tags may be very small, such as ones inserted beneath the skin for animal tracking. Larger tags have been encased in plastic and attached to store merchandise to prevent theft. Similar tags are mounted on windshields inside automobiles and used for automated toll collection. Tags have also been embedded in automobile tires for identification and tracking purposes. More advanced versions of these tire tags also contain a pressure sensor. Such an RFID sensor tag can alert the driver if the tire loses pressure.

The frequency of RFID systems can vary from the lower ranges of the spectrum around 135 kHz to the higher frequency range at 5.875 GHz. The most commonly used is the 13.56 MHz ISM (industrial scientific medical) band. If an on-board battery powers the tag, it is termed an active tag. Passive tags draw power from radiation emitted by the reader and are less expensive and generally preferred over active ones. Their only drawback is a much-reduced reading range compared to an active tag. This disadvantage is offset by the very long lifetimes and durability of passive tags compared to active ones.

Figure 10.49 shows the major components of an RFID sensor tag system. The reader transmits radiation to the tag. In this case the tag is passive so it contains a circuit that rectifies received radiation to supply power to the rest of the tag circuitry. The microcontroller contains identification information unique to the particular tag and may also receive input from an on-board sensor. The controller exports this information to a switch on the antenna. The switch determines whether the antenna is tuned to its resonant frequency and affects how much of the incident radiation is reflected. The result is a modulated signal at the reader that can then be decoded. This mode of detection is referred to as *backscatter modulation*.

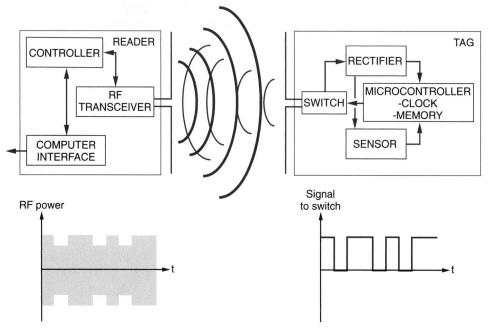

Figure 10.49 Block diagram of an RFID sensor system.

▷ SUMMARY

- Common impedance matching networks employ shunt T-line stubs or lumped-element capacitors and inductors. The Smith Chart is used in designing these matching networks. With lumped elements, a more compact matching network may be constructed but care must be taken to ensure the devices operate well below their self-resonant frequency.

- Scattering parameters, or S-parameters, are a very useful way to represent a circuit device or network and are extensively used in microwave circuit design. When a network is terminated at all ports in the same characteristic impedance Z_o, the scattering parameters represent ratios of voltage waves entering and leaving the circuit. Formally,

$$S_{ij} = \frac{V_i^-}{V_j^+}\bigg|_{V_k^+ = 0, k \neq j}$$

- Any network constructed of passive components is reciprocal. In terms of S-parameters, for a reciprocal network the scattering matrix equals its transpose:

$$[S] = [S]^t$$

- No power is dissipated in a lossless network. A network is lossless if

$$[S]^t[S]^* = [U]$$

where $[S]^*$ is the complex conjugate of $[S]$ and $[U]$ is the unitary matrix. Since a squared S-parameter represents a power ratio, the squared S-parameters of any column of a scattering matrix will sum to 1 for a lossless network.

- Return loss RL at the ith port of a network is related to the reflection coefficient Γ_i looking into the ith port and is defined as

$$RL_i = -20 \log \left(|\Gamma_i| \right)$$

If all ports are terminated in matched loads,

$$RL_i = -20 \log \left(|S_{ii}| \right)$$

- Insertion loss IL is a measure of loss from one port to another in a network. For a signal entering port j and exiting port i,

$$IL_{ij} = -20 \log \left(\left| \frac{V_i^-}{V_j^+} \right| \right)$$

If all ports are terminated in matched loads,

$$IL_{ij} = -20 \log (|S_{ij}|)$$

- Microwave power can be coupled, combined, or redirected using three- or four-port couplers and dividers. These devices are most commonly defined by their scattering matrix.

- For a microwave circulator

$$[S] = \begin{bmatrix} S_{11} & S_{12} & S_{13} \\ S_{21} & S_{22} & S_{23} \\ S_{31} & S_{32} & S_{33} \end{bmatrix} \overset{\text{ideal}}{=} \begin{bmatrix} 0 & 0 & 1 \\ 1 & 0 & 0 \\ 0 & 1 & 0 \end{bmatrix}$$

The insertion loss *IL* between the input port and the desired coupled port is given by

$$IL_{21} = -20 \log (|S_{21}|)$$

whereas the isolation *I*, a measure of how much signal makes it to the wrong port, is given by

$$I = -20 \log (|S_{31}|)$$

The amount of signal reflected at the input port is given by the VSWR. These definitions assume all ports are terminated in matched loads.

- A typical four-port coupler consists of an input port (1), a through port (2), a coupled port (3), and an isolated port (4). The ideal scattering matrix for such a coupler is

$$[S] = \begin{bmatrix} 0 & S_{21} & S_{31} & 0 \\ S_{21} & 0 & 0 & S_{42} \\ S_{31} & 0 & 0 & S_{43} \\ 0 & S_{42} & S_{43} & 0 \end{bmatrix}$$

The amount directed to the coupled port is given by the coupling

$$C = -20 \log (|S_{31}|)$$

whereas the amount directed to the through port is represented by the main-line loss,

$$IL = -20 \log (|S_{21}|)$$

The power to the isolated port is given by the isolation

$$I = -20 \log (|S_{41}|)$$

and the amount reflected at the input port is represented by the VSWR. These definitions assume all ports are terminated in matched loads.

- Some of the most common coupler types used in microwave circuits are the ring hybrid, the quadrature hybrid, and the Lange coupler. Whereas the ring and quadrature hybrid couplers are narrow-bandwidth devices, the Lange coupler performs over an octave bandwidth.

- The four basic filter types are low-pass, high-pass, bandpass, and bandstop. A filter is characterized by its passband ripple and insertion loss, as well as by the sharpness of its roll off.

- Multisection filters made with a number of inductors and capacitors can have small passband ripple and sharp roll off. The Butterworth and Chebyshev filter design approaches are most popular. Tables of filter coefficients simplify the design procedure.

- The gain of an amplifier network is given by

$$G = 20 \log (|S_{21}|)$$

The amplifier is also characterized by its dynamic range (the range over which gain is linear), by its noise figure, and by the VSWR at each port. An amplifier's 1-dB output compression point is the output power level corresponding to a 1-dB decrease in gain.

- Amplifiers are designed using S-parameters and Smith Charts. By assuming unconditionally stable operation of a unilateral ($S_{12} = 0$) transistor, design of an amplifier for maximum gain consists of designing the input and output matching networks such that

$$\Gamma_S = S_{11}^*$$

and

$$\Gamma_L = S_{22}^*$$

- Broader bandwidth amplifiers are realized by sacrificing gain or by constructing balanced amplifiers. A balanced amplifier delivers twice the power of a single amplifier and is impedance matched at the input and the output.

- Oscillators convert DC power to RF power, and ideally they will deliver a very narrow bandwidth signal. They are characterized by their power output, frequency, and phase noise.

- Mixers use the nonlinearity of a diode or transistor element to multiply a pair of input signals. The output consists of a number of frequency components. For down-conversion, the difference or intermediate frequency component is retained.

- A mixer is characterized by its conversion loss,

$$CL = 10 \log \left(\frac{P_{RF}}{P_{IF}} \right)$$

by its isolation

$$I = -10\log\left(\frac{P_{RF_{out}}}{P_{RF_{in}}}\right)$$

or

$$I = -10\log\left(\frac{P_{LO_{out}}}{P_{LO_{in}}}\right)$$

and by its dynamic range. The mixer's 1-dB compression point is the RF input power level where CL increases by 1 dB.

- Microwave CAD software is required in the design and analysis of modern wireless transceivers.

▶ SUGGESTED REFERENCES

Golio, M., ed., *The RF and Microwave Handbook,* CRC Press, 2001.

Gonzalez, G., *Microwave Transistor Amplifiers: Analysis and Design,* 2nd ed., Prentice–Hall, 1997.

Ludwig, R., and Bretchko, P., *RF Circuit Design: Theory and Applications,* Prentice–Hall, 2000.

Pozar, D. M., *Microwave Engineering,* 2nd ed., Wiley, 1998.

▶ PROBLEMS

10.1 Lumped-Element Matching Networks

10.1 A matching network consists of a length of a T-line through section in series with a capacitor. Determine the length (in wavelengths) required of the through section and the capacitor value needed (at 1.0 GHz) to match a $10 - j35$ Ω load impedance to the 50-Ω line.

10.2 Design an L-section matching network to match a $10 + j15$ Ω load to a 50-Ω line. Determine specific values of the lumped elements at a 1.0-GHz operating frequency.

10.3 Design an L-section matching network to match an $80 - j50$ Ω load to a 50-Ω line. Determine specific values of the lumped elements at a 1.0-GHz operating frequency.

10.4 Design an L-section matching network to match a $30 + j70$ Ω load to a 50-Ω line. Determine specific values of the lumped elements at a 2.5-GHz operating frequency.

EMAG SOLUTIONS **10.5** Suppose you want to match a $20 + j50$ Ω load to a 50-Ω line. For the design of an L-section matching network, you notice the normalized load impedance lies outside both the $1 \pm jx$ circle and the rotated $1 \pm jx$ circle. Find all four possible solutions, and for each one determine specific values of the lumped elements at a 2.5-GHz operating frequency.

10.6 Suppose you want to match a 100-Ω line to a load $Z_L = 200 - j100$ Ω (a resistor in series with a capacitor) at a frequency of 500 MHz. (a) Determine the element values for the load. (b) Design a shorted shunt-stub matching network. (c) Design an L-section matching network.

EMAG SOLUTIONS **10.7** Design an L-section matching network to match a load $Z_L = 100 + j80$ Ω to a 50-Ω line. Find the lumped-element values at an operating frequency of 11.18 GHz.

Further, your design should allow for DC biasing the load element through the matching network.

10.8 Design an L-section matching network for a load that has a 25.0-Ω resistor in series with a 1.061-pF capacitor. Assume a 50-Ω system impedance at 3.0 GHz.

10.9 There are two fundamental solutions for the L-section matching network of Problem 10.2. Develop a routine to plot the $|\Gamma|$ versus frequency for both solutions from 0.1 to 10 GHz. (*Hint: This is somewhat similar to MATLAB 9.1.*)

10.10 There are two fundamental solutions for the L-section matching network of Problem 10.3. Develop a routine to plot the $|\Gamma|$ versus frequency for both solutions from 0.1 to 10 GHz.

10.11 Suppose the L-section matching network of Example 10.1 is realized with a capacitor that can be characterized by the circuit model of Figure 9.13b, where $R_x = 0.010$ Ω and $L_L = 7.2$ nH, and an inductor that can be characterized by the circuit model of Figure 9.10, where $R_x = 0.10$ Ω and $C_x = 5.2$ pF. Compare plots of $|\Gamma|$ versus frequency for the ideal case to the case where parasitics are included. The frequency range is from 0.1 to 10 GHz.

10.2 Scattering Parameters

10.12 Find the scattering matrices for the simple two-port networks shown in Figure 10.50.

10.13 Cut a 50-Ω T-line and insert a series 50-Ω resistor followed by a shunt 50-Ω resistor. Determine the scattering matrix for this two-port network. Is the network lossless? Is it reciprocal? Calculate the insertion loss.

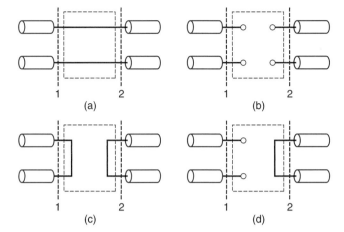

1   (a)   2   1   (b)   2

1   (c)   2   1   (d)   2

Figure 10.50   Two-port networks for finding $[S]$ in Problem 10.12.

10.14   In a 50-Ω system, a two-port network consists of a 25-Ω series resistor followed by a 50-Ω shunt resistor (see Figure 10.11a). Calculate the return loss looking into port 1 of this network if port 2 is terminated in a 100-Ω resistor.

10.15   A series capacitor of value $C = 2.0$ pF is inserted in a 50-Ω T-line. At 1.0 GHz, determine $[S]$, the return loss, and the insertion loss.

10.16   A series inductor of value $L = 3.5$ nH is inserted in a 50-Ω T-line. At 1.0 GHz, determine $[S]$, the return loss, and the insertion loss.

10.17   The scattering matrix for a three-port network is

$$[S] = \begin{bmatrix} 0.60 & 0 & j0.80 \\ 0 & 1.0e^{j90°} & 0 \\ j0.80 & 0 & 0.60 \end{bmatrix}$$

(a) Is this network reciprocal? (b) Is it lossless? (c) Determine the return loss at port 1 if ports 2 and 3 are connected together by a matched T-line of electrical length 45°.

10.18   The scattering matrix (assuming a 50-Ω impedance system) for a two-port network is

$$[S] = \begin{bmatrix} 0.5 & 0.5e^{j45°} \\ 0.5e^{j45°} & 0.5e^{j90°} \end{bmatrix}$$

(a) Is this network reciprocal? (b) Is this network lossless? (c) Determine the return loss looking into port 1 if port 2 is terminated in an open-ended Z_o stub of electrical length 45°.

10.19   Three T-lines with the same characteristic impedance Z_o are connected as shown in Figure 10.51. (a) Determine the scattering matrix that represents this three-port network. (b) Is this network reciprocal? (c) Is it lossless?

10.3   Couplers and Dividers

10.20   Consider a three-port network that is matched at all ports ($S_{11} = S_{22} = S_{33} = 0$). Show that it is impossible to construct a reciprocal network that is lossless for this case.

10.21   A circulator referenced to a 50-Ω impedance is characterized by

$$[S] = \begin{bmatrix} 0.50 & 0.050e^{j60°} & 0.75e^{j60°} \\ 0.75e^{j60°} & 0.50 & 0.050e^{j60°} \\ 0.050e^{j60°} & 0.75e^{j60°} & 0.50 \end{bmatrix}$$

(a) Is this network reciprocal? (b) Is it lossless? Calculate (c) insertion loss, (d) isolation, and (e) VSWR.

10.22   Calculate the insertion loss and the VSWR for the previous problem if the isolated port is terminated in a short circuit.

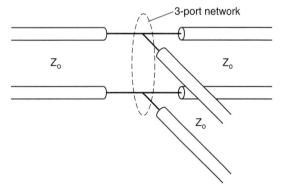

Figure 10.51   Finding $[S]$ for the three-port network for Problem 10.19.

10.23 The following information is supplied for a commercial L-band circulator: $IL_{max} = 0.60$ dB, $I_{min} = 18$ dB, and $VSWR_{max} = 1.35$. Calculate the worst-case magnitudes for the scattering matrix. Assume a symmetrical circulator.

10.24 Verify the scattering matrix (10.27) for the resistive power divider of Figure 10.16.

10.25 Suppose 10.0 mW of microwave power is fed into port 1 of the resistive divider shown in Figure 10.16. With ports 2 and 3 terminated in matched loads, determine how much power is transmitted to each port and how much is dissipated in the divider.

10.26 Repeat Problem 10.25 for a Wilkinson power divider.

10.27 A four-port "20-dB coupler" is specified as having 20-dB coupling, 50-dB isolation, and 0.25 dB of insertion loss. If 100 mW of power is input, calculate the power out of the other three ports. Assume all ports are terminated in matched loads.

10.28 Suppose the coupling for an ideal symmetrical four-port coupler is 3 dB. Find the scattering matrix and determine the insertion loss.

10.29 Suppose to port 1 of an ideal ring hybrid coupler we apply the appropriate frequency voltage

$$V_1^+ = 1.0e^{j0°} \text{ V}$$

If port 2 is terminated in a short circuit, determine the voltage exiting ports 1, 3, and 4.

10.30 Consider an ideal ring hybrid coupler, with all ports terminated in matched loads. A signal $5.0e^{j30°}$ V is injected into port 2, and $3.0e^{j30°}$ V is injected into port 3. Determine the signals exiting ports 1 and 4.

10.31 Suppose to port 1 of a quadrature hybrid coupler we apply the appropriate frequency voltage

$$V_1^+ = 1.0e^{j0°} \text{ V}$$

If port 2 is terminated in a short circuit, determine the voltage exiting ports 1, 3, and 4.

10.32 Given a 50.0-mil-thick Teflon substrate, design a quadrature hybrid coupler for 2.50-GHz operation.

10.33 Suppose you join a pair of quadrature hybrid couplers (port 2 of coupler 1 attached to port 1 of coupler 2, port 3 of coupler 1 attached to port 2 of coupler 2). The resulting network will have four ports: ports 1 and 3 from the first coupler, and ports 2 and 4 from the second coupler. Determine the overall scattering matrix.

10.4 Filters[10.7]

10.34 Derive an insertion loss expression for a high-pass filter realized using a shunt inductor inserted in a $Z_o = R_o$ system. What inductance value is needed for a 1.0-GHz cutoff frequency if $R_o = 50$ Ω?

10.35 Derive an insertion loss expression for a high-pass filter realized using a series capacitor inserted in a $Z_o = R_o$ system. What capacitance value is needed for a 1.0-GHz cutoff frequency if $R_o = 50$ Ω?

10.36 Design a fifth-order low-pass filter with $f_c = 2.0$ GHz for a 50-Ω system where we will allow only 0.1 dB of ripple. Use the Figure 10.28b circuit configuration, and compare your insertion loss plot with that of Figure 10.31.

10.37 Design a third-order low-pass filter with $f_c = 1.0$ GHz for a 50-Ω system starting with the Figure 10.28a circuit configuration. Determine component values for each amount of ripple (0.1, 1, and 3 dB) and compare the three insertion loss responses.

10.38 Starting with the Figure 10.28b circuit configuration, design a third-order high-pass filter with $f_c = 2.4$ GHz for a 50-Ω system where we will allow only 1 dB of ripple. Plot the insertion loss.

10.39 Starting with the Figure 10.28a circuit configuration, design a fifth-order high-pass filter with $f_c = 1.0$ GHz for a 50-Ω system where we will allow 3 dB of ripple. Plot the insertion loss.

10.40 Starting with the Figure 10.28b circuit configuration, design a third-order Chebyshev bandpass filter for a 50-Ω system. The passband is to be from 900 to 1100 MHz with only 1 dB of ripple allowed.

10.41 Starting with the Figure 10.28b circuit configuration, design a third-order Chebyshev bandpass filter for a 50-Ω system. The passband is to be from 900 to 1100 MHz with 3 dB of ripple allowed. Plot the insertion loss and calculate the shape factor.

10.42 Starting with the Figure 10.28b circuit configuration, design a third-order Chebyshev bandpass filter for a 50-Ω system. The passband is to be from 2.2 to 2.6 GHz with only 1 dB of ripple allowed.

10.43 Starting with the Figure 10.28b circuit configuration, design a third-order Chebyshev bandstop filter for a 50-Ω system. The stopband is to be from 900 to 1100 MHz with 3 dB of ripple allowed. Plot the insertion loss.

10.44 Starting with the Figure 10.28a circuit configuration, design a fifth-order Chebyshev bandstop filter for a

---

[10.7]Most of the problems in this section require an insertion loss plot that may best be created using MATLAB.

50-Ω system. The stopband is to be from 2.3 to 2.5 GHz with 1 dB of ripple allowed. Plot the insertion loss.

10.5 Amplifiers

10.45 The following S-parameters were measured at 2.0 GHz in a 50-Ω system:

$$S_{11} = 0.68e^{j125°}, \quad S_{12} = 0$$

$$S_{21} = 3.6e^{j40°}, \quad S_{22} = 0.86e^{-j74°}$$

(a) Determine the gain, in decibels, without any matching networks. (b) Determine the maximum gain, assuming optimized matching networks.

10.46 For Problem 10.45, (a) design open-ended shunt-stub matching networks. In the sketch of your solution, indicate line lengths in terms of wavelengths. (b) You are to realize the matching networks in microstrip constructed on 25.0-mil-thick Teflon. Determine the required microstrip width, and provide a labeled sketch of your network similar to Figure 10.42.

10.47 For Problem 10.45, design a matching network using lumped elements.

10.48 The following S-parameters were measured at 10 GHz in a 50-Ω system:

$$S_{11} = 0.72e^{j76°}, \quad S_{12} = 0$$

$$S_{21} = 4.4e^{j\,125°}, \quad S_{22} = 0.58e^{-j30°}$$

(a) Determine the gain, in decibels, without any matching networks. (b) Determine the maximum gain, assuming optimized matching networks.

10.49 For Problem 10.48, design shorted shunt-stub matching networks with the overall line lengths minimized. In the sketch of your solution, indicate line lengths in terms of wavelength.

10.50 For Problem 10.48, design the matching networks using lumped elements.

10.6 Receiver Design

10.51 Determine the IF power, in watts, exiting a mixer that has a 6.0-dB conversion loss if 0 dB$_m$ of RF power and of LO power enters the mixer.

10.52 Referring to Example 10.21 and Figure 10.48, suppose you require a 100-μW output power level and the antenna receives –80 dB$_m$. If you have several of each amplifier available, design the receiver. You are also allowed to insert a fixed-value attenuator.

Vector Relations

Vector Algebra

Given two vectors $\mathbf{A} = A_x\mathbf{a}_x + A_y\mathbf{a}_y + A_z\mathbf{a}_z$ and $\mathbf{B} = = B_x\mathbf{a}_x + B_y\mathbf{a}_y + B_z\mathbf{a}_z$, the following relationships apply:

Addition

$$\mathbf{A} + \mathbf{B} = (A_x + B_x)\,\mathbf{a}_x + (A_y + B_y)\,\mathbf{a}_y + (A_z + B_z)\mathbf{a}_z$$

Subtraction

$$\mathbf{A} - \mathbf{B} = (A_x - B_x)\,\mathbf{a}_x + (A_y - B_y)\,\mathbf{a}_y + (A_z - B_z)\mathbf{a}_z$$

Commutative Property

$$\mathbf{A} + \mathbf{B} = \mathbf{B} + \mathbf{A}$$

Dot Product

$$\mathbf{A} \cdot \mathbf{B} = |\mathbf{A}||\mathbf{B}| \cos\theta_{AB} = A_xB_x + A_yB_y + A1_zB_z$$

Cross Product

$$\mathbf{A} \times \mathbf{B} = |\mathbf{A}||\mathbf{B}|\sin\theta_{AB} = \begin{vmatrix} \mathbf{a}_x & \mathbf{a}_y & \mathbf{a}_z \\ A_x & A_y & A_z \\ B_x & B_y & B_z \end{vmatrix}$$

Vector Operations

Divergence

Cartesian

$$\nabla \cdot \mathbf{A} = \frac{\partial A_x}{\partial x} + \frac{\partial A_y}{\partial y} + \frac{\partial A_z}{\partial z}$$

Cylindrical

$$\nabla \cdot \mathbf{A} = \frac{1}{\rho}\frac{\partial}{\partial\rho}\left(\rho A_\rho\right) + \frac{1}{\rho}\frac{\partial A_\phi}{\partial\phi} + \frac{\partial A_z}{\partial z}$$

Spherical

$$\nabla \cdot \mathbf{A} = \frac{1}{r^2}\frac{\partial}{\partial r}\left(r^2 A_r\right) + \frac{1}{r\sin\theta}\frac{\partial}{\partial\theta}\left(A_\theta\sin\theta\right) + \frac{1}{r\sin\theta}\frac{\partial A_\phi}{\partial\phi}$$

Gradient

Cartesian

$$\nabla V = \frac{\partial V}{\partial x}\mathbf{a}_x + \frac{\partial V}{\partial y}\mathbf{a}_y + \frac{\partial V}{\partial z}\mathbf{a}_z$$

Cylindrical

$$\nabla V = \frac{\partial V}{\partial \rho}\mathbf{a}_\rho + \frac{1}{\rho}\frac{\partial V}{\partial \phi}\mathbf{a}_\phi + \frac{\partial V}{\partial z}\mathbf{a}_z$$

Spherical

$$\nabla V = \frac{\partial V}{\partial r}\mathbf{a}_r + \frac{1}{r}\frac{\partial V}{\partial \theta}\mathbf{a}_\theta + \frac{1}{r\sin\theta}\frac{\partial V}{\partial \phi}\mathbf{a}_\phi$$

Curl

Cartesian

$$\nabla \times \mathbf{A} = \begin{vmatrix} \mathbf{a}_x & \mathbf{a}_y & \mathbf{a}_z \\ \dfrac{\partial}{\partial x} & \dfrac{\partial}{\partial y} & \dfrac{\partial}{\partial z} \\ A_x & A_y & A_z \end{vmatrix} = \left(\frac{\partial A_z}{\partial y} - \frac{\partial A_y}{\partial z}\right)\mathbf{a}_x + \left(\frac{\partial A_x}{\partial z} - \frac{\partial A_z}{\partial x}\right)\mathbf{a}_y + \left(\frac{\partial A_y}{\partial x} - \frac{\partial A_x}{\partial y}\right)\mathbf{a}_z$$

Cylindrical

$$\nabla \times \mathbf{A} = \frac{1}{\rho}\begin{vmatrix} \mathbf{a}_\rho & \mathbf{a}_\phi & \mathbf{a}_z \\ \dfrac{\partial}{\partial \rho} & \dfrac{\partial}{\partial \phi} & \dfrac{\partial}{\partial z} \\ A_\rho & A_\phi & A_z \end{vmatrix} = \left[\frac{1}{\rho}\frac{\partial A_z}{\partial \phi} - \frac{\partial A_\phi}{\partial z}\right]\mathbf{a}_\rho + \left[\frac{\partial A_\rho}{\partial z} - \frac{\partial A_z}{\partial \rho}\right]\mathbf{a}_\phi + \frac{1}{\rho}\left[\frac{\partial(\rho A_\phi)}{\partial \rho} - \frac{\partial A_\rho}{\partial \phi}\right]\mathbf{a}_z$$

Spherical

$$\nabla \times \mathbf{A} = \frac{1}{r^2\sin\theta}\begin{vmatrix} \mathbf{a}_r & \mathbf{a}_\theta & \mathbf{a}_\phi \\ \dfrac{\partial}{\partial r} & \dfrac{\partial}{\partial \theta} & \dfrac{\partial}{\partial \phi} \\ A_r & rA_\theta & (r\sin\theta)A_\phi \end{vmatrix}$$

$$= \frac{1}{r\sin\theta}\left[\frac{\partial(\sin\theta A_\phi)}{\partial \theta} - \frac{\partial A_\theta}{\partial \phi}\right]\mathbf{a}_r + \frac{1}{r}\left[\frac{1}{\sin\theta}\frac{\partial A_r}{\partial \phi} - \frac{\partial(rA_\phi)}{\partial r}\right]\mathbf{a}_\theta + \frac{1}{r}\left[\frac{\partial(rA_\theta)}{\partial r} - \frac{\partial(A_r)}{\partial \theta}\right]\mathbf{a}_\phi$$

Laplacian

Cartesian

$$\nabla^2 V = \frac{\partial^2 V}{\partial x^2} + \frac{\partial^2 V}{\partial y^2} + \frac{\partial^2 V}{\partial z^2}$$

Cylindrical

$$\nabla^2 V = \frac{1}{\rho}\frac{\partial}{\partial \rho}\left(\rho\frac{\partial V}{\partial \rho}\right) + \frac{1}{\rho^2}\frac{\partial^2 V}{\partial \phi^2} + \frac{\partial^2 V}{\partial z^2}$$

Spherical

$$\nabla^2 V = \frac{1}{r^2}\frac{\partial}{\partial r}\left(r^2\frac{\partial V}{\partial r}\right) + \frac{1}{r^2\sin\theta}\frac{\partial}{\partial \theta}\left(\sin\theta\frac{\partial V}{\partial \theta}\right) + \frac{1}{r^2\sin^2\theta}\frac{\partial^2 V}{\partial \phi^2}$$

Vector Identities

For vectors **A** and **B** with scalar field V:

$$\nabla \cdot (\mathbf{A} + \mathbf{B}) = \nabla \cdot \mathbf{A} + \nabla \cdot \mathbf{B}$$

$$\nabla \times (\mathbf{A} + \mathbf{B}) = \nabla \times \mathbf{A} + \nabla \times \mathbf{B}$$

$$\nabla \cdot (V\mathbf{A}) = \mathbf{A} \cdot \nabla V + V\nabla \cdot \mathbf{A}$$

$$\nabla \times (V\mathbf{A}) = \nabla V \times \mathbf{A} + V(\nabla \times \mathbf{A})$$

$$\nabla \cdot \nabla V = \nabla^2 V$$

$$\nabla \times \nabla \times \mathbf{A} = \nabla(\nabla \cdot \mathbf{A}) - \nabla^2 \mathbf{A}$$

$$\nabla \cdot (\nabla \times \mathbf{A}) = 0$$

$$\mathbf{A} \times \nabla V = V\nabla \times \mathbf{A} - \nabla \times V\mathbf{A}$$

B

Coordinate System Transformations

Vectors in Cartesian, cylindrical, and spherical coordinate systems are represented by

$$\mathbf{A}_{\text{Cart}} = A_x\mathbf{a}_x + A_y\mathbf{a}_y + A_z\mathbf{a}_z$$

$$\mathbf{A}_{\text{cyl}} = A_\rho\mathbf{a}_\rho + A_\phi\mathbf{a}_\phi + A_z\mathbf{a}_z$$

$$\mathbf{A}_{\text{spher}} = A_r\mathbf{a}_r + A_\theta\mathbf{a}_\theta + A_\phi\mathbf{a}_\phi$$

Rectangular—Cylindrical Transformations

Coordinates

$$\rho = \sqrt{x^2 + y^2} \qquad\qquad x = \rho \cos \phi$$

$$\tan \phi = \frac{y}{x} \qquad\qquad y = \rho \sin \phi$$

$$z = z \qquad\qquad z = z$$

Unit Vectors

$$\mathbf{a}_\rho = \cos \phi\, \mathbf{a}_x + \sin\phi\,\mathbf{a}_y \qquad \mathbf{a}_x = \cos \phi\, \mathbf{a}_\rho - \sin \phi\, \mathbf{a}_\phi$$

$$\mathbf{a}_\phi = -\sin \phi\, \mathbf{a}_x + \cos\phi\,\mathbf{a}_y \qquad \mathbf{a}_y = \sin \phi\, \mathbf{a}_\rho + \cos \phi\, \mathbf{a}_\phi$$

$$\mathbf{a}_z = \mathbf{a}_z \qquad\qquad\qquad \mathbf{a}_z = \mathbf{a}_z$$

Vector Components

$$A_\rho = A_x \cos\phi + A_y \sin\phi \qquad A_x = A_\rho \cos\phi - A_\phi \sin\phi$$

$$A_\phi = -A_x \sin\phi + A_y \cos\phi \qquad A_y = A_\rho \sin\phi + A_\phi \cos\phi$$

$$A_z = A_z \qquad\qquad\qquad A_z = A_z$$

Rectangular—Spherical Transformations

Coordinates

$$r = \sqrt{x^2 + y^2 + z^2} \qquad\qquad x = r\sin\theta\cos\phi$$

$$\tan\theta = \frac{\sqrt{x^2 + y^2}}{z} \qquad\qquad y = r\sin\theta\sin\phi$$

$$\tan\phi = \frac{y}{x} \qquad\qquad z = r\cos\theta$$

Unit Vectors

$$\mathbf{a}_r = \sin\theta\cos\phi\,\mathbf{a}_x + \sin\theta\sin\phi\,\mathbf{a}_y - \cos\theta\mathbf{a}_z \qquad \mathbf{a}_x = r\sin\theta\cos\phi\,\mathbf{a}_r + \cos\theta\cos\phi\,\mathbf{a}_\theta - \sin\phi\,\mathbf{a}_\phi$$

$$\mathbf{a}_\theta = -\cos\theta\cos\phi\,\mathbf{a}_x + \cos\theta\sin\phi\,\mathbf{a}_y - \sin\theta\mathbf{a}_z \qquad \mathbf{a}_y = r\sin\theta\sin\phi\,\mathbf{a}_r + \cos\theta\sin\phi\,\mathbf{a}_\theta + \cos\phi\,\mathbf{a}_\phi$$

$$\mathbf{a}_\phi = -\sin\phi\,\mathbf{a}_x + \cos\phi\,\mathbf{a}_y \qquad\qquad \mathbf{a}_z = \cos\theta\mathbf{a}_r - \sin\theta\mathbf{a}_\theta$$

Vector Components

$$A_r = A_x \sin\theta\cos\phi + A_y \sin\theta\sin\phi + A_z\cos\theta \qquad A_x = A_r\sin\theta\cos\phi + A_\theta\cos\theta\cos\phi - A_\phi\sin\phi$$

$$A_\theta = A_x \cos\theta\cos\phi + A_y \cos\theta\sin\phi - A_z \sin\theta \qquad A_y = A_r \sin\theta\sin\phi + A_\theta \cos\theta\sin\phi + A_\phi \cos\phi$$

$$A_\phi = -A_x \sin\phi + A_y \cos\phi \qquad\qquad A_z = A_r \cos\theta - A_\theta \sin\theta$$

▷ **EXAMPLE B.1**

Given a vector $\mathbf{A} = \mathbf{a}_x + 2\mathbf{a}_y + 3\mathbf{a}_z$ located at the point P(4, 3, 1), transform this vector to its equivalent cylindrical vector.

First we must find the cylindrical coordinates for the point. Using the preceding equations we find

$$\rho = \sqrt{x^2 + y^2} = \sqrt{4^2 + 3^2} = 5$$

$$\phi = \tan^{-1}\left(\frac{y}{x}\right) = \tan^{-1}\left(\frac{3}{4}\right) = 0.6435 \text{ radians}$$

$$z = z = 1$$

Now we transform each unit vector into its cylindrical coordinate unit vector:

$$\mathbf{a}_x = \cos\phi\,\mathbf{a}_\rho - \sin\phi\,\mathbf{a}_\phi = 0.8\mathbf{a}_\rho - 0.6\mathbf{a}_\phi$$

$$\mathbf{a}_y = \sin\phi\,\mathbf{a}_\rho + \cos\phi\,\mathbf{a}_\phi = 0.6\mathbf{a}_\rho + 0.8\mathbf{a}_\phi$$

$$\mathbf{a}_z = \mathbf{a}_z$$

We can then insert these equivalent unit vectors into the given Cartesian vector:

$$\mathbf{A}_{cyl} = 1(0.8\mathbf{a}_\rho - 0.6\mathbf{a}_\phi) + 2(0.6\mathbf{a}_\rho + 0.8\mathbf{a}_\phi) + 3\mathbf{a}_z = 2.0\mathbf{a}_\rho + 1.0\mathbf{a}_\phi + 3\mathbf{a}_z$$

Differential Lengths

Whereas the differential length dr shown in Figure B.1 is apparent, the differential lengths $rd\theta$ and $r\sin\theta d\phi$ are not so obvious. Figure B.2 details the geometry used to find the differential length $rd\theta$. Consider that, for a very small angle $d\theta$, the triangle formed appears almost as a right triangle. The length $d\ell$ that we wish to find can then be related by geometry to the hypotenuse r and the angle $d\theta$ as $r\sin d\theta$. Also, for very small angles, $\sin d\theta = d\theta$. Therefore, $d\ell = rd\theta$.

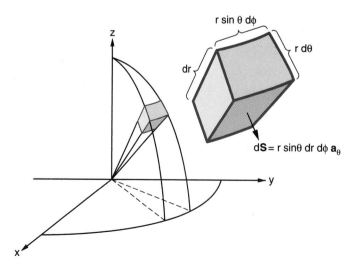

Figure B.1 A differential element in the spherical coordinate system (Figure 2.14).

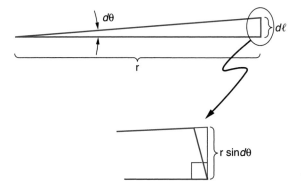

Figure B.2 Detail of geometry used to find the differential length $rd\theta$.

This procedure is also employed to find the differential length $r\sin\theta d\phi$. Here we project the length onto the x–y plane and discover that the hypotenuse is $r\sin\theta$, leading to $r\sin\theta d\phi$.

C

Complex Numbers

The imaginary quantity $\sqrt{-1}$ is represented by the letter j, or $j = \sqrt{-1}$ and $j^2 = -1$. A complex quantity z is the sum of a real part ($x = \text{Re}(z)$) and an imaginary part ($y = \text{Im}(z)$), that is,

$$z = x + jy$$

This is the *rectangular form* of z. By applying Euler's identity ($e^{j\theta} = \cos\theta + j\sin\theta$), the complex number can also be written in *polar form* as

$$z = |z| e^{j\theta}$$

where $|z| = \sqrt{x^2 + y^2}$, and $\tan\theta = y/x$. Figure C.1 illustrates the concept of a complex number.

The *complex conjugate* of z, written z^*, is related to z as

$$z^* = x - jy = |z| e^{-j\theta}$$

and $\sqrt{zz^*} = z$.

Another useful complex relationship is

$$\sqrt{j} = \frac{1 - j}{\sqrt{2}}$$

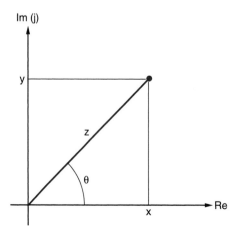

Figure C.1   Graphical representation of a complex number.

► **MATLAB C.1**

The value $z = x + jy$ can be expressed in MATLAB as either

```
z=complex(x,y)
```

or

```
z=x+i*y
```

The magnitude of z (magz) and angle of z (angz) are found by

```
magz=abs(z)
```

and

```
angz=ang(z)
```

D

Integrals, Conversions, and Constants

Useful Integrals

$$\int u\,dv = uv - \int v\,du$$

$$\int \frac{dx}{x} = \ln x$$

$$\int e^x dx = e^x$$

$$\int \sin ax\,dx = -\frac{1}{a}\cos ax$$

$$\int \cos ax\,dx = \frac{1}{a}\sin ax$$

$$\int \frac{dx}{\sqrt{x^2 + a^2}} = \ln\left(x + \sqrt{x^2 + a^2}\right)$$

$$\int \frac{dx}{\left(x^2 + a^2\right)^{3/2}} = \frac{x}{a^2\sqrt{x^2 + a^2}}$$

$$\int \frac{dx}{x^2 + a^2} = \frac{1}{a}\tan^{-1}\frac{x}{a}$$

$$\int e^{ax}\cos bx\,dx = \frac{e^{ax}}{a^2 + b^2}(a\cos bx + b\sin bx)$$

$$\int e^{ax}\cos(c + bx)\,dx = \frac{e^{ax}}{a^2 + b^2}\left[a\sin(c + bx) - b\cos(c + bx)\right]$$

Quadratic Equation

Given $ax^2 + bx + c = 0$

$$x = \frac{-b \pm \sqrt{b^2 - 4ac}}{2a}$$

Half-Angle Formulas

$$\sin^2 \theta = \frac{1}{2}(1 - \cos 2\theta)$$

$$\cos^2 \theta = \frac{1}{2}(1 + \cos 2\theta)$$

Hyperbolic Functions

$$\sinh x = \frac{1}{2}\left(e^x - e^{-x}\right)$$

$$\cosh x = \frac{1}{2}\left(e^x + e^{-x}\right)$$

$$\tanh x = \frac{\sinh x}{\cosh x}$$

Conversions and Constants

1 inch = 1000 mils
1 mil = 25.4 μm

Table D.1 Physical Constants

Constant	Symbol	Value
Avogadro's number	N_A	6.02×10^{23} atoms/mole
Boltzmann's constant	k	1.38×10^{-23} J/K
Charge of an electron	q	-1.602×10^{-19} C
Earth's gravitational acceleration	g	9.78 m/s^2
Free space permeability	μ_o	$4\pi \times 10^{-7}$ H/m
Free space permittivity	ε_o	8.854×10^{-12} F/m $\approx 10^{-9}/(36\pi)$ F/m
Free space intrinsic impedance	η_o	120π Ω
Planck's constant	h	6.63×10^{-34} J-s
Speed of light in vacuum	c	2.998×10^8 m/s

Table D.2 Unit Conversions

Parameter	Unit	Equivalence
Capacitance	F	C/V
Conductivity	S/m	1/(Ω-m)
Current	A	C/s
Electromotive force	V	Wb/s = J/C
Electric field intensity	V/m	N/C = J/(C-m)
Force	N	kg-m/s^2
Inductance	H	Wb/A = J/A^2 H-A = V-s
Magnetic field intensity	A/m	—
Magnetic flux	Wb	V-s
Power	W	J/s
Resistance	Ω	V/A
Work (energy)	J	N-m = W-s = V-C

APPENDIX **E**

Material Properties

TABLE E.1 Conductors at Room Temperature

Material	Symbol	σ (S/m)[a]
Aluminum	Al	3.8×10^7
Carbon	C	3×10^4
Copper	Cu	5.8×10^7
Gold	Au	4.1×10^7
Graphite	—	7×10^4
Iron	Fe	1×10^7
Lead	Pb	5×10^6
Nichrome	—	1×10^6
Nickel	Ni	1.5×10^7
Silver	Ag	6.2×10^7
Solder	—	7×10^6
Stainless steel	—	1.1×10^6
Tin	Sn	8.8×10^6
Tungsten	W	1.8×10^7

[a]Conductivity is approximate since it depends on impurities, moisture, and temperature.

TABLE E.2 Properties for Selected Dielectrics[a]

	ε_r	E_{br} (V/m)	$\tan\delta$ at 1 MHz	σ (S/m)
Air	1.0005	3×10^6	~0	~0
Alumina	9.9	—	0.0001	—
Barium titanate	1200	7.5×10^6	—	—
Glass	10	30×10^6	0.004	~10^{-12}
Ice	4.2	—	0.12	—
Mica	5.4	200×10^6	0.0003	10^{-15}
Polyethylene	2.26	47×10^6	—	10^{-16}
Polystyrene	2.56	20×10^6	—	10^{-17}
Quartz (fused)	3.8	30×10^6	0.0002	10^{-17}
Silicon (pure)	11.8	—	—	4.4×10^{-4}
Soil (dry)	3–4	—	0.017	2×10^{-3}
Teflon	2.1	60×10^6	< 0.0002	10^{-15}
Water (distilled)	81	—	0.04	10^{-4}
Seawater	72	—	0.9	5

[a]Dielectric permittivity results from the polarization of fixed molecules and their bound electrons. Conductors, with their free electrons, do not have polarized molecules and hence have free space permittivities ($\varepsilon_{r(metals)} = 1$).

Table E.3   Magnetic Materials

Substance	μ_r
Diamagnetic	
Bismuth	0.99983
Gold	0.99986
Mercury	0.99997
Silver	0.99998
Lead	0.999983
Copper	0.999991
Water	0.999991
Nonmagnetic	
Vacuum	1.000000
Paramagnetic	
Air	1.0000004
Aluminum	1.00002
Palladium	1.0008
Tungsten	1.00008
Titanium	1.0002
Platinum	1.0003
Ferromagnetic	
Cobalt	250
Nickel	600
Silicon iron	3500
99.8% pure iron	5000
Mumetal	
(75% Ni, 5% Cu, 2% Cr)	100,000
99.96% pure iron	280,000
"Supermalloy"	
(79% Ni, 5% Mo)	1,000,000

Table E.4   American Wire Gauge (AWG) Diameters

AWG	Diameter (mil)[a]	AWG	Diameter (mil)[a]
8	128.5	26	15.94
10	101.9	28	12.64
12	80.81	30	10.03
14	64.08	32	7.950
16	50.82	34	6.305
18	40.30	36	5.000
20	31.96	38	3.965
22	25.35	40	3.145
24	20.10		

[a]1 mil = 25.4 µm.

Answers to Selected Problems

Chapter 2: Electrostatics

P2.1 $Q(6, 6, 7)$

P2.5 $\mathbf{F} = 0.89\mathbf{a}_x + 1.8\mathbf{a}_y\ \mu\mathrm{N}$

P2.7 $\mathbf{F}_{\mathrm{tot}} = 3.2\mathbf{a}_x\ \mathrm{nN}$

P2.9 (a) P(8.7, 46°, 18°), (b) P(5.0, 53°, −90°), (c) P(6.5, 130°, 190°)

P2.11 (b) $V = 14\ \mathrm{m}^3$, (c) $S = 35\ \mathrm{m}^2$

P2.13 (a) P(2.00, 2.00, 2.00), (b) P(−3.00, 5.20, −3.00), (c) P(0.00, −10.0, 6.00)

P2.15 (a) Q(0, 2.0m, 0), (b) $Q = 72\ \mathrm{nC}$

P2.19 $\mathbf{E} = 18\mathbf{a}_x + 20\mathbf{a}_y + 680\mathbf{a}_z\ \mathrm{V/m}$

P2.23 $\mathbf{E} = \dfrac{\rho_s}{\pi\varepsilon_o}\tan^{-1}\left(\dfrac{a}{d}\right)\mathbf{a}_x.$

P2.25 $Q_{\mathrm{tot}} = 160\pi\ \mathrm{nC}$

P2.29 $\psi = 6.0\ \mathrm{C}$

P2.33

```
%HwCh2Pr33
clear
a=.04; %inner cyl radius in m
b=.05; %outer cyl radius
Q=3; %charge per unit length, C/m
N=100; %number of data points
maxrad=.10; %max radius for plot, m

for i=1:N/2.5
    rho(i)=i*maxrad/N;
    D(i)=0.01;%actual value is zero
    %put .01 in to show the line in the plot
```

```
end
for i=(N/2.5)+1:N/2;
    rho(i)=i*maxrad/N;
    K=Q/(2*pi*(b^2-a^2));
    D(i)=K*(rho(i).^2-a^2)/rho(i);
end

for i=(N/2+1):N
    rho(i)=i*maxrad/N;
    D(i)=Q/(2*pi*rho(i));
end

plot(rho,D)
xlabel('radius (cm)')
ylabel('magnitude of D (C/m^2)')
grid on
```

See Figure F.1.

P2.39 (a) $\rho_v = 2\ \mathrm{C/m}^3$, (b) = 16 C, (c) = 16C

P2.41 (a) $\rho_v = 1.83\ \mathrm{C/m}^3$, (b) = 35.5 C, (c) = 35.5 C

P2.43 $\mathbf{a} = 0.124\mathbf{a}_x + 0.990\mathbf{a}_y - 0.062\mathbf{a}_z$

P2.45 (a) 300V, (b) 300 nJ

P2.47 $\mathrm{V}_{\mathrm{BA}} = -250\ \mathrm{V}$

P2.49 $\mathrm{V}_{\mathrm{ho}} = -36\pi\ \mathrm{V}$

P2.51 $R = \dfrac{1}{4\pi\sigma}\left(\dfrac{1}{a} - \dfrac{1}{b}\right)$

P2.53 7.9 Ω

P2.55 1.0 mΩ/m

P2.57 $\mathbf{D} = 2.3\ \mathbf{a}_x\ \mathrm{nC/m}^2$

P2.59 5

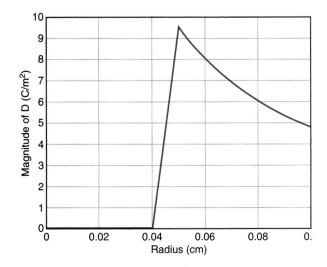

Figure F.1

P2.61 yes, at $x = 2$ m

P2.63 14°

P2.67 5.4 $\mu\Omega$

P2.69 $V(z) = \dfrac{2\rho_o d^2}{\pi^2 \varepsilon} \sin\left(\dfrac{\pi z}{2d}\right) + \left(\dfrac{V_d}{d} - \dfrac{2\rho_o d}{\pi^2 \varepsilon}\right) z$

$\mathbf{E} = \left(-\dfrac{\rho_o d}{\pi \varepsilon} \cos\left(\dfrac{\pi z}{2d}\right) - \dfrac{V_d}{d} + \dfrac{2\rho_o d}{\pi^2 \varepsilon}\right)\mathbf{a}_z$

P2.71 (a) > 6 kV/m, (b) 5.3 μF, (c) 64 μC

P2.73 19 pF

P2.75 (a) 0.52 nF, (b) $W_{E1} = 14.6$ nJ, $W_{E2} = 6.5$ nJ

Chapter 3: Magnetostatics

P3.1 (a) $-17\mathbf{a}_x + 2\mathbf{a}_y + 10\mathbf{a}_z$, (b) $12\mathbf{a}_\rho + 2\mathbf{a}_\phi - 4\mathbf{a}_z$,
(c) $15\mathbf{a}_r - 5\mathbf{a}_\theta - 5\mathbf{a}_\phi$

P3.3 (a) $\theta_P = \theta_Q = 78°$, $\theta_R = 24°$, (b) $0.93\mathbf{a}_x - 0.31\mathbf{a}_y + 0.22\mathbf{a}_z$, (c) 11.4 m²

P3.5 $\mathbf{H} = 160\mathbf{a}_x + 51\mathbf{a}_y + 38\mathbf{a}_z$ mA/m

P3.7 $\mathbf{H} = -10.4\ \mathbf{a}_z$ mA/m

P3.11 exact 1960 A/m; approx 2000 A/m

P3.13 (a) $1.57\mathbf{a}_x$ A/m, (b) $1.1\ \mathbf{a}_x - 0.80\ \mathbf{a}_z$ A/m

P3.15 6A, clockwise viewed from +x-axis

P3.17 $\rho \le a: \mathbf{H} = \dfrac{I\rho}{2\pi a^2}\mathbf{a}_\phi$

$\rho \ge a: \mathbf{H} = \dfrac{I}{2\pi\rho}\mathbf{a}_\phi$

P3.19 (a) 40 mA/m, and (b) for $a = 0.04$m,

$\rho \le a: \mathbf{H} = 0$

$\rho \ge a: \mathbf{H} = \dfrac{I}{2\pi\rho}\mathbf{a}_\phi$.

P3.21 $\rho < a: \mathbf{H} = \dfrac{I}{2\pi\rho}\mathbf{a}_\phi$

$\rho > a: \mathbf{H} = 0$.

P3.23 $H = \dfrac{NI}{2\pi b}a_\phi$

P3.25 $\mathbf{J} = -10\mathbf{a}_\rho + 13\mathbf{a}_\phi + 0.89\mathbf{a}_z$ A/m²

P3.29 (a) $-86 \times 10^{-9}\ \mathbf{a}_z$T, (b) $-86 \times 10^{-9}\ \mathbf{a}_z$(Wb/m²), (c) = $-860 \times 10^{-6}\ \mathbf{a}_z$G

P3.31 $\phi = \dfrac{\mu_o I h}{2\pi} \ln\left(\dfrac{b}{a}\right)$

P3.33 $-4.0\ \mu$N $\mathbf{a}_x$

P3.35 $\mathbf{E} = 5\mathbf{a}_x - 3\mathbf{a}_z$ kV/m

P3.37 $\mathbf{F} = 0.63$ pN $\mathbf{a}_z$

P3.39 $B_o = 0.490$ T, directed north

P3.41 line 1 = line 3 = 0,

line 2 $\mathbf{F} = -0.20\ \mathbf{a}_z$nN

line 4 $\mathbf{F} = +0.20\ \mathbf{a}_z$nN

P3.45 $\tau = -20\ \mathbf{a}_x$pNm

P3.47 $\mathbf{B} = -20 \times 10^{-3}\ \mathbf{a}_x$Wb/m²

P3.49 $\alpha_2 = \tan^{-1}\left(\dfrac{\mu_{r_2}}{\mu_{r_1}} \tan\alpha_1\right)$, $\alpha_2 = 5.5°$

P3.51 (a) $\mathbf{B}_2 = 4\mathbf{a}_y + 27\mathbf{a}_z$ Wb/m², (b) $\alpha_1 = 56°$, $\alpha_2 = 82°$

P3.53 $\mathbf{H}_2 = 4\mathbf{a}_x + 2\mathbf{a}_y + 8\mathbf{a}_z$ A/m

P3.55 $\left.\dfrac{L}{h}\right|_{total} = \dfrac{\mu}{\pi}\left[\dfrac{1}{4} + \ln\left(\dfrac{d-a}{a}\right)\right]$

P3.59 $M_{12} = \mu\pi a^2/2b$

P3.61 (a) $\phi = 40$ μWb, (b) $W_m = 3.2$ mJ

P3.63 $H = 190$ kA/m

P3.65 $F = 1.6$ N

Chapter 4 Dynamic Fields

P4.1 120 days

P4.3 −19 C/m³

P4.5 (a) $\alpha = 0.01$ Np/m, $f = 5$ MHz, $\lambda = 2$ m, $u_p = 10^7$ m/s, $\phi = -\pi/4$ radians, (b) 460 m

P4.9 10 mA clockwise viewed from the +z-axis

P4.11 17.8μV, 1.2μV, 2.25×10^{-18} V

P4.13 (a) 0.46 A clockwise viewed from +z-axis, (b) from induced current we have $\mathbf{B} = -13\ \mathbf{a}_z\mu$Wb/m²

P4.15 $I = 8$ A, and see Figure F.2

P4.17 0.18A (clockwise when viewed from the +z-axis)

P4.19 $V_{emf} = 0$

P4.21 68 mV

P4.27 $i_d = 1.5 \cos(2\pi \times 10^3 t)$ μA

P4.29 $i_d = -97 \sin(6\pi \times 10^6 t)$ μA

P4.31 (a) $E_o = 100$ V/m, $f = 2$ MHz, $u_p = 10^8$ m/s, $\lambda = 50$ m, $\varepsilon_r = 9$, (b) $\mathbf{H}(y, t) = 0.796 \cos(4\pi \times 10^6 t - 0.1257y)\mathbf{a}_x$ A/m

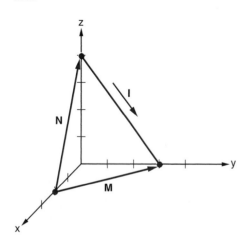

Figure F.2

P4.33 $\mathbf{E}(y, t) = 257e^{-0.14y}\cos(20\pi \times 10^6 t - 0.2\pi y - 12.6°)$ $\mathbf{a}_z$ V/m

P4.35 $\varepsilon_r = 100$, $f = 48$ MHz, $\beta = 10$ radians/m, $\mathbf{E}(\rho,t) = 226 \sin(3 \times 10^8 t + 10\rho)\mathbf{a}_z$ V/m

P4.41 with $\beta = \pi/300$ radians/m, we have

$$\mathbf{H}(z,t) = -53\cos(\omega t - \beta z)\mathbf{a}_x + 27\cos(\omega t - \beta z)\mathbf{a}_y\ \dfrac{mA}{m}$$

Chapter 5: Plane Waves

P5.3 (a), (b) 5 GHz, (c) $\varepsilon_r = 36$

P5.5 $\gamma = 9.4 \times 10^{-3} + j2.1$ 1/m, $\alpha = 9.4 \times 10^{-3}$ Np/m, $\beta = 2.1$ rad/m, $\eta = 1890e^{j257°}$ Ω

P5.7 $\mathbf{H}(x,y,t) =$

$$-2.53\cos(\pi \times 10^6 t - 3x + 2y)\mathbf{a}_x$$

$$-3.80\cos(\pi \times 10^6 t - 3x + 2y)\mathbf{a}_y\ \dfrac{A}{m}$$

$$\mathbf{a}_p = 0.83\mathbf{a}_x - 0.55\mathbf{a}_y$$

P5.11 $u_p = 0.75 \times 10^8\ \dfrac{m}{s}$, $\omega = 7.5 \times 10^8\ \dfrac{rad}{s}$,

$$\mathbf{E}(y,t) = -9.4\cos(7.5 \times 10^8 t - 10y)\mathbf{a}_x\ \dfrac{V}{m}$$

P5.15 See Figure F.3

P5.19

$$\therefore \alpha = \beta = 4.4\dfrac{1}{m},$$

$$\mathbf{H}(z,t) = -16e^{-4.4z}\cos(2\pi \times 10^6 t - 4.4z - 16°)\mathbf{a}_x\ \dfrac{A}{m}$$

P5.23

$$\mathbf{H}(z,t) = 360e^{-200z}\cos(2\pi \times 10^9 t - 200z - 45°)\mathbf{a}_y\ \dfrac{mA}{m}$$

P5.25 0.66 Ω/m

P5.27 (a) $\mathbf{E}(z,t) = 1\cos(1.2\pi \times 10^9 t - 18.2z)\mathbf{a}_y\ \dfrac{V}{m}$

(b) $\mathbf{H}(z,t) = -3.8\cos(1.2\pi \times 10^9 t - 18.2z)\mathbf{a}_x\ \dfrac{mA}{m}$

(c) $\mathbf{P}_{avg} = 1.9\dfrac{mW}{m^2}\mathbf{a}_z$

P5.29 144 W, 102 W

P5.31 linear polarization with 27° tilt angle

P5.33 RHEP

P5.37 $\Gamma = 0.60$, $\tau = 1.60$

P5.39 28.8 mW/m², 71.2 mW/m²

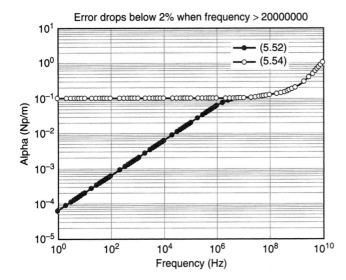

Figure F.3

P5.41 $E^r = -100\cos(\pi \times 10^6 t + \beta_1 z + \pi/4)\, \mathbf{a}_x$ V/m

$E^t = 0$

P5.43 $E^r = -10.\cos(2\pi \times 10^8 t + \beta_1 z)\, \mathbf{a}_x$ V/m

$E^t = 196 e^{-\alpha_2 z}\cos\left(wt - \beta_2 z + 45^\circ\right)\mathbf{a}_x\,\dfrac{\mu V}{m}$,

$P^t_{avg} = 3.7\dfrac{\mu W}{m^2}\mathbf{a}_z$

P5.47 Incident:

$H^i(z,t) = 0.300\cos\left(2\pi \times 10^7 t - 0.628 z + \dfrac{\pi}{4}\right)\mathbf{a}_y\,\dfrac{A}{m}.$

$P^i_{avg} = 5.655\dfrac{W}{m^2}\mathbf{a}_z$

Reflected:

$E^r(z,t) = 13.3\cos\left(2\pi \times 10^7 t + 0.628 z + 171^\circ\right)\mathbf{a}_x\,\dfrac{V}{m}.$

$H^r(z,t) = -0.106\cos\left(2\pi \times 10^7 t + 0.628 z + 171^\circ\right)\mathbf{a}_y\,\dfrac{A}{m}.$

$P^r_{avg} = 0.704\dfrac{W}{m^2}(-\mathbf{a}_z)$

Transmitted:

$E^t(z,t) = 31.7\cos\left(2\pi \times 10^7 t - 1.56 z + 64.8^\circ\right)\mathbf{a}_x\,\dfrac{V}{m}.$

$H^t(z,t) = 0.373\cos\left(2\pi \times 10^7 t - 1.56 z + 31.8^\circ\right)\mathbf{a}_y\,\dfrac{A}{m}.$

$P^t_{avg} = 4.954\dfrac{W}{m^2}\mathbf{a}_z$

$\left(\text{Check: } 5.655\ \text{W}/\text{m}^2 = 0.704\ \text{W}/\text{m}^2 + 4.954\ \text{W}/\text{m}^2\right)$

P5.49

$\theta_t = 7.4^\circ$, $\Gamma_{TM} = -0.589$, $\tau_{TM} = 0.318$,

Incident:

$E^i(z,t) = \left(0.766\mathbf{a}_x - 0.643\mathbf{a}_z\right)\cos(\omega t - 1.34 x - 1.60 z)\dfrac{V}{m}$

$H^i(z,t) = 2.65\cos(\omega t - 1.34 x - 1.60 z)\mathbf{a}_y\,\dfrac{mA}{m}$

Reflected:

$E^r(z,t) = \left(-0.452\mathbf{a}_x + 0.379\mathbf{a}_z\right)\cos(\omega t - 1.34 x + 1.60 z)\dfrac{V}{m}$

$H^r(z,t) = -1.56\cos(\omega t - 1.34 x + 1.60 z)\mathbf{a}_y\,\dfrac{mA}{m}$

Transmitted:

$E^t(z,t) = \left(0.316\mathbf{a}_x - 0.041\mathbf{a}_z\right)\cos(\omega t - 1.35 x - 10.4 z)\dfrac{V}{m}$

$H^t(z,t) = 4.22\cos(\omega t - 1.35 x - 10.4 z)\mathbf{a}_y\,\dfrac{mA}{m}$

Chapter 6: Transmission Lines

P6.1 $R' = 3.32\ \Omega/\text{m}$, $L' = 223\ \text{nH}/\text{m}$, $G' = 560 \times 10^{-18}\text{S}/\text{m}$, $C' = 112\ \text{pF}/\text{m}$

P6.3 $L' = \dfrac{\mu_o}{2\pi}\ln\left(\dfrac{b}{a}\right) + \dfrac{\mu_o}{8\pi} + \dfrac{\mu_o}{2\pi}$

$\left[\left(\dfrac{c^2}{c^2-b^2}\right)^2\ln\left(\dfrac{c}{b}\right) - \left(\dfrac{c^2}{c^2-b^2}\right) + \dfrac{1}{4}\left(\dfrac{c^2+b^2}{c^2-b^2}\right)\right]$

$L' = 330\ \text{nH}/\text{m}$

P6.5 $R' = 2.37\ \Omega/\text{m}$, $L' = 139\ \text{nH/m}$, $G' = 7.63\ \mu\text{S/m}$, $C' = 401\ \text{pF/m}$

P6.7 2.4λ

P6.9 (a) 18 mm, (b) 1.3×10^8 m/s, (c)5.1×10^{-3} dB/m

P6.13 $\gamma = 0.092 + j62.8$ /m

P6.15 $\Gamma_L = 0.12e^{j168°}$, $Z_{in} = 34 - j7.5\ \Omega$

P6.17 $Z_L = 50\ \Omega$

P6.19 (a) $Z_{in} = 22 - j28\ \Omega$, (b) $v_{in} = 2.1\cos(\omega t - 36°)$ V, (c) $v_L = 4.5\cos(\omega t + 106°)$ V

P6.23 (a) $\Gamma_L = 0.80e^{j95°}$, (b) VSWR = 9.0, (c) $Z_{in} = 60 - j180\ \Omega$, (d) 0.058λ

P6.25 $Z_L = 15 - j70\ \Omega$

P6.27 (a) 2.4 GHz, (b) 2.0, (c) $Z_L = 28 - j12\ \Omega$

P6.29 $Z_L = 12 + j35\ \Omega$

P6.31 (a) $d = 0.254\lambda$, $C = 1.14$ pF, (b) $d = 0.408\lambda$, $L = 22.3$ nH

P6.33 (a) $Z_s = 92.2\ \Omega$, (b) through: 0.170λ, stub: $0.104\ \lambda$, or through: 0.329λ, stub: 0.396λ

P6.35 through: 0.372λ, stub: 0.146λ

P6.37 through: 0.379λ, stub: 0.191λ

P6.39 $Z_{in} = 65 + j3.8\ \Omega$; $f_{max} = 100$ GHz

P6.41 $Z_{in} = 32 + j48\ \Omega$

P6.45 $f_{max} = 18$ GHz

	1 GHz	10 GHz	20 GHz
α_c (dB/m)	1.3	4.0	5.7
α_d (dB/m)	0.65	6.5	13.0
α_{tot} (dB/m)	2.0	10.5	18.7

P6.47 (a) 38.6 mils, (b) 0.058 m (2.29 in), (c) through: 9.0 mm (354 mils), stub: 19.5 mm (768 mils)

P6.53 $v_L(t) = 2V_o^i\left(1 - e^{-\tau/z_o c}\right)U(\tau)$ and

$$i_L(t) = \frac{2V_o^i}{Z_o}e^{-\tau/z_o c}U(\tau)$$

P6.55 $12.5\ \Omega$ resistor located at 0.525 m

Chapter 7   Waveguide

P7.1 TE_{10}:   1.374 GHz,
$\quad\quad TE_{01}$:   2.747 GHz,
$\quad\quad TE_{20}$:   2.747 GHz,
$\quad\quad TE_{11}$:   3.07 GHz,
$\quad\quad TM_{11}$:   3.07 GHz,

$\quad\quad TE_{21}$:   3.885 GHz,
$\quad\quad TM_{21}$:   3.885 GHz,
$\quad\quad TE_{30}$:   4.121 GHz,

P7.3 TE_{10}:   6.56 GHz,
$\quad\quad TE_{20}$:   13.1 GHz,
$\quad\quad TE_{30}$:   19.7 GHz,
$\quad\quad TE_{01}$:   19.7 GHz,
$\quad\quad TE_{11}$:   20.8 GHz,
$\quad\quad TM_{11}$:   20.8 GHz,
$\quad\quad TE_{21}$:   23.7 GHz,
$\quad\quad TM_{21}$:   23.7 GHz,

P7.5 $u_u = 2 \times 10^8$ m/s, $u_p = 2.7 \times 10^8$ m/s, $u_G = 1.5 \times 10^8$ m/s.

P7.7

Mode	f_c (GHz)	λ (m)	u_p (m/s)	u_G (m/s)	Z (Ω)
TE_{10}	6.56	.0206	3.3×10^8	2.7×10^8	410
TE_{20}	13.1	.0328	5.2×10^8	1.7×10^8	660
TE_{11}	14.7	.0470	7.5×10^8	1.2×10^8	945
TM_{11}	14.7	.0470	7.5×10^8	1.2×10^8	150

P7.13

$$H = \frac{-j\beta}{\beta_u^2 - \beta^2}\frac{\pi}{a}H_o\sin\left(\frac{\pi x}{a}\right)\cos\left(\frac{\pi y}{b}\right)\cos(\omega t - \beta z)a_x$$

$$+ \frac{j\beta}{\beta_u^2 - \beta^2}\frac{\pi}{b}H_o\cos\left(\frac{\pi x}{a}\right)\sin\left(\frac{\pi y}{b}\right)\cos(\omega t - \beta z)a_y$$

$$+ H_o\cos\left(\frac{\pi x}{a}\right)\cos\left(\frac{\pi y}{b}\right)\cos(\omega t - \beta z)a_z$$

P7.21   2.4 GHz

P7.25   (a)1990, (b) 850, (c) 598

P7.27   8.2 μm

P7.29   1.451

P7.31   850 nm: 4.7 μW, 1300 nm: 13 μW

P7.33   power margin = 23.9 dB, $f_{max} = 984$ MHz

Chapter 8   Antennas

P8.1   (a) $\mathbf{E}_s = \dfrac{\eta_o I_s}{r}\cos^2\theta\,\mathbf{a}_\phi$

(b) $\mathbf{P}(r,\theta,\phi) = \dfrac{1}{2}\eta_o\dfrac{I_o^2}{r^2}\cos^4\theta\,\mathbf{a}_r$

(c) $R_{rad} = 96\pi^2\ \Omega = 950\ \Omega$

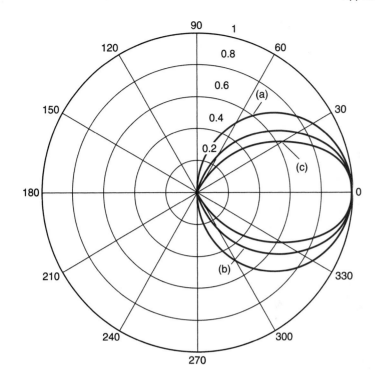

Figure F.4

P8.3 (a) Beamwidth = 120°, $\Omega_p = \pi$, $D_{max} = 4$, (b) Beamwidth = 90°, $\Omega_p = 2\pi/3$ sr, $D_{max} = 6$, (c) $BW = 75°$, $\Omega_p = \pi/2$ sr, $D_{max} = 8$ (see Figure F. 4 for plot)

P8.5 (a) Beamwidth = 82.5°, (b) $\Omega_p = 1.78$ sr, (c) $D_{max} = 7.1$

P8.9 (a) $P_{max} = 0.052$ pW/m², (b) $R_{rad} = 8.8$ mΩ

P8.13 44 Ω

P8.17 $R_{diss} = 1.46$ Ω, e = 0.986, $G_{max} = 1.53$

P8.19 19 µW/m²

P8.23 VSWR = 2.2 and see Figure F.5

P.8.25 (a) 3.6 cm, (b) 60 Ω

P8.29 39°

P8.31 $\Omega_p = 2.6$ sr, $D_{max} = 4.8$, $R_{rad} = 100$ Ω

P8.35 $P_{max} = 76$ µW/m², and see Figure F.6

P8.37 $P_{max} = 26$ µW/m², and see Figure F.7

P8.39 $P_{max} = 57$ pW/m², same pattern as P8.35

P8.43 $P_{max} = 480$ µW/m² and see Figure F.8

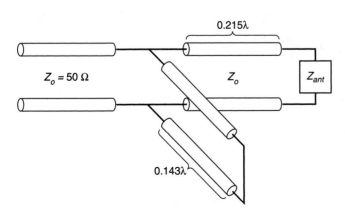

Figure F.5

P8.45 $P_{\text{rad}} = 35$ dBm

P8.47 $P_{\text{rec}}/P_{\text{rad}} = -77$ dB

P8.49 $Z_{\text{in}} = 73 - j42\ \Omega$

P8.53 (a) 0.67 ms, (b) 1.5 pW, (c) 93 km

P8.55 (a) 384×10^6 m, (b) 1 MW

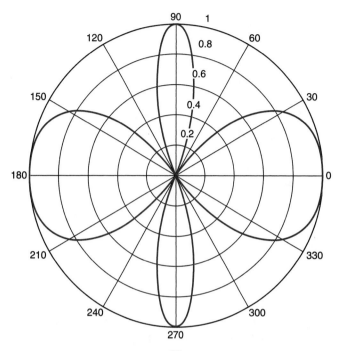

Figure F.6

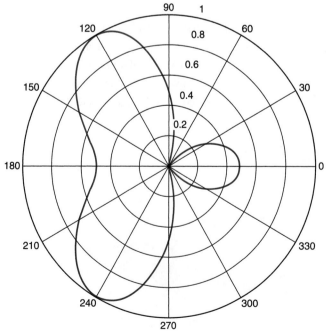

Figure F.7

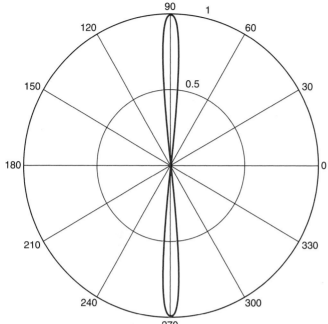

Figure F.8

Chapter 9:   Electromagnetic Interference

P9.1   (a) $R_{dc} = 670\ \mu\Omega$, (b) $R_{ac} = 58\ m\Omega$, (c) $L = 14\ nH$

P9.3 $L = 1.3\ \mu H$, $C_x = 670\ fF$, $f_{SRF} = 170\ MHz$

P9.5 $L = 520\ \mu H$, $f_{SRF} = 8.4\ MHz$

P9.7   186 $\mu F/m^2$, 6 V

P9.9 $1e^{j\,179°}$, $1e^{j\,168°}$, $1e^{j\,87°}$

P9.11   2 GHz

P9.13   A: 135 mV, B: 225 mV, C: 270 mV

P9.15 $SE = 262$ dB

P9.19 SE(Ag paint) = 80 dB, SE(Ag) = 156 dB

P9.23   (a) $IL = 0.36$ dB, (b) $IL = 49$ dB

P9.25   41 dB

Chapter 10   Microwave Engineering

P10.1   See Figure F.9.

P10.3   See Figure F.10.

P10.7   See Figure F.11.

P10.13 $[S] = \begin{bmatrix} 0.200 & 0.400 \\ 0.400 & -0.200 \end{bmatrix}$

The circuit is reciprocal but not lossless, and IL = 8 dB.

P10.15 $[S] = \begin{bmatrix} 0.625e^{-j51°} & 0.780e^{j39°} \\ 0.780e^{j39°} & 0.625e^{-j51°} \end{bmatrix}$,

$IL = 2.2$ dB, $RL = 4.1$ dB

P10.17   (a) yes, (b) yes, (c) $RL = 0$ dB

P10.19   (a) $[S] = \dfrac{1}{3}\begin{bmatrix} -1 & 2 & 2 \\ 2 & -1 & 2 \\ 2 & 2 & -1 \end{bmatrix}$,   (b) yes,   (c) yes

P10.21   (a) no, (b) no, (c) $IL = 2.5$ dB, (d) $I = 26$ dB, (e) VSWR = 3

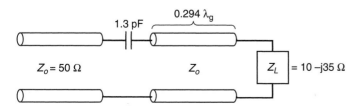

Figure F.9

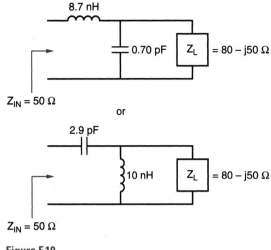

or

Figure F.10

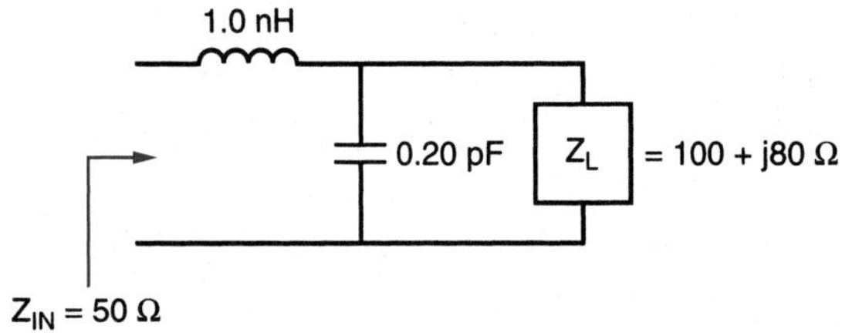

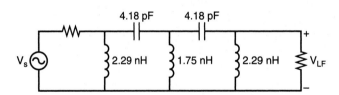

Figure F.11

P10.23 $[S] = \begin{bmatrix} 0.15 & 0.13 & 0.93 \\ 0.93 & 0.15 & 0.13 \\ 0.13 & 0.93 & 0.15 \end{bmatrix}$

P10.25 $P_2 = P_3 = 2.5$ mW, 5 mW dissipated

P10.27 $P_2 = 94$ mW, $P_3 = 1$ mW, $P_4 = 1$ μW

P10.29 $V_1^- = 0.50e^{j0°}$V, $V_3^- = 0.71e^{-j90°}$V,
$V_4^- = 0.50e^{j180°}$V

P10.31 $V_1^- = 0.50e^{j0°}$V, $V_3^- = 0.71e^{j180°}$V, $V_4^- = 0.50e^{-j90°}$V

P10.33 $[S_{overall}] = \begin{bmatrix} 0 & 0 & -j & 0 \\ 0 & 0 & 0 & +j \\ -j & 0 & 0 & 0 \\ 0 & +j & 0 & 0 \end{bmatrix}$

P10.35 $IL = 10\log\left(1 + \left(\dfrac{1}{4\pi f R_o C}\right)^2\right)$, $C = 1.6 pF$

P10.39 See Figure F.12 for one solution.

P10.45 (a) $G_T = 11.1$ dB, (b) $G_{Tmax} = 19.7$ dB

P10.47 See Figure F.13 for one solution.

P10.49 See Figure F.14.

P10.51 0.25 mW

Figure F.12

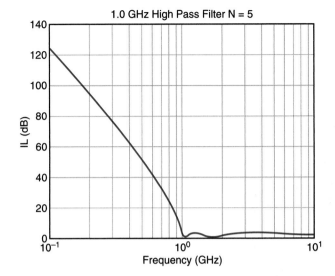

Figure F.12 (*continued*)

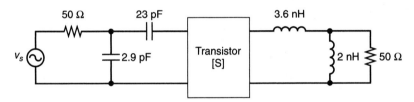

Figure F.13

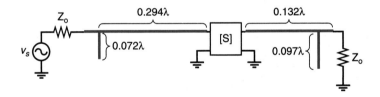

Figure F.14

Index

Table D.1   Physical Constants

Constant	Symbol	Value
Avogadro's number	N_A	6.02×10^{23} atoms/mole
Boltzmann's constant	k	1.38×10^{-23} J/K
Charge of an electron	q	-1.602×10^{-19} C
Earth's gravitational acceleration	g	9.78 m/s^2
Free space permeability	μ_o	$4\pi \times 10^{-7}$ H/m
Free space permittivity	ε_o	8.854×10^{-12} F/m $\approx 10^{-9}/(36\pi)$ F/m
Free space intrinsic impedance	η_o	120π Ω
Planck's constant	h	6.63×10^{-34} J-s
Speed of light in vacuum	c	2.998×10^8 m/s

Table D.2   Unit Conversions

Parameter	Unit	Equivalence
Capacitance	F	C/V
Conductivity	S/m	$1/(\Omega\text{-m})$
Current	A	C/s
Electromotive force	V	Wb/s = J/C
Electric field intensity	V/m	N/C = J/(C-m)
Force	N	kg-m/s^2
Inductance	H	Wb/A = J/A^2 H-A = V-s
Magnetic field intensity	A/m	—
Magnetic flux	Wb	V-s
Power	W	J/s
Resistance	Ω	V/A
Work (energy)	J	N-m = W-s = V-C

Vector Operations

Divergence

Cartesian

$$\nabla \cdot \mathbf{A} = \frac{\partial A_x}{\partial x} + \frac{\partial A_y}{\partial y} + \frac{\partial A_z}{\partial z}$$

Cylindrical

$$\nabla \cdot \mathbf{A} = \frac{1}{\rho}\frac{\partial}{\partial \rho}\left(\rho A_\rho\right) + \frac{1}{\rho}\frac{\partial A_\phi}{\partial \phi} + \frac{\partial A_z}{\partial z}$$

Spherical

$$\nabla \cdot \mathbf{A} = \frac{1}{r^2}\frac{\partial}{\partial r}\left(r^2 A_r\right) + \frac{1}{r\sin\theta}\frac{\partial}{\partial \theta}\left(A_\theta \sin\theta\right) + \frac{1}{r\sin\theta}\frac{\partial A_\phi}{\partial \phi}$$

Gradient

Cartesian

$$\nabla V = \frac{\partial V}{\partial x}\mathbf{a}_x + \frac{\partial V}{\partial y}\mathbf{a}_y + \frac{\partial V}{\partial z}\mathbf{a}_z$$

Cylindrical

$$\nabla V = \frac{\partial V}{\partial \rho}\mathbf{a}_\rho + \frac{1}{\rho}\frac{\partial V}{\partial \phi}\mathbf{a}_\phi + \frac{\partial V}{\partial z}\mathbf{a}_z$$

Spherical

$$\nabla V = \frac{\partial V}{\partial r}\mathbf{a}_r + \frac{1}{r}\frac{\partial V}{\partial \theta}\mathbf{a}_\theta + \frac{1}{r\sin\theta}\frac{\partial V}{\partial \phi}\mathbf{a}_\phi$$

Curl

Cartesian

$$\nabla \times \mathbf{A} = \begin{vmatrix} \mathbf{a}_x & \mathbf{a}_y & \mathbf{a}_z \\ \dfrac{\partial}{\partial x} & \dfrac{\partial}{\partial y} & \dfrac{\partial}{\partial z} \\ A_x & A_y & A_z \end{vmatrix} = \left(\frac{\partial A_z}{\partial y} - \frac{\partial A_y}{\partial z}\right)\mathbf{a}_x + \left(\frac{\partial A_x}{\partial z} - \frac{\partial A_z}{\partial x}\right)\mathbf{a}_y + \left(\frac{\partial A_y}{\partial x} - \frac{\partial A_x}{\partial y}\right)\mathbf{a}_z$$